Microbiology: Concepts and Applications

Microbiology: Concepts and Applications

Edited by **Lucy Phillip**

SYRAWOOD
PUBLISHING HOUSE

New York

Published by Syrawood Publishing House,
750 Third Avenue, 9th Floor,
New York, NY 10017, USA
www.syrawoodpublishinghouse.com

Microbiology: Concepts and Applications
Edited by Lucy Phillip

© 2016 Syrawood Publishing House

International Standard Book Number: 978-1-68286-156-1 (Hardback)

Printed in the United States of America.

Contents

Preface

Microbiology is a rapidly emerging branch of biology that focuses on study of microorganisms. Encompassing numerous sub disciplines; microbiology has a wide scope of research and application. It has evolved as a discipline of great interest all around the world with its diverse applications in biotechnology, medical and pharmaceutical science. The book aims to present a comprehensive account of various concepts of microbiology and its applications in different industrial processes and other fields. It focuses on microbiological principles and techniques to provide a better understanding of the field. This book is beneficial for students, researchers and professionals engaged in the field of microbiology.

This book is a comprehensive compilation of works of different researchers from varied parts of the world. It includes valuable experiences of the researchers with the sole objective of providing the readers (learners) with a proper knowledge of the concerned field. This book will be beneficial in evoking inspiration and enhancing the knowledge of the interested readers.

In the end, I would like to extend my heartiest thanks to the authors who worked with great determination on their chapters. I also appreciate the publisher's support in the course of the book. I would also like to deeply acknowledge my family who stood by me as a source of inspiration during the project.

Editor

Molecular strategies of microbial adaptation to xenobiotics in natural environment

Olusola Abayomi Ojo

Department of Microbiology, Lagos State University, Badagry Expressway, P.O. BOX 12142 Ikeja, Lagos, Nigeria. E-mail: solayom@yahooo.com.

The unprecedented population increase and industrial development during the twentieth century has increased conventional solid and liquid waste pollutants to critical levels as well as produced a range of previously unknown strange synthetic chemicals for which society was unprepared. Increasing pollution of the environment by xenobiotic compounds has provoked the need for understanding the impact of toxic compounds on microbial populations, the catabolic degradation pathway of xenobiotics and upgrade in bioremediation processes. Adaptation of native microbial community to xenobiotic substrates is thus crucial for any mineralization to occur in polluted environment. Enzymes which catalyze the biodegradation of xenobiotics are often produced by induction process and this subsequently determine the acclimation time to xenobiotic substrates. Microbial degraders are adapted to xenobiotic substrates via various genetic mechanisms that subsequently determine the evolution of functional degradative pathways. The ultimate goal of these genetic mechanisms is to creating novel genetic combinations in microorganisms that facilitates mineralization of xenobiotics. Moreover, recent development of high-throughput molecular techniques such as polymerase chain reaction (PCR), microarrays and metagenomic libraries have helped to uncover issues of genetic diversity among environmentally relevant microorganisms as well as identification of new functional genes which would enhance pollution abatement management in the twenty-first century.

Key words: Biodegradation, bioremediation, DNA, metagenomics, microarrays and xenobiotics.

TABLE OF CONTENTS

INTRODUCTION

Xenobiotics are chemically synthesized organic compounds most of which do not occur in nature (Schlegel, 1995). Xenobiotics are defined as compounds that are foreign to a living organism. Where these compounds are not easily recognized by existing degra-dative enzymes, they accumulate in soil and water (Esteve-Nunez et al., 2001). Xenobiotics include fungi-cides, pesticides, herbi-cides, insecticides, nematicides, and so on. Most of which are substituted hydrocarbons, phenyl carbonates, and similar compounds. Some of these substances of which great quantities are applied to crops and soil are very recalcitrant and are degraded only very slowly or not at all. Therefore, the discovery of a new catabolic path-way to complete mineralization of the pollu-tant would be much desirable. Synthetic fibres like polyethylene and polypropylene, though harmless, are practically non-degradable. Whilst the plastizers and softeners contained in textiles are gradually oxidized, the polymer skeleton remains intact (Schlegel, 1995). Among the xenobiotics, polyaromatic, chlorinated and nitro-aromatic compounds were reported to be toxic, mutagenic and carcinogenic for living organisms. However, microbial diversity and versa-tility for adaptation to xenobiotics makes them the best candidates among all living organisms to convey xeno-biotic compounds into natural biogeochemical cycles. Although, more microorganisms are being described as able to degrade these anthropogenic molecules, some xenobiotics have been shown to be unusually recalcitrant (Esteve-Nunez et al., 2001).

The discovery of new catabolic pathways leading to mineralization of this pollutant would be more valuable and afford a better knowledge of the diversity of catabolic pathways for the degradation of xenobiotics as well as bring valuable information for bioremediation processes (Black, 1999). The large majority of the earth's micro-organisms remain uncharacterized because of the ina-bility to isolate and cultivate them on appropriate media. Although, cultivation techniques are improving, the scien-tific knowledge of their growth conditions in nature (that is, chemistry of the original environment, life in complex communities, obligate interactions with other organisms, etc.) remains insufficient to cultivate most of these micro-organisms (Leadbetter, 2003). This is particularly true in complex biological systems like soils, where, despite a huge bacteria diversity (up to 10^{10} bacteria and probably thousands of different species per gram of soil) (Rosello-Mora and Amann, 2001).Whereas less than 1% of bact-eria has been cultured so far (Torsvik and Qvreas, 2002).

Several different molecular methods independent of cultivation have been developed to explore the diversity of microorganisms, cultivable or not, in natural environ-ment.

Most of these methods are based on PCR amplification and subsequent analysis of bacterial rRNA genes by sequencing and fingerprint methods (clone libraries, res-tricttion fragment length polymorphism (RFLP), random amplified polymorphic DNA (RAPD), denaturing gradient gel electrophoresis (DGGE, etc.). The discovery of many new bacterial lineages and the reassignment to the most ecologically significant group when using these methods have led to a dramatic change in our perception of micro-

bial diversity and phylogenetic tree of life (Ward et al., 1990; Amann et al., 1995).

Venter et al. (2004) using a cultivation-independent molecular approach, found thousands of new bacterial species and more than one million new protein-coding genes in 2001 of Sargasso seawater. Thus, corroborating the fact that there are millions of genes, uncharacterized microorganisms and other protein-coding genes yet to be discovered, thus, presenting a tremendous potential for the discovery of new antibiotics, secondary metabolites or xenobiotic degradation pathways. El–Fantroussi (2000) also reported similar discovery while studying the biodegradation of Linuron, a herbicide where the majority of the microbial species involved in the biodegradation were difficult to culture, but were detectable by denatu-ring gradient gel electrophoresis (DGGE).

The development of species-specific oligonucleotide primers and polymerase chain reaction (PCR) can now be used as a confirmatory assay for microbial isolates, since it's highly sensitive if conducted with one of the real-time technologies (Schaad et al., 1999). DNA array technology, essentially a reverse dot blot technique, is an emerging methodology useful for rapid identification of DNA fragments and may be applicable for rapid iden-tification and detection of plant pathogens (Levesque, 1997, 2001). An array of species-specific oligonucleotide probes representing the various pathogens of potato, built on a solid support such as a nylon membrane or microscope slide, could be probed readily with labelled PCR products amplified from a potato sample. This is done by using conserved primers to amplify common bacterial genome fragments from extracts of potato tu-bers that might contain the bacterial pathogens, the presence of DNA sequences indicative of pathogenic species would be revealed by hybridization to species-specific oligonucleotide probes within the array (Fesse-haie, 2003). This has been done using conserved ribo-somal primers and labelled simultaneously with digo-xigenin-dUTP. Hybridization of amplicons to the array and subsequent serological detection of digoxigenin label revealed different hybridization patterns that were distinct for each species and subspecies tested (Fessehaie, 2003).

These recently developed molecular methods have the capacity to explore simultaneously the astonishing taxo-nomic and functional variety among microorganisms. The metagenomic libraries are constructed from the environ-mental genome and clones are screened either for a desired trait ("function –driven" approach) or for a specific sequence ("sequence – driven" approach) (Schloss and Handelsman, 2003). These molecular techniques would enhance several aspects of environmental biotechnology, spanning the spectrum from environmental monitoring (Guschin et al., 1997) to bioremediation and biodegrad-ation (Dennis et al., 2003).

The objective of this review is to emphasize the poten-cy of application of new molecular biology techniques such as polymerase chain reaction (PCR), micro-arrays and metagenomic libraries for assessment of the genetic diversity among environmentally relevant micro-organisms and identification of new functional gene involved in the catabolism of xenobiotics.

FACTORS AFFECTING XENOBIOTIC DEGRADATION

Microorganisms are ubiquitous; hence they have the capacity to adapt to xenobiotic compounds as novel growth and energy substrates. The pollution of the environment with synthetic organic compounds has become an issue of public health concern. Many harmful synthetic organic compounds, which are slowly degradable, have been identified; this includes halogennated aromatics (such as benzenes, biphenyls and anilines), halogenated aliphatics and several pesticides (Spain and Van Veld, 1983). Several factors may be responsible for the slow biodegradation of these compounds in the natural environment; this may include unfavourable physio-chemical conditions (such as temperature, pH, redox potential, salinity and oxygen concentration), presence of alternative nutrients, the accessibility of the sub-strates, or predation (Goldstein et al., 1985). Slow biode-gradation of xenobiotics may also occur due to absence of genetic information coding for appropriate catabolic enzymes or proteins in native microbiota.

However, microbial communities exposed to strange synthetic organic compounds over a long period often metabolize them completely (Rieger et al., 2002). Although, adaptive acquisition of degradative abilities by bacteria for some organic compounds or the resistance to heavy metals induced after long acclimation period in laboratory-simulated ecosystems has been observed (Aelion et al., 1987). Acclimation to xenobiotics may be due to;

(i) Induction of specific enzymes among microbial community which enhanced the degradative capacity of the entire community (Spain and Van Veld, 1983),
(ii) Development of a specific sub-population of microbial community with capacity for co-metabolic process with the main microbial population,
(iii) Adaptation can also be due to the selection of mutants which acquired altered enzymatic specificities or novel metabolic activities and which were not present at the onset of the exposure of the community to the introduced compounds (Barkay and Pritchard, 1988; Timmis and Pieper, 1999).

Such a selective process (that is, induction, growth and mutation) may be responsible for the adaptation observed in mineralization of recalcitrant xenobiotics. (Haigler et al., 1988; Timmis and Pieper, 1999).
Van der Meer et al. (1992) critically analyzed some of these genetic mechanisms by which microorganisms ad-

apt to xenobiotics that includes genetic recombination, transposition, mutational drift and gene transfer. These genetic strategies accelerate the processes of evolution of catabolic pathways in bacteria. The analysis of sequence information showed divergence of micro-organisms isolated from geographically separated areas of the world but harbouring the catabolic genes for xeno-biotics. Haigler et al. (1988), Timmis and Pieper, (1999) attributed the genetically influenced selective process as being the underlying principle behind mineralization of recalcitrant halogenated aromatics.

GENES AND DEGRADATION OF AROMATICS

Aromatic compounds carrying substituents forms a special class of xenobiotics because of their recalcitrance. Often reported are the aerobic processes of mineralization whereas anaerobic processes of biode-gradation do occur in natural environment (Reineke, 1984; Becker et al., 1999). A general comparison of the major pathways for catabolism of aromatic compounds in bacteria has revealed that the initial biotransformation steps are mediated by different enzymes that subse-quently produce limited number of central intermediates such as proto-catechuates and substituted cathecols (Reineke, 1984). These dihydroxylated intermediates are channelled into either a 'meta cleavage or ortho cleavage' pathway (Haigler et al., 1988).The ortho cleavage pathways are involved in the degradation of catechol and protocatechu-ate (Doten et al., 1987). Moreover, the enzymes involved in the mineralization of chlorocathecols (that is, substituted catechols) have wider substrates specificities; hence they are rather referred to as the modified ortho cleavage pathway (Figure1). This parti-cular pathway has been detected in Pseudomonas sp. Strain B13; Alcaligenes eutrophus JMP134 among many others which metabolize chlorinated benzenes (Haigler et al., 1988). The Modified ortho cleavage pathway genes for three bacteria species were extensively studied:

(i) The clc ABD operon of Pseudomonas putida (pAC27) (Ghosal and You, 1989).
(ii) The tfdCDEF operon of A. eutrophus JMP134 (Pjp4) (Don et al., 1985).
(iii) The tcbCDEF operon of Pseudomonas sp. strain P51 (pP51) (Van der Meer et al., 1991).

The outcome of these studies and many others corroborated the fact that the genes for the modified ortho cleavage pathways are generally located on catabolic plasmids and their organization into operon structures was contrary to that of the chromosomally encoded cat and pca genes (Don et al., 1985).The cat and pca genes encodes for the ortho cleavage pathway enzymes, they are located on the chromosomes (Doten et al., 1987; Neidle and Ornston, 1986).

STRATEGIES OF ADAPTATION TO XENOBIOTICS

The spontaneous occurrence of DNA rearrangements in xenobiotic-degraders that resulted in evolution of different pathways for mineralization of synthetic compounds in natural environment is one of the principal mechanisms of adaptation to xenobiotic substrates. The evolution of catabolic pathways (that is, modified *ortho* cleavage pathway, *meta* cleavage pathway and others) in micro-organisms for xenobiotic substrates often involves different gene clusters encoding for the aromatic path-way enzymes.

RECOMBINATION AND TRANSPOSITION

Recombination is the combining of genes (DNA) from two or more different cells. This is principally based on molecular methods involving cutting of DNA fragments from different cells harbouring desired catabolic traits. These DNA fragments through hybridization in host cells that are known as recombinants, is seeded onto polluted environment where the expression of the catabolic trait is desired (Black, 1999). This strategy is often practiced in vitro than in vivo. El-Fantroussi (2000) in the soil enrichment method for the degradation of an herbicide (Linuron) engaged a modified strategy which can be extended for bioremediation process in soil polluted by this herbicide.

In another biodegradation involving the use of *Acinetobacter* and *Pseudomonas* species, DNA rearrangement strategy was used to achieve mineralization. The orders of the genes encoding the *ortho* cleavage pathways of *Acinetobacter calcoaceticus* and *P. putida* differ from one another and from those of other organisms, suggesting that various DNA rearrangements have also occurred (Van der Meer et al., 1991). Gene rearrangements are also evident even between the different operons for the modified *ortho* pathways enzymes (Figure 1). There are as yet no clear indications of what mechanisms may direct these rearrangements. Rearrangement of DNA fragments is believed to be due to insertion elements which subsequently enhance gene transfer as well as activation or inactivation of silent gene (Tsuda et al., 1989).

GENE DUPLICATION

This is an important mechanism for the evolution of different strains of microorganisms of the same species. Once a gene becomes duplicated, the extra gene copy thus becomes independent of selective pressures and subsequently imbibes mutations with speed. These mutations could eventually lead to full inactivation, rendering this copy silent. Reactivation of the silent gene copy could then occur through the action of insertion elements. This occurred in *Flavobacterium* sp. Strain K172 that produced two isozymes of 6-aminohexanoate dim-mer hydrolase, one of the enzymes involved in the degradation of nylon oligomers (Okada et al., 1983).

MUTATIONAL DRIFT

Mutational drift in terms of point mutation is of much relevance in xenobiotic degradation. It is possible that a number of stress factors such as chemical pollutants induce error-prone DNA replication that subsequently accelerates DNA evolution.

Point mutation involves base substitution, or nucleotide replacement, in which one base is substituted for another at a specific location in a gene (Black, 1999). This kind of mutation changes a single codon in mRNA, and it may or may not change the amino acid sequence in a protein. Several examples have illustrated that single-site-mutations can alter substrate specificities of enzymes or effector specificities. Clarke (1984) isolated mutants with altered substrate specificities of the AmiE amidase of *Pseudomonas aeruginosa*, which were provoked by single-base-pair changes. Sequential mutations in the cryptic *ebg* genes of *Escherichia coli* were shown to result in active enzymes capable of metabolizing lactose and other sugars.

Single-site-mutations are believed to arise continuously and at random as a result of errors in DNA replication or repair. Although, the important effects of single-base-pair mutation on the adaptive process have been demonstrated experimentally, the accumulation of the single-base-pair changes may not be the main reason for the differences in the properties of the catabolic enzymes elicited by xenobiotic-degraders. There are other factors that would precipitate changes in DNA sequences, this included gene conversion or slipped-strand mispairing (Niedle et al., 1988).

GENE TRANSFER

Gene transfer is a process of movement of genetic information between organisms (Black, 1999). The importance of gene transfer for adaptation of host cells to new compounds has been explicitly demonstrated in many studies on experimental evolution of novel meta-bolic activities. Such studies identified biochemical block-ades in natural pathways that prevented the degradation of novel substrates and these barriers scaled by transferring appropriate genes (Reineke et al., 1982).

Genetic interactions in microbial communities are effected by several mechanisms such as conjugation via plasmid replicons, transduction and transformation (Saye et al., 1990). The occurrence of plasmids in bacteria in the natural environment is certainly a general phenolmenon and an important pool of genetic information residing on plasmid vehicles may flow among indigenous organisms. The self transmissible plasmids that carry

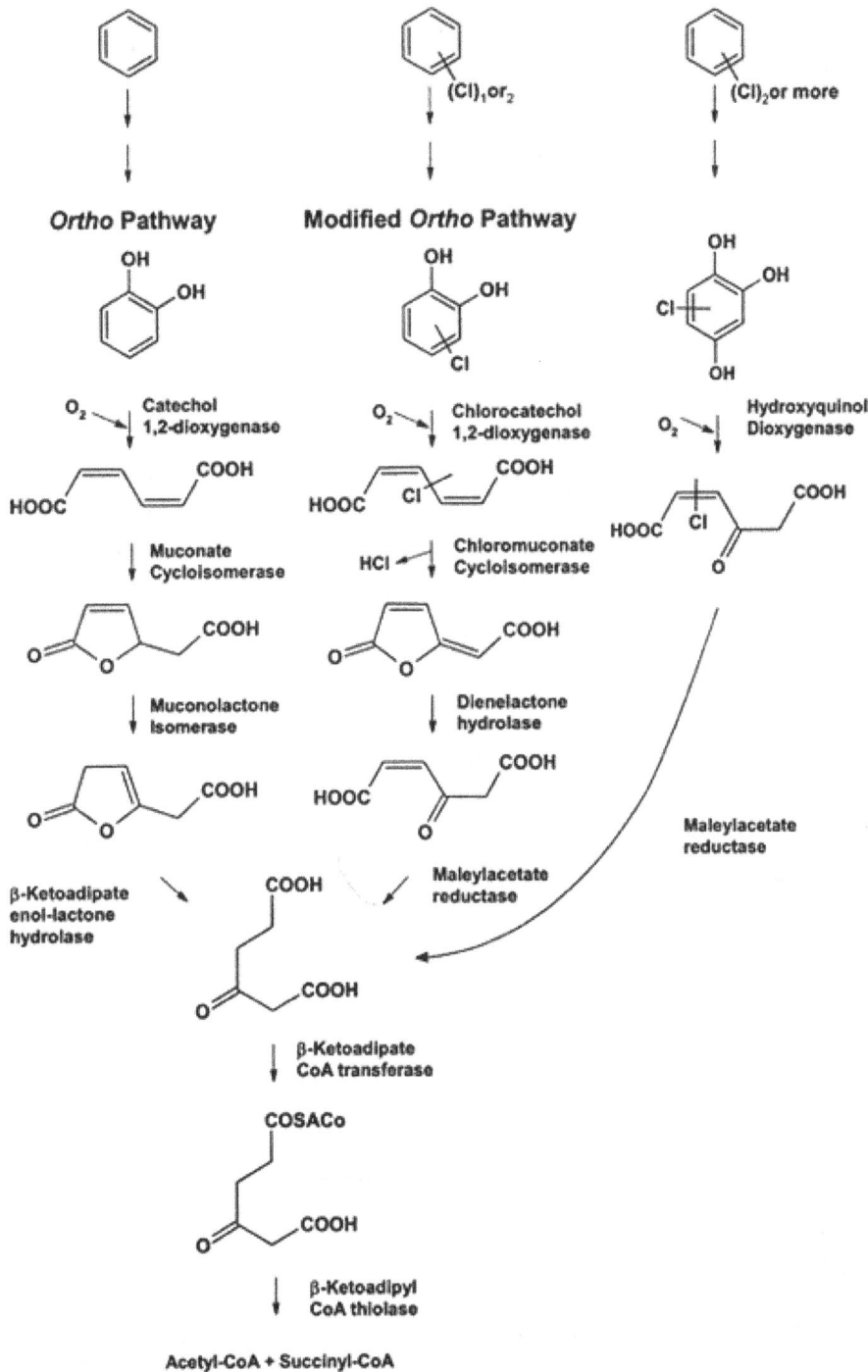

Figure 1. Modified *ortho* cleavage pathway. Adapted from Ferraroni et al. (2004).

genes for degradation of aromatic or of other organic compounds are known and their roles in spreading these genes to other organisms is predictable (Assinder and Williams, 1988).

Although, the transfer of catabolic plasmids can lead to regulatory and / or metabolic problems for the cells and therefore additional mutations in the primary transconju-gants are often needed to construct strains with the desi-red metabolic activities (Reineke et al., 1982).

NEW MOLECULAR TECHNIQUES FOR DETECTING XENOBIOTIC- DEGRADER

There are many microorganisms in natural environment

that can degrade biphenyls, halogenated aromatics, naphthalene and xylenes. However, the assessment of the distribution of these microorganisms exhibiting specific genetic traits has been handicapped due to the fact that a large proportion are not culturable and some genes are latent.The discovery of DNA-DNA hybridization technique that is a relatively novel experimental app-roach in environmental biotechnology provided the solution to the problems of culturability and gene expression among microorganisms (Sayler and Layton, 1990; Leadbetter, 2003). Fessehaie et al. (2003) obtained oligonucleotides from bacteria pathogenic on potato which he designed and formatted into an array by pin spotting on nylon membranes. Genomic DNA from bacterial cultures was amplified by polymerase chain reaction (PCR) using conserved ribosomal primers and labeled simultaneously with digoxigenin-dUTP. Hybridization of amplicons to the array as well as detection of digoxigenin label showed different hybridization patterns that were distinct for each species and subspecies tested. DNA array technology is essentially a reverse dot blot technique useful for identification of DNA fragments and this was applied for rapid identification and detection of bacteria pathogenic on potato. An array of species-specific oligonucleotide pro-bes representing the various pathogens of potato was constructed on a solid support (that is, nylon memb-ranes). This was probed with labelled PCR products amplified from a potato sample, mean-while, conserved primers to amplify common bacterial genome fragments from extracts of potato tubers that had previously been infected by different bacterial pathogens was generated. The presence of DNA sequences indicative of patho-genic species would be shown by the hybridization to species-specific oligonucleotide probes within the array. This discriminatory technology identifies genomic DNA fragments of bacterial pathogens via the distinct species-specific hybridization patterns shown using the gel elec-trophoresis. Specific DNA sequences of native micro-organisms could also be detected in environmental sou-rces by hybridization with probes after amplification of those sequences using PCR technique (Bej et al., 1990; Weisburg et al., 1991; Eyers et al., 2004). Theoretically, the use of conserved DNA sequences in a gene family as universal primers in polymerase chain reaction amplifica-tion and consequent cloning of the amplified fragments would facilitate the detection and isolation of a wider variation of genotypes from the environment (Van der Meer et al., 1991). Weisburg et al. (1991) used this method for the characterization of variations in 16S rRNA genes from microorganisms in natural communities.

METAGENOMIC LIBRARIES

Metagenomics is the culture-independent genomic analy-sis of entire microbial communities (Schloss and Handel-sman, 2003). In other words, metagenomics provides access to the pool of genomes of a given environment.

While direct genomic cloning gives access to retrieve unknown sequences or functions in a given ecosystem that may be used for the design of primers. The PCR amplification requires prior knowledge of the sequences of genes for the design of primer. Total genomic DNA is extracted from the environment (Figure 2A) before meta-genomic libraries can be constructed. The genomic DNA is enzymatically or mechanically fragmented. Fragments can be separated on the basis of their size by pulsed field gel electrophoresis (PFGE). This methodology per-mits fragments of an appropriate size to be isolated from the gel and inserted into host cell by cloning vectors (bacteriophage lambda, cosmid, fosmid or bacterial artifi-cial chromosome (BAC) vectors). BAC vectors are quite efficient in maintaining stably large DNA inserts (up to 300 kb) in low copy numbers in the host cells (1-2 per cell) (Shizuya and Simon, 1992; Rondon et al., 2000). Metagenomic libraries can be screened for functional and / or genetic diversity. New catabolic genes for the degra-dation of xenobiotics are discovered via the "functional approach". Clones are screened for a desired trait on appropriate media. For example, haloaromatic com-pounds could be used as sole electron acceptors since it has been reported that bacteria can metabolize them. (El-Fantroussi et al., 1998; Van de Pas et al., 2001). Whereas polyaromatic hydro-carbons could be utilized as sole C-source and energy-source, this often occurs after several years of adaptation that has led to a selection of a bacterial consortium cap-able of completely minerali-zing such compounds. The biodegradation of linuron, a commonly used herbicide was monitored by enrichment process, using reverse transcription-PCR and denaturing gradient gel electrophoresis (DGGE), it was revealed that a mixture of bacterial species was involved in the mine-ralization. Although, these bacterial species appear to be difficult to culture since they were detectable by DGGE but were not cultivable on agar plates (El-Fantroussi, 2000).

Consequently, growth measurements could identify clones bearing catabolic genes. This function-driven screening remains a straightforward and successful met-hod for the discovery of catabolic genes, as opposed to inferring the function of cloned genes by searching for homologous sequences available in database (Rondon et al., 2000). Indeed, sequences coding for important metabolic functions are frequently poorly conserved, making the comparison of clone sequences with homolo-gous ones very difficult. This functional screening appro-ach has been successfully used to identify novel and previously undescribed genes coding for antibiotics, lipa-ses, enzymes for the metabolism of 4-hydroxybutyrate and genes encoding biotin synthetic pathways (Schloss and Handelsman, 2003). Most of the catabolic genes known to date were isolated from cultivable microorg-anisms (Table 1) (Eyers et al., 2004). However, no genes for TNT denitration have been discovered up till now. But high denitration activities were obtained under

Table 1. Properties of selected identified genes involved in degradative pathways of recalcitrant substituted aromatic compounds.

Target compound	Function	Name	Target metabolite in pathway	Mode of isolation	Size	Microorganism of origin	References
Naphthalene, Toluene, Anthracene Polycyclic aromatic hydrocarbon(PAH)	Reductase	NL1	PAH	Genomic library expressed in Sphingomonas strains	160 – 195kb	Sphingomonas subterranean, S. aromaticivorans F199,B0695 S. xenophaga BN6, S. sp. HH69	Basta et al., (2005)
2-Nitrotoluene	Reductase Ferrodoxin Iron-sulphur protein α Iron-sulphur protein β	ntdAa ntdAb ntdAc ntdAd	2-Nitrotoluene	Genomic library expressed in E. coli and screening for metabolic activities	4.9 kb	Pseudomonas sp. JS42	Parales et al., (1996)
2,4-Dinitrotoluene	Dioxygenase Monooxygenase Dioxygenase Isomerase/hydrolase Dehydrogenase	dntA dntB dntD dntG dntE	2,4-Dinitrotoluene 4-Methyl-5-nitrocatechol 2,4,5-Trihydroxytoluene 2,4-Dihydroxy-5-methyl-6-oxo-2,4-hexadecanoic acid Methylmalonic acid semialdehyde	Genomic library expressed in E. coli and screening for metabolic activities	27 kb	Burkholderia cepacia R34	Johnson et al., (2002)
2,4,6-Trinitrophenol	Hydride transferase Hydride transferase NADPH reductase	npdl npdG npdC	2,4,6-Trinitrophenol Hydride-Meisenheimer complex of 2,4,6-trinitrophenol 2,4,6-Trinitrophenol and its hydride-Meisenheimer	Genomic library and selection of clones thanks to mRNA differential display experiments	12.5kb	Rhodococcus erythropolis HL PM-1	Walters et al., (2001) ; Heiss et al., (2002)
2,4,6-Trinitrotoluene	Reductase	xenB	2,4,6-Trinitrotoluene	Genomic library expressed in E. coli and screening for metabolic activities	1.05kb	P. fluorescens I-C	Blehert et al., (1999)
2,4,6-Trinitrophenol	NADPH F420 Reductase Hydride transferase II	npdG npdl	2,4,6-Trinitrophenol Hydride-Meisenheimer complex of 2,4,6-trinitrophenol	Genomic library and selection of clones***	12.5kb	Rhodococcus erythropolis HL PM-1	Heiss et al., (2003)

Table 1. contd.

4-Nitrotoluene	Monooxygenase	ntnMA	4-Nitrotoluene	Genomic library screened with designated probes; activities confirmed by cloning and expression in *Escherichia coli.*	14.8kb	*Pseudomonas* sp. TW3	James and Williams (1998); James et al. (2000)
	Nitrobenzyl alcohol dehudrogenase	ntnD	4-Nitrobenzyl alcohol			
	Nitrobenzadehyde dehudrogenase	ntnC	4-Nitrobenzadehyde	Genomic library expressed in *P. putida*PaW340 and screening for metabolic activities		
	Nitrobenzoate reductase	pnbA	4-Nitrobenzoate		6 kb	*Pseudomonas* sp. TW3	Hughes and Williams (2001)
	Hydroxylaminobenzoate lyase	pnbB	4-Hydroxylaminobenzoate			

defined conditions with a TNT-contaminated soil sample (Eyers et al., 2004). Further more, the presence of a specific bacterial consortium in this polluted soil was demonstrated by DGGE (Eyers et al., 2004).Therefore, metagenomic libraries are particularly promising for identifying denitration genes, compared with methods based on isolation of pure cultures.

Comparatively, the sequence–driven approach of the metagenomic libraries is based on conserved regions in microbial genes. Clone libraries are screened for specific DNA sequences by means of hybridization probes and PCR primers, whose design is based on the information available in databases. Such hybridization probes may use, in the case of denitration of 2,4,6-trinitrophenol or other electron-deficient aromatics, the *npdG* and *npdI* sequences of *Rhodococcus erythropolis* HL-PM1(Table 1), as homologous sequences that have been identified in other nitroaromatic compounds-degrading *Rhodococcus* strains (Heiss et al., 2003). These sequences are clustered separately from the related enzymes. Hence, it was suggested that they may be suitable as genes probes for finding bacteria in the environment with the ability to "hydrogenate" electron-deficient aromatic ring systems (Heiss et al., 2003).

LIMITATIONS OF METAGENOMIC LIBRARIES

Although, metagenomic libraries constitute at present the most powerful tool to assess the functional diversity of natural microbial communities, they do not cover genomes of low abundance in complex environment like soils, which can be responsible for an important degradation or related process. Hence, the frequency of clones of a desired nature in a library can be very low. This implies screening thousands of clones, which can be laborious and time-consuming. High- throughput equipment is now available to facilitate colony picking, inoculation in microtitre plates and screening numerous clones at the same time. However, an enrichment strategy is required before the construction of the library to select a specific feature and in that way ensure better cover for a subset of the community (Entcheva et al., 2001). Communities enriched naturally by long-term exposure to high concentrations of xenobiotics may host the genes of interest at a high frequency. In this context, metagenomics might shorten the time required to understand the genetics of degradation.

Moreover, the catabolism of specific xenobiotics may be achieved by two or more bacteria via co-metabolic process (Abraham et al., 2002). In this case, it is not possible to isolate a contiguous piece of DNA containing all the genes involved in the catabolic pathway. Therefore, Schloss and Handelsman (2003) suggested studying multiple clones simultaneously on liquid media in which substrates and products can diffuse freely among members of the mixture. Finally, it is important for functional screening that catabolic genes are effectively expr-

essed. Hence, another challenge lies in choosing an appropriate host of expression. Moreover, the host has to be both relatively insensitive to the toxic xenobiotic and unable to catabolize it in the absence of the vector.

MICROARRAYS

DNA chip is an emerging technique making it possible to analyze hundreds and even thousand of genes at the same time, thus getting away from the "one gene at a time" analysis. This allows extremely small amount of biomolecules (RNA, cDNA, etc) to be printed at high density on a substrate. Comparatively, microarrays present the advantage of miniaturization (thousands of probes can be spotted on a chip), high sensitivity and rapid detection which is not obtainable with the traditional nucleic acid membrane hybridization. Zhou and Thompson (2002) has reported that microarray –based genomic technologies would revolutionize the analysis of microbial community structure, function and dynamics. DNA microarrays are coated glass microscope slides onto which thousands of target DNA sample are spotted in a precise pattern. There are two principal microarrays types based on the nature of the target (DNA, e.g., cDNA, PCR products, oligonucleotides or plasmids, or RNA) and on the method of spotting (mechanical microspotting or photolithography). In the first method, purified DNA is spotted onto the membrane or coated glass slide. Although, DNA will stick to glass, aminosilane-coated and poly-L-lysine (PLL)-coated slides are predominantly used. Silanized slides ($RSiX_3$, where X is typically alkoxy) attach the DNA by forming covalent bonds between primary amines on the surface and the phosphate backbone (Celis et al., 2000; Ye et al., 2001). The photolithography method uses an ultraviolet light source that passes through a mask where a photochemical reaction takes place on a siliconized glass surface (Affymetrix). This is by far the most efficient method of generating high-density oligonucleotide chips, but has practical limitations in terms of fragment length and affordability (Kumar et al., 2000). After printing the microarrays, therefore, extraction of DNA or messenger RNA from pure cultures or the environment (Figure 2B), labelling it with specific fluorescent molecules and hybridizing it to target DNA spotted on the glass slide. The resulting image of fluorescent spots is visualized by confocal laser scanning and it's digitized for quantitative analysis.

Fessehaie et al. (2003) designed and formatted into an array by pin spotting on nylon membranes, genomic DNA from bacterial cultures which was amplified by PCR using conserved ribosomal primers and labelled simultaneously with digoxigenin –dUTP. Hybridization of amplicons to the array and subsequent serological detection distinctly confirmed the identity of each species of potato pathogens within the mixed cultures and inoculated potato tissues.

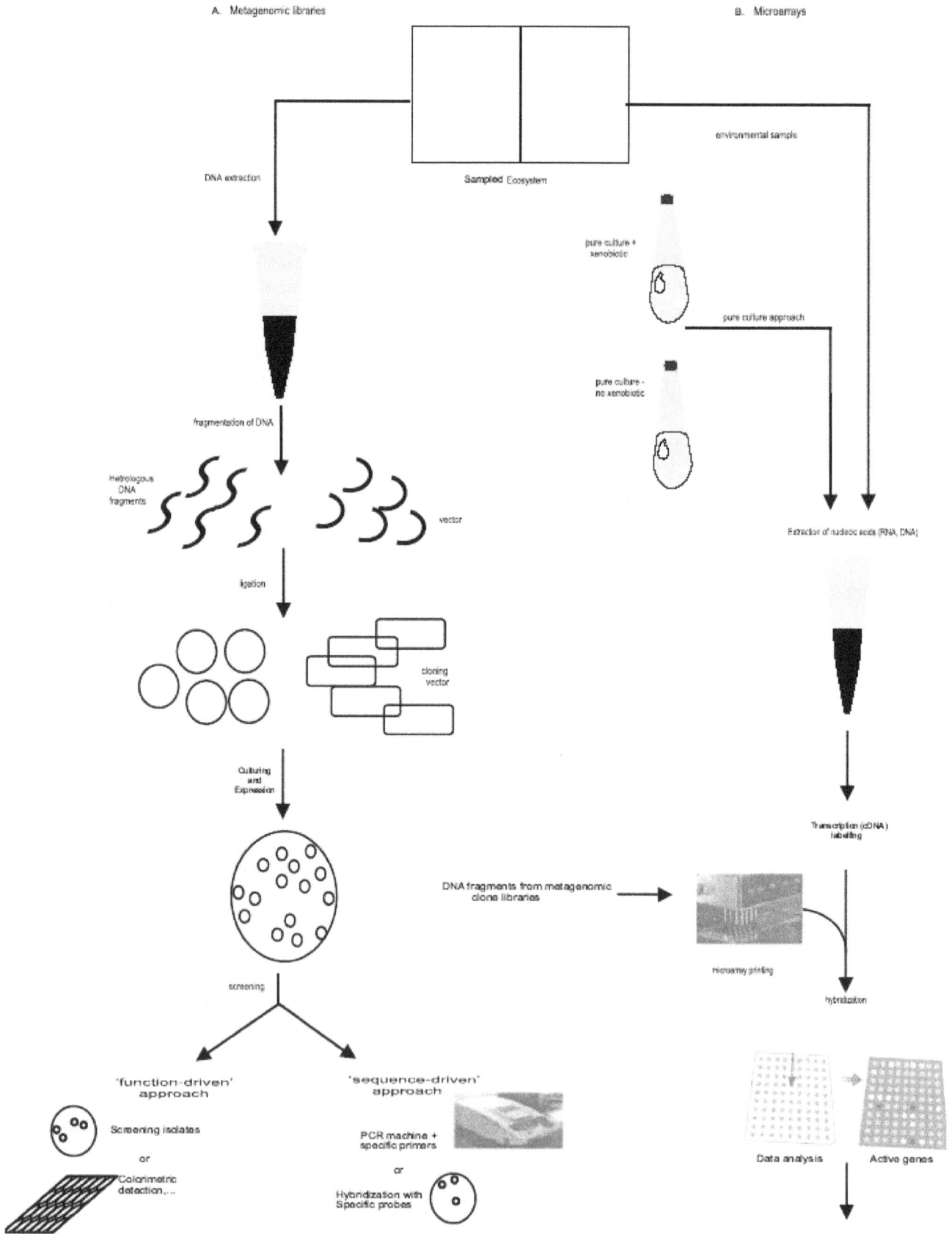

Figure 2. (A). Construction and analysis of metagenomic libraries from sampled ecosystems. (B). Construction of microarrays and hybridization with samples from the environment or pure cultures.

DNA microarrays is applied in research for gene expression profiling, that is, identification of changes in mRNA expression of strains exposed to a particular substrate, for example, a specific xenobiotic. Schut et al. (2001) constructed a DNA microarray with probes targeting 271 open reading frames (ORFs) from the genome sequence of the hyperthermophile *Pyrococcus furiosus*. When this strain was grown with elemental sulphur (S), two previously uncharacterized operons were identified and their products were proposed to be part of a novel S-reducing, membrane-associated, iron-sulphur cluster-containing complex. DNA microarray was also used to assess the mRNA expression levels in *Bacillus subtilis* grown under anaerobic condition (Ye et al., 2000). Transcriptional activities of more than 100 genes affected by the oxygen–limiting conditions were identified (Ye et al., 2000). Microarray has also been used for physiological studies of environment samples. Wu et al. (2001) detected genes involved in nitrogen –cycling (nirS) using 1ng of labelled genomic DNA of a *Pseudomonas stutzeri* strain and 25 ng of bulk community DNA extracted from soil samples. Catabolic genes involved in degradation of xenobiotics were assessed with microarrays by Dennis et al. (2003).

In microarray technology, sequence information is needed to design probes. However, this approach cannot be applicable for discovering new catabolic genes for which no sequences are available in databases. Moreover, knowledge of the entire sequence is not necessary for the construction of microarrays; and PCR products of a random genomic library constructed from a microorganism of interest may be used. It is expected that to a toxic substrate, differential gene expression would result at the transcript level. This is reflected by differential hybridization patterns in the presence or absence of the toxic pollutant. Afterward, clones of the library associated with differentially hybridized probes can be picked up for sequencing. This methodology applied in identifying genes involved in N$_2$ –fixation in *Leptospirillum ferrooxidans* by Parro and Moreno-Paz (2003). Sebat et al. (2003) used metagenomes as microarrays by deriving cosmid library from a microcosm of groundwater and used this library as probes for microarrays. Afterwards, they hybridized the microarrays with cDNA of individual strains isolated from the microcosm and cDNA of the microcosm itself. Comparisons of the hybridization profiles of the microcosm with isolated strains, clones were identified in the library corresponding to uncultured members of the microcosm. Sequencing of these clones revealed ORFs assigned to functions that have potential ecological importance, including hydrogen oxidation, NO$_3$-reduction and transposition (Sebat et al., 2003).

Therefore a random genomic library approach associated with microarrays offers a high potential for the discovery of novel genes and operons. This is desirable since it offers additional information apart from the biology of the microorganisms but precise knowledge of bio-

degradation processes of xenobiotics for application in pollution control and prevention under field conditions.

LIMITATIONS OF MICROARRAY TECHNOLOGY

Presence of humic matter, organic contaminants and metals in environmental samples that may interfere with RNA and DNA hybridization (Zhou and Thompson, 2002). Possibility of extraction of undegraded mRNA is additional problem (Burgmann et al., 2003). In cases of sequences of poor abundance, microarrays are not superior in sensitivity to PCR that is 100 to 10,000 fold more than that of microarrays (Zhou and Thompson, 2002). The success of the application of microarray technology in a study lies in the possibility of determining, in complex environments, the relative abundance of a microorganism bearing a specific functional gene. Therefore, it is important to be able to differentiate between differences in hybridization signals due to population abundance from those due to sequence divergence (Wu et al., 2001).

COMPLEMENTARY ROLES OF FUNCTION-DRIVEN AND SEQUENCE-DRIVEN APPROACHES

In the case of metagenomic libraries, sequencing clones with interesting functional properties may reveal a sequence that can be used to confirm the phylogenetic affiliation of the organism from which the DNA was isolated. Where the sequencing clones harbours an rRNA sequence in a big fragment that can lead to functional information about the microorganism from which the fragment originated (Beja et al., 2000; Liles et al., 2003; Quaiser et al., 2003). Environmental systems often contain a high diversity of bacteria, the use of labelled xenobiotics can provide information about active bacteria within a complex environmental system such bacteria incorporates radio labelled atoms into their DNA, making it denser than non-labelled DNA. By centrifugation, separation of labelled from non-labelled DNA can be achieved (Radajewski et al., 2000) and thereby distinguish the bacteria involved in the catabolic process from those which are not. This labelled DNA can be used afterwards to construct metagenomic libraries. This forms an excellent method for finding clones carrying catabolic genes.

Conclusion

It is clear that polluted sites, particularly those that are or become sources of contamination of surface and groundwater, have to be remediated. Bioremediation can be a cost-effective and ecologically suitable alternative to physical methods.

The genetic characterization of an increasing number of aerobic pathways for degradation of (substituted) aromatic compounds in different bacteria has made it possible to compare the similarities in genetic organization and in sequence, which exist between genes and proteins of these specialized catabolic routes and more common pathways. Sequence information provided the scientific evidence of the occurrence of catabolic genes coding for specialized enzymes in the degradation of xenobiotic chemicals. Moreover, molecular biology evidences corroborated the fact that a range of genetic mechanisms, such as gene transfer, mutational drift, genetic recombination and transposition can accelerate the evolution of catabolic pathways in microorganisms.

Metagenomic libraries open access to the world of uncultivated microorganisms and their undescribed catabolic genes for the degradation of xenobiotics. Microarrays are also useful for the discovering new catabolic genes as well as provide the opportunity of easily monitoring catabolic genes. In both approaches, the use of radio-labelled molecules could improve the recovery and identification of microorganisms involved in biodegradation of xenobiotics. The technical challenges associated with metagenomic libraries and microarrays not withstanding, these methods presents an exceptional opportunity for discovering the scientific basis of microbial degradation of xenobiotics. The application of both metagenomic libraries and DNA microarrays in bioremediation processes will facilitate rapid transfer of genetic information between axenic and native microbial communities, thus enhancing microbial adaptation to metabolism of xenobiotics in the environment.

In order to minimize future environmental impact by xenobiotics it may be economical to develop new synthetic compounds that fit in the naturally existing catabolic potential of the microorganisms.

REFERENCES

Abraham WR, Nogales B, Golyshin PN, Pieper DH, Timmis KN (2002). Polychlorinated biphenyl-degrading microbial communities in soil and sediments. Curr. Opin. Microbiol. 5: 246-253.

Aelion CM, Swindoll CM, Pfaender FK (1987). Adaptation to and biodegradation of xenobiotic compounds by microbial communities from pristine aquifer. Appl. Environ. Microbiol. 53: 2212-2217.

Amann RI, Ludwig W, Schlefer KH (1995). Phylogenetic identification and in situ detection of individual cells without cultivation. Microbiol. Rev. **59**: 143-169.

Assinder SJ, Williams PA (1988). Comparison of the *meta* pathway operons on NAH plasmids *pWW60-22* and *TOL* plasmid *pWW53-4* and its evolutionary significance. J. Gen. Microbiol. 134: 2769-2778.

Barkay T, Pritchard H (1988). Adaptation of aquatic microbial communities to pollutant stress. Microbiol. Sci. 5:165-169.

Basta T, Buerger S, Stolz A (2005). Structural and replicative diversity of large plasmids from sphingomonads that degrade polycyclic aromatic compounds and xenobiotics. Microbiology 151: 2025-2037.

Bej AK, Steffan RJ, Di Cesare J, Haff L, Atlas RM (1990). Detection of coliform bacteria in water by polymerase chain reaction and gene probes. Appl. Environ. Microbiol. 56: 307-314.

Beja O, Aravind L, Koonin EV, Suzuki MT, Hadd A, Nguyen LP, Jovanovich SB, Gates CM, Feldmanc RA, Spudich JL, Spudich EN, Delong EF (2000). Bacterial rhodopsin: evidence for a new type of phototrophy in the sea. Science 289: 1902-1906.

Black JG (1999). Bioremediation In: *Microbiol. Principles and explorations.*Pp. 751-752.

Blehert DS, Fox BG, Chambliss GH (1999). Cloning and sequence analysis of two *Pseudomonas* flavoprotein xenobiotic reductases. J. Bacteriol. 181: 6254-6263.

Burgmann H, Widmer F, Sigler WV, Zeyer J (2003). mRNA extraction and reverse transcription-PCR protocol for detection of *nifH* gene expression by *Azotobacter vinelandii* in soil. Appl. Environ. Microbiol. 69:1928-1935

Celis JE, Kruhoffer M, Gromova I, Frederiksen C, Ostergaard M, Thykjaer T, Gromov P, Yu H, Palsdottir H, Magnusson N, Orntoft TF (2000). Gene expression profiling: monitoring transcription and translation products using DNA microarrays and proteomics. FEBS Lett. 480: 2-16.

Clarke PH (1984). The evolution of degradative pathways. In: DT Gibson (ed.), Microbial degradation of organic compounds. Marcel Dekker, Inc. New York. pp. 11-27.

Dennis P, Edwards EA, Liss SN, Fulthorpe R (2003). Monitoring gene expression in mixed microbial communities by using DNA microarrays. Appl. Environ. Microbiol. 69: 769-778.

Don RH, Weightman AJ, Knackmuss HJ, Timmis KN (1985). Transposons mutagenesis and cloning analysis of the pathways for degradation of 2, 4-dichlorophenoxyacetic acids and 3-chlorobenzoate in *Alcalgenes eutorphus* JMP134 (pJP4). J. Bacteriol. 161: 85-90.

Doten RC, Ngai KL, Mitchell DJ, Ornston LN (1987). Cloning and genetic organization of the pca gene cluster from Acinetobacter calcoaceticus. J. Bacteriol. 169: 3168-3174.

El- Fantroussi S, Naveau H, Agathos SS (1998). Anaerobic dechlorinating Bacteria. Biotechnol. Prog. 14:167-188.

El-Fantroussi S (2000). Enrichment and molecular characterization of a bacterial culture that degrades methoxy-methyl urea herbicides and their aniline derivatives. Appl. Environ. Microbiol. 66(12): 5110-5115.

Entcheva P, Liebl W, Johann A, Hartsch T, Streit WR (2001). Direct cloning from enrichment cultures, a reliable strategy for isolation of complete operons and genes from microbial consortia. Appl. Environ. Microbiol. 67: 89-99.

Esteve-Nenez A, Caballero A, Ramos JL (2001). Biological degradation of 2, 4, 6-trinitrotoluene. Microbiol. Mol. Biol. Rev. 65:335-352.

Eyers L, Stenuit B, El fantroussi S, Agathos SN (2004). Microbial characterization of TNT-contaminated soils and anaerobic TNT degradation: high and unusual denitration activity. In: W. Verstraete (ed.) *Eur. Symp. Environ. Biotechnol.* 5, Oostende Belgium, pp. 51-54.

Ferraroni M, Solyanikova IP, Kolomytseva MP, Scozzafava A, Golovleva L, Briganti F (2004). Crystal Structure of 4-Chlorocatechol 1,2-Dioxygenase from the Chlorophenol-utilizing Gram-positive *Rhodococcus opacus* 1CP. J. Biol. Chem. 279(26): 27646-27655.

Fessehaie A, De Boer SH, Levesque CA (2003). An oligonucleotide array for the identification and differentiation of bacteria pathogenic on potato. Phytopathology 93: 262-269.

Guschin DY, Mobarry BK, Proudnikov D, Stahl DA, Rittman BE, Mirzabekov AD (1997). Oligonucleotide microchips as genosensors for determinative and environmental studies in microbiology. Appl. Environ. Microbiol. 63: 2397-2402.

Ghosal D, You S (1989). Operon structure and nucleotide homology of the chlorocatecol oxidation genes of plasmids pAC27 and pJP4. Gene 83: 225-232.

Goldstein RM, Mallory LM, Alexander M (1985). Reasons for possible failure of inoculation to enhance biodegradation. Appl. Environ. Microbiol. 50: 977-983.

Haigler BE, Nishino SF, Spain JC (1988). Degradation of 1, 2-dichlorobenzene by a *Pseudomonas* sp. Strain JS6. Appl. Environ. Microbiol. 54: 294-301.

Haigler BE, Spain JC (1989). Degradation of p-chlorotoluene by a mut ant of *Pseudomonas* sp. Strain JS6. Appl. Environ. Microbiol. 55: 372-379.

Heiss G, Hofmann KW, Trachtmann N, Walters DM, Rouviere P, Knackmuss H-J (2002). npd gene functions of Rhodococcus (opacus) erythropolis HL PM-1 in the initial steps of 2,4,6-trinitrophenol degradation. Microbiology 148:799-806.

Heiss G, Trachtmann N, Abe Y, Takeo M, Knackmuss H-J (2003). Homologous npdGI genes in 2, 4-Dinitrophenol-and 4-nitrophenol-degrading Rhodococcus spp. Appl. Environ. Microbiol. 69(5): 2748-2754.

Hughes MA, Williams PA (2001). Cloning and characterization of *pnb* genes, encoding enzymes for 4-nitrobenzoate catabolism in *Pseudomonas putida* TW3. J. Bacteriol. 183:1225-1232.

James KD, Williams PA (1998). *ntn* genes determining the early steps in the divergent catabolism of 4-nitrotoluene and toluene in *Pseudomonas* sp. Strain TW3. J. Bacteriol. 180: 2043-2049.

James KD, Hughes MA, Williams PA (2000). Cloning and expression of *ntnD, encoding a novel NAD (P)*[+]*-independent* 4-nitrobenzyl alcohol dehydrogenase from *Pseudomonas* sp. Strain TW3. J. Bacteriol. 182: 3136-3141.

Johnson GR, Jain RK, Spain JC (2002). Origins of the 2,4-dinitrotoulene pathway. J. Bacteriol. 184: 4219-4232.

Kumar A, Larson O, Parodi　D, Liang Z (2000). Silinized nucleic acids: a general platform for DNA immobilization. Nuc. Acids Res. 28: E71.

Leadbetter JR (2003). Cultivation of recalcitrant microbes: Cells are alive, well and revealing their secrets in the 21st. century laboratory. Curr. Opin. Microbiol. 6: 274-281.

Levesque CA (1997). Molecular detection tools in integrated disease management: Overcoming current limitations. Phytoparasitica 25:3-7.

Levesque CA (2001). Molecular methods for detection of plant pathogens-what is the future? Can. J. Plant Pathol. 24: 333-336.

Liles MR, Manske BF, Bintrim SB, Handelsman J, Goodman RM (2003). A census of rRNA genes and linked genomic sequences within a soil metagenomic library. Appl. Environ. Microbiol. 69: 2684-2691.

Niedle EL, Harnett C, Bonitz S, Ornston LN (1988). DNA sequence of *Acinetobacter calcoaceticus catechol* 1,2-dioxygenase-I structural gene *cat*A: evidence for evolutionary divergence of intradiol dioxygenases by acquisition of DNA sequence repetitions. J. Bacteriol. 170: 4874-4880.

Niedle EL, Ornston LN (1986).Cloning and expression of *Acinetobacter calcoaceticus* catechol 1, 2-dioxygenase structural gene *cat*A in *Escherichia coli*. J. Bacteriol. 168: 815-820.

Okada H, Negoro S, Kimura H, Nakamura S (1983). Evolutionary adaptation of plasmid-encoded enzymes for degrading nylon oligomers. Nature (London) 306: 203-206.

Parales JV, Kumar A, Parales RE, Gibson DT (1996). Cloning and sequencing of genes encoding 2-nitrotoluene dioxygenase from *Pseudomonas* sp. JS42. Gene 181: 57-61.

Parro V, Moreno-Paz M (2003). Gene function analysis in environmental isolates: the *nif* regulon of the strict iron oxidizing bacterium *Leptospirillum ferrooxidans*. Proc. Natl. Acad. Sci. USA 100: 7883-7888.

Quaiser A, Ochsenreiter T, Lanz　C, Schuster SC, Treusch AH, Eck J, Schleper C (2003). Acidobacteria form a coherent but highly diverse group within the bacterial domain: evidence from environmental genomics. Mol. Microbiol. 50:563-575.

Radajewski S, Ineson P, Parekh NR, Murrell JC (2000). Stable-isotope probing as a tool in microbial ecology. Nature 403: 646-649.

Reiger PG, Meier H-M, Gerle M, Vogt U, Groth T, Knackmuss H-J (2002). Xenobiotics in the environment: present and future strategies to obviate the problem of biological persistence. J. Biotech. 94: 101-123.

Reineke W (1984). Microbial degradation of halogenated aromatic compounds, In: DT. Gibson (ed.), Microbial degradation of organic compounds. Marcel Dekker, Inc. New York. pp. 319-360.

Reineke W, Jeenes DJ, Williams PA, Knackmuss HA (1982). TOL plasmid in constructed halo benzoate-degrading *Pseudomonas* strains: prevention of *meta* pathway. J. Bacteriol. 150: 195-201.

Rondon MR, August PR, Bettermann AD, Brady SF, Grossman TH, Liles MR, Loiacono　KA, Lynch BA, Macneil IA, Micor C, Tiong CL, Gilman M, Osburne MS, Clardy J, Handelsman J, Goodman RM (2000). Cloning the soil metagenome: a strategy for accessing the genetic and functional diversity of uncultured microorganisms. Appl. Environ. Microbiol. 66: 2541-2547.

Saye DJ, Ogunseitan OA, Sayler GS, Miller RV (1990).Transduction of linked chromosomal genes between *Pseudomonas aeruginosa* strains during incubation *in situ* in a fresh water habitat. Appl. Environ. Microbiol. 56: 140-145.

Sayler GS, Layton AC (1990). Environmental application of nucleic acid Hybridization. Annual Rev. Microbiol. 44: 625-648.

Schaad NW, Berthier SY, Sechler A, Knorr D (1999). Detection of *Clavibacter michiganensis* subsp. *sepedonicus* in potato tubers by

BIO-PCR and an automated real-time fluorescence detection system. Plant Dis. 83:1095-1100.

Schlegel HG (1986). Xenobiotics In: *General Microbiology* 6th edition Cambridge University Press N.Y. p. 433.

Schloss PD, Handelsman J (2003). Biotechnological prospects from metagenomics. Curr. Opin. Microbiol. 14:303-310

Sebat JL, Colwell FS, Crawford RL (2003). Metagenomic profiling: microarray analysis of an environmental genomic library. Appl. Environ. Microbiol. 69: 4927-4934.

Shizuya H, Simon M (1992). Cloning and stable maintenance of 300-kilobase-pair fragments of human DNA in *Escherichia coli* using an F-factor-based vector. Proc. Natl. Acad. Sci. USA. 89: 8794-8797.

Schut G, Zhou J, Adams MW (2001). DNA microarray analysis of the hyperthermophilic archaeon *Pyrococcus furiosus*: evidence for a new type of sulfur-reducing enzyme complex. J. Bacteriol. 183: 7027-7036.

Spain JC, Van Veld PA (1983). Adaptation of natural microbial communities to degradation of xenobiotic compounds: effects of concentration, exposure, time, inoculum, and chemical structure. Appl. Environ. Microbiol. 45: 428-435.

Timmis KN, Pieper DH (1999). Bacteria designed for bioremediation. *TIBTECH* 17: 201-204.

Torsvik V, Qvreas L (2002). Microbial diversity and function in soil: from genes to ecosystem. Curr. Opin. Microbiol. 5:240-245.

Tsuda M, Minegishi KI, Lino T (1989). Toluene transposons Tn4561 and Tn4563 are class II transposons. J. Bacteriol. 171: 1386-1393.

Van der Meer JR, Eggen　RIL, Zehnder　AJB, De Vos WM (1991). Sequence analysis of *Pseudomonas* sp. Strain P51 *tcb* gene cluster, which encodes metabolism of chlorinated catecols: evidence for specialization of catecol 1, 2-dioxygenases for chlorinated substrates. J. Bacteriol. 173: 2425-2434.

Venter JC, Remington K, Heidelberg JF, Halpern AL, Rusch D, Eisen JA, Wu DY, Paulsen I, Nelson KE, Nelson W, Fouts DE, Levy S, Knap AH, Lomas MW, Nealson K, White O, Peterson J, Hoffman J, Parsons R, Baden-Tillson H, Pfannkoch C, Rogers YH, Smith HO (2004). Environmental genome shotgun sequencing of the Sargasso Sea. Science 304: 66-74.

Walters DM, Russ R, Knackmuss H, Rouviere PE (2001). High-density sampling of a bacterial operon using mRNA differential display. Gene 273: 305-315.

Ward DM, Weller R, Bateson MM (1990). 16s rRNA sequences reveal numerous uncultured microorganisms in a natural community. Nature 344: 63-65.

Weisburg WG, Barns SM, Pelletier DA, Lane DJ (1991). 16S ribosomal DNA amplification for phylogenetic study. J. Bacteriol. 173: 697-703.

Wu L, Thompson DK, Li G, Hurt RA, Tiedje JM, Zhou J (2001). Development and evaluation of functional gene arrays for detection of selected genes in the environment. Appl. Environ. Microbiol. 67: 5780-5790.

Ye RW, Tao W, Bedzyk L, Young T, Chen M, Li G (2000). Global gene expression profiles of *bacillus subtilis* grown under anaerobic conditions. J. Bacteriol. 183: 7027-7036.

Ye RW, Wang T, Bedzyk L, Croker KM (2001). Applications of DNA microarrays in microbial systems. J. Microbiol. Methods. 47: 257-272.

Zhou J, Thompson DK (2002). Challenges in applying microarrays to environmental studies. Curr. Opin. Biotechnol. 13: 204-207.

Kojic acid: Applications and development of fermentation process for production

Rosfarizan Mohamad[1], Mohd Shamzi Mohamed[1], Nurashikin Suhaili[1], Madihah Mohd Salleh [2] and Arbakariya B. Ariff [1*]

[1]Department of Bioprocess Technology, Faculty of Biotechnology and Biomolecular Sciences, Universiti Putra Malaysia, 43400 UPM Serdang, Selangor, Malaysia.
[2]Faculty of Bioscience and Bioengineering, Universiti Teknologi Malaysia, Skudai, 81310 Johor, Malaysia.

Kojic acid, 5-hydroxy-2-hydroxymethyl-γ-pyrone, has many potential industrial applications. In this review, the properties and diverse applications of kojic acid in industries are described. The review also discusses the advance in kojic acid fermentation, focusing on the process development in microorganisms and strain selection, medium and culture optimization, as well as fermentation modes for commercially viable industrial scale production. The performances of various fermentation techniques that have been applied for the production of kojic acid are compared, while the advantages and disadvantages of each technique are discussed in this paper.

Key words: Kojic acid, mild antibiotic, anti-browning agent, tyrosinase inhibitor, submerged fermentation, resuspended cell system, surface culture.

INTRODUCTION

Kojic acid, which is an organic acid, is produced biologically by different types of fungi during aerobic fermentation using various substrates (Kitada et al., 1967; Ariff et al., 1996; Wakisaka et al., 1998; El-Aasar, 2006). The name 'kojic acid' (which was originally known as Koji acid) was derived from "Koji", the fungus starter or inoculum used in oriental food fermentations for many centuries. Its chemical structure was then extensively investigated and defined as 5-hydroxy-2-hydroxymethyl-γ-pyrone (Yabuta, 1924). Chemical synonyms of kojic acid are known as 5-hydroxy-2- hydroxymethyl-4H-pyran-4-one (Nandan and Polasa, 1985) and 5-hydroxy-2-hydroxymethyl-4-pyrone (Kahn et al., 1995).

The market for kojic acid has been developed for some 40 years since 1955, where Charles Pfizer and Company, USA, announced the first attempt to manufacture this organic acid. The company patented the methods for the production of kojic acid and its recovery, as well as the preparation of derivatives usable as pesticides. However, there was no urgent commercial use for this compound at that time until a rapid growth occurred in various industries

recently. In short, the interest in kojic acid is increasing enormously with a growing presence in industries related to its applications, especially in cosmetic industry (Brtko et al., 2004; Bentley, 2006). Although, kojic acid has been industrially produced and applied for some time, it is still extensively studied. Two main areas are normally considered; those associated with the strain development, and those concerned with the development of fermentation process. Kojic acid is produced by *Aspergillus* spp. and *Penicillium* spp., belonging mainly to the *flavus-oryzae-tamarii* groups. Among them, *A. flavus* (Basappa et al., 1970; Ariff et al., 1996), *A. oryzae* (Kwak and Rhee, 1992; Takamizawa et al., 1996), *A. tamarii* (Gould, 1938) and *A. parasiticus* (Nandan and Polasa, 1985; Coupland and Niehaus, 1987; El-Aasar, 2006) were reported to have the ability to produce large amounts of kojic acid. Although, several potential kojic acid producing strains have been isolated, very little attention has been paid to the improvement of the strains, either through mutation or genetic engineering techniques. Details of industrial techniques of kojic acid fermentation are rarely revealed since they comprise the proprietary know-how of each producing company. This review describes and discusses the properties and potential applications of kojic acid, as well as the development of kojic acid fermentation through both approaches, namely microbiology and pro-

*Corresponding author. E-mail: arbarif@biotech.upm.edu.my.

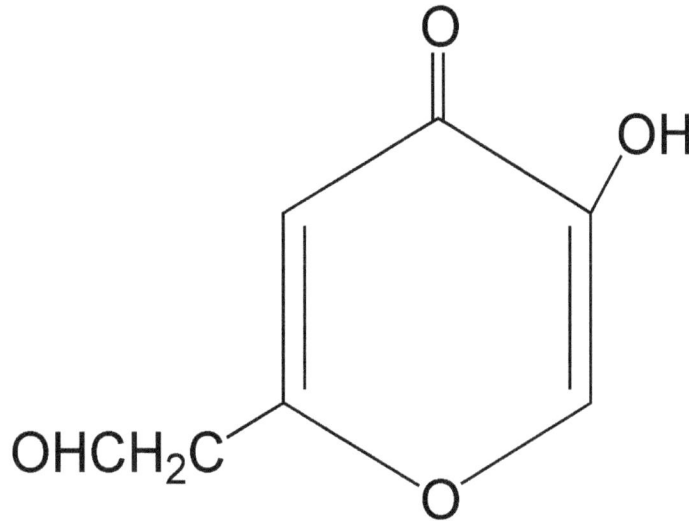

Figure 1. The chemical structure of kojic acid.

cess engineering.

The properties of kojic acid

Kojic acid crystallises in the form of colourless, prismatic needles that sublime in vacuum without any changes. Meanwhile, the melting point of kojic acid ranges from 151°C - 154°C (Ohyama and Mishima, 1990). Kojic acid is soluble in water, ethanol and ethyl acetate. On the contrary, it is less soluble in ether, alcohol ether mixture, chloroform and pyridine (Wilson, 1971). The molecular weight of kojic acid, as determined by the cryoscopic method for a formula of $C_6H_6O_4$, is 142.1 (Uchino et al., 1988). Kojic acid has a maximum peak of ultraviolet absorption spectra at 280 -284 nm (Choi et al., 2002; Watanabe-Akanuma et al., 2007).

The chemical structure of kojic acid is shown in Figure 1. Kojic acid is classified as a multifunctional, reactive γ-pyrone with weakly acidic properties. It is reactive at every position on the ring and a number of products which have values in industrial chemistry, such as metal chelates, pyridones, pyridines, ethers, azodyes, mannich base, and the products of cyanoethylation can be formed from kojic acid (Ichimoto et al., 1965; Wilson, 1971). Numerous chemical reactions of kojic acid have been studied over the decades since its isolation. At carbon 5 positions, the hydroxyl group acts as a weak acid, which is capable to form salts with few metals such as sodium, zinc, copper, calcium, nickel and cadmium (Crueger and Crueger, 1984).

THE APPLICATIONS OF KOJIC ACID

The most striking benefit of kojic acid is found in cosmetic and health care industries. It primarily functions as the basic material for the production of skin whitening creams, skin protective lotions, whitening soaps and tooth care products. Kojic acid has the ability to act as the ultra violet protector, whereby, it suppresses hyper-pigmentation in human skins by restraining the formation of melanin through the inhibition of tyrosinase formation, the enzyme that is responsible for skin pigmentation (Ohyama and Mishima, 1990; Noh et al., 2009). Kojic acid has a melanogenesis-inhibitory effect on *in-vitro* living pigment cells. Kojic acid induces a distinct reduction of eumelanin content and its essential precursor monomer, 5,6DHI 2C in hyper-pigmented B16 cells. The melanogenesis inhibitory effect of kojic acid on the ultraviolet-induced hyper-pigmentation and pigmentary disorders of human skins has been reported in some studies (Ohyama and Mishima, 1990; Lee et al., 2006). At present, kojic acid is primarily used as the basic ingredient for excellent skin lightener in cosmetic creams, where it is used to block the formation of pigment by the deep cells on the skins (Masse et al., 2001). Since the incident of skin cancer is increasing rapidly due to exposure towards high ultraviolet radiation of sunlight, currently, this acid is also widely used in cosmetic industry as a skin protective lotion. It is normally used in combination with alpha-hydroxy acid in the formulation of skin whiteners to control lightened freckles and age spots. Hydroquinone has been banned for cosmetic usage in Asia, and it is noted as a possible carcinogenic compound by the Food and Drug Authority (FDA) of USA. This has led to a significant increase in the use of kojic acid as a replacement for hydroquinone that bleaches and possibly damages skins in cosmetic products. In addition, kojic acid and its manganese and zinc complexes can potentially be used as radio protective agents, particularly against γ-ray (Emami et al., 2007).

Recently, methods for the synthesis of various kojic

Table 1. Applications of kojic acid.

Fields	Functions	References
Medical	Antibacterial	Kotani et al. (1976); Nohynek et al. (2004).
	Antifungal	Kayahara et al. (1990); Balaz et al. (1993).
	Pain killer	Beelik (1956).
Food	Flavour enhancers	Wood (1998); Burdock et al. (2001)
	Antioxidant	Chen et al. (1991); Niwa and Miyachi (1986).
	Maltol and ethyl maltol	Tatsumi et al. (1969).
Agriculture	Anti melanosis	Chen et al. (1991); Saruno et al. (1978).
	Insecticide activator	Buchta (1982); Dowd (1988); Dobias et al. (1977).
Cosmetic	Whitening agent	Niwa and Akamatsu (1991); Masse et al. (2001); Bently (2006); Ohyama and Mishima (1990).
	Ultra violet filter	Noh et al. (2009).
	Tyrosinase inhibitor	
	Radical scavenging activity	Niwa and Akamatsu (1991).
	Radioprotective agent	Emami et al. (2007).
Chemistry	Reagent for iron determination	Bentley (1957).
	Synthesis of 2-methyl-4-pyrone	Hasizume et al. (1968).
	Iron chelator	Mitani et al. (2001).
	Kojic acid-chitosan conjugates	Guibal (2004), Synytsya et al. (2008).

acid derivatives, such as kojic acid ester, kojic acid laureate and kojic acid dipalmitate have been reported in many studies (e.g. Brtko et al., 2004; Lee et al., 2006; Khamaruddin et al., 2008; Ashari et al., 2009). These derivatives have been found to improve both the stability and solubility of kojic acid in oily cosmetic products. In addition, the tyrosinase inhibitory activity of kojic acid derivatives, which are synthesized through an ethylene linkage of phosphonate with aldehyde using interme-diates derived from kojic acid, is about 8 times more potent than kojic acid (Lee et al., 2006).

Apart from its main application in cosmetics, kojic acid also has many potential industrial applications (Table 1). It is the first pyrone derivative that is chemically used for analytical iron determinations in ores. Metal chelates of kojic acid have been advocated as the source materials giving the controlled release of metallic ions in curing agents or catalysts (Wilson, 1971).

Kojic acid gives a deep red colour with as little as 0.1 ppm ferric ion, and the ferric complex cannot be reversibly oxidised or reduced (Buchta, 1982).

The natural origin of kojic acid, which ensures its non-problematic biodegradation, makes it an attractive basic skeleton for the development of biologically active compound via derivatization. Kojic acid and its derivatives possess antibiotic properties against gram-negative as well as gram-positive micro-organisms (Kotani et al.,

1976). The bacterial growth is generally inhibited in the presence of more than 0.5% (w/v) of kojic acid (Beelik, 1956). It is also active against human tubercle bacilli in *in vitro* technique under a variety of conditions, whereby 45 mg/100 mL of kojic acid completely inhibits the surface growth of bacilli (Lee et al., 1950). In addition, kojic acid in the form of azidometalkojates also shows antibacterial and antifungal effects on several species of *Bacillus*, *Staphylacoccus*, *Saccharomyces*, *Aspergillus*, *Rhizopus* and *Fusarium*. Azidometalkojates, in the form of zinc derivatives, also exhibits a certain cytotoxic activity on HeLa tumour cells (Hudecova et al., 1996). Recently, Nohynek et al. (2004) reported that the antibacterial feature of kojic acid, over a few common bacterial strains, is found to be significant only in dilutions of 1:1000 - 1:2000.

Kojic acid and its derivatives (mostly at the 7-iodo kojic acid) have a potent activity against bacteria such as *Staphylococcus aureus* 209 (Kotani et al., 1976). The compound of kojic acid derivatives was also tested for antifungal activities against *Phythium graminicola*, *Fusarium oxysporum* and *Rhizoctonia solani*, which cause seedling blight, fusarium wilt and sheath blight, respectively (Kayahara et al. 1990; Balaz et al., 1993). Besides its antibiotic functions, kojic acid also shows a certain insecticidal activity against *Heliothis zea* and *Spodoptera frugiperda* insects. In addition, it has been employed as a chelating agent for the production of insecticides (Buchta, 1982; Dowd, 1988). Kojic acid has been shown to inhibit the growth of *Trogoderma* larvae and induce sterility in males and females of this genus (Sehgal, 1976). Furthermore, it also inhibits the development of *Musca domestica* (Beard and Walton, 1969), as well as that of *Drosophila melanogaster* (Dobias et al., 1977). The potential use of kojic acid and its derivatives in humans and/or veterinary medicines as biological active compounds has also been reviewed by Brtko et al. (2004).

Kojic acid is distributed naturally in traditional Japanese food such as miso, soy sauce and sake, thus, endowing these various food types with special tastes, colours and flavours (Wood, 1998). Not only that, kojic acid also acts as a precursor for flavour enhancers (that is, maltol and ethyl maltol). It is used in the production of comenic acid, which is an intermediate for the synthesis of maltol (3-hydroxy-2-methyl-γ-pyrone) and its derivatives (Tatsumi et al., 1969). The synthesis of maltol is useful for improving the flavours of various food products and as the ingredient in perfumes and flavours (Ichimoto et al., 1965). The health aspects of kojic acid in food have been evaluated by Burdock et al. (2001). In fact, the researchers suggested that the consumption of kojic acid, at levels normally found in food, does not require a concern for safety.

Uchino et al. (1988) recognised kojic acid for its antispeck activity. It can be used to prevent browning formation (speck) during the storage and processing of raw noodles (uncooked). Besides, it also has an inhibitory effect on polyphenol oxidase in mushrooms (Saruno et al., 1978), apples, potatoes and crustaceans including white shrimps, grass prawns and Florida spiny lobsters (Chen et al., 1991). The inhibitory effect of kojic acid on polyphenol oxidase is associated to the inhibition of melanosis by interfering the uptake of oxygen required for enzymatic browning, and reduction of o-quinones to diphenols to prevent the formation of the final pigment (melanin) or the combination of the above actions.

Kojic acid is recognised as an important intermediate in the production of chemicals that can be used as pharmaceuticals. Novotny et al. (1999) claimed that kojic acid could be used in the preparation of compounds with an anti-neoplastic potential. In addition, the anti-neoplastic activity of kojic acid derivatives is based on various mechanisms of actions on different levels of cellular metabolism and functions, which make this compound useful as a cytotoxic agent. Garcia and Fulton (1996) reported that kojic acid, in combination with glycolic acid and hydroquinone, can be used for the treatment of melasma and related conditions.

The formulation is now available to dermatologists to satisfy the patient's preferences. Although, hydroquinone alone is effective and has been available for years, kojic acid has the advantages of being pharmaceutically more stable and also as a tyrosinase inhibitor. Recently, it has been reported that kojic acid can be easily conjugated

with chitosan to produce kojic acid-chitosan conjugates, suggesting that kojic acid has a potential use in chemical industry (Guibal 2004; Synytsya et al., 2008). At present, chitosan is well-known for its wide applications in various industries such as food, pharmaceuticals, cosmetics, agriculture and environment (Ravi-Kumar, 2000). Moreover, kojic acid has also been conjugated with amino acids to form conjugates which exhibit a higher tyrosinase inhibition activity and stability than kojic acid alone (Noh et al., 2009).

THE DEVELOPMENT OF KOJIC ACID FERMENTATION

Micro-organisms and strain improvement

In general, *Aspergillus* spp., which belongs mainly to the *A. tamarii* group, is widely used in fermentation of kojic acid. Kitada et al. (1967); Kwak and Rhee (1991) produced kojic acid using *A. oryzae* and yielded 0.26 g of kojic acid/g glucose. In addition, kojic acid production by *A. parasiticus* and *A. candidus* has also been reported by El-Aasar (2006) and Wei et al. (1991), respectively. In these two cases, the yield was 0.089 g kojic acid/g glucose and 0.3 g kojic acid/g sucrose, respectively. On the other hand, a very high yield (that is, 0.453 g kojic acid/g glucose) was also obtained in the fermentation using *A. flavus* (Ariff et al., 1997; Rosfarizan et al., 2007). Having mentioned this, the risk of aflatoxin production by this strain cannot be ignored. However, Madihah (1996) reported that aflatoxin production by kojic acid producing *A. flavus* can be inhibited by a suitable medium formulation and appropriate culture conditions. In addition, kojic acid and aflatoxin synthesis follow different pathways and therefore, kojic acid is not an intermediate for the synthesis of aflatoxin by *Aspergillus* spp. (Basappa et al., 1970).

Kojic acid producing strains, *A. flavus* has been improved by the monospore isolation method to obtain a stable monokaryotic strain capable of producing a substantially higher amount of kojic acid (19.2 g/L) as compared to the unstable heterokaryons (10.5 g/L) (Rosfarizan et al., 1998). Kojic acid producing strain could also be improved through mutation and genetic recombination techniques (Fantini, 1975; Crueger and Crueger, 1984). Moreover, the improvement of kojic acid production by phototropic *A. oryzae* (green conidia) through mutation has been reported by Demain (1973). The mutation was achieved by producing lysine auxotrophy with yellow conidia, and this was found to produce kojic acid five times higher than the parent strain. A mutant strain of *A. oryzae* MK107-39 was also capable to produce kojic acid by about 7.7 times than its parent strain (Futamura et al., 2001). The mutation of *A. oryzae* ATCC 22788 via NTG treatment and UV irradiation, followed by protoplast was also found to improve kojic acid production (41 g/L) of about 100 times been reported that kojic acid can be easily conjugated

Table 2. Carbon sources for kojic acid production by various types of micro-organism.

Carbon source	Number of carbon atom	Micro-organism	References
Ethanol	2	A. oryzae	Basappa et al. (1970)
Glycine			
Sodium acetate		A. flavus	Arnstein and Bentley (1956)
1,3-Dihydroxy-2-propanone	3	A. oryzae	Arnstein and Bentley (1956)
Glycerol		A. tamarii	Gould (1938)
Tartaric acid	4	A. oryzae	Tamiya (1932)
Arabinose	5	A. flavus	Arnstein and Bentley (1953a),
Xylose			Basappa et al. (1970)
Fructose	6	A. tamarii	Gould (1938)
Glucose		A. luteo-virescens	Morton (1945)
		A. oryzae	Arnstein and Bentley (1956), Kitada et al. (1967), Takamizawa et al. (1996)
		A. flavus	Ariff et al. (1996), Kwak and Rhee (1991), Bajpai et al. (1982), Wan et al. (2005)
		A. candidus	Wei et al. (1991)
		A. parasiticus	El-Aasar (2006).
Shikimic acid	7	A. oryzae	Katagiri and Kitahara (1933)
Maltose	12	A. oryzae	Kitada et al. (1967)
Sucrose		A. tamari	Marston (1949)
		A. flavus	Rosfarizan and Ariff (2007)
Starch	-	A. flavus	Rosfarizan et al. (1998)

OPTIMISATION OF MEDIUM COMPOSITION

Carbon sources

A variety of carbon containing substrates may be used as carbon sources for kojic acid fermentation (Table 2). These substrates include starch, sucrose, maltose, glucose, fructose, mannose, galactose, xylose, arabinose, sorbitol, acetate, ethanol, glycerol and arabinose (Arnstein and Bentley, 1956; Wilson, 1971; Burdock et al., 2001; Rosfarizan and Ariff 2007). The use of various carbon sources such as starch, sucrose, fructose, glucose and xylose for kojic acid fermentation by A. oryzae had also been investigated by Kitada et al. (1967). Excellent growth was obtained in all types of carbon sources investigated. Nevertheless, kojic acid was not detected in the fermentation using starch, fructose and xylose. The highest yield of kojic acid was obtained in the fermentation using glucose as the carbon source, followed by sucrose and fructose. Yields ranging from 0.5 - 0.6 g kojic acid/g glucose that can be obtained from fermentation by various kojic acid-producing strains (Kitada et al., 1967). Thus, glucose is not only used as a carbon source for biomass built-up, but it is also used as a precursor for kojic acid synthesis (Arnstein and Bentley, 1956; Kitada and Fukimbara, 1971).

On the other hand, Wei et al. (1991) reported that sucrose was a better carbon source than glucose for the production of kojic acid by A. candidus, in which the acid yielded 0.5 g kojic acid/g sucrose for the fermentation using sucrose that produced about two times higher than the yield obtained from glucose, that is, 0.25 g kojic acid/g glucose. Generally, polysaccharides with a long chain of carbon sources, such as starch, are considered as poor carbon sources for fermentation of kojic acid. A. flavus strain (S33-2), which is capable of growing on cooked starch and producing substantially high level of kojic acid, has been isolated (Rosfarizan et al., 1998). The yield of 0.25 g kojic acid/g starch was obtained in the fermentation using corn starch.

The concentration of glucose, as a carbon source, also greatly influences kojic acid production. Kitada et al. (1967) investigated the effect of glucose concentrations ranging from 25 - 150 g/L on the production of kojic acid by A. oryzae. Although, the maximum cell yield was not significantly different at above 50 g/L glucose, the highest kojic acid production (24.2 g/L) was obtained using 100 g/L glucose, giving a yield of 0.24 kojic acid/g glucose. At 25 and 50 g/L glucose, all glucose supplied was consumed for biomass built-up and kojic acid production was delayed. About 50% of the supplied glucose was not consumed during the fermentation using 150 g/L glucose. El-Aasar (2006) also reported that the highest production of kojic acid was obtained in the fermentations which employed either 60 g/L glucose, 40 g/L sucrose and 60 g/L beet molasses with the yields of 0.43 g kojic acid/g

Table 3. Nitrogen sources for kojic acid production by various types of micro-organisms.

Micro-organism	Nitrogen source (g/L)		References
A. oryzae	$(NH_4)_2SO_4$	0.5	Katagiri and Kitahara (1933)
	Yeast extract	10	Arnstein and Bentley (1953c)
		5	Ogawa et al. (1995)
		5	Kitada et al. (1967)
A. flavus	NH_4NO_3	1.1	May et al. (1931)
	Yeast extract	1	Ariff et al. (1996)
		5	Megalla et al. (1987)
		6	Wan et al. (2005)
A. tamarii	$NaNO_3$	2	Gould (1938)
A. albus	Polypeptone	7	Saruno et al. (1978)
A. candidus	Yeast extract	1	Wei et al. (1991)
A. parasiticus	Peptone	2	Coupland and Niehaus (1987)
	Yeast extract	10	El-Aasar (2006)

glucose, or 0.475 g kojic acid/g sucrose, 0.35 kojic acid g/g molasses, respectively.

Nitrogen sources

Table 3 shows the variation in nitrogen sources chosen for kojic acid fermentation from several strains of *Aspergillus* spp. Kitada et al. (1967) reported that organic nitrogen sources are generally better than inorganic nitrogen sources for kojic acid fermentation. Complex organic nitrogen sources such as peptone and yeast extract may contain vitamins, which act as a precursor for kojic acid production. Furthermore, some organic nitrogen sources have a good buffering system, while inorganic nitrogen sources, such as ammonia, excessively reduce the culture pH during NH_4^+ absorption. Low pH may influence the synthesis of kojic acid during fermentation and inhibit its growth. Yeast extract has been reported to be the most favourable organic nitrogen source for kojic acid production as compared to peptone and polypeptone (Wei et al., 1991; Ariff et al., 1996). However, Kitada et al. (1967) and Coupland and Niehaus (1987) claimed that peptone is better than yeast extract for kojic acid production. The presence of important growth factors, such as vitamins and oligoelements in specific nitrogen sources, also plays an important role in enhancing kojic acid production (Gad, 2003). Coupland and Niehaus (1987) pointed out that the addition of 10 mM $(NH_4)_2SO_4$ and glycerine to a medium containing 2 g/L peptone repressed kojic acid synthesis significantly.

On the other hand, high production of kojic acid (83 g/L), using a mixture of nitrogen source (0.5 g/L yeast extract + 0.75 g/L $(NH_4)_2SO_4$), had been reported by Kwak and Rhee (1992). Megalla et al. (1987) also used a mixture of nitrogen sources (2.0 g/L $(NH_4)_2SO_4$ + 1 g/L

yeast extract) to produce a fairly high kojic acid yield (25 g/L) by *A. flavus*. A high kojic acid production (28.4 g/L) was obtained by *A. parasiticus* fermentation utilizing 10 g/L yeast extract, whereas, the amount of acid production was greatly reduced with $(NH_4)_2SO_4$ and $NaNO_3$ (El-Aasar, 2006). Hence, NH_4^+ ion has the tendency to lower the culture pH excessively (Kitada et al., 1967).

A medium containing 100 g/L glucose as a carbon source and yeast extract concentrations ranging from 0.5 - 2.5 g/L or peptone concentrations ranging from 1 - 5 g/L as the nitrogen source is normally used in kojic acid fermentation by *Aspergillus* spp. Due to the presence of high concentration of yeast extract or peptone in a medium, glucose cannot to be converted to kojic acid but it is utilised for cell growth instead. Limitation of nitrogen supply is required to limit the growth so that more glucose can be converted to kojic acid. Using 100 g/L glucose, the optimum production of kojic acid was obtained at 0.3 g/L total amino nitrogen (Kwak and Rhee, 1992) and 2 g/L peptone (Kitada et al., 1967). Therefore, nitrogen limited fermentation is essential to limit the growth, so that the excess glucose which still remains in the culture can be converted to kojic acid by the activity of non-growing mycelial cells (Ariff et al., 1996).

C/N ratio

It is important to quantify the variations of carbon to nitrogen (C/N) ratio as it leads to the inhibition and enhancement of kojic acid synthesis, as well as to obtain optimal production. The effects of C/N ratio on the production of kojic acid have been reported by various researchers, and these are summarized in Figure 2. It is apparent that there is a critical C/N ratio for kojic acid production. It can be seen that C/N ratio higher than 100

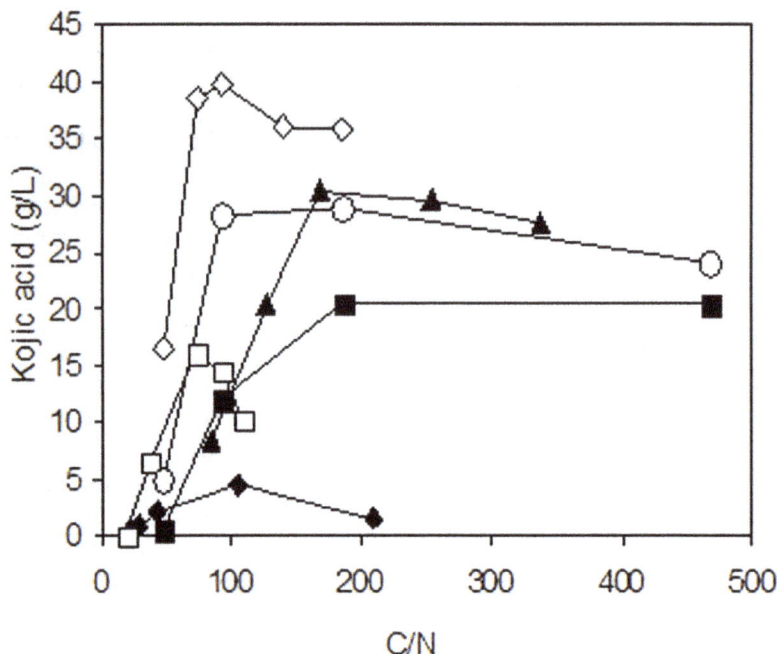

Figure 2. Effects of C/N ratio on the performance of kojic acid fermentation as reported by various researchers.
Note: (▲) Kwak and Rhee (1992); (♦) Coupland and Niehaus (1987); (■) Ogawa et al. (1995); (◊) Madihah et al. (1996); (□) Kitada et al. (1967); (O) Ogawa et al. (1995) MSLC.

is essential for the enhancement of kojic acid production. In most cases, the C/N ratio which is higher than 100 was not found to significantly influence kojic acid production. However, kojic acid production was reduced greatly at C/N ratio lower than 100. From this observation, it can be concluded that nitrogen limited fermentation is essential for kojic acid production. A large amount of carbon sources should be available during the production phase for the conversion to kojic acid by the cell bound enzymes and other activities of non-growing cells.

Minerals

Czapek-Dox medium containing KH_2PO_4, KCl, $NaNO_3$, $MgSO_4$ and $FeSO_4.7H_2O$ is favourable for the growth of kojic acid producing fungus (*A. flavus* and *A. oryzae*) and kojic acid production (Basappa et al., 1970; Wei et al., 1991). However, the medium with carbon and nitrogen sources containing only KH_2PO_4 and $ZnCl_2$ as minerals can also be used for a good growth of *A. parasiticus* and substantially high kojic acid production (Coupland and Niehaus, 1987).

Phosphate is an important nutrient for the growth of most fungi. It is incorporated in molecules such as nucleic acids, phospholipids and sugar phosphate; and plays an essential role in energy metabolism. The concentration of phosphate in the culture broth gives a significant influence on kojic acid production by *A. oryzae* (Arnstein and Bentley, 1953b). Previous studies revealed that high phosphate concentration in the culture broth (that is, 0.55 -13.72 mM) resulted in rapid kojic acid production. Conversely, at lower concentrations of phosphate (that is, 0.0055 - 0.055 mM), the rate of kojic acid was very much lower, while the maximum concentration of kojic acid obtained was about 2 times lower than those obtained in fermentation using high concentrations of phosphate. However, Coupland and Niehaus (1987) reported that the variation in phosphate concentrations ranging from (0.1 - 100 mM) gave no effects on kojic acid production by *A. flavus*.

Inhibitors and stimulants

Some components in a fermentation medium assist to regulate the production of products rather than to support the growth of micro-organisms. Such additives include inhibitors and stimulators, both of which may be used to manipulate the progress of fermentation. Past studies showed that the production of kojic acid by *A. oryzae* was inhibited by metabolic inhibitors such as sodium fluoride, monoiodoacetic acid, sodium arsenate, malonate, potassium cyanide, sodium azide dinitrophenol and pentachlorophenol at concentrations ranging from 10^{-2} to 10^{-4} M (Kitada and Fukimbara, 1971). However, no inhibition of kojic acid biosynthesis occurred when sodium sulphite was added to the medium. Furthermore, Basappa et al. (1970) reported that malonate and $Na_2S_2O_4$ inhibited the production of kojic acid by A. flavus. The inhibition of kojic acid production and glucose consumption by the above mentioned inhibitors was recovered by the addition of intermediates such as succinate, pyruvate and citrate at a concentration of 10^{-2} M to the inhibited culture (Kitada and Fukimbara, 1971). The growth, kojic acid production and glucose consumption by kojic acid producing fungus (A. flavus, A. oryzae and A. parasiticus) were inhibited by some chlorinated hydrocarbons such as aldrin, DDT and Lindane, which are normally used as pesticides (Nandan and Polasa, 1985).

Megalla et al. (1987) discovered that the addition of copper-monovales-nicotinic acid complex into A. flavus culture enhanced kojic acid production by about 47%. In the fermentation using 100 g/L glucose, the highest production of kojic acid was obtained when 75 µg/100 mL copper (I) – B2 complex was added. This compound, which is also known as a chelating agent, is considered to be a heavy metal vitamin derivative, which precipitates in certain enzymatic reactions. Moreover, the researchers also investigated the biochemical effects of the chelating agent on kojic acid biosynthesis and related the results on the model basis of the biosynthesis pathway of kojic acid, as proposed by Bajpai et al. (1981). The emphasis was placed on the enzymes that were dependent on NAD and NADP. The presence of highly reactive NAD and NADP like structures in the medium promoted the action of dehydrogen, which is the main enzyme involved in the biosynthesis of kojic acid.

The synthesis of kojic acid was unaffected by the changes in the concentrations of zinc and ferric ion, although, high concentrations of NaCl (> 100 mM) and NaOAc (1 - 2 mM) inhibited the synthesis (Coupland and Niehaus, 1987). The addition of 1 - 2 g/L cycasin (methyl-azoxymethyl-β-D-glucose), extracted from the Cycas revoluta plant, inhibited the spore formation of A. oryzae and increased the kojic acid production by about 6-fold (Tadera et al., 1985). Enhanced kojic acid production by A. flavus with the addition of methanol in the culture broth had been reported by Madihah et al. (1996). The presence of methanol in the culture might reduce the bubble size in the stirred tank fermenter and increase the oxygen transfer rate, which in turn, enhanced the fermentation performance.

OPTIMISATION OF CULTURE CONDITION

Culture pH

Most studies conducted on the effects of culture pH towards the growth and production of kojic acid was based on the initial culture pH (Lin et al., 1976; Clevstrom and Lundjgren, 1985). The effects of initial pH on kojic

acid production by A. parasiticus were studied by Lin et al. (1976) using yeast extract as the nitrogen source. The researchers reported that the kojic acid production was optimal at two pH values (4.5 and 6.2). Katagiri and Kitahara (1933) found that an initial pH 5 was favourable for the growth of A. oryzae, but an initial pH 2.4 was required to enhance the synthesis of kojic acid. This means that, the optimal pH for kojic acid synthesis was different with the optimal pH for the growth of kojic acid producing strain. In the replacement culture, the highest conversion of glucose to kojic acid was obtained at a very low pH (1.9) (Wood, 1998). At slightly higher pH than 1.9 (pH 2.2), the conversion of glucose to kojic acid was reduced significantly. Conversely, at very low pH, the metabolism of the fungus might be shunted into another pathway, which led to kojic acid synthesis. Clevstrom and Lundjgren (1985) also pointed out that the formation of kojic acid occurred most readily at low pH values (2 - 3). In addition, Wilson (1971) reported that the optimal pH for kojic acid formation was in the range of pH 2 - 3, and a small deviation of pH from this range tended to reduce the production sharply. On the other hand, El-Aasar (2006) claimed that the kojic acid production by A. parasiticus was optimal at pH 5.

It is important to note that the optimum pH for kojic acid production also depends on the types of carbon or nitrogen used. Kitada et al. (1967) identified that the optimum pH for kojic acid production was in the range of pH 2 - 3, when a combination of glucose or sucrose as the carbon source and peptone or yeast extract as the nitrogen source was used. On the other hand, when ammonium nitrate was used as the nitrogen source, the maximum kojic acid production was achieved at pH 3.08. In addition, Basappa et al. (1970) also reported that the pH range for the maximum kojic acid production by A. flavus was between 6 - 7, when acetate was used as the carbon source.

Aeration and agitation

Fermentations of kojic acid at pilot and industrial scales are usually conducted using stirred tank fermenters to ensure efficient oxygen transfer into the culture. Some studies on the effects of aeration and agitation on kojic acid production in stirred tank fermenters at industrial scale had been conducted by Kitada et al. (1971). The highest kojic acid production (32 g/L) in a 300 L stirred tank fermenter was obtained at 1 vvm and 240 rpm (impeller tip speed = 8.04 m/s) which gave the value of oxygen transfer rate coefficient of 11.2×10^{-6} g/mol O_2/min.atm.mL. High kojic acid production was obtained in the fermentation where dissolved oxygen tension (DOT) was controlled at very high level (80% saturation) during an active growth phase, and DOT was then switched to 30% saturation during the production phase (Ariff et al., 1996). High DOT (80% saturation) was required during the growth phase for the production enhancement of enzymes responsible for kojic acid synthesis. Glucose was converted to kojic acid by these cell-bound enzymes during the production phase. Very low DOT (30% saturation) should be controlled during the production phase to avoid excessive degradation of kojic acid to other compounds.

FERMENTATION TECHNIQUES

Various fermentation techniques, such as submerged, solid state and surface cultures, and modes of fermenter operation such as batch, fed-batch and continuous, have been developed and used for the improvement of various fermentation processes (Table 4). Nevertheless, very few attempts had been made to produce kojic acid using solid-state fermentation. For example, Kharchenko (1999) reported that a comparatively high yield of kojic acid (8.5 - 9.5 g kojic acid/kg substrate) was obtained in a solid-state fermentation using grains and grain-forages with high amount of proteins and carbohydrates such as maizes, oats, ryes and barley grains. Most reports on kojic acid production available in the literature are related to submerged fermentation. The feasibility of using different types of submerged fermentation techniques and modes of fermenter operation on the improvement of kojic acid fermentation are outlined and discussed below.

Surface culture

Surface culture is known as the technique where molds are grown on a solid or liquid medium without any agitation or shaking of the culture vessel (Wei et al., 1991). Kojic acid production by surface culture of A. parasiticus had been reported in several studies (Lin et al., 1976; Nandan and Polasa, 1985). Meanwhile, A. candidus ATCC 44054 produced more kojic acid in the surface culture as compared to the cultures shaken at 100 rpm using the same medium (Wei et al., 1991). However, the process parameters such as pH, temperature and dissolved oxygen tension are difficult to control in the surface culture system.

Submerged culture

Submerged culture is widely used for high performance of kojic acid production (Kitada et al., 1967; Wei et al., 1991; Ariff et al., 1996; Ariff et al., 1997). The growth of aerobic micro-organisms in a submerged culture is controlled by the availability of substrates, energy and enzymes. Cultures are always of a heterogeneous nature, whereby the rates of reactions can be limited by the rate of substrate or product transfer at a particular interface. Different techniques of fermentation, such as

batch, fed-batch and continuous for the improvement of kojic acid production, are also possible to be applied in submerged fermentation in order to achieve an optimal and economic fermentation process (Kitada et al., 1967; Ariff et al., 1996).

A typical time course of batch kojic acid fermentation by A. flavus is shown in Figure 3 (Ariff et al., 1996; Ariff et al., 1997). The kojic acid fermentation by A. flavus was classified as a non-growth associated process, where the fermentation could be divided into two phases, namely the growth phase and the production phase. The growth normally reached the maximum after 120 h and the production of kojic acid would start after about 48 h - 72 h, whereby the production continued almost linearly until the exhaustion of glucose. On the other hand, Kitada et al. (1967) found that kojic acid fermentation by A. oryzae was a mixed process, whereby, kojic acid was produced during the growth and non-growth phases. After all supplies of glucose had been consumed, kojic acid accumulated in the culture may be utilised by microorganisms to produce other substances such as oxalic and other acids (Clevstrom and Ljunggren, 1985), resulting in the decrease of kojic acid production. Excessive degradation of kojic acid, at certain conditions, was also observed towards the end of batch fermentation (Kitada et al., 1967; Ogawa et al., 1995; Ariff et al., 1996).

The use of two-stage continuous culture for kojic acid production by A. oryzae using peptone as the growth-limiting nutrient had been reported by Kitada and Fukimbara (1970). Slightly lower concentration of kojic acid at steady-state (6 - 7 g/L) in the first and second vessels was obtained as compared to the batch fermentation (9 g/L). The continuous culture is an ideal method of fermentation for the production of microbial biomass and other growth associated processes such as ethanol production rather than for the production of secondary metabolites or non-growth associated processes such as kojic acid.

In kojic acid fermentation, the fungi initially utilize the carbon source for growth and then synthesize the kojic acid in subsequent declining phase and early stationary phase (Kitada et al., 1967). Extending the culture time while the cells are still actively excreting the metabolite or containing a stable mycelial-bound enzyme that is responsible for kojic acid synthesis (Bajpai et al., 1982) could improve kojic acid production. Fed-batch culture had been applied for the production of kojic acid by A. oryzae using the membrane surface liquid culture, where powdered glucose was added intermittently to the initial batch fermentation after glucose supply was exhausted (Ogawa et al., 1995). In this technique of fermentation, the production of kojic acid was increased almost linearly up to about 80 g/L after 384 h. In addition, the yield (0.56 g kojic acid/g glucose) and productivity (0.208 g/L.h) obtained from the fed-batch fermentation were higher than that obtained from the batch fermentation (0.45 g kojic acid/g glucose and 0.121 g/L.h, respectively). Fed-

Table 4. Kojic acid production using different fermentation techniques by various kojic acid-producing micro-organisms.

Method	Micro-organism	Carbon source (g/L)	Nitrogen source (g/L)	pH	Yield (g kojic acid/g carbon source)	Productivity (g/L.h)	References
Surface culture	A. candidus	Sucrose (200)	Yeast extract (20)	6.5	0.3	0.208	Wei et al. (1991)
	A. tamarii	Glucose (50)	NaNO$_3$ (2)	4.85	0.24	0.081	Gould (1938)
	A. oryzae	Maltose (50)	(NH4)$_2$SO$_4$ (0.5)	4.5	0.41	0.043	Katagiri and Kitahara (1933)
	A. flavus	Glucose (150)	NH$_4$NO$_3$ (28)	-	0.57	0.135	May et al. (1931)
		Glucose (100)	Yeast extract (5)	3.5	0.29	0.13	Ariff et al. (1996)
Submerged culture	A. oryzae	Glucose (100)	Yeast extract (6)	5.0	0.41	0.22	Wan et al. (2005)
	A. oryzae	Glucose (100)	Peptone (5)	4.0	0.24	0.143	Kitada et al. (1967)
	A. parasiticus	Glucose (100)	Yeast extract (0.5)	6.0	0.51	0.177	Ogawa et al. (1995)
		Glucose (60)	Yeast extract (10)	5.0	0.57		El-Aasar (2006)
Resuspended cell system	A. flavus	Glucose (200)	Yeast extract (20)	6.5	0.41	0.107	Bajpai et al. (1982) Rosfarizan and Ariff (2007)
Immobilised cell system	A. oryzae	Glucose (100)	Yeast extract (1)	6.0	0.38	0.154	Kwak and Rhee (1991)
Membrane surface liquid culture	A. oryzae	Glucose (100)	Yeast extract (1)	6.0	0.31	0.137	Wakisaka et al. (1998)
	A. oryzae	Glucose (100)	Yeast extract (0.5)	6.0	0.46	0.121	Ogawa et al. (1995)

batch fermentation was also performed by Ariff et al. (1996) in order to investigate the requirement of yeast extract during kojic acid production by A. flavus. Rapid kojic acid production, together with rapid glucose consumption, occurred even though there was a growth and a high nitrogen level was still present in the culture. However, kojic acid production was very much lower than that obtained in nitrogen limited batch fermentation. Reduced kojic acid production in the fed-batch fermentation was due to the high consumption of glucose for the growth during the production phase and less glucose was converted to kojic acid. This result also showed that the growing cells and unlimited nitrogen did not repress the metabolic pathway of kojic acid production. How-

ever, active growth limited a turnover of biomass, which in turn, reduced the ability of mycelia to synthesize kojic acid. Limitation of the nitrogen supply was required to limit the growth so that, more glucose could be converted to kojic acid. Since mycelial-bound enzyme system involved in kojic acid biosynthetic pathway was stable for a long starvation period (Kwak and Rhee, 1992), efficient utilization of the remaining activities from batch fermentation might be the extension of production with glucose feeding alone. Fed-batch culture by feeding glucose alone should be started after glucose in the culture is almost exhausted. The glucose feeding should be continued until no more glucose could be converted to kojic acid. The application of fed-batch culture for improve-

ment of kojic acid production by A. flavus using sago starch as the carbon source had been reported by Rosfarizan et al. (2002). Improved kojic acid production, in terms of total productivity, could also be achieved using repeated batch fermentation technique (Wan et al., 2005), where the cultivation period could be extended by culture withdrawal (75% of the total volume) and substitution with a fresh medium.

Membrane surface liquid culture

In submerged culture, molds like A. flavus grow either in the form of pulpy or pellet-like morphology (Braun and Vecht-Lifshitz, 1994). In the pulpy

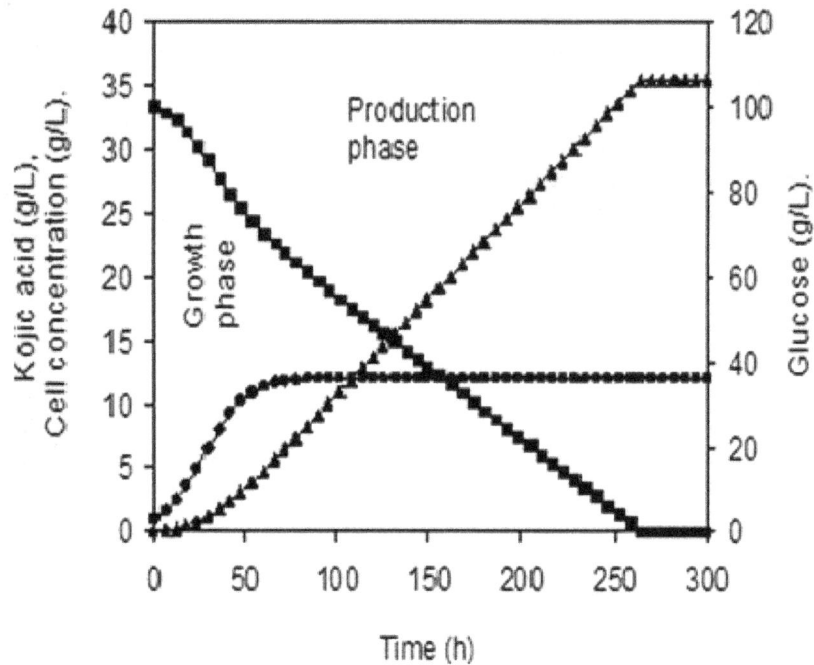

Figure 3. A typical time course of batch kojic acid fermentation by *A. flavus*.
Note: (●) Cell concentration; (■) Glucose; (▲) Kojic acid.

state, the molds may be damaged by shear stress while the cells in the interior of pellets are prone to undergo autolysis due to oxygen deficiency. On the other hand, the production rate of kojic acid in the surface culture is usually limited by mass transfer rate where oxygen uptake by the cells is not efficient. In order to overcome problems and disadvantageous exist in the submerged and surface cultures, Yasuhara et al. (1994) proposed a novel cultivation method termed as the membrane surface liquid culture. In this culture system, molds are grown on the surface of a microporous membrane that faces the air with its opposite side in contact with the liquid medium. The membrane surface liquid culture has both the advantageous features of surface and submerged cultures. In the membrane surface liquid culture, molds are grown on the surface of a porous membrane facing the air, similar to the surface culture, and no agitation. However, the liquid medium is used as in a conventional submerged culture, where the culture conditions can be easily controlled during fermentation.

The membrane surface liquid culture had been applied for the production of kojic acid by *A. oryzae* NRRL484 (Ogawa et al., 1995). The researchers reported that the maximum concentration of kojic acid (30 g/L) and the production rate obtained in the membrane surface liquid culture were higher than those obtained in the shaking culture (20 g/L). Since the amount of energy required for agitation and aeration is negligible and downstream processing is simple because the medium is not contaminated with cells, it can be suggested that the membrane surface liquid culture is an energy-saving

process. Nonetheless, some technical problems related to scaling-up of the process such as the method of spore inoculation, removal of cells and sterilization of apparatus should be resolved before it can be used as an industrial process.

Resuspended cell system

Kojic acid production phase could be extended by the addition of glucose to the replenishment culture after growth almost ceased under high aeration rate (Kitada et al., 1967), suggesting that, the mycelia cells were still active in synthesizing kojic acid even after prolonged incubation. The cell-bound enzyme system that is responsible in kojic acid biosynthesis is stable for prolonged incubation in glucose solution (Bajpai et al., 1982). This special characteristic of the cell-bound enzyme has led to the development of resuspended cell system for production of kojic acid. The researchers also found that mycelia, grown in the yeast extract sucrose (YES) medium and resuspended in a buffer containing only carbohydrates, produced kojic acid almost to the same extent as in the case when a complete growth medium was used. These observations suggest that, limitation of nitrogen source is required to suppress the growth during resuspended cell system, so that, more glucose can be converted to kojic acid. The use of resuspended cell system for biotransformation of glucose to kojic acid has recently gathered a new momentum of interest in a view of its capability to produce pigment-free kojic acid, which will make the purification process much easier. The kinetics of biotransformation of glucose and sucrose to kojic acid by resuspended cell system of *A. flavus* has recently been reported by Rosfarizan and Ariff (2007).

Immobilised cell system

Immobilised spores of *A. oryzae* in calcium alginate beads have been used as biocatalyst for kojic acid production (Kwak and Rhee, 1992). In order to maximize the metabolic activity of the immobilised *A. oryzae*, the amount of growth limiting nutrients and bead size was found to be a very important factor affecting mycelial distribution on polymeric gel beads and the overall reaction rates of immobilised cell cultures. Regardless of the size of the immobilised particles, there existed an optimal nitrogen concentration for the maximum production rate of kojic acid, at which smaller bead sizes resulted in a higher kojic acid production rate. Higher specific oxygen uptake rate in fermentation, with smaller bead size, might be one of the factors which increased the productivity, which was required to enhance the production of enzymes involved in kojic acid biosynthesis pathway (Woodhead and Walker, 1975). As mentioned earlier, high oxygen demand was required during the active growth phase of *A. flavus* for the enhancement of enzyme synthesis relevant to kojic acid production (Ariff et al. 1996).

The immobilised cells of *A. oryzae* in calcium alginate beads could also be applied in repeated batch fermentation under long-starvation conditions (Kwak and Rhee, 1992). The immobilised cells were capable of producing kojic acid linearly up to 83 g/L with fermentation time, while maintaining the stable metabolic activity for a prolonged production period (528 h). In freely suspended cells, the maximum concentration of kojic acid obtained was only 18 g/L.

CONCLUSION

Kojic acid has many industrial applications and its demand is increasing enormously with the growing industries related to its applications. Although, various fermentation techniques can be applied for kojic acid production, it was proven that, the batch submerged fermentation gives the highest efficiency in terms of yield, maximum kojic acid concentration and also overall productivities. In general, it can also be concluded that glucose and yeast extract are the preferred carbon and nitrogen sources for kojic acid production by various fungal strains.

REFERENCES

Ashari SE, Mohamad S, Ariff A, Basri M, Salleh AB (2009) Optimization of enzymatic synthesis of palm-based kojic acid ester using response surface methodology. J. Oleo. Sci. 58: 503-510.

Ariff AB, Rosfarizan M, Herng LS, Madihah S, Karim MIA (1997). Kinetics and modelling of kojic acid production by Aspergillus flavus Link in batch fermentation and resuspended cell mycelial. World J. Microbiol. Biotechnol. 13: 195-201.

Ariff AB, Salleh, MS, Ghani, B, Hassan, MA, Rusul G, Karim MIA (1996). Aeration and yeast extract requirements for kojic acid production by Aspergillus flavus Link. Enzyme Microbiol. Technol. 19: 545-550.

Arnstein HRV, Bentley R (1953a). The biosynthesis of kojic acid 1. Production from [1-14] and [3: 4-14C] glucose and [2-14C]-1: 3-Dihydroxyacetone. Biochem. J. 54: 493-508.

Arnstein HRV, Bentley R (1953b). The biosynthesis of kojic acid 2. The occurrence of aldolase and trioephosphate isomerase in Aspergillus species and their relationship to kojic acid biosynthesis. Biochem. J. 5: 508-516.

Arnstein HRV, Bentley R (1953c). The biosynthesis of kojic acid 3. The incorporation of labeled small molecules into kojic acid. Biochem. J. 54: 517-522.

Arnstein HRV, Bentley R (1956). The biosynthesis of kojic acid 4. Production from pentoses and methyl pentoses. Biochem. J. 62: 403-411.

Bajpai P, Agrawala PK, Vishwanathan L (1981). Enzymes relevant to kojic acid biosynthesis in Aspergillus flavus. J. Gen. Microbiol. 127: 131-136.

Bajpai P, Agrawala PK, Vishwanathan L (1982). Production of kojic acid by resuspended mycelial of Aspergillus flavus. Can. J. Microbiol. 28: 1340-1346.

Balaz S, Michal U, Julius B, Jona B, Jozzef D (1993). Relationship between antifungal activity and physicochemical properties of kojic acid derivatives. Folia Microbiol. 9: 387-391.

Basappa SC, Sreenivasamurthy V, Parpia HAB (1970). Aflatoxin and kojic acid production by resting cells of Aspergillus flavus Link. J. Gen. Microbiol. 61: 81-86.

Beard RL, Walton GS (1969). Kojic acid as an insecticidal mycotoxin. J. Invertebr. Pathol. 14: 53-59.

Beelik A (1956). Kojic acid. Adv. Carbohydr. Chem. 11: 145-183.

Bentley R (1957). Preparation and analysis of kojic acid. Methods Enzymol. 3: 238-241.

Bentley R (2006) From miso, sake and shoyu to cosmetics: a century of science for kojic acid. Nat. Prod. Rep. 23: 1046-1062.

Braun S, Vecht-Lifshitz SE (1994). Mycelial morphology and metabolite production. Trends Biotechnol. 9: 63-68.

Brtko J, Rondahl L, Fickova M, Hudecova D, Eybl V, Uher M (2004) Kojic acid and its derivatives: history and present state of art. Cent. Eur. J. Public Health. 12: 16-18.

Buchta K (1982). Organic acids of minor importance. In Rehm HJ, Reed G, H Dellweg H (eds.) Biotechnology: A Comprehensive Treaties Vol. 3: Biomass, Microorganisms for Special Applications, Microbial Products, Energy from Renewable Resources. Ingelheim: Federal Republic of Germany.

Burdock GA, Soni MG, Carabin GI (2001) Evaluation of health aspects of kojic acid in food. Reg. Toxicol. Pharmacol. 33: 80-101.

Chen JS, Wei CI, Rolle RS, Balaban MO, Otwell SW, Marshall MR (1991). Inhibitory effect of kojie acid on some plant and crustacean polyphenol oxidases . J. Agric. Food Chem. 39: 1396-1401.

Choi SH, Kim S, Kim H, Suk K, Hwang JS, Lee BG, Kim AA, Kim SY (2002). (4-Methoxy-benzylidene)-(3-methoxy-phenyl)-amine, a nitrogen analog of silbene as a potent inhibitor of melanin production. Chem. Pharm. Bull. 50: 450-452.

Clevstrom G, Ljunggren H (1985). Aflatoxin formation and the dual phenomenon in Aspergillus flavus Link. Mycopathol. 92: 129-139.

Coupland K, Niehaus Jr WG (1987). Effect of nitrogen supply, Zn^{2+} and salt concentration on kojic acid and versicolorin biosynthesis by Aspergillus parasiticus. Exp. Mycol. 11: 206-213.

Crueger W, Crueger A (1984). A Textbook of Industrial Microbiology. Sci. Tech. and Sinauer Associates Inc, London.

Demain AL (1973). Mutation and the production of secondary metabolites. Adv. Appl. Microbiol. 16: 177-199.

Dobias J, Nemec P, Brtko JK (1977). Inhibitory effect of kojic acid and its two derivatives on the development of Drosophila melanogaster. Biologia 32: 417-421.

Dowd PF (1988). Toxicological and biochemical interactions of the fungal metabolites fusaric acid and kojic acid with xenobiotics in Heliothis zea (F.) and Spodoptera frugiperda (J.E. Smith). Pestic. Biochem. Physiol. 32: 123-134.

El-Aasar SA (2006). Cultural conditions studies on kojic acid production by *Aspergillus parasiticus*. Int. J. Agric. Biol. 8: 468-473.

Emami S, Hosseinimehr SJ, Taghdisi SM, Akhlaghpoor S (2007). Kojic acid and its manganese and zinc complexes as potential radioprotective agents. Bioorg. Med. Chem. Lett. 17: 45-48.

Fantini AA (1975). Strain development. Methods Enzymol. 43: 24-41.

Futamura T, Okabe M, Tamura T, Toda K, Matsunobu T, Park YS (2001). Improvement of production of kojic acid by a mutant strain *Aspergillus oryzae*, MK 107-39. J. Biosci. Bioeng. 3: 272-276.

Gad AS (2003). Modification of molasses for kojic acid production by *Aspergillus parasiticus*. N. Egyptian J. Microbiol. 5: 14-26.

Garcia A, Fulton Jr. JE (1996). The combination of glycolic acid and hydroquinone or kojic acid for the treatment of melasma and related conditions. Dermatol. Surg. 22: 443-447.

Gould BS (1938). The metabolism of Aspergillus tamarii Kita, Kojic acid production. Biochem. J. 32: 797.

Guibal E (2004). Interactions of metal ions with chitosan-based sorbents: A review. Sep. Purif. Technol. 38: 43-74.

Hasizume K, Yamamura S, Inoue S (1968). A convenient synthesis of 2-methyl-4-pyrone from kojic acid. Chem. Pharm. Bull. 16: 2292-2296.

Hudecova D, Jantavo S, Melnik M, Uher M (1996). New azidome-talkojates and their biological activity. Folia Microbiol. 41: 473-476.

Ichimoto I, Fujii K, Tatsumi C (1965). Studies on kojic acid and its related γ-pyrone compounds. Part IX. Synthesis of maltol from kojic acid (Synthesis of maltol (3)). Agric. Biol. Chem. 2: 325-330.

Kahn V, Lindner P, Zakin V (1995). Effect of kojic acid on the oxidation of o-dihydrohyphenols by mushroom tyrosinase. J. Agric. Food Chem. 18: 253-271.

Kharchenko SM (1999). The biosynthesis of kojic acid by Aspergillus flavus Link strains isolated from feed. Microbiol. Zh. 61: 15-21.

Katagiri H, Kitahara K (1933). The formation of kojic acid by *Aspergillus oryzae*, Kyoto Imperial University, Kyoto.

Kayahara H, Shibata N, Tadasa K, Maedu H, Kotani T, Ichimoto I (1990). Amino acids and peptide derivatives of kojic acid and their antifungal properties. Agric. Biol. Chem. 54: 2441-2442.

Khamaruddin NR, Basri M, Lian GEC, Salleh AB, Rahman RNZ, Ariff AB, Mohamad R, Awang R (2008) Enzymatic synthesis and characterization of palm-based kojic acid ester. J. Oil Palm Res. 20: 461-469.

Kitada M, Fukimbara T (1970). Studies on kojic acid fermentation (IV). Kojic acid production by continuous culture. J. Ferment. Technol. 48: 411-415.

Kitada M, Fukimbara T (1971). Studies on kojic acid fermentation (VII). The mechanism of the conversion of glucose to kojic acid. J. Ferment. Technol. 49: 847-851.

Kitada M, Kenaeda J, Miyazaki K, Fukimbara T (1971). Studies on kojic acid fermentation (VI). Production and recovery of kojic acid fermentation on industrial scale. J. Ferment. Technol. 49: 343-349.

Kitada M, Ueyama H, Fukimbara T (1967). Studies on kojic acid fermentation (I) Cultural condition in submerged culture. J. Ferment. Technol. 45: 1101-1107.

Kotani T, Ichimoto I, Tatsumi C, Fujita T (1976). Bacteriostatic activities and metal chelation of kojic acid analogs. Agric. Biol. Chem. 40: 765-770.

Kwak MY, Rhee JS (1991). Cultivation characteristics of immobilized *Aspergillus oryzae* for kojic acid production. Biotechnol. Bioeng. 39: 903-906.

Kwak MY, Rhee JS (1992). Control mycelial growth for kojic acid production using ca-alginate immobilized fungal cells. Appl. Microbiol. Biotechnol. 36: 578-583.

Lee FH, Boltjes B, William E (1950). Kojic acid as an inhibitor of tubercle bacilli. Am. Rev. Tuberculosis. 61: 738-741.

Lee YS, Park JH, Kim MH, Seo SH, Kim HJ (2006) Syntheisis of tyrosinase inhibitory kojic acid derivative. Archiv. Pharm. 339: 111-114.

Lin MT, Mahajan JR, Dianese JC, Takatsu A (1976). High production of kojic acid crystals by *Aspergillus parasiticus* UNBF A12 in liquid medium. Appl. Environ. Microbiol. 32: 298-299.

Madihah MS, Ariff AB, Hassan MA, Rusul G, Karim MIA (1996). Enhanced kojic acid production by Aspergillus flavus Link in Growth medium containing methanol. ASEAN Food J. 11: 158-162.

Marston RQ (1949). Production of kojic acid from *Aspergillus lutescens* [12]. Nat. 164: 961.

Masse MO, Duvallet V, Borremans M, Goeyens L (2001) Identification and quantitative analysis of kojic acid and arbutine in skin-whitening cosmetics. Int. J. Cosmet. Sci. 23: 219-232.

May OE, Moyer AJ, Wells PA, Herbik H (1931). The production of kojic acid by *Aspergillus flavus*. J. Am. Chem. Soc. 53: 774-782.

Megalla SE, Nassar AY, Gohar MAS (1987). The role of copper (I) nicotinic acid complex on kojic acid biosynthesis by *Aspergillus flavus*. J. Basic Microbiol. 27: 29-33.

Mitani H, Koshiishi I, Sumita T, Imanari T (2001). Prevention of the photodamage in the hairless mouse dorsal skin by kojic acid as an iron chelator. Eur. J. Pharmacol. 411: 169-174.

Morton HE, Kocholaty R, Junowicz K, Kelner A (1945). Toxicity and antibiotic activity of kojic acid produced by *Aspergillus luteovirescens*. J. Bacteriol. 50: 579-584.

Nandan R, Polasa H (1985). Inhibition of growth of kojic acid biosynthesis in *Aspergillus* by some chlorinated hydrocarbons. Indian J. Microbiol. 25: 21-25.

Niwa Y, Akamatsu H (1991). Kojic acid scavenges free radicals while potentiating leukocyte functions including free radical generation. Inflammation 15: 303-315.

Niwa Y, Miyachi Y (1986). Antioxidant action of natural health products and Chinese herbs. Inflammation 10: 79-91.

Noh JM, Kwak SY, Seo HS, Seo JH, Kim BG, Lee YS (2009) Kojic acid-amino acid conjugates as tyrosinase inhibitors. Bioorg. Med. Chem. Lett. 19: 5586-5589.

Nohynek GJ, Kirkland D, Marzin D, Toutain H, Leclerc-Ribaud C, Jinnai H (2004). An assessment of the genotoxicity and human health risk of topical use of kojic acid [5-hydroxy-2-(hydroxymethyl)-4H-pyran-4-one]. Food Chem. Toxicol. 42: 93-105.

Novotny L, Rauko P, Abdel-Hamid M, Vachalkova A (1999). Kojic acid - A new leading molecule for a preparation of compounds with an anti-neoplastic potential. Neoplasma 46: 89-92.

Ogawa A, Wakisaka Y, Tanaka T, Sakiyama T, Nakanishi K (1995). Production of kojic acid by membrane-surface liquid culture of *Aspergillus oryzae* NRRL484. J. Ferment. Bioeng. 80: 41-45.

Ohyama Y, Mishima Y (1990). Melanosis-inhibitory effect of kojic acid and its action mechanism. Fragrance J. 6: 53-58.

Ravi-Kumar MNV (2000). A review of chitin and chitosan applications. React. Funct. Polym. 46: 1-27.

Rosfarizan M, Ariff AB (2007). Biotransformation of various carbon sources to kojic acid by cell-bound enzyme system of A. flavus Link 44-1. Biochem. Eng. J. 35: 203-209.

Rosfarizan M, Ariff AB, Hassan MA, Karim MIA, Shimizu H, Shioya S (2002). Important of carbon source feeding and pH control strategies for maximum kojic acid production from sago starch by A. flavus. J. Biosci. Bioeng. 94(2): 99-105.

Rosfarizan M, Madihah S, Ariff AB (1998). Isolation of a kojic acid producing fungus capable of using starches as a carbon source. Lett. Appl. Microbiol. 26: 27-30.

Saruno R, Kato F, Ikeno T (1978). Kojic acid, a tyrosinase inhibitor from *Aspergillus albus*. Agric. Biol. Chem. 43: 1337-1338.

Sehgal SS (1976). Effectiveness of kojic acid in inducing sterility in *Tragoderma granarium* Everts (Coleoptera). Cienc. Cultura. 28: 777-779.

Synytsya A, Blafkova P, Synytsya A, Copikova J, Spevacek J, Uher M (2008). Conjugation of kojic acid with chitosan. Carbohydr. Polym. 72: 21-31.

Tadera K, Yagi F, Kobayashi A (1985). Effects of cycasin on kojic acid-producing molds. Agric. Biol. Chem. 49: 203-205.

Takamizawa K, Nakashima S, Yahashi Y, Kubata BK, Suzuki T, Kawai K, Horitsu H (1996). Optimization of kojic acid production rate using the Box-Wilson method. J. Ferment. Bioeng. 82: 414-416.

Tamiya H (1932). A remark on the importance of buffering the culture solution for the metabolism of Aspergillus oryzae. Arch. Microbiol. 3: 559-560.

Tatsumi C, Ichimoto I, Uchida S, Nonomura S (1969). Production of comenic acid from kojic acid by microorganism. J. Ferment. Technol. 47: 178-184.

Uchino K, Nagawa M, Tonosaki Y, Oda M, Fukuchi A (1988). Kojic acid as an anti-speck agent. Agric. Biol. Chem. 52: 2609-2610.

Wakisaka Y, Segawa T, Imamura K, Sakiyama T, Nakanishi K (1998). Development of a cylindrical apparatus for membrane-surface liquid culture and production of kojic acid using *Aspergillus oryzae* NRRL 484. J. Ferment. Bioeng. 85: 488-494.

Wan HM, Chen CC, Giridhar R, Chang TS (2005). Repeated-batch production of kojic acid in a cell-retention fermenter using *Aspergillus oryzae* M3B9. J. Ind. Microbiol. Biotechnol. 32: 227-233.

Watanabe-Akanuma M, Inaba Y, Ohta T (2007) Mutagenicity of UV-irradiated maltol in *Salmonella typhimurium*. Mutagenesis 22: 43-47.

Wei CI, Huang TS, Fernando SY, Chung KT (1991). Mutagenicity studies of kojic acid. Toxicol. Lett. 59: 213-220.

Wilson BJ (1971). Miscellaneous *Aspergillus* toxins. In Ciegler A (ed.) Microbes Toxins, Fungal Toxins VI, Academic Press, New York.

Wood BJB (1998). Microbiology of Fermented Food (2nd Edition), Springer, London.

Woodhead S, Walker ARL (1975). The effect of aeration on glucose catabolism in *Penicillium expansum*. J. Gen. Microbiol. 89:327-336.

Yabuta T (1924). The constitution of kojic acid, a δ-pyrone derivative formed by *Aspergillus flavus* from carbohydrates. J. Chem. Soc. Trans. 125: 575-587.

Yasuhara A, Ogawa A, Tanaka T, Sakiyama T, Nakanishi K (1994). Production of neutral protease from *Aspergillus oryzae* by a novel cultivation method on a microporous membrane. Biotechnol. Tech. 8: 249-254.

The role, isolation and identification of *Vibrio* species on the quality and safety of seafood

Shikongo-Nambabi M. N. N. N.[1]*, Petrus N. P[2] and M. B. Schneider[3]

[1]Department of Food Science and Technology, Neudamm Campus, University of Namibia, Windhoek, Namibia.
[2]Department of Animal Science, Neudamm Campus, University of Namibia, Windhoek, Namibia.
[3]Faculty of Agriculture and Natural Resources, University of Namibia, Namibia.

Seafoods in their natural environments are associated with a variety of microorganisms. Fish shelf life reduction results from microbial metabolism, mainly by Gram negative bacteria that produce chemical compounds responsible for bad odour, texture and taste. Shelflife is estimated by performing total viable bacterial counts at ambient and refrigeration temperatures. The type and number of bacteria present on seafood depends on the microbial composition of the surrounding waters, on the intrinsic factors, extrinsic factors, processing, and implicit factors and on the microbial interactions within the fish itself. Although, sea food safety assessment is preferably determined by detecting indicator organisms; such as Enterobacteriaceae and coliforms, none of these groups fulfil all requirements that guarantee food safety necessitating direct detecting of relevant pathogens. *Vibrio* species are part of the bacteria genera associated with seafoods borne diseases. Prompt and accurate detection and identification methods of pathogens are imperative to determine the product compliance with seafood microbiological criteria. Although cultural methods have long been used in detecting human pathogens including *Vibrio* species in fish, these methods are time consuming and sometimes inaccurate. Also some pathogens have the propensity to change into the Viable but non culturable (VBNC) state in unfavourable environments. The use of molecular methods is hampered by drawbacks, such as inter species 16S rRNA sequence similarity and that some strains carry multiple copies of the 16S rRNA gene. A combination of classical, numerical taxonomy and Multi locus sequence analysis (MLSA) methods are promising to give absolute resolution between closely related *Vibrio* species.

Key words: *Vibrios* spp., seafood, spoilage, pathogens, detection, identification.

INTRODUCTION

Seafood spoilage is of biochemical and /or microbial origin and results in limited shelf life and the eventual sensory rejection of the food (Gennari et al., 1999; Gram, 1992; Gram and Dalgaard, 2002; Huis in't Veld, 1996). Bacterial counts performed at different incubation temperatures and detection of indicator organisms usually indicate the degree of contamination and are used to predict the sanitary conditions under which the food is produced and to estimate the microbiological quality of a food item (ICMSF, 1978). The specific bacteria that carry out metabolism in the fish muscle after death depend on the type of fish and its chemical composition, the feeding habit, the area where the fish is harvested and the type of fishing gear used during harvesting (Françoise, 2010; Huis in't Veld, 1996). The storage conditions dictate the microbial successions and the final predominant groups (Gram and Huss, 1996; Huis in't Veld, 1996). *Vibrio, Shewanella, Aeromonas,* Enterobacteriaceae and *Photobacterium* are psychrotrophic and play important roles in the quality and shelf life of marine fish during storage (Koutsomanis and Nychas, 1999). *Aeromonas, Bacillus, Campylobacters, Clostridium, Mycobacterium, Legionella, Edwardsiella, Plesiomonas, Salmonella,*

Yersinia and Vibrios are common seafood borne human pathogens. They cause diseases including gastroenteritis, septicaemia and wound infections that are sometimes fatal, botulism and scombrotoxicosis (Novotny et al., 2004). Vibrio species bacteria are ubiquitous in aquatic environments including fresh, coastal and marine habitats. They are also found as commensals on the surfaces and in the digestive tracts of fish and in zooplanktons (Drake et al., 2007; Montanari et al., 1999). They are transmitted to humans via raw or improperly cooked fish or contaminated water. The resultant diseases are either toxigenic or infectious. However, scombrotoxicosis caused by some species of Vibrios is a Type I hypersensitivity reaction mediated by histamine the metabolic end product of the amino acid histidine found abundant in scombroidea fish species.

Vibrio species testing is part of the microbiological criteria for marine fish in international trade. Apart from their ubiquity in aquatic environments some Vibrio species are emerging pathogens that cause up to more than 50% deaths of all clinical cases (Hsueh et al., 2004; Novotny et al., 2004). Vibrio cholera and Vibrio parahaemolyticus are sensitive to moderate heat treatment 45 to 60°C while Vibrio vulnificus responds well to acid treatment.

Laboratory efforts of detecting Vibrios have for many decades depended on cultural means followed by identification by biochemical tests. This approach was suitable as most Vibrio species are non fastidious, although a few have specific nutritional requirements (Thompson et al., 2004) while others may be undetectable during dormant states (Wong and Wang, 2004). The initial step in Vibrio isolation is resuscitation in alkaline peptone water (APW) pH 8.6, followed by plating on thiosulphate citrate bile salts sucrose (TCBS) agar as the differential selective media on which sucrose fermenting and non sucrose fermenting species are differentiated. Other media such as brain heart infision (BHI) broth and cellobiose polymyxin B colistin (CPC) agar for V. vulnificus have been designed. Problems encountered during Vibrio isolation have led to refinement of cultural methods and to the design of molecular based methods including protein profiling; the detection of species specific genes and in some instances a polyphasic approach.

The present document is a review of the literature about microbiological quality issues encountered in fish factories with specific emphasis on three most important human pathogenic Vibrio species; their biology and survival patterns, the available methods of detection and identification are highlighted. The uniqueness of this paper is that it presents the current state of the art, summarises the challenges met and it uses information from various research groups to bridge the gaps, shortfalls and uncertainties commonly experienced in the field. The first part discusses the consequence of metabolic activities of these microorganisms in fish and the interpretation of microbial load to public health. The second part explores the importance of human pathogens transmitted via the consumption of contaminated fish. Their ecology, clinical manifestations in humans, methods of isolation and identification are discussed. Problems encountered and progress made with the use of different methods has also been reviewed.

SPOILAGE OF SEAFOOD

Fish spoilage occurs as a result of autolysis and lipolysis due to the activity of endogenous enzymes or contamination by metabolically active microorganisms (Chang et al., 1998; Ordóñez et al., 2000; Chytiri et al., 2004). Bacteria are pivotal in the process of seafood spoilage by either initiating or accelerating the spoilage process (Gram, 1992; Gennari et al., 1999; Tryfinopoulon et al., 2002).

The number of total viable bacteria is a measure of the general microbiological quality of the food. High mesophilic counts indicate poor sanitary practise and temperature abuse during processing hence signaling a health hazard. Mesophilic counts however depend on the type of food and number present during harvesting; it may also indicate that a slow spoilage process is taking place. The pshychrotrophic count may indicate the shelf life under refrigeration storage (ICMSF, 1978; Mol et al., 2007). Psychrotrophic bacteria belonging to the class γ-Proteobacteria group are well documented as spoilage organisms in fresh and preserved fish (Chytiri et al., 2004; Françoise, 2010; Gennari et al., 1999; Gram and Huss, 1996; González et al., 1999; Himelbloom et al., 1991; Koutsoumanis and Nychas, 1999; Ordóñez et al., 2000; Tryfinopoulou et al., 2002). The total viable aerobic plate count (TVC) typically ranges between 10^2-10^6 cfu/g/cm^2 in fresh fish (Chang et al., 1998; Giménez et al., 2002; Mahmoud et al., 2004; Popovic et al., 2010). Fish spoilage is usually associated with total counts of 10^7 to 10^8 of the specific spoilage organisms (Gram and Huss, 1996; Ordóñez et al., 2000).

Fresh fish contain 10^4 to 10^6 cfu/g on the skin, 10^4 to 10^7 in the gills (Gennari et al., 1999), and 10^3 to 10^5 in the intestines (Nickelson and Finne, 1992). Studies carried out on fresh fish and ice-stored fish from temperate regions show that the predominant microflora include Shewanella putrefaciens, Photobacterium phosphoreum, Brochotrix thermosphacta, Pseudomonas species, Aeromonas species and lactic acid bacteria (LAB) (Chytiri et al., 2004; Françoise, 2010; Tryfinopoulou et al., 2002). Similar studies on initial microflora on sardines from the Adriatic Sea found Pseudomonas fluorescence, Pseudomonas putida, Shewanella putrefaciens, Achromobacter, Acinetobacter, Psychrobacter and Flavobacterium (Gennari et al., 1999). Fish from warm waters mainly carries Gram-positive bacteria such as coryneforms, micrococci and Enterobacteriaceae as the predominant microflora (Gennari et al., 1999). *Shewanella* and *Pseudomonas* are the predominant microorganisms in seafood stored on ice under aerobic conditions regardless of its origin (Gram and Huss, 1996; Ordóñez et al., 2000). Kyrana and Lougovois (2002) found *Vibrio* species and Enterobacteriacea as the major spoilage organism on sea bass stored at ambient temperatures. However, only 0.01 to 2% of the the total microflora are culturable, hence total viable counts do not represent absolute numbers and identities of the bacteria present (Françoise, 2010).

Factors that influence these microbial identities including contamination from the environment are the intrinsic factors (water activity, pH, nutritional composition of the fish, and redox potential), the extrinsic factors (temperature, surrounding environment in packages), the processing factors (lightly versus heavily preserved fish, slicing and grinding, the methods of preservation) and implicit factors (conditions of storage, transport and distribution, the biochemical reactions within the residual groups, and their interactions; synergism, antagonism) (Gram and Huss, 1996; Gram and Dalgaard, 2002; Huis in't Veld, 1996; Tryfinopoulou et al., 2002). Seafood spoilage may be minimised through adequate temperature control. Lactic acid bacteria may be used to inhibit spoilage bacteria during preservation and can significantly prevent spoilage by putrefaction (Sudalayandi and Manja, 2011). However, this approach still needs to be verified by further studies due to possible associated side effects including the selective growth promoting effect on psychrotrophic pathogens including *Listeria monocytogenes* and *Clostridium botulinum* type E (Françoise, 2010).

MICROBIAL SAFETY OF SEAFOOD

Apart from spoilage the safety of seafood also has to be controlled in terms of the presence of possible food–borne pathogens such as the human pathogenic *Vibrio* species, *C. botulinum*, *Aeromonas hydrophylla*, *B. cereus*, *Salmonella* spp., *Y. enterocolitica*, *L. monocytogenes* (Françoise, 2010). In 1978, 10.5% of all disease outbreaks and 3.6% of all case of seafood diseases in the United States of America were linked to the consumption of both shellfish and fin fish (Wekell et al., 1994).

Enterobacteriaceae counts are used as a measure of the degree of sanitation in food. High counts are a result of unsanitary handling or temperature abuse. The numbers do not always correlate with the extent of contamination from the original source, due to their ability to grow at varying rates in different food commodities (ICMSF, 1978). The coliform group is not well defined

and results obtained can vary depending on the specimen, growth medium used, incubation temperature and methods used to read results. Human pathogens of exogenous origin in seafood include *Escherichia coli*, *Salmonella*, *Shigella*, *Yersinia enterocolitica*, *Campylobacter* spp., *Staphylococcus aureus* and *Bacillus cereus* introduced through poor personal hygiene or cross contamination with other contaminated foods (Popovic et al., 2010; Wekell et al., 1994)

Biogenic amines

Biogenic amines occur in a wide range of fresh and processed foods including cheese, sauerkraut, wine, liver, leafy vegetables, fruits, milk, chocolates, and meats (Chong et al., 2011; Karovičová and Kohanjdová, 2005; Santos, 1996). Some biogenic amines are natural constituents of the foods; others are a result of endogenous enzymes and or microbial metabolism. The nature and amount of the biogenic amine formed in a particular food depends of the specific food chemical composition and the types of microorganisms present (Santos, 1996). The proteinaceous nature of fish favours formation of biogenic amines some of which may cause intoxication in humans (Karovičová and Kohanjdová, 2005; Santos, 1996). Biogenic amines have been detected in seafoods including fresh fish, fermented fish and fish pastes. The most common of these amines include agmantine, cadaverine, putrecine, phenylethylamine, histamine, serotonin, spermine, spermidine and tryptamine, tyramine (Naila et al., 2011; Rabie et al., 2011).

In living tissues under normal conditions biogenic amine have a diverse role of synthetic and metabolic functions; as a source of nitrogen, precursors for the synthesis of hormones, alkaloids, nucleic acids and proteins, amines and as a source of aroma in foods. In humans they are also important as chemical messengers, and in regulating the blood pressure. Putrecine, cadaverine and spermine are important as free radical scavengers and antioxidants. However, some biogenic amines (agmatine, spermine and spermidine) can react with nitrites to form pro-carcinogenic nitrosamines (Santos, 1996) while others (histamine) in high concentrations trigger allergic reactions known as scombrotoxicosis in humans (Davis and Henry, 2007) associated with scombroid fish species (tuna and horse mackerel, and sardines) (Karovičová and Kohanjdová, 2003; Santos, 1996) and with other fish species such as salmon (Auerswald et al., 2006) as well as other food types such as cheese (Karovičová and Kohanjdová, 2003; Santos, 1996). Histamine is the most toxic biogenic amine known. Putrecine, cadaverine and agmatine suppress oxidation of histamine enhancing histamine toxicity (Chong et al., 2011; Santos, 1993). Biogenic amines are formed mostly through decarboxylation of amino acids, through amination and trans-amination of ketones and aldehydes mainly by bacteria decarboxylases than by food endogenous enzymes (Karovičová and Kohanjdová, 2003; Santos, 1996). Bacteria of the genera Enterobacteriaceae, *Lactobacillus*, *Bacillus*, *Citrobacter*, *Proteus*, *Pseudomonas*, *Salmonella*, *Shigella*, *Klebsiella*, *Escherichia*, *Pediococcus*, *Streptococcus*, *Staphylococcus*, *Vibrio* species, *Raoultella planticolla* and some *Clostridia* are known to produce biogenic amines from their precursor molecules (Kanki et al., 2007; Santos, 1996). *Proteus morganii*, *Klebsiella pneumoniae*, *Klebsiella oxytoxa* *Hafnia alvei*, *Staphylococcus hominis* and *Enterococcus hirae* produce histamine at a fast rate and are important in the microbiological quality of fish, while *Photobacterium phosphoreum* and *Photobacterium damsela* are psychrotolerant and mesophiles respectively that produce biogenic amines in scombroid fish species at low temperatures and at ambient temperature (Economou et al., 2007; Kanki et al., 2007; Santos, 1996). Higher levels of histamine are found in improperly stored scombroid fish species due to high levels of histidine in these fish muscles (Auerswald et al., 2006; Karovičová and Kohanjdová, 2005). Apart from histamine, putrecine, cadaverine, tyramine, spermine and spermidine have also been detected in mackerel, herring, tuna, and sardines (Santos, 1996), tryptamine, 2-phenylethylamine, agmatine and serotonin in sea bass (Öxogul et al., 2006).

Its accumulation in fish muscle is commonly caused by temperature abuse during fish harvesting, processing, transport and storage, for example, when the fish is held at temperatures above 7°C for several hours (Auersw ald et al., 2006; CDC, 2007; Economou et al., 2007) the rate of formation also depends on bacterial count (Takahashi et al., 2003). The amine is resistant to freezing and cooking (Chong, 2011). Formation of biogenic amines may be controlled by controling bacterial growth and by a number of food processing techniques including modified atmosphere packaging, food irradiation, high temperature treatment, addition of starter cultures that break down histamine and by hydrostatic presure (Chog et al., 2011; Naila et al., 2010). The levels of allowable histamine concentrations in food range from 50 to 200 ppm in Australia and USA respectively (Auerswald et al., 2006).

Vibrio species as food-borne pathogens

Vibrios are responsible for a number of clinical conditions such as cholera, gastroenteritis, septicaemia and wound infections (Jay et al., 2005; Oliver and Kaper, 1997; Thompson et al., 2004).Twelve *Vibrio* species have been documented as potential food-borne disease agents in humans: *V. cholerae*, *V. parahaemolyticus*, *V. vulnificus*,

Vibrio alginolyticus, *Vibrio funissii*, *Vibrio fluvialis*, *Vibrio damsela*, *Vibrio mimicus*, *Vibrio hollisae*, *Vibrio cincinatiencis*, *Vibrio harveyi* and *Vibrio metchnikovii* (Adams and Moss, 2008; ICMSF, 1996; Thompson and Swings, 2006). A few species are pathogens of fish, while some other species are involved in coral bleaching (Thompson et al., 2004).

Vibrio species are transmitted to humans mostly via sewage contaminated water or seafood (finfish, molluscs and crustaceans) when consumed raw or partially cooked (De Paola et al., 2000; ICMSF, 1996; Oliver and Kaper, 1997). Though *Vibrio* species have been isolated from marine environments, poor processing practises are regarded as the major cause of the food contamination (Kaysner et al., 1992). The bacteria may persist in the food depending on storage temperatures, pH and the product water activity (ICMSF, 1996) until the food is consumed, thereby causing disease. Pathogenic *Vibrio* species are a health concern especially in fish harvested from poor quality waters (ICMSF, 1986). The level of these pathogens in shellfish and in water does not correlate with the level of indicator organisms. In addition human pathogenic *Vibrio* species can acquire survival strategies that enable them to evade detection by conventional monitoring techniques (Odeyemi, 2012), necessitating direct detection of each species in order to ensure public health and food safety (Harwood et al., 2004). The level of *Vibrios* in sea water is generally high during summer months, but very low or undetectable during winter when cultural methods of detection are used due to the ability of these organisms to revert into a dormant unculturable state (Whitehouse et al., 2010).

Pathogenic *Vibrio* species associated with seafood food

Vibrio cholerae

V. cholerae is the most commonly occurring pathogenic *Vibrio* species, followed by *V. parahaemolyticus*. Cholera is characterised by profuse watery diarrhoea with flakes and mucus, dehydration and sometimes death when adequate medical intervention is not instituted (Jay et al., 2005; Kaper et al., 1995; Talkington et al., 2011). Huq et al. (1990) have suggested that *V. cholerae* is wide spread in estuarine and marine waters around the world, although the numbers may be low in sea water throughout the year. They are found in areas where salinity is between 4 to 17% and their presence does not correlate with either *E. coli* or *Salmonella*. They establish symbioses with planktons as a means of overcoming low temperatures that prevail during winter in temperate regions (Huq et al., 1990; Montanari et al., 1999). *V. cholerae* is sensitive to temperatures, higher than 45°C, and to many disinfectants used in the food industry.

Studies elsewhere on the survival patterns of *V. cholerae* have shown that the bacteria remained viable for 14 days in refrigerated raw vegetables and at room temperature (28 to 30℃) for 28 days. On dry cereals (maize, rice and biscuits) the survival time was 1 to 5 days at 4℃. On the same cooked food item survival times were 14 and 24 days for room temperature and refrigeration storage respectively. It survives longer under refrigeration conditions, with 4 to 9 days in raw, and 2 to 21 days in cooked fish (ICMSF, 1996). Infections by *V. cholera* can lead to epidemics, pandemics, or may be endemic in specific areas. Since its discovery in 1817 there have been seven cholera pandemics. The bacteria are carried around the world mostly by asymptomatic carriers. The most recent outbreak occurred in Haiti and it was linked to a clone in Nepal (Talkington et al., 2011; Hendriksen et al., 2011).

Vibrio parahaemolyticus

V. parahaemolyticus associated gastroenteritis manifested by profuse, watery diarrhoea free from blood and mucus, abdominal cramps, nausea, and vomiting. Outbreaks of *V. parahaemolyticus* food poisoning are associated with consumption of raw molluscs (Oysters, clams), cooked crustaceans (shrimp, crab and lobsters) in America and Europe, but in Japan, South East Asia, India and Africa raw fish is always implicated as a vehicle (Jay et al., 2005).

V. parahaemolyticus is usually resident in coastal waters (Adams and Moss, 2008), but recently it has been isolated from fresh water fish (Noorlis et al., 2011). It is halophylic requiring 3 to 10% NaCl for growth. It is distinguished from other *Vibrio* species by its inability to ferment sucrose. It is mesophilic hence isolated only in summer in temperate regions, but in warm water (19 to 20℃), it is detected throughout the year (ICMSF, 1996). It is associated with crustaceans; shrimp and crab, and molluscan shell fish (Popovic et al., 2010) and free swimming fish at a concentration of 10^2 to 10^3 cells per gram, while in very warm water there may be 10^6 cfu per gram (ICMSF, 1996). About 98 of the bacteria isolated from marine animals and seawater are Kanagawa negative and were regarded to be non pathogenic (ICMSF, 1996). However Pal and Das (2010), Rojas et al. (2011) and Jun et al. (2012) showed that up to 35% of 60 *V. parahaemolyticus* bacteria isolated from shell fish and fin fish samples collected from markets, aquatic environments and restaurants carried the Thermostable Direct Haemolysisn (*tdh*) gene some of which were resistant to antibiotics. Velazquez-Roman et al. (2012) confirmed the prevalence of pandemic strains of *V. parahaemolyticus* (O3:K6) in sediment, seawater and shrimp samples from the Pacific coast that coincided with recurrent sporadic cases of gastroenteritis in Northwestern Mexico between 2004 and 2010. During this study 52% of the environmental strains were found to contain virulent genes. *V. parahaemolyticus* is rarely isolated in water where temperatures are below 15℃ (Matches et al., 1971). The bacteria are moderately sensitive to freezing and can persist in frozen food for long periods (Vasudevan et al., 2002). *V. parahaemolyticus* is very sensitive to heat (killed at 47 to 60℃) to ionizing radiation, and to halogens (Adams and Moss 2008). Food poisoning associated with this organism arises from gross mishandling during preparation and from temperature abuse (Nickelson and Finne, 1992).

V. parahaemolyticus has been isolated from shellfish (Bauer et al., 2006; De Paola et al., 2000; Drake et al., 2007; Hervio-Heath et al., 2002; Pal and Das, 2010; Vieira and Iaria, 1993). The levels in oysters correlated with the levels in corresponding waters (De Paola et al., 1990) pointing at the filter feeding habit as the way shell fish pick up the bacteria from waters. *V. parahaemolyticus* was also detected in lobster samples collected from a supermarket and from a fish processing factory at 3 to 21 cfu/g). Some of environmentally isolated *V. parahaemolyticus* were Kanagawa test positive (Vieira and Iaria 1993). The minimum levels of detection for *V. parahaemolyticus* in chilled or raw crustaceans are 10^3/g (Vieira and Iaria, 1993). *V. parahaemolyticus* and *V. vulnificus* have been isolated from bivalve molluscs, but their numbers did not correlate with the number of coliforms (Normanno et al., 2005).

Vibrio vulnificus

V. vulnificus is regarded as an emerging pathogen, infection in humans was first reported in 1964 in the USA and in 1987 in Taiwan (Harwood et al., 2004; Hsueh et al., 2004). It is an opportunistic pathogen in the elderly, immunocompromised or in individuals with impaired liver function, or underlying disease such as cirrhosis, diabetes mellitus or those on steroid therapy (Drake et al., 2007; Harwood et al., 2004). Infections are usually acquired through consumption of raw or improperly cooked shellfish or through contact with seawater (Cazorla et al., 2011; Hsueh et al., 2004). *V. vulnificus* causes three important disease syndromes; septicaemia, necrotising wound infections and gastroenteritis with a mortality rate of 40 to 50% occurring one to two days after onset of the symptoms (Cazorla et al., 2011; Harwood et al., 2004; Hsueh et al., 2004). *V. vulnificus* is a common inhabitant of seawater, but the levels are not correlated with those of indicator organisms (Harwood et al., 2004).Hsueh et al. (2004) have shown that the bacteria are most prevalent in seawater during summer when the temperatures are between 26 to 29℃. It occurs in environments with salinity of 0.5 to 2.5% (Harwood et al., 2004), but the concentration was found to increase in oil polluted seawater (Tao et al., 2011). *V. vulnificus* is sensitive to low pH's and acid treatment is suggested for effective control of the survival of this pathogen in seafood (Lee et al., 1997). Tamplin and Capers (1992) have demonstrated the presence of *V. vulnificus* (10^3 to 10^5 cfu /ml) in sea water and (10^2 and 10^3 cfu/g) on oyster samples collected from the Gulf of Mexico. They are also found associated with other shellfish (clams, and mussels), sediments and planktons which are believed to act as reservoirs (Harwood et al., 2004). The USA Interstate Shellfish Sanitation Conference has set a limit of 30 *V. vulnificus* per gram oyster (Harwood et al., 2004). Tamplin and Capers (1992) showed that the organisms could not be cleared from the oyster tissues by normal depuration procedures when UV treated and filtered water was used.

Vibrio alginolyticus

Vibrio alginolyticus is largely opportunistic pathogen causing systemic infections in persons with underlying diseases such as the immunocompromised individuals, those with severe burns, cancers or with a history of alcohol abuse (Oliver and Kaper, 1997), though it has occasionally been associated with cases of gastroenteritis and diarrhoea. In healthy individuals *V. alginolyticus* is associated with extra intestinal infections such as wound or ear infections (Novotny et al., 2004). The bacterium was also isolated from the blood of a leukaemia patient alongside *Pseudomonas aeruginosa* (Oliver and Kaper, 1997). *V. alginolyticus* is also important food spoilage organism producing histamine by the decarbolylation of histidine and is responsible for scombroid poisoning characterised by nausea, vomiting, abdominal cramps, neurological disorders and skin irritations (Novotny et al., 2004; Ray and Bhunia, 2008).

V. alginolyticus is the most commonly isolated *Vibrio* species in marine environments from all over the world. Its numbers correlate with increases in temperatures (Oliver and Kaper, 1997). It has been isolated from both fin fish and shell fish. Hervio–Heath et al. (2002) isolated *V. alginolyticus* as the most predominant *Vibrio* species from mussels and water samples from the coastal areas in France. Di Pinto et al. (2006) analysed 38 shellfish samples and detected *V. alginolyticus* from 76% of those samples while only 42% of their samples were positive for *V. parahaemolyticus.* Gonzales–Escanola et al. (2006) and Xie et al. (2005); detected *V. parahaemolyticus* virulence associated genes in some *V. alginolyticus* strains.

Pathogenic strains of *V. alginolyticus* carry the collagenase and ToxR genes and can be identified through dection of these genes (Cai et al., 2009). Detection of *Vibrio* species in food and water typically relied on isolation of the bacteria followed by identification by means of classical biochemical tests (Croci et al., 2007; Harwood et al., 2004). *Vibrio* species are non fastidious and grow readily on basic laboratory media, but some need supplementation of vitamins, amino acids and minerals (Farmer and Hickmann- Brenner, 1991; Thompson et al., 2004). They grow better at alkaline pH (7.5 to 8.5) and require added NaCl. The optimum growth temperature ranges from 15 to 30°C (Thompson et al. , 2004).

Most *Vibrio* species grow on Mac Conkey agar, but do not ferment lactose (Farmer et al., 2004). Isolation of *Vibrio* species from environmental sources usually is done by a pre-enrichment step in Alkaline Peptone Water (APW), pH 8.6 supplemented with 1 to 2% NaCl (Harwood et al. 2004; Kaysner et al. 1992), followed by plating on a solid medium such as thiosulfate citrate bile salts sucrose (TCBS) agar. Enrichment media are normally incubated at room temperatures (18 to 22°C), while solid media are incubated at 25°C. TCBS is a selective differential media that incorporates bile salts, alkaline pH (8.6) and 1% NaCl as selective agents, sucrose as a fermentable sugar and bromothymol blue as the pH indicator (Farmer and Hickmann-Brenner, 1991; Harwood et al., 2004). On TCBS sucrose fermenters form yellow colonies, while non-sucrose fermenters are green (Farmer and Hickmann-Brenner, 1991; Kaysner et al., 1992). Enterobacteriaceae, *Pseudomonas* and Gram positive bacteria are inhibited on TCBS (Harwood et al., 2004). The problems encountered with TCBS are that some species do not grow well on it, the selectivity and performance of the medium may vary from batch to batch or between manufacturers, bacteria other than vibrios may grow on it (Farmer and Hickmann-Brenner, 1991; Harwood et al., 2004; Shikongo-Nambabi et al., 2010) and sometimes TCBS may be too inhibitory for some species (especially *V. vulnificus*). Other media used in the isolation of vibrios include Tryptone Soy Agar to which 1 to 2% NaCl is added and Marine Agar (MA). Luria-Bertani broth is used for the enrichment of psychrotrophic species (Thompson et al., 2004). Lee et al. (1997) showed that Brain Heart Infusion Broth (BHI) was a better enrichment medium for *V. vulnificus* than Luria-Bertani (LB) broth, Cellobiose Polymyxin–B Colistin (CPC) broth or Alkaline Peptone Water (APW).

The media that best suit isolation of *V. vulnificus* from shellfish and other environmental sources include amongst others *V. vulnificus* (VV) agar, CPC agar and its modification, as well as Sodium Dodecyl Sulfate-Polymyxin-B - Sucrose (SPS) agar (Harwood et al., 2004).

Williams et al. (2011) showed that using CPC[+] can help isolate *V. vulnificus* with a high rate of recovery from the environment only when present in high numbers, while CHROMagar *Vibrio* medium can clearly distinguish between four medically important *Vibrio* species. However, the researchers recommended that the two media be used simultaneously to avoid false positive results.

The viable but non culturable (VBNC) state

The mystery behind the survival of pathogenic *Vibrio species* in sea water that could not be detected by cultural methods at low temperatures was broken when a dormant state of the bacteria was investigated and discovered (Jiang and Chai, 1996; Oliver et al., 1995). This dormant state, termed the Viable but Non culturable (VBNC) state is associated with a number of pathogenic bacteria including *Vibrio* species regulated at the level of gene expression that enhances resistance to stress (Baffone et al., 2006; Wong and Wang, 2004). The VBNC differs from stress responses induced by nutrient limitation (Coutard et al., 2005). Apart from nutrient deprivation low temperature is an essential factor for VBNC state (Coutard et al., 2005; Jiang and Chai, 1996).

The VBNC state is accompanied by alterations in metabolic activities, morphological changes and changes in genes expression. Jiang and Chai (1996) found that starved pathogenic and non pathogenic strains of *V. parahaemolyticus* maintained at low temperatures changed shapes from rods to cocci quicker than the cells that were kept at room temperatures. After one week *V. parahaemolyticus* cells also changed from smooth rod like shapes with a single flagellum to spheroid cells without flagellae, but with polymer like filaments and wrinkled cell walls. Similar morphological changes (rods to cocci) were observed in *V. vulnificus* cells kept at 5°C in ASW microcosms (Smith and Oliver, 2006). Baffone et al. (2006) observed reduced cell size, spherical shapes, formation of blebs and polymer–like filaments in VBNC states whereas no such changes have been reported during stressful conditions at normal growth temperatures.

Alam et al. (2007) showed that after incubation free swimming coccoid *V. cholera* O1 cells in the VBNC could not be cultured, while normal shaped cells recovered from attached biofilms could be grown on culture media. The cell size reduction from rods to cocci is regarded as part of the survival strategy to optimise nutrient uptake and utilisation during the adverse environmental conditions such as those encountered during winters in temperate regions and marine environments (Jiang and Chai, 1996; Wong and Wang, 2004).

Griffitt et al. (2011) using recognition of individual gene fluorescence *in situ* hybridisation (RING-FISH) method showed that culture based methods were unable to detect *V. parahaemolyticus* cells in VBNC hence counts obtained in environmental samples were lower than the actual population density in samples in agreement with Thongchankaew et al. (2011) who used the denaturant gradient gel electrophoresis (DGGE) method to detect

V. prahaemolyticus in water samples from Tarutao Island, Thailand that could not be detected by cultural methods. Brauns et al. (1991) demonstrated that the (*V vulnificus* haemolysin gene) *vvh* decreased significantly during the VBNC state and was barely detectable using similar conditions that were optimal for detecting the same gene in viable, culturable cells. Oliver and Bockian (1995) demonstrated that though pathogenicity decreased during the VBNC state, *V. vulnificus* strains expressed virulence factors during the dormant state and were able to resuscitate in vivo. Saux et al. (2002) using a Reverse Transcriptase seminested PCR detected the *vvh* gene in VNBC cells induced in Artificial Sea Water (ASW).

Similarly, Coutard et al. (2005) demonstrated that Nonculturable viable *V. parahaemolyticus* cells maintain albeit at reduced levels expression of housekeeping genes (rpoS and 16S-23S rRNA), while the synthesis of virulent associated proteins (Thermostable Direct Haemolysin) was suspended. However, Smith and Oliver (2006) did not detect expression the haemolysin gene (*vvhA*) during the *V. vulnificus* VBNC state induced in ASW microcosms. Using membrane diffusion chambers dipped in the North Carolina estuarine waters however the same researchers (Smith and Oliver, 2006) showed that expression of the *vvhA* was strain dependent. Smith and Oliver (2006) also detected the gene encoding the periplasmic catalase (*katG*); however the levels decreased during the VBNC state and concluded that a decrease of the catalase production triggers cells to adopt the VBNC state since the cells are no longer able to protect themselves against the H_2O_2 that is present in culture media.

Entry into the VNBC may (Coutard et al., 2005) or may not (Oliver et al. 1995) depend on the physiological and nutritional state of the cells. Although Coutard et al. (2005) found that logarithmic phase cells are more readily inducible into the VBNC than stationary phase cells, Oliver et al. (1991) and Oliver et al. (1995) studying *V. vulnificus* observed that both logarithmic and stationary phase cells entered the VBNC state at the same time.

In the VBNC state cells are more resistant to heat, salinity, acid and pH than dividing cells (reviewed by Drake et al., 2007). The resistance of VBNC pathogenic *Vibrio* species to mild treatments is a matter of concern in seafood that is consumed raw or partially cooked Wong and Wang (2004).

The VNBC state is inducible by a temperature downshift to 3 to 4°C (Coutard et al., 2005; Oliver and Bockian, 1994; Smith and Oliver, 2006; Wong and Wang, 2004) but higher temperatures (10 to 15°C) were required to enter this state under natural conditions (Oliver et al., 1995). In both *in vitro* and *in situ* studies viable counts remained high throughout the VBNC state, and the culturability was reversed within 24 h by temperature upshift to 22°C, for a more extensive review, refer to Drake et al., (2007).

Table 1. Phenotypic traits used to differentiate between *V. parahaemolyticus* and *V. alginolyticus* (Farmer et al., 2004; Oliver and Kaper, 1997).

Phenotypic test	*Vibrio parahaemolyticus* (%)	*Vibrio alginolyticus* (%)
Voges-Proscauer (VP) test in1% NaCl	80-95	0
Urea hydrolysis	15	0
Cellobiose	5	3
Dulcitol	3	0
Sucrose	1	99
ONPG	5	0
L- Arabinose	80-89	0-1
Growth in nutrient broth with		
10%	0-2	69-100
12%	0-1	17-100

IDENTIFICATION OF *VIBRIO* SPECIES

General taxonomy

The taxonomy of vibrios was initially based on the classical methods of classification; identification and nomenclature where morphological features (cell shape and presence of extracellular appendages such as flagellae) and biochemical reaction played an important role (Thompson and Swings, 2006). The current taxonomy of vibrios is based on the polyphasic approach that includes phenotypic and molecular methods (Arias et al., 1997; Thompson and Swings, 2006).

The DNA–DNA hybridisation and phylogenetic relationship studies based on the 16S rRNA comparison have been extensively used in *Vibrio* classification (Thompson and Swings, 2006), but with little success in delineating vibrios to species level; most *Vibrio* species have more than 90% 16S rDNA similarities (Aznar et al., 1994).

Multilocus Sequence Analysis (MLSA) is currently used to provide better differentiation of *Vibrio* isolates into respective species (Thompson and Swings, 2006). Although the current family *Vibrionaceae* comprises eight genera the phylogenetic analysis of this group based on the concatenated genes of the three genes, 16S rRNA, *recA* and *rpoA* has proposed four different families e.g. *Vibrionaceae*, *Photobacteriaceae*, *Enterovibrionaceae* and *Salinivibrionaceae* (Thompson and Swings, 2006). The species pathogenic to humans will, however, remain within the genus *Vibrio*.

Phenotypic identification

Phenotypic traits used for the identification of *Vibrio* species are the Gram reaction, oxidase test where vibrios are always positive and the oxidation/fermentation (OF) test in which *Vibrio* species are facultatively fermentative. *Vibrio* species can be differentiated from one another and

from *Aeromonas* species by the sensitivity test to the vibriostatic agent O/129 (Famer et al., 2004). Biochemical test systems such as API 20E, and Biolog can also be used for final identification. These methods are, however, slow and unreliable, since some strains exhibit atypical phenotypic characteristics (Thompson et al., 2004). One of the main obstacles is to correctly differentiate *V. alginolyticus* from *V. parahaemolyticus* that have 60 to 70% DNA homology. *V. alginolyticus* was initially classified as a biotype of *V. parahaemolyticus*. The two species can only be differentiated on the basis of a few phenotypic characters (Farmer et al., 2004; Oliver and Kaper, 1997) (Table 1). In addition the two species have have up to 99% 16S rRNA DNA similarity. Determining nucleotide sequences of certain gene such as the hsp60 gave better resolution between the two species (Kwok et al., 2002). A polyphasic approach using phenotypic and molecular biology traits is most reliable, most importantly is the numerical taxonomy and the detection of *V. alginolyticus* species specific collagenase gene (Shikongo-Nambabi et al., 2010).

Immunological based methods

Immunological methods are based on the reaction of antibodies with specific antigens to form immune complexes. Serotyping has been used as a tool for the terminal confirmation during the identification of human pathogenic *Vibrio* species, especially *V. cholerae*. *V. cholerae* is divided into serovars with O1 being the most important and have caused seven cholera pandemics. Strain O139 Bengal is a non O1 strain first isolated in 1992 from the coastal waters of the Bay of Bengal during a cholera epidemic in India, Bangladesh, and in Thailand is also of concern (ICMSF, 1996). The same technique was used in typing the pathogenic strains of *V. parahaemolyticus* O3:K6 and O1: K UT (Iida et al., 2001; Myers et al., 2003). De Paola et al. (2003) used species specific antisera to differentiate pathogenic strains of *V. parahaemolyticus* into 27 serotypes. Serotyping is of epidemiological significance and can be used to differentiate pandemic from non pandemic strains (Velazquez-Roman et al., 2012).

DNA based methods

Hybridisation

Oligonucleotide probes are used to detect complementary genes or gene fragments as a means of identifying cultures. Probes directed to the variable region of the 16S rRNA gene have been developed, but these were not very useful for *Vibrio* species identification partly due to cross reactions of some probes that reacted with strains other than their specific targets and the specificity of some probes had not yet been tested across the whole *Vibrio* genus. This approach is likely to face problems as some *Vibrio* species may share 100% 16S rDNA homology (Thompson et al., 2004).With the advent of PCR technology the use of hybridisation has largely been replaced.

PCR detection of unique gene fragments

Several methods employing *in vitro* amplification of specific gene fragments by the Polymerase Chain Reaction (PCR) and derivatives of this method have been used for the identification of *Vibrio* species. PCR technique exploits the specificity of short synthetic DNA fragments to bind to complementary sequences and the ability of the DNA polymerase enzymes to directly synthesize the opposite strand under a defined set of conditions using the available DNA as a template. The process is robust, specific and fast hence enabling detection of target genes, gene sequences or specific DNA sequences in test samples.

Hoshino et al. (1998) developed a multiplex PCR consisting of three primer pairs targeting the *rfb*) (gene region specific for O1 and O139) and the cholera toxin (*ctxA*) gene. The *rfb* gene based PCR could detect up to 65 and 200 O1 and O139 cfu per assay in clinical samples respectively. Keasler and Hall (1993) designed a multiplex PCR simultaneously detecting the cholera toxin (*ctxA*) gene in pathogenic and environmental *V. cholerae* O1 Classical and El Tor biotypes and differentiating the two biotypes through their differences in the toxic coregulated pilus (*tcpA)* genes. Theron et al. (2000) developed and evaluated the performance of a seminested *ctxAB* gene specific PCR for the detection of pathogenic *V. cholerae* in environmental water and drinking water sources. This protocol was shown to be highly sensitive, specific and rapid producing results

within 10 h. In a quest for a broader spectrum detection protocol due to the fact that non-epidemic strains could also cause disease, the outer membrane protein (*ompW*) gene was targetted. This gene forms part of the *tox*R regulon and was shown to be present in all *V. cholerae* strains and conserved across different biotypes and serogroups, but absent from all other *Vibrio* species studied (Nandi et al, 2000). Oligonucleotide primers specific to the *ompW* gene were designed and tested for their ability to amplify the specific gene in both clinical and environmental strains. The specificity of these primers was confirmed using DNA probes (Nandi et al., 2000). Le Roux et al. (2004) have evaluated and identified a combination of three primers for the detection of *V. cholerae ompW* gene in environmental isolates. Their work has shown that the PCR approach is more specific than the API 20E and VITEK 32 systems in identifying environmental *V. cholerae* strains.

Other approaches that have been followed include a multiplex Real Time (RT- PCR) targeting four *V. cholerae* virulence genes (Gubala, 2006) and a similar fourplex Real Time PCR targeting *Vibrio choleare* specific genes, for example, repeat in toxin (*rtxA*), extracellular secretory protein (*epsM*), the toxic coregulated pilus A (*tcpA*) and *ompW* gene with a view to enabling detection of both toxigenic and non toxigenic strains (Gubala and Proll, 2006). More recently Fykse et al. (2007) designed real time nucleic acid sequence based amplification (NASBA) PCR that amplifies specific RNA for a number of virulence factors and house keeping genes. The NASBA detects only actively metabolising cells, as opposed to DNA amplification that might have originated from dead cells.A number of *V. parahaemolyticus* genes have been used as targets to develop species specific PCR based detection. These genes include the thermolabile direct haemolysin, *tl* (Baffone et al., 2006; Croci et al., 2007), the thermostable direct haemolysin, *tdh*, the thermostable direct haemolysin related haemolysin, *trh (*Baffone et al., 2006*)*, the phosphatidyl serine synthetase gene (pR72H fragment) (Lee et al., 1995), gyrase B gene (Venkateswaran et al., 1998), the metalloprotease gene (Luan et al., 2007) and the collagenase gene (Di Pinto et al., 2006). Venkateswaran et al. (1998) cloned and sequenced the gyrase B (*gyrB*) gene of *V. parahaemolyticus* and of its close genetic relative, *V. alginolyticus*. They (Venkateswaran et al., 1998) subsequently developed oligonucleotide primers (Vp-1 and Vp-2r) that amplify a 285bp fragment from the *V. parahaemolyticus gyrB* gene by PCR. All *V. parahaemolyticus* strains were recognised by this primer set, and false positives were not detected. Lee et al. (1995) developed a pair of oligonucleotide sequences (Vp32 and VP33) that bind to opposite ends of a 320-387bp DNA fragment termed pR72H,from the chromosome of *V. parahaemolyticus* that was shown to be found only in this species. This is a fragment of unknown function located after an rRNA operon and composed of a non coding region and a phosphatidylserine synthetase gene that was found conserved in *V. parahaemolyticus* (Lee et al., 1995; Robert-Pillot et al., 2002). Hervio-Heath et al. (2002) used the *V. parahaemolyticus* (Vp32/Vp33), the *tdh* and the *trh* specific primers to identify suspect isolates and to determine their pathogenicity respectively.

Kim et al. (1999) developed *V. parahaemolyticus* specific primers that detected the species specific Tox–R gene. Low amplification signals were, however, also obtained with closely related species, *V. alginolyticus* and *V. vulnificus*. They (Kim et al., 1999) recommended that detection of this gene as an identification tool should be supplemented with screening the suspect isolates for *V. parahaemolyticus* virulence specific genes (*tdh* and *trh*) so as to confirm the results. Luan et al. (2007) developed PCR primers specific for the *V. parahaemolyticus* metalloprotease gene and used 101 bacterial strains, 85 of which were identified by phenotypic methods as *V. parahaemolyticus* to assess the specificity of their new primers. When the specificity and sensitivity of the new primer *VPM1* and *VPM2* were compared to three other primer sets that were already in use (including primers pairs directed against three other known virulence genes; (*tl, tdh* and *trh*) the designed metalloprotease specific primer pair gave the best results with a sensitivity of up to 4pg DNA. Unlike the L-*tdh*/R-*tdh* and L-*trh*/R-*trh* primer sets that gave false negatives, *VPM1*/*VPM2* detected all the *V. parahaemolyticus* strains tested, and did not react with other bacteria. Reverse transcriptase (R-T) PCR for *tdh, trh1* and *trh2* have also been designed (Mothershed and Witney, 2006).

Di Pinto et al. (2006) used three oligonucleotide primer pairs specific for either *V. parahaemolyticus*, *V. cholerae* or *V. alginolyticus* collagenase gene and have demonstrated a simultaneous detection of the two species (*V. parahaemolyticus* and *V. alginolyticus*) that were present in alkaline peptone water (APW) enriched shellfish tissue homogenates. In addition the researchers recommended use of these primer pairs in discriminating between *V. alginolyticus* and *V. parahaemolyticus*. This multiplex-PCR was able to detect the presence of these two bacterial species in some culture negative samples circumventing the low sensitivity inherent in culturing and the inability of phenotypic tests to identify isolates with atypical biochemical profiles. Qian et al. 2008 cloned two proteins from *V. alginolyticus* (*OmpK* and *OmpW*) expressed them in *E. coli* and designed specific primers to the genes. More recently Dalmasso et al. (2009) designed a multiplex PER (Primer extension reaction) PRC protocol directed against *rpoA* gene to simultaneously detect and identify six human pathogenic Vibrio species (*V. cholerae*, *V. parahaemolyticus*, *V. vulnificus*, *V. mimicus*, *V. alginolyticus* and *V. fluvialis*) in fishery products. At the same time Cai et al. (2009)

designed a multiplex PCR against collagenase, ompK and toxR genes to differentiate virulent and avirulent strains of *V. alginolyticus*. Virulent strains were recognised by the presence of the collagenase and toxR genes that were absent from avirulent strains. The test also improved the rapidity, sensitivity and specificity for *V. alginolyticus* detection and could detect up to 8.8×10^2 cfu.The gyrase B gene was also targeted (Kumar et al., 2006) for *V. vulnificus* identification using the primer set *gyr-vv1* and *gry-vv2*. Arias et al. (1995) developed a highly sensitive nested PCR specific for the 23S rDNA of *V. vulnificus* and Lee et al. (1998) developed another nested PCR directed against the *V. vulnificus* haemolysin gene (*vvh*). Real Time PCR (RT-PCR) was later developed to detect *V. vulnificus* in sea water and oyster tissue homogenates targeting the species specific *vvh* gene (Panicker et al., 2004). Chakraborty et al. (2006) developed a species specific PCR targeting the ToxR gene of the less characterised human pathogen, *V. fluvialis* facilitating a successful differentiation of this pathogen from the closely related *Aeromonas* species.

Molecular typing techniques

Various molecular techniques have been used to type the strains belonging to the various Vibrio species. These methods include amplified fragment length polymorphism (AFLP), rapidly amplified polymorphic DNA (RAPD), restriction fragment length polymorphism (RFLP), ribotyping, repetitive extragenic palindromic (Rep) sequences, pulsed field gel electrophoresis, Rep (Enterobacterial repetetive intergenic consensus and extragenic palindromic) -PCR, multilocus sequence typing (MLST) and whole genome sequence typing (WGST). Most of these techniques could also be used for the identification of specific Vibrio species; they are important tools in epidemiological investigations and have been recommended for determining the source of virulent strains and their temporal and spatial distributions (Caburlotto et al., 2011; Cazorla et al., 2011; Hendriksen et al., 2011; Ju et al., 2009; Thompson and Swings, 2006).

Sequencing of 16S rDNA

Modern methods of identification that compare nucleotide sequences of group specific genes have gained popularity within the last three decades. The 16S rRNA gene is used for both phylogenetic studies and as a taxonomic marker (Thompson et al., 2005).The whole gene or variable regions of the 16S rRNA gene are sequenced through the use of various primer pair combinations. Ideally comparison of the obtained nucleotide sequences with the sequences available in the

Table 2. Copy numbers of 16S rRNA genes detected in bacteria.

Bacterial species	Number of 16S gene copies	Reference
Bacillus subtilis	10	(Conville and Witebsky, 2007)
E. coli	7	(Conville and Witebsky, 2007)
Salmonella Typhimurium	7	(Conville and Witebsky, 2007)
Clostridium perfringens	9	(Conville and Witebsky, 2007)
Clostridium perfringens	9	(Conville and Witebsky, 2007)
Vibrio cholerae	9	(Conville and Witebsky, 2007)
Nocardia species	3-5	(Conville and Witebsky, 2007)
V. parahaemolyticus	11	(Harth et al., 2007)

GenBank, BCCM or other gene banks facilitates identification of unknown isolates. The 16S rDNA PCR method increases the efficiency of bacterial identification due to its rapid, reproducible nature. This notion was supported by Petti et al. (2005) who showed that the 16S rDNA sequencing was able to correctly identify bacteria including pathogens that had been misidentified by traditional methods. Harris and Hartley (2003) developed a broad range 16S rDNA PCR for use in identifying bacteria isolated from various clinical specimens and compared the results with cultural and serological methods. PCR amplification of the 16S rDNA detected many potentially pathogenic organisms from culture negative samples implying that the 16S rDNA PCR improves the identification process as compared to cultural methods. Complete 16S RNA gene sequences of many Vibrio species have been determined and Vibrio species 16S rRNA gene specific primers and probes have been developed, as cited by Maeda et al. (2003) who developed a clustering scheme based on Vibrio species 16S rRNA gene specific PCR that clustered 46 Vibrio species into 16 groups.

There are also several problems associated with the use of the 16S rDNA for identifying bacteria. No universally accepted criteria exist for the required level of homology to delineate isolates of the same species or genus. Some bacteria with different phenotypic characters may share up to 100% 16S rDNA sequences whereas less than 99% 16S rDNA sequence homology has been observed in bacteria belonging to the same species (Harris and Hartley, 2003). In some species multiple heterogenous copies of the 16S rRNA gene operons exist (Case et al., 2007; Pontes et al., 2007) (Table 2). Some closely related Vibrio species share up to 99% 16S rDNA sequence homology (Shikongo-Nambabi et al., 2010). It is therefore recommended to combine the 16S rRNA gene sequence with other species specific morphological and physiological characteristics to correctly identify Vibrio species (Mienda, 2012). The 16S-23S Intergenic spacer (IGS) region nucleotide sequences complement the 16S rDNA in Vibrio species identification (Hoffmann et al., 2010). Using the capillary gel electrophoresis technology to analyse the 16S-23S IGS region nucleotide sequences after PCR quickly generates strain and species specific patterns useful in epidemiological and idenfication of Vibrio species.

Also the databases used in sequence comparison for identification are public facilities hence sequences that are incorrectly identified may also be erroneously published and could lead to misidentification of accurately read sequences.

Multilocus sequence analysis (MLSA)

With the problems experienced with the prokaryotic identification schemes (phenotypic traits) and 16S rRNA gene sequences, as well as with the DNA- DNA hybridisation (DDH), multilocus sequence analysis (MLSA) is gaining popularity as a promising taxonomic tool to differentiate between closely related bacterial species. MLSA uses gene sequences from more than one locus, generally of house keeping genes that are widely distributed among bacteria to be studies and have single copy in the genome (Pontes et al., 2007; Thompson et al., 2005). The genes should also be long enough to give sufficient information, but should be of the length that permits easy sequencing (Thompson et al., 2005).

The genetic loci that were initially found suitable in MLSA for taxonomic studies of Vibrio species included the 16S rRNA gene, *rpoA* (RNA polymerase alpha subunit) and *recA* (recombinant repair protein) and *pyrH* (uridylate kinase gene) (Thompson et al., 2005). Recently Thompson et al. (2007) have developed an MLSA scheme to distinguish V. harveyi from the closely related Vibrio campbellii species using seven housekeeping genes; *recA*, *topA* (topoisomerase I), *pyrH*, *ftsZ* (a cell division protein), *mreB* (actin like cytoskeleton protein), *gapA* (glyceraldehydes-3 phosphate dehydrogenase) and the *gyrB* DNA (gyrase B gene beta subunit). The resulting 3596 nucleotide long DNA gave a better resolution than when only three genes, *gyrB, reA* and *gapA* were used. MLSA therefore acts as a buffer against mutations and horizontal gene transfer (HGT) problems associated with the 16S rDNA sequences (Thompson et al., 2005). Whitehouse et al. (2010) coupled Electrospray Ionization Mass Spectrometry to Multi Locus Sequence PCR (PCR/ESI-MS) using 8 primer pairs that react with the whole *Vibrionaceae* family to simultaneously detect, quantify and identify *V. cholera*, *V. parahaemolyticus* and *V. vulnificus* species in sea water and fresh water in Georgia. The novel scheme was robust and highly specific able to discriminate between closely related *V. cholera* from *V. mimicus*.

CONCLUSION

The initial microbiological quality of marine fish is depended upon the microbial load of the marine water where the fish is harvested as governed by the prevailing physicochemical parameters (pH, salinity, nutrient content and temperature). The predominant micro-organisms that persist in the finished products to cause spoilage and or food borne diseases depend on the intrinsic factors of the fish in question; the extrinsic factors of the food processing, storage and distribution environment, the explicit factors of the microbial population, and the processing factors. Bacterial meta-bolism of the final product can cause significant quality deterioration while the presence of pathogens can be hazardous to human health. Stringent quality assurance regimes are required to ensure optimal seafood safety and shelf life.

The current laboratory methods used in quality assurance and quality control are sometimes ineffective due to low sensitivity or atypical phenotypic or genotypic characters of target organisms. A polyphasic approach that encompasses both phenotypic and genotypic traits is often best for accurate identification of sefood borne pathogens particularly human pathogenic *Vibrio* species. Novel methods such as Electrospray Ionization Mass Spectrometry Multi Locus Sequence PCR or the 16S-23S IGS region nucleotide sequences can be used to enhance the discriminative power of *Vibrio* species identification protocols.

ACKNOWLEDGEMENT

The authors would like to thank the Ministry of Education of Namibia for funding this research.

REFERENCES

Adams MR, Moss MO (2008). Bacterial Agents of Foodborne Illness. In: Food Microbiology, 3rd ed., RSC, Cambridge UK, pp. 182-269.

Alam M, Sultana M, Nair GB, Siddique AK, Hasan NA, Sack RB, Sack DA, Ahmed KU, Sadique A, Watanabe H, Grim CJ, Huq A, Colwell RR (2007). Viable but nonculturable *Vibrio* cholera O1 in biofilms in the aquatic environment and their role in cholera transmission.

PNAS, 104: 17801-17805

Arias CR, Garay E, Aznar R (1995). Nested PCR Method for Rapid and Sensitive Detection of *Vibrio vulnificus* in Fish, Sediments and Water. Appl. Environ. Microbiol., 61: 3476-3478.

Arias CR, Verdonck L, Swings J, Aznar R, Garay E (1997). A Polyphasic approach to Study the Intraspecific Diversity Among *Vibrio vulnificus* isolates. Syst. Appl. Microbiol., 20: 622-633.

Auerswald L, Morren C, Lopata A (2006). Histamine levels in seventeen species of fresh and processed South African seafood. Food Chem., 98: 231-239.

Aznar R, Ludwig W, Amann RI, Schleifer KH (1994). Sequence determination of rRNA Genes of Pathogenic *Vibrio* Species and Whole–Cell Identification of *Vibrio vulnificus* with rRNA–Targeted Oligonucleotide Probes. Int. J. Syst. Bacteriol., 44: 330-337.

Baffone W, Tarsi R, Pane L, Campana R. Repetto B, Mariottini GL, Pruzo C (2006). Detection of free living and plankton- bound *Vibrios* in coastal waters of the Adriatic Sea (Italy) and study of their pathogenicity associated properties. Environ. Microbiol., 8: 1299-1305.

Brauns LA, Hudson MC, Oliver JD (1991). Use of the Polymerase Chain Reaction in Detection of Culturable and Nonculturable *Vibrio vulnificus* Cells. Appl. Environ. Microbiol., 57: 2651-2655.

Cai SH, Lu YS, Wu ZH, Jian JC, Huang YC (2009). A novel multiplex PCR method for detecting virulent strains of *Vibrio alginolyticus*. Aquac. Res., 41 27-4134.

Case RJ, Boucher Y, Dahllöf I, Holmström C, Doolittle W F, Kjelleberg S (2007) Use of 16S rRNA and *rpoB* Genes as Molecular Markers for Microbial Ecology Studies. Appl. Environ. Microbiol., 73: 278-288.

Cazorla C, Guigon A, Noel M, Quilici M-L, Lacassin F (2011). Fatal *Vibrio vulnificus* Infections associated with Eating Raw Oysters, New Caledonia. Emerg. Infect. Dis., 17: 136-137.

CDC: Morbidity and Mortality Weekly Report (2007). Scombroid Fish Poisoning Associated With Tuna Steaks-Louisiana and Tennessee, 2006 JAMA Weekly Report, 298: 1-4.

Chakraborty R, Sinha S Mukhopadhyay K, Asakura M, Yamasaki S, Bhattacharya SK, Nair GB, Ramamurthy T (2006). Species specific identification of *Vibrio fluvialis* by PCR targeted to the conserved transcriptional activation and variable membrane tether regions of the toxR gene. J. Med. Microbiol. Correspond., pp. 805-808.

Chang KL, Chang J, Shiau C-Y, Pan BS (1998). Biochemical, Microbiological and Sensory Changes of Sea Bass (*Lateolabrax japonicus*) under Partial Freezing and Refrigerated Storage. J. Agric. Food Chem., 46: 682-686.

Chong CY, Abu Bakar F, Russly AR, Jamilah B, Mahyudin NA (2011). MiniReview. The effect of food processing on biogenic amines formation. Int. Food Res. J., 18: 867-876.

Chytiri S, Chouliara I, Savvaidis IN, Kontominas MG (2004). Microbiological, chemical, and sensory assessment of iced whole and filleted aquacultured rainbow trout. Food Microbiol., 21: 157-165.

Conville PS, Witebsky FG (2007). Analysis of Multiple Differing Copies of the 16S rRNA gene in Five Clinical isolates and Three Type Strains of *Nocardia* Species and Implications for Species Assignment. J. Clin. Microbiol., 45: 1146-1151.

Croci L, Suffredini E, Cozzi L, Toti L, Ottaviani D, Pruzzo C, Serratore P, Fischetti R, Goffredo E, Loffredo G, Mioni R, the *Vibrio parahaemolyticus* Working Group (2007) Comparison of different biochemical and molecular methods for the identification of *Vibrio parahaemolyticus*. J. Appl. Microbiol., 102: 229-237.

Coutard F, Pommepuy M, Loaec S, Hervio-Heath D (2005). mRNA detection by reverse transcription-PCR for monitoring viability and potential virulence in a pathogenic strain of *Vibrio parahaemolyticus* in viable but non culturable state. J. Appl. Microbiol., 94: 951-961.

Dalmasso A, Civera T, Bottero MT (2009). Multiplex primer-extension assay for identification of six pathogenic *Vibrios*. Int. J. Food Microbiol., 129: 21-25.

Davis J, Henry MPH (2007). Scombroid Fish Poisoning Associated With Tuna Steaks-Louisiana andTennessee, 2006 JAMA 298: 1269-1270.

De Paola A, Hopkins IH, Peeler JT, Wentz B, Mcphearson RM (1990). Incidence of *Vibrio parahaemolyticus* in US coastal waters and oysters. Appl. Environ. Microbiol., 56: 2299-2302.

De Paola A, Kaysner CA, Bowers JC, Cook DW (2000). Environmental investigations of *Vibrio parahaemolyticus* in oysters following outbreaks in Washington, Texas, and New York (1997 and 1998). Appl. Environ. Microbiol., 66: 4649-4654.

De Paola A, Ulaszek J, Kaysner CA, Tenge BJ, Nordstrom JI, Wells J, Puhr N, Gendel SM (2003). Molecular, Serological, and Virulence Characteristics of *Vibrio parahaemolyticus* Isolated from environmental, Food, and Clinical Sources in North America and Asia. Appl. Environ. Microbiol., 69: 3999-4005.

Di Pinto A, Giuseppina C, Fontanarosa M, Teria V, Tantillo G, (2006). Detection of *Vibrio alginolyticus* and *Vibrio parahaemolyticus* in shellfish samples using Collagenase - Targeted Multiplex PCR. J. Food Saf., 26: 150-159.

Drake SL, DePaola A, Jaykus LA (2007). An Overview of *Vibrio vulnificus* and *Vibrio parahaemolyticus*. Compr. Rev. Food Sci. Food Saf., 6: 120-144.

Economou V, Brett MM, Papadopoulou C, Frillingos S, Nichols T (2007). Changes in histamine and microbiological analyses in fresh and frozen tuna muscle during temperature abuse. Food Addit. Contam., 8: 820-832.

Farmer JJ, Hickman-Brenner FW (1991). The Genera *Vibrio* and *Photobacterium* In: Ballows et al., (eds) The Prokaryotes (2nd edn.) Springer–Verlag, New York, pp. 2952-3011.

Farmer JJ, Janda JM, Brenner FW, Cameron DN, Birkhead KM (2004). In: Garrity et al., (eds.) Bergy's Manual of Systematic Bacteriology (2nd edn) Springer-Verlag, New York, pp. 494-545.

Françoise I (2010). Occurence and role of lactic acid bacteria in seafood product. Food microbial., 27: 698-701.

Fykse EM, Skogan G, Davies W, Olsen JS, Blatny JM (2007). Detection of *Vibrio cholerae* by Real–Time Nucleic Acid Sequence-Based Amplification. Appl. Environ. Microbiol., 73:1457-1466.

Gennari M, Tomaselli S, Cotrona V (1999). The Microflora of fresh and spoiled sardines (*Sardina pilchardus*) caught in Adriatic (Mediterranean) Sea and stored in ice. Food Microbiol., 16: 15-28.

Giménez B, Roncalés P, Beltrán JA (2002). Modified atmosphere packaging of filleted rainbow trout. J. Sci. Food Agr., 82: 1154-1159.

González CJ, López-Díaz TM, García-López, ML, Prieto M, Otero A (1999). Bacterial Microflora of wild Brown trout (*Salmon trutta*), Wild pike (*Esox lucius),* and Aquacultured Rainbow Trout (*Onchorhynchus mykiss*). J. Food Protect, 62: 1270-1277.

Gram L (1992). Evaluation of the Bacteriological quality of seafood. Int. J. Food Microbiol., 16: 25-39.

Gram L, Dalgaard P (2002). Fish spoilage bacteria-problems and solutions. Curr. Opin. Biotech., 13: 262-266.

Gram L, Huss HH (1996). Microbiological spoilage of fish and fish products. Int. J. Food Microbiol., 33: 121-137.

Griffitt KJ, Noriella III F, Johnson CN, Grimes DJ (2011). Enumeration of *Vibrio parahaemolyticus* in the viable but nonculturable state using direct plate counts and recognition on individual gene fluorescence *in situ* hybridisation. J. Micobiol. Meth., 85: 114-118

Gubala AJ (2006). Multiplex real–time PCR detection of *Vibrio cholerae*. J. Microbiol. Meth. 65: 278-293.

Gubala AJ, Proll DF (2006). Molecular–Beacon Multiplex Real–Time PCR Assay for Detection of *Vibrio cholerae*. Appl. Environ. Microbiol., 72: 6424-6428.

Harris KA, Hartley JC (2003). Development of broad-range 16S rDNA PCR for use in the routine diagnostic clinical microbiology service. J. Med. Microbiol., 52: 682-691.

Harth E, Romeo J, Torres RR, Espejo RT (2007). Intragenomic heterogeneity and intergenomic recombination among *Vibrio parahaemolyticus* 16S rRNA genes. Microbiology 153: 2640-2647.

Harwood VJ, Gandhi JP, Wright AC (2004). Methods for Isolation and confirmation of *Vibrio vulnificus* from oysters and environmental sources: a review. J. Microbiol. Meth., 59: 301-316.

Hendiksen RS, Price LB, Schupp JM, Gillece JD, Kaas, RS, Engelthaller DM, Bortolia V, Pearson T, Wters AE, Updhyay BP, Shrestha SD, Adhikari S, Shakya D, Keim PS, Aarestrup FM (2011). Population Genetics of *Vibrio cholera* from Nepal in 2010: Evidence on the Origin of the Haitian Outbreak. mBio., 2: 1-6.

Hervio-Heath D, Colwell RR, Derrien A, Robert-Pillot A, Fournier JM, Pommepuy M (2002). Occurrence of pathogenic vibrios in coastal areas of France. J. Appl. Microbiol., 92: 1123-1135.

Himelbloom BH, Brown EK, Lee JS (1991). Microorganisms on Commercially Processed Alaskan Finfish. J. Food Sci., 56: 1279-1281.

Hoffmann M, Brown EW, Feng PCH, Keys CE, Fischer M, Monday SR (2010). PCR-based method for targeting 16S-23S rRNA intergenic spacer regions among *Vibrio* species. Microbiol., 10: 2-14.

Hoshino K, Yamasaki S, Mukhopadhyay AK, Chakraborty S, Basu A, Bhattacharya SK, Nair GB, Shimada T, Takeda Y (1998). Development and evaluation of a multiplex PCR assay for rapid detection of toxigenic *Vibrio cholera* O1 and O139. FEMS Immunol Med. Mic., 20: 201-207.

Hsueh PR, Lin CY, Tang HJ, Lee HC, Liu JW, Liu YC, Chuang YC (2004). *Vibrio vulnificus* in Taiwan. Emerg. Infect. Dis., 10: 1363-1368.

Huis in't Veld JHJ (1996). Microbial and Biochemical spoilage of foods: An overview. Int. J Food. Microbiol., 33: 1-18.

Huq A, Colwell RR, Rahman R, Ali A, Chowdhurry MAR, Parveen S, Sack DA, Russek-Chen E (1990). Detection of *Vibrio cholerae* O1 in the aquatic environment by fluorescent monoclonal antibody and culture methods. Appl. Environ. Microbiol., 56: 2370-2373.

CMSF (1978). Indicator Microorganisms; Important consideration for the food analyst; Preparation and dilution of the food homogenate; Enumeration of mesophilic aerobes: the agar plate methods In: Elliot et al. (eds) Microorganisms in Foods 1 Their significance and Methods of Enumeration (2nd edn) ICMSF, Toronto, pp. 3-124.

ICMSF (1986). Meaningful microbiological criteria for foods; Sampling plans for fish and shellfish In: Roberts et al. (eds) Microorganisms in Foods 2 Sampling for microbiological analysis: Principles and specific applications (2nd edn) Blackwell Scientific Publications, Oxford, pp. 3-15: 181-196.

ICMSF (1996). *Vibrio cholerae*; *V. parahaemolyticus; V. vulnificus* In: Roberts et al. (eds) Microorganisms in Foods 5 Characteristics of Microbial Pathogens (2nd edn) Blackie Academic and Professional Publishers, London, pp. 414-439.

Iida T, Hattori A, Tagomori K (2001). Filamentous phage Associated with recent Pandemic Strains of *Vibrio parahaemolyticus*. Emerg. Infect. Dis., 7: 477-478.

Jay JM, Loessner MJ, Golden DA (2005). Foodborne gastroenteritis caused by *Vibrio, Yersinia* and *Campylobacter* species. In: Modern Food Microbiology. (7th edn). Springer Science, New York, pp. 657-664.

Jiang X, Chai TJ (1996). Survival of *Vibrio parahaemolyticus* at Low Temperatures under Starvation Conditions and Subsequent Resuscitation of Viable, Nonculturable Cells. Appl. Environ. Microbiol., 62: 1300-1305.

Jun JW, Kim JH, Shin CHC Jr. SP, Han SY, Chai JY, Park SC (2012). Foodborne Pathogens and Disease. doi: 10.1089/fpd.2011.1018.

Kanki M, Yoda T, Tsukamoto T, Baba E (2007). Histidine Decarboxylases and Their Role in Accumulation of Histamine in Tuna and Dried Saury. Appl. Environ. Microbiol., 73: 1467-1473.

Kaper JB, Morris JG, Levine MM (1995). Cholera. Clin. Microbiol. Rev., 8: 48-86.

Karovičová J, Kohajdová Z (2005). Biogenic Amines in Food. Chemical Pap. 59: 70-79.

Kaysner CA, Tamplin ML, Twedt RM (1992). *Vibrio* In: Vanderzant C, Splittstoesser DF (eds) Compendium of methods for the microbiological examination of foods (3rd edn) APHA, Washington D.C., pp. 451-447.

Keasler SP, Hall RH (1993). Detecting and biotyping *Vibrio cholerae* O1 with multiplex polymerase chain reaction. Lancet, 341: 1661.

Kim YB, Okuda J, Matsumoto C, Takahasi N, Hashimoto S, Nishibuchi M (1999). Identification of *Vibrio parahaemolyticus* Strains at the Species Level by PCR Targeted to the toxR Gene. J. Clin. Microbiol. 37: 1173-1177.

Koutsoumanis K, Nychas GJE (1999). Chemical and Sensory Changes Associated with Microbial Flora of Mediterranean Boque (*Boops boops*) Stored Aerobically at 0, 3, 7, and 10°C. Appl. En viron. Microbiol. 56: 698-706.

Kumar HS, Parvathi A, Karunasagar I, Karunasagar I (2006). A gyrB–based PCR for the detection of Vibrio vulnificus and its application for the direct detection of this pathogen in oyster enrichment broths. Int. J. Food Microbiol., 111: 216-220.

Kwok AYC, Wilson JT, Coulthart M, Ng LK, Mutharia L, Chow AW (2002). Phylogenetic study and identification of human pathogenic Vibrio species based on partial hsp60 gene sequences. Can. J. Microbiol., 48: 903-910.

Kyrana VR, Lougovois VP (2002). Sensory chemical and microbiological assessment of farm-raised European sea bass (Dicentrarchus labrax) stored in melting ice. Int. J. Food Sci. Tech., 37: 319-328.

Lee CY, Pan SF, Chen CH (1995). Sequence of a cloned pR72H Fragment and Its Use for Detection of Vibrio parahaemolyticus in Shellfish with the PCR. Appl. Environ. Microbiol., 61: 1311-1317.

Lee YL, Eun JB, Choi SH (1997). Improving detection of Vibrio vulnificus in Octopus variabilis by PCR. J. Food Sci., 62: 179-182.

Lee SE, Kim SY, Kim S J, Kim HS, Shin JH, Choi SH, Chung SS, Rhee JH (1998). Direct Identification of Vibrio vulnificus in Clinical Specimens by Nested PCR. J. Clin. Microbiol., 36: 2887-2892.

Le Roux WJ, Masoabi D, De Wet CME, Venter SN (2004). Evaluation of a rapid polymerase chain reaction based identification technique for Vibrio cholerae isolates. Water Sci. Technol., 50: 229-232.

Caburlotto G, Lleo MM, Gennari M, Balboa S, Romalde JL (2011). The use of multiple typing methods allows a more accurate molecular chacterization of Vibrio parahaemolyticus strains isolated from the Italian Adriatic Sea. FEMS Microbiol. Ecol., 77: 611-622. doi: 10.1111/j.1574-6941.2011.01142.x

Luan XY, Chen JX, Zhang XH, Jia J-T, Sun FR, LI Y (2007). Comparison of different primers for rapid detection of Vibrio parahaemolyticus using the polymerase chain reaction. Lett. Appl. Microbiol., 44: 242-247.

Maeda T, Matsuo Y, Furushita M, Shiba T (2003). Seasonal dynamics in coastal Vibrio community examined by rapid and clustering method based on 16S rDNA. Fisheries Sci., 69: 358-394.

Mahmoud BSM, Yamazaki K, Miyashita K, Shik IL S, Dong-Suk C, Suzuki T (2004) Bacterial micro flora of carp (Cyprinus carpio) and its shelf-life extension by essential oil compounds. Food Microbiol., 21: 657-666.

Matches JR, Liston J, Daneault LP (1971). Survival of Vibrio parahaemolyticus in Fish Homogenates During storage at Low Temperatures. Appl. Environ. Microbiol., 21: 951-952.

Mienda BS (2012). Proteolytic Activity of Vibrio parahaemolyticus Isolated from Epinephilus spp. A Preliminary Report. Res. Biotech., 3: 36-40.

Mol S, Erkan N, Üçok D, Tosum Ş (2007). Effect of Psychrophilic bacteria to Estimate Fish Quality. J. Muscle Foods, 18: 120-128.

Montanari MP, Pruzzo C, Pane L, Colwell RR (1999). Vibrios associated with plankton in a coastal zone of the Adriatic Sea (Italy). FEMS Microbiol. Ecol., 29: 241-147.

Mothershed EA, Witney AM (2006). Nucleic acid based methods for the detection of bacterial pathogens. Present and future considerations for the clinical laboratory. Clin. Chim. Acta, 363: 206-220.

Myers ML, Panicker G, Bej AK (2003). PCR Detection of a Newly Emerged Pandemic Vibrio parahaemolyticus O3: K6 Pathogen in Pure Cultures and Seeded Waters from the Gulf of Mexico. Appl. Environ. Microbiol., 69: 2194-2200.

Nandi B, Nandy RK, Mukhopadhyay S, Nair GB, Shimada T, Ghose A (2000). Rapid Method for Species–Specific Identification of Vibrio cholerae Using Primers Targeted to the Gene of Outer Membrane Protein OmpW. J. Clin. Microbiol., 38: 4145-4151.

Naila A, Flint S, Fletcher G, Bremer P, Meerdink G (2010). Control of Biogenic Amines in Food-existing and Emerging Approaches. J. Food Sci., 75: R139-R150.

Naila A, Flint S, Fletcher GC, Bremer PJ, Meerdink G (2011). Biogenic amines and potential histamine-Forming bscteria in Rihaakuru (acooked fish paste). Food Chem., 128: 479-484.

Nickelson R, Finne G (1992). Fish, crustaceans, and precooked Seafoods In: Vanderzant C, Splittstoesser DF (eds) Compendium of Methods for the Microbiological Examination of Foods. APHA, Washington D.C., pp. 875-895.

Noorlis A, Ghazali FM, Cheah YK, Tuan ZTC, Ponniah J, Tunung R, Tang JYH, Nishibuchi M, Nakaguchi Y, Son R (2011). Prevalence and quantification of Vibrio species and Vibrio parahaemolyticus in freshwater fish at hypermarket level. Int. Food Res. J., 18: 689-695.

Normanno G, Parisia L, Addante N, Quaglia NC, Dambrosio A, Montagna C, Chiocco D (2005). Vibrio parahaemolyticus, Vibrio vulnificus, and microorganisms of fecal origin in mussels (Mytilus galloprovincialis) sold in Puglia region (Italy). Int. J. Food Microbiol., 24: 549-558.

Novotny L. Dvorska L, Lorencova A, Beran V Pavlik I (2004). Fish: A potential source of bacterial pathogens for human beings. Vet-Med. Check, 49: 343-358.

Odeyemi OA (2012). Biofilm producing Vibrio species Isolated from Siloso Beach, Singapore: A Preliminary Study. Wemed Central Microbiology, 2012: 3: WMC002631. Accessed: 20th January 2012.

Oliver JD, Nilsson L, Kjelleberg S (1991). The formation of nonculturable cells of Vibrio vulnificus and its relationship to the starvation state. Appl. Environ. Microbiol., 57: 2640-2644.

Oliver JD, Hite F, Mcdougal D Andon NL, Simpson LM (1995). Entry into, and Resuscitation from, the Viable but Nonculturable State by Vibrio vulnificus in an Estuarine Environment. Appl. Environ. Microbiol., 61: 2624-2630.

Oliver JD, Kaper JB (1997). Vibrio species In: Doyle et al. (eds) Food Microbiology Fundamentals and Frontiers. ASM Press, Washington D.C., pp. 228-264 .

Oliver JD, Bockian R (1995). In vivo resuscitation, and virulence towards mice, of viable but nonculturable cells of Vibrio vulnificus. Appl. Environ. Microbiol., 61: 2620-2623.

Ordóñez JA, López-Gálvez DE, Fernández M, Hierro E, de la Hoz L (2000). Microbial and physicochemical modifications of hake (Merluccius merluccius) steaks stored under carbon dioxide enriched atmosphere. J. Sci. Food Agric., 80: 1831-1840.

Pal D and Das N (2010). Isolation, identification and molecular characterisation of Vibrio parahaemolyticus from fish samples in Kolkata. Eur. Rev. Med. Pharmacol. Sci., 14: 545-549.

Panicker G, Myers ML, Bej AK (2004). Rapid Detection Vibrio vulnificus in Shellfish and Gulf of Mexico Water by Real-Time PCR. Appl. Environ. Microbiol., 70: 498-507.

Petti CA, Polage CR, Schreckenberger P (2005). The Role of 16S rRNA gene sequencing in Identification of Microorganisms Misidentified by Conventional Methods. J. Clin. Microbiol., 43: 6123-6125.

Pontes DS, Lima-Bittencout CI, Chartone-Souza E, Nascimento AMA (2007). Molecular approaches: Advantages and artefacts in assessing bacterial diversity. J. Indian Microbiol. Biotech., 34: 463-473.

Popovic T, Skukan AB, Dzidara P, Coz-Rakovac R, Strunjak-Perovic I, Kozacinski L, Jadan M, Brlek-Gorski D (2010). Microbiological quality of marketed fresh and frozen seafood caught off the Adriatic coast of Croatia. Vet. Med-Czech., 55: 233-241

Qian R, Xiao Z, Zhang C, Chu W, Mao Z, Yu L (2008). Expression of two major outer membrane proteins from Vibrio alginolyticus. World J. Microb. Biotechnol. 24: 245-251.

Rabie MA, Elsaidy S, el-Badawy AA, Siliha H Malcata FX (2011). Biogenic amine contents in selected Egyptian fermented foods as determined by ion-exchange chromatography. J. Food Protect., 74: 681-685.

Ray B (2003). Control of Microorganisms in Foods. In: Stern RA, Kaplan L (eds) Fundamental Food Microbiology CRC, London, pp. 439-534.

Ray B, Bhunia A (2008). Microbial Foodborne Diseases Opportunistic Pathogens, Parasite, and Algal Toxins. In: Fundamental Food Microbiology (4th edn) CRC, London, pp. 315-347.

Robert-Pillot A, Guénolé A, Fournier JM (2002). Usefulness of pR72H PCR assay for differentiation between Vibrio parahaemolyticus and Vibrio alginolyticus species: validation by DNA-DNA hybridisation. FEMS Microbiol. Lett ., 215: 1-6.

Rojas MVR, Mattè MH, Dropa M, da Silva ML, Mattè GR (2011). Characterization of Vibrio parahaemolyticus isolated from oysters and mussels in São Paulo, Brazil. Rev. Inst. Med. Trp. São Paulo., 53: 201-205.

Santos MHS (1996). Biogenic amines: Their importance in foods. Int. J.

Food Microbiol., 29: 213-231.

Saux MFL, Hervio-Heath D, Loaec S, Colwell RR, Pommepuy M (2002). Detection of Cytotoxin-Hemolysin mRNA in Nonculturable Populations of Environmental and Clinical *Vibrio vulnificus* Strains in Artificial Seawater. Appl. Environ. Microbiol., 68: 5641-5646.

Shikongo-Nambabi MNNN; Chimwamurombe PM, Venter SN (2010). Molecular and Phenotypic Identification Of *Vibro* Spp. Isolated from Processed Marine Fish. Proceedings of the National Research Symposium 15-17 September Safari Hotel Windhoek, Namibia (in print).

Smith BE, Oliver JD (2006). *In situ* and in *vitro* gene expression by Vibrio vulnificus during entry into, persistence within, and resuscitation from the viable but nonculturable state. Appl. Environ. Microbiol., 72: 1445-1451.

Sudalayandi K and Mnja (2011). Efficacy of lactic acid bacteria in the reduction of timethylamine-nitrogen and related spoilage derivatives of fresh Indian mackerel fish chunks. Afr. J. Biotechnol., 10: 42-47.

Talkington D, Bopp C, Tarr C, Parsons MB, Dahourou G, Freeman M, Joyce K, Turnsek M, Garret N, Humphrys M, Gomez G, Stroika S, Boncy J, Ochieng B, Oundo J, Klena J, Smith A,Keddy K, Gerner-Smit P (2011). Characterisation of Toxigenic Vibrio cholera from Haiti, 2010-2011. Emerg. Infect. Dis., 17: 2122-2129.

Takahashi H, Kimura B, Yoshikawa M, Fujii T (2003). Cloning and Sequencing of the Histidine Decarboxylase Genes of Gram-Negative, Histamine –Producing Bacteria and Their Application in Detection and Identification of These organisms in Fish. Appl. Environ. Microbiol., 69: 2568-2579.

Tamplin ML, Capers GM (1992). Persistence of *Vibrio vulnificus* in Tissues of Gulf Coast Oysters, *Crassostrea Virginia*, Exposed to Seawater Disinfection with UV Light. Appl. Environ. Microbiol., 58: 1506-1510.

Tao Z, Bullard S, Arias C (2011) High Numbers of *Vibrio vulnificus* in Tar Balls from Oiled Areas of the North-Central Gulf of Mexico Following the BP Deepwater Horizon Oil Spill. Eco.Health

Theron J, Cilliers J, Du Preez M, Brözel VS, Venter SN (2000). Detection of toxigenic *Vibrio cholerae* from environmental water samples by enrichment broth cultivation-pit-stop semi-nested PCR procedure. J. Appl. Microbiol., 89: 539-546.

Thongchankaew U, Sukhumungoon P, Mitraparp-arthorn P Srinitiwarawong K, Plathong S, Vuddhakul V (2011). Diversity of *Vibrio* spp. at the Andaman Tarutao Island, Thailand. Asian J. Biotechnol., 3: 530-539.

Thompson FL, Iida T and Swings J (2004). Biodiversity of *Vibrios*. Microbiol. Mol. Biol. Rev., 68: 403-431.

Thompson FL, Gevers D, Thompson CC, Dawyndt P, Naser S, Hoste B, Munn CB, Swings J (2005). Phylogeny and Molecular Identification of *Vibrios* on the Basis of Multilocus Sequence Analysis. Appl. Environ. Microbiol., 71: 5107-5115.

Thompson FL, Swings J (2006). Taxonomy of the *Vibrios*. In: Thompson et al J (eds) The Biology of Vibrios. ASM Press, Washington D.C., pp. 29-43.

Thompson FL, Gomez-Gil B, Vasconcellos ATR and Sawabe T (2007). Multilocus Sequence Analysis Reveals that *Vibrio harveyi* and *V. campbellii* Are Distinct Species. Appl. Environ. Microbiol., 73: 4279-4285.

Tryfinopoulou P, Tsakalidou E, Nychas GJE (2002). Characterisation of *Pseudomonas* spp. Associated with Spoilage of Gilt-Head Sea Bream Stored under Various Conditions. Appl. Environ. Microbiol., 68: 65-72.

Vasudevan P, Marek P, Daigle S, Hoagland T, Venkitanarayan AN (2002). Effects of chilling and Freezing on survival of *Vibrio parahaemolyticus* on fish fillets. J. Food Safety, 22: 209-217.

Velazquez-Roman J, León-Sicairos N, Flores-Villaseñor H, Villafaña-Rauda1 S, Canizalez-Roman A (2012). Pandemic *Vibrio parahaemolyticus* O3:K6 present in the Coastal Environment of Northwest Mexico is associated with recurrent diarrheal cases from 2004 to 2010. Appl. Environ. Microbiol., 87: doi: 10.1128/AEM.06953-11.

Venkateswaran K, Dohmoto N, Harayama N (1998). Cloning and Nucleotide Sequence of the *gyrB* Gene of *Vibrio parahaemolyticus* and Its Application in Detection of This Pathogen in Shrimp. Appl. Environ. Microbiol., 64: 681-687.

Vieira RHSF, Iaria STI (1993). *Vibrio parahaemolyticus* in Lobster *Panulirus laevicauda* (Latreille). Rev. Microbiol., 24: 16-21.

Wekell MM, Manger R, Kolburn K, Adams A, Hill W (1994). Microbiological quality of Seafoods: viruses, bacteria and parasites. In: Shahidi F, Botta RJ (eds) Seafoods Chemistry, Processing Technology and Quality. Chapman and Hall, Glasgow, pp. 220-232.

Whitehouse CA, Baldwin C, Sampath R, Blyn LB, Melton R, Li F, Hall TA, Harpin V, Matthews H, Tediashvili M, Jaiani E, Kokashvili T, Janelidze N, Grim C, Colwell RR, Huq A (2010). Identification of Pathogenic *Vibrio* Species by Multilocus PCR-Electrospray Ionization Mass Spectrometry and Its Application to Aquatic Environments of the Former Soviet Republic of Georgia. Appl. Environ. Microbiol., 1996-2001.

Williams TC, Froelich BA, Oliver JD (2011). Comparison of two selective and differential media for the isolation of *Vibrio vulnificus* from the environment. University of North Carolina. at Chalotte, 9201 University City Blvd, Charlotte, NC 28223.

Wong HC, Wang P (2004) Induction of viable but nonculturable state in *Vibrio parahaemolyticus* and its susceptibility to environmental stresses. J. Appl. Microbiol., 96: 359-366.

Xie ZY, Hu, CQ, Chen C, Zhang LP, Ren CH (2005). Investigation ofseven vibrio virulence genes among *Vibrio alginolyticus* and *Vibrio parahaemolyticus* strains from the coastal mariculture systems in Guangdong, China. Lett. Appl. Microbiol., 41: 202-207.

4

New Targets for Antibacterial Agents

Fatma Abdelaziz Amer, Eman Mohamed El-Behedy and Heba Ali Mohtady

Microbiology and Immunology Department, Faculty of Medicine, Zagazig University, Zagazig, Egypt.

The alarming increase and spread of resistance among emerging and re-emerging bacterial pathogens to all clinically useful antibiotics is one of the most serious public health problems of the last decade. Thus, the search for new antibacterials directed toward new targets is not only a continuous process but also, at this time, an urgent necessity. Recent advances in molecular biological technologies have significantly increased the ability to discover new antibacterial targets and quickly predict their spectrum and selectivity. The most extensively evaluated bacterial targets for drug development are: quorum sensor biosynthesis; the two component signal transduction(TCST) systems; bacteria division machinery; the shikimate pathway; isoprenoid biosynthesis and fatty acid biosynthesis.

Key words: new targets, quorum sensor biosynthesis, TCST systems, bacteria division machinery, shikimate pathway, isoprenoid biosynthesis, fatty acid biosynthesis, antibiotic resistance.

TABLE OF CONTENTS

INTRODUCTION

Bacterial infection is a ubiquitous health hazard. There are a number of very good clinically efficacious antibiotics in use today; however, the development of bacterial resistance has rendered almost all of them less effective. This critical situation necessitates the design of novel antibacterial agents. These agents must target essential bacterial pathways, but may have new modes of action or even interfere with novel bacterial targets. Many essential bacterial proteins have been identified as potential drug targets. However, an ideal target is recognized as that different from existing targets, essential for microbial cell survival, highly conserved in a clinically relevant spectrum of species, absent or radically different in man, easy to assay, and has an understood biochemistry. This review highlights the progress in the search for new antibacterial targets.

IMPACT OF GENOMICS, BIOINFORMATICS AND RELATED TECHNOLOGIES ON THE SEARCH FOR NEW TARGETS

With advancements in genomics and bioinformatics, it is

*Corresponding author. E-mail: egyamer@yahoo.com.

now possible to search through a bacterial genome to identify potential antibacterial targets. Selection of an appropriate target begins with bioinformatics to look for open reading frames (ORFs) conserved across the potential bacterial target organisms. Genes or gene products that can actually be used, are chosen by various approaches. These include automated comparisons of bacterial genomes to categorize genes and the encoded proteins. Primary sequence comparison programs, like BLAST (The Basic Local Alignment Search Tool) or PSI-BLAST (Position-Specific Iterated BLAST) determine gene functions by sequence homology, which is also used to determine clusters of orthologous groups (COGs). COGs are groups of genes shared by evolutionarily distant organisms. These orthologous families of genes are prime candidates for broad-spectrum antimicrobial agents (Hood 1999). Another approach is gene expression profiling with cluster analysis which uses microarray technology to analyze gene expression, in order to organize genes into functional groups (Eisen et al. 1998). Unknown genes functions can be estimated based on the general pathways or metabolic functions of nearby clusters. A better method of antibacterial target selection is the structural genomics, which determines the three-dimensional structures of proteins. Function is more directly a consequence of its structure than its sequence (Gerstein and Jansen 2000). Typically, 30% to 50% of the several thousand genes that make up each bacterial genome as yet have no apparent function. Direct comparison of the three-dimensional protein structure to a protein structure database can be used to assign unknown gene function. Advances already made in this area have been described that apply the sequence-structure-function model to function prediction (Fetrow et al. 1998). Moreover, phylogenetic groups that are based on the specific folds shared by organisms can be used. These folds and sequence families in bacterial pathogens can be useful antibiotic targets (Gerstein 2000). Motif analysis is another strategy to identify potential antibiotic targets among genes with unknown functions. Many databases, including PROSITE database, can search for motifs in a sequence (PROSITE is a method of determining what is the function of uncharacterized proteins translated from genomic or cDNA sequences). The motifs may show the approximate biochemical function of the gene. Moreover, gene fusion is a computational method to infer protein interactions from genome sequences. Proteins that interact with each other tend to have homologs in other organisms that are joined into a single protein chain. This method would give additional functional information for target proteins (Hood 1999).

Other tools were also developed in order to identify and study potential targets which although unnecessary *in vitro*, are essential for disease production. Examples include *In Vivo* Expression Technology (IVET), Differential Fluoresence Induction (DFI), and Signature-Tagged Mutagenesis (STM) (Strauss and Falkow 1977),

Each of these methods provide a different type of information. IVET identifies genes specifically induced during an active infection and can be used to generate temporal information (Camilli and Mekalanos 1995); DFI uncouples metabolic requirements from selection parameters, thus focusing on infection specific process, while STM identifies genes required for the establishment of an infection (Heithoff et al. 1997).

The drug discovery process

Genes that pass these filters are the best candidates for going to high-throughput screens to generate initial hits (compounds that interact with a target). High-throughput screening can be considered the process in which batches or "libraries" of compounds are tested rapidly and in parallel, for binding activity or biological activity against target molecules. Test compounds act as inhibitors of target enzymes if one has been established, as competitors for binding of a natural ligand to its receptor, as agonists or antagonists for receptor-mediated intracellular processes, and so forth. Then chemical modification of hits are carried out by repeated cycles of synthesis and testing of analogs to produce "leads," which are compounds with improved chemical characteristics, thereby increasing their suitability as potential drugs. Finally, leads are further optimized by additional repeated modification to produce drug development candidates with optimized characteristics for further pre-clinical and clinical development (Allsop and Illingworth 2002).

NOVEL ANTIBACTERIAL TARGETS

QUORUM SENSOR BIOSYNTHESIS

Bacteria are sensitive to an increase in population density and respond quickly and coordinately by inducting certain sets of genes. This mode of regulation, known as quorum sensing (QS), is based on the interaction of low-molecular-weight signal molecules called autoinducers (AIs) or pheromones with a sensor kinase and response regulator to activate or repress gene expression. QS systems are considered to be global regulators and play a key role in controlling many metabolic processes in the cell, including bacterial virulence. These systems offer attractive targets for a novel class of antibacterial drugs, capable of inducing chemical attenuation of pathogenicity.

Three types of autoinducers have been identified: acylated homoserine lactones, autoinducing peptides and autoinducer-2 compounds (Raffa et al. 2005). Acylated homoserine lactones (AHLs, acyl-HSLs, or HSLs) such as *N*-3-oxohexanoyl-$_L$-homoserine lactones (AI-1), are present in a wide spectrum of Gram-negative organisms. The AHL molecules are produced by LuxI homologues and constitute, in complex with LuxR homologues, transcripttional regulators. On the other hand, autoinducing peptides (AIPs) are amino acids or short peptides synthe-

Table 1. Synthetic quorum-sensing inhibitors (QSIs)

Development of QSI compound(s)	Mode of action	References
Substitutions in the AHL acyl side chain		
- Extended acyl side chains	inhibition of the LuxR homologues	Chhabra et al. 1993 Passador et al. 1996 Schaefer et al. 1996 Zhu et al. 1998
- Introducing ramified alkyl, cycloalkyl or aryl/phenyl substituents at the C-4 position resulting in both inducers (analogues with non-aromatic substitutions) and antagonists (analogues with phenyl substitutions)	Antagonistic activity of the phenyl compounds may result from the interaction between the aryl group and aromatic amino acids of the LuxR receptor, preventing it from adopting the active dimeric form.	Reverchon et al. 2002
- Replacing the C-3 atom with sulphur	inhibition of transcriptional regulators LuxR and LasR*.	Persson et al. 2005
Substitutions and alteration in the AHL lactone ring		
- To C-3	AHL agonist	Olsen et al. 2002
- To C-4	Unable to interact with LuxR homologues.	
- Exchanging the homoserine ring with a 5- or 6- membered alchohol or ketone ring (activators and inhibitors)	Some blocked the two *Ps. aeruginosa* QS systems *in vitro*. Their target specificity for the QS regulon was not verified by transcriptomics.	Smith et al. 2003a, b
Extensive modification in both AHL acyl side chain and lactone ring		
- 4- nitro-pyridine-*N*-oxide (4-NPO)	- Block the *E. coli* established hybrid LuxR QS system. - Target the LasR and RhlR* receptors in *Ps. aeruginosa*.	Rasmussen et al. 2005a Hentzer et al. 2003 Schuster et al. 2003

Adapted from Rasmussen and Givskov (2006).
*LasR and RhlR: transcriptional regulators of *Ps. aeruginosa* QS systems.

sized in Gram-positive bacteria and are processed, modified, and excreted by the ATP-binding cassette export systems. AIPs bind to cell surface-bound histidine protein kinase, which autophosphorylates and in turn phosphorylates a response regulator that activates transcription of one or more target genes. Lastly, autoinducer-2 compounds (AI-2) are common to both Gram-negative and Gram–positive bacteria. They are derived from furanones. LuxS is an enzyme which produces 4,5-dihydroxy-2,3-pentanedione (DPD) acylhomoserine lactones a forerunner of AI-2. This autoinducer binds to LuxP protein (a LuxR homolog). The AI-2/LuxP complex then binds to membrane-bound histidine protein kinase, and a signal transduction occurs by multistep phosphorylation similar to that of AIPs. In other bacteria, extracellular AI-2 is transported back into the cell through a LuxS-regulated (Lsr) transporter.

Based on this information, many QS inhibitor or block-ing strategies can be proposed. Many include the block-ing of signal synthesis. The LuxI family of synthases could be a target. The x-ray structure of the lactone synthase Esa1 from *Ps. stewartii* was solved and could be a starting point for structure-based design of drugs (Watson et al. 2002). If LuxS is the target, inhibitors could well function as broad-spectrum antibiotics. Other strategies are blockade of the autoinducer receptor site of the LuxR homologues, histidine protein kinase, or Lsr transporter. Other strategies include, blocking the formation of active dimers that are required for binding and expression of target genes as well as enhancement of signal molecule degradation. A combination of mechanisms would be expected to be more effective than single mechanism approaches. A number of studies have identified several molecules that function as QS inhibitors (QSI) both *in vitro* and *in vivo*. These included both synthetic and natu-ral QSIs (Tables 1 and 2).

QS systems are also important determinants of morphology and communication when bacteria grow as aggregates in biofilms (Vuong et al. 2003). Biofilm-associated microorganisms appear to contribute to cystic fibrosis, native valve endocarditis, otitis media, period ontitis, and chronic prostatitis. A spectrum of indwelling medical devices and other devices used in the healthcare environment have been shown to harbor biofilms, result-

Table 2. Natural quorum-sensing inhibitors (QSIs).

QSI compound(s)	Source	Quorum-sensing system affected	References
Agrocinopine B	Crown gall cells of host plants.	Tra system of *A. tumefaciens.*	Oger and Farrand 2001.
Furanone	*D. pulchra.*	Swr system of S. liquefaciens and other Gram-negative bacteria.	Givskov et al. 1996.
Canavanine	*M. sativa*	Sin/ExpR system of *S. meliloti*	Keshavan et al. 2005.
Norepinephrine, epinephrine	Human hormones	AI-3 system of EHEC	Sperandio et al. 2003.
Penicillic acid, Patulin	*Penicillium* spp.	Las and Rhl systems of *Ps. aeruginosa*	Rasmussen et al. 2005b

Modified from González and Keshavan (2006)

Table 3. Microorganisms commonly associated with biofilm on indwelling medical devices

Microorganism	Associated with biofilms on
C. albicans	- Artificial voice prosthesis
	- Central venous catheters
	- Intrauterine device
Coagulase negative Staphylococci	- Artificial hip prosthesis
	- Artificial voice prosthesis
	- Central venous catheters
	- Intrauterine device
	- Prosthetic heart valves
	- Urinary catheters
Enterococcus spp.	- Artificial hip prosthesis
	- Central venous catheters
	- Intrauterine device
	- Prosthetic heart valves
	- Urinary catheters
Kl. Pneumoniae	- Central venous catheters
	- Urinary catheters
Ps. aeruginosa	- Artificial hip prosthesis
	- Central venous catheters
	- Urinary catheters
Staph. aureus	- Artificial hip prosthesis
	- Central venous catheters
	- Intrauterine device
	- Prosthetic heart valve

Modified from Donlan (2001).

ing in device-associated infections (Donlan 2001).

Table (3) provides a listing of microorganisms commonly associated with biofilms on indwelling medical device. A very important clinical requirement is to block biosynthesis of the quorum signalers that initiate biofilm development. Calfee et al. (2001), reported that an anthranilate analog (methyl anthranilate) inhibited signal production of the QS system of *Ps. aeruginosa* and decre-

ased the expression of cellular virulence factors in a dose-dependent fashion. Hentzer et al. (2003), showed that the target of a synthetic furanone, which prevented biofilm formation in *Ps. aeruginosa* were the genes of QS system. That the effect was specific to biofilms was shown by the lack of activity against planktonic cultures of the same bacteria.

THE TWO-COMPONENT SIGNAL TRANSDUCTION SYSTEMS

The two-component signal transduction (TCST) systems are the principal means for coordinating responses to environmental changes in bacteria as well as some plants, fungi, protozoa, and archaea. These systems typically consist of a receptor histidine kinase (HK), which reacts to an extracellular signal by phosphorylating a cytoplasmic response regulator, causing a change in cellular behavior. The two-component family presents an excellent novel target for antibacterial drug discovery due to many reasons. First, significant homology is shared among different genera of bacteria, particularly in those amino acid residues located near active sites. Second, pathogenic bacteria use TCST to regulate expression of essential virulence factors required for *in vivo* survival. Moreover, bacteria contain many TCST systems and in several different bacteria, at least one TCST is essential for *in vitro* growth. Lastly, a different mechanism is responsible for signal transduction in mammals.

Antibacterials inhibitors of TCSTs have been identified. The most common inhibitors reported to date are hydrophobic compounds that inhibit HK- autokinase activity, noncompetitively with respect to ATP; however, in the majority of cases these compounds do not appear to be selective for signal transduction pathways and exert their effect by multiple mechanisms of action (Hilliard et al. 1999; Kitayama et al. 2004; Stephenson and Hoch, 2002).

Studies conducted in subsequent years (Stephenson and Hoch 2004), concluded that designing competitive inhibitors specific to HKs presents a huge challenge, because the Bergerat ATP binding fold in HKs is not unique for prokaryotes. Nevertheless, a novel thienopyridine (TEP) compound has been reported. TEP inhibits bacterial histidine kinases competitively (with ATP) but does not comparably inhibit mammalian serine/threonine kinases. TEP could serve as a starting compound for a new class of histidine kinase inhibitors with antibacterial activity (Gilmour et al. 2005).

BACTERIAL DIVISION MACHINERY

The division machinery of bacteria is an attractive target because it comprises seven or more essential proteins that are conserved almost throughout the bacterial kingdom but are absent from humans.

Most of the effort done to exploit this machinary as a new target has been directed at the FtsZ protein which is a GTPase. That is due to its essential role in prokaryotic cell division, its widespread conservation in the Bacterial Kingdom, its absence in the mitochondria of higher eukaryotes, its evolutionary distance from tubulin, its known biochemical activity and atomic structure and its predominance among all bacterial cell division proteins, totalling 10,000–20,000 copies per single bacterium (Pinho and Errington 2003; Stokes et al. 2005).

Localization of the FtsZ protein at the site of cell division is the first stage in the cell replication process followed by self-polymerization. The polymerized FtsZ recruits other cell division proteins, including FtsA, ZipA, FtsK, FtsQ, FtsL, FtsW, FtsI, and FtsN, leading to the formation of a Z-ring and the initiation of the complex process of septation. All of these cell division proteins are localized at mid-cell, and work in concert to constrict the cell and produce cell division.

Ftsz, and thus represent potential classes of new antibiotics. Ftsz may also be a feasible novel target for antituberculous agents (Reynolds et al. 2004; White et al. 2002).

Another target that has been exploited to screen and analyze inhibitory molecules, is the FtsA. This highly conserved protein presumably constitutes a key bacterial component because of its ATPase enzymatic activity and its essential protein–protein interaction with FtsZ. The FtsZ–FtsA protein–protein interaction and the FtsZ:FtsA ratio are crucial for the progress of bacterial cell division. Accumulating evidence suggests that FtsA plays the role of a motor protein in providing energy for constriction by way of its ATPase activity (Errington et al., 2003). Lately, Paradis-Bleau et al. (2005), reported the identification of peptide sequences which showed specific inhibition of ATPase activity of FtsA. They considered this the first step for the future development of antimicrobial agents via peptidomimetism.

One more interaction which can also be an elegant tool with the aim of developing antibacterials with novel modes of action is the FtsZ-ZipA protein–protein interac-

tion. Briefly these proteins are essential components of the septal ring which forms at the site of cell division. Inhibition of this interaction between the two proteins results in inhibition of cell division, leading to filamenttation and ultimately cell death. Development of inhibitors of FtsZ-ZipA protein-protein interaction has been focused upon only recently (Jennings et al. 2004; Sutherland et al. 2003). It is expected that efforts in this area will not diminish in the future.

THE SHIKIMATE PATHWAY

The shikimic acid pathway (the aromatic biosynthetic pathway), is conserved in bacteria, fungi, plants and apicomplexan parasites but is absent in mammals. Analysis of the genomes of Strep. pneumoniae and Staph. aureus has illustrated the presence and conservation of the chorismate biosynthetic genes in key Gram-positive pathogens, which can facilitate their characterization and essentiality testing (McDevitt et al. 2002).

This pathway effects the conversion of two simple products of carbohydrate metabolism (phosphoenolpyruvate and erythrose 4- phosphate into the unstable diene chorismate (Figure 1). Chorismate represents a major bifurcation point of the pathway and is a common non-aromatic precursor for the biosynthesis of a range of important aromatic metabolites (Daugherty et al. 2001).

The enzymes of the shikimate pathway constitute an excellent target for the design of new antibacterial agents. Such agents may prove beneficial for immunocom-promised patients who are suffering multiple infection from bacterial and parasitic organisms.

Shikimate kinase represents an attractive target for the development of new antimicrobial agents, herbicides, and antiparasitic agents. The shikimate kinase structure of many bacteria has been determined. These include, Erw. chrysanthemi (Krell et al. 2001), E. coli (Romanowski and Burley 2002), Myco. tuberculosis (Pereira et al. 2004) and Camp. jejuni and H. pylori (Cheng et al. 2005). These structures provide shikimate-binding information as a rational basis for further investigation towards structure-guided inhibitors.

EPSP synthase can also be an attractive target. Knockout mutations of aroA, which encodes EPSP synthase, in both Gram negative and Gram positive bacteria were found to lead to attenuation of bacterial virulence, supporting the utility of this targeting approach (Izhar et al. 1990).

Chorismate Synthase (CS) is the most unusual of the entire pathway and is unique in nature. Studies recently conducted to understanding the structure of Myco-.tuberculosis CS together with its cofactor and substrate binding modes compared to Strep. pneumoniae reported a degree of similarity (Fernandes et al. 2005). Such studies may facilitate the search for inhibitors of this enzyme as alternative agents to treat tuberculosis.

Furthermore, fluorinated analogues of shikimate, block-

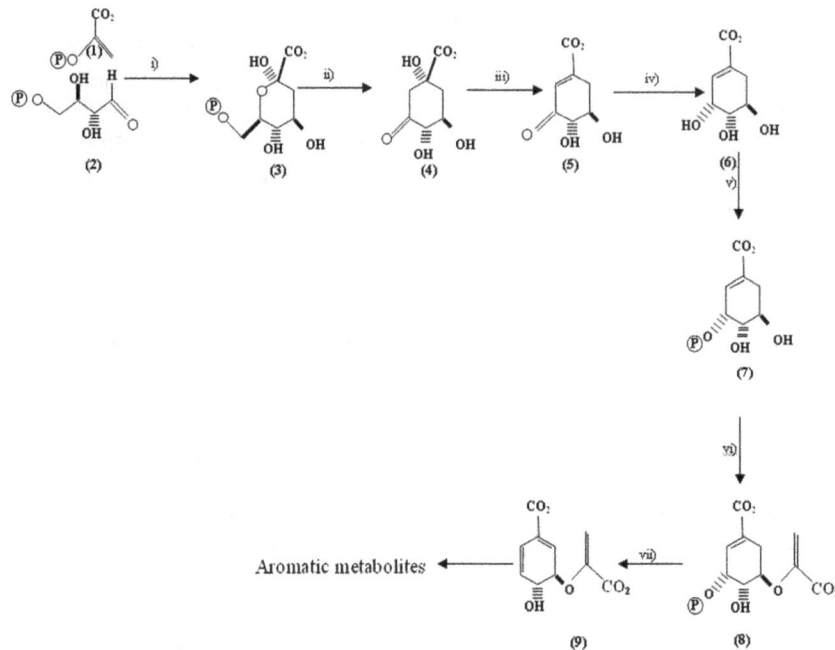

Figure 1. Shikimic acid pathway, modified from Daughery et al. (2001).
(1) Phosphoenolpyruvate (2) Erythrose-4-phosphate i) 3-deoxy-D-arabino-heptulosonate 7-phosphate (DAHP) synthase (3) 3-deoxy-D-arabino-heptulosonate 7-phosphate ii) 3-dehydroquinate synthase (4) 3-dehydroquinate iii) 3-dehydroquinate dehydratase (5) 3-dehydroshikmate iv) Shikimate dehydrogenase (6) Shikimate v) Shikimate kinase (7) Shikimate-3-phosphate vi) 5-enolpyruvyl-shikimate-3-phosphate (EPSP) synthase (8) EPSP vii) Chorismate synthase (9) Chorismate.

ed the growth of *Plasm. falciparum in vitro*, demonstrating that the shikimate pathway is a valid target for development of new broad-spectrum antimicrobial and antiparasitic agents (McConkey 1999).

ISOPRENIOD BIOSYNTHESIS

Isoprenoids are known in having invaluable role in various biological processes such as cell-wall biosynthesis, electron transport, photosynthetic light harvesting, lipid membrane structure and intracellular signaling. Isopentenyl diphosphate (IPP) and its isomer; dimethylallyl diphosphate (DMAPP), act as a universal precursor in biosynthesis of isoprenoids. The reactions IPP $\leftrightarrow$ DMAPP is catalyzed by isopentenyl diphosphate (IPP) isomerase enzyme. These two isomers are produced by either classical (mevalonate) or non-classical (non-mevalonate or deoxyxylulose phosphate) pathway, as shown in Figures 2 A & 2 B, respectively.

The classical `mevalonate' pathway involves the construction of the six-carbon branched chain mevalonate skeleton from three molecules of acetyl-CoA (Figure 2 A). However, the recently identified `non-mevalonate' pathway involves the combination of pyruvate and glyceroldehydes-3-phosphate to make 1-deoxy-D-xylulose-5-phosphate (Figure 2 B). Many Gram-negative bacteria

and apicoplast type protozoa including the *Plasmodium* spp. utilize the non-mevalonate pathway. Whereas, eukaryotes, archaea, Gram-positive cocci, *Bor. burgdorferi* and *Cox. burnetii_*employ exclusively the mey-alonate pathway (Hedi and Rodwell, 2004).

The distribution of the new non-mevalonate pathway made it an attractive target for the design of antibiotics against pathogenic bacteria and the malaria parasite *Plasm. falciparum* (Rohmer et al. 2004). Perhaps, any of the enzymes involved in this pathway may be considered as a good antibacterial target. The 1-deoxy-D-xylulose-5-phosphate (DOXP) synthetase is a promising enzyme. The inhibition of this enzyme will block the key step for essential metabolites (Walsh 2003). Another enzyme is the 1-deoxy-D-xylulose 5-phosphate reducto-isomerase (DXR). Fosmidomycin is an active antibiotic against many Gram-negative and some Gram-positive bacteria; specifically inhibits DXR. Recently, two novel inhibitors of DXR are showing high activity similar to fosmidomycin and are significantly active against *E. coli in vitro* (Kuntz et al. 2005). The genetic code of mevalonate pathway enzymes in Gram-positive cocci are of particular interest in the light of the recent development of multidrug-resistant strains of these group of pathogenic bacteria and conesquently may representing some concern for human health and cost effectiveness. Comprehensive study of the six enzymes of the mevalonate pathway should lead

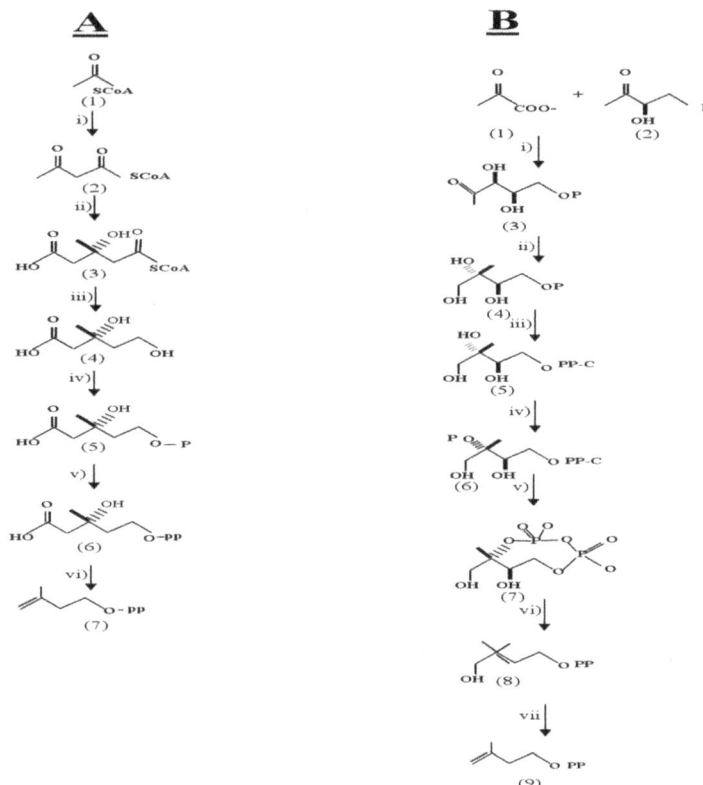

Figure 2. (A) the classical (mevalonate); (B) the non-classical (non-mevalonate) pathways for isoprenoid synthesis in bacteria, modified from Walsh (2003).

(A). (1) Acetyl CoA; i) Acetoacetyl-CoA thiolase; (2) Acetoacetyl-CoA; ii) 3-hydroxy-3-methylglutaryl co-enzyme A (HMG-CoA) synthase; (3) HMG-CoA; iii) HMG-CoA reductase (HMGR); (4) Mevalonate; iv) Mevalonate kinase (MK); (5) Mevalonate-5-phosphate; v) Phosphomevalonate kinase (PMK); (6) Mevalonate -5-diphosphate; vi) Diphosphomevalonate decarboxylase (DPMD); (7) Isopentenyl diphosphate (IPP).

(B). (1) Pyruvate; (2) D-glyceraldehyde-3-phosphate; i) 1-deoxy-D-xylulose-5-phosphate synthase (DXS); (3) 1-deoxy-D-xylulose-5-phosphate; ii) 1-deoxy-D-xylulose-5-phosphate reductoisomerase (DXR); (4) 2C-methyl-D-erythritol-4-phosphate; iii) 4-diphosphocytidyl-2C-methyl-D-erythritol synthase (YgbP, IspD); (5) 4-diphosphocytidyl-2C-methyl-D-erythritol; iv) 4-diphosphocytidyl-2C-methyl-D-erythritol kinase (YchB, IspE); (6) 4-diphosphocytidyl-2C-methyl-D-erythritol-2-phosphate; v) 2C-methyl-D-erythritol 2,4 cyclodiphosphate synthase (YgbB, IspF); (7) 2C-methyl-D-erythritol 2,4 cyclodiphosphate (MEcPP); vi) 4-hydroxy-3-methyl-but-2-enylpyrophosphate (HMB-PP) synthase (GcpE, IspG); (8) = HMB-PP; vii) = HMB-PP reductase (LytB, IspH); (9) = Isopentenyl diphosphate (IPP).

to a synthetic design for inhibitors with therapeutic effect against Gram positive pathogens. It is well known fact that mevalonate pathway is also essential for humans. Therefore, any chemotherapeutic approach targeting this pathway must exploit differences between the bacterial and human enzymes. Comparative genome analysis has identified the complete set of mevalonate pathway enzymes in *Streptococci*, *Staphylococci* and *Enterococci*, which were found to be heterologous with the coding sequences for comparable enzymes in higher organisms,

meaning that there is a difference between isoprenoid metabolism using mevalonate pathway in prokaryotes and in eukaryotes (Voynova et al. 2004). The pathway enzyme HMG-CoA reductase is of particular interest in this context. It exhibits significant differences in the three-dimensional structure and in the sensitivity to inhibition by the active-site inhibitors known as statins (Tabernero et al. 2003). Isopentenyl diphosphate (IPP) isomerase is another important enzyme. Recently, an unrelated IPP isomerase (type II) has been discovered (Laupitz et al.,

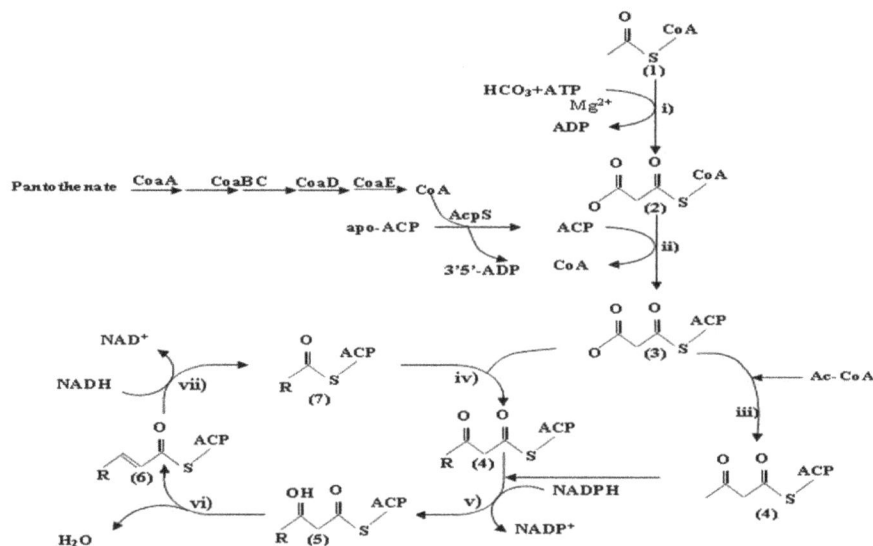

Figure 3. Fatty acid biosynthsis (type II), modified from Zhang et al. (2006).
(1) Acetyl CoA; i) AccABCD (acetyl-CoA carboxylase complex); (2) Malonyl CoA; ii) FabD (Malonyl CoA:ACP transacylase; (3) Malonyl-ACP; iii) FabH (β-ketoacyl-ACP synthase III); (4) β-ketoacyl-ACP; iv) FabB/F (β-ketoacyl-ACP synthase I/II); v) FabG (β-ketoacyl-ACP reductase); (5) β-hydroxyacyl-ACP; vi) FabA/Z (β-hydroxydecanoyl-ACP dehydratase/ β-hydroxyacyl-ACP dehydratase); (6) Trans-2-enoyl-ACP; vii) FabI/K (enoyl-ACP reductase I/II); (7) Acyl-ACP.
CoaA) Pantothenate kinase; CoaBC) Phosphopantothenoylcystein synthetase/decarboxylase complex; CoaD) Phosphopantetheine adenyltransferase; CoaE) Phospho-CoA kinase; AcpS) ACP synthase.

2004).

This type II enzyme invariably accompanies the mevalonate pathway in human pathogens of the Gram-positive coccus group. Since type II isomerase is absent in human, the enzyme appears to be an attractive target for therapy against Gram-positive cocci infection. Currently, the structure of the enzyme is determined by X-ray crystallography this should facilitate the development of cognate inhibitors (Steinbacher et al. 2003). Comparable differences may also recognize other bacterial enzymes of mevalonate pathway from their human counterparts. This will provide a rationale for the potential selectivity of target therapy.

FATTY ACID BIOSYNTHESIS

The fatty acid synthesis (FAS), required for the membrane building phospholipids, in living organisms comprises a repeated cycle of reactions involving the condensation, reduction, dehydration, and subsequent reduction of carbon-carbon bonds. Higher eukaryotes carry out these reactions by a large multifunctional protein (type I pathway). Whereas, in bacteria, plant chloroplasts and *Plasm. falciparum* each reaction is catalyzed by discrete enzymes; type II pathway (Figure 3), this allows the prospects of selective inhibition (Zhang et al. 2006). Most of the FAS II enzymes are essential for bacterial viability

and consequently they are under investigation as targets for antibacterial drug discovery.

The bacterial acetyl coenzyme A (acetyl-CoA) carboxylase complex (AccABCD) catalyzes the first step in fatty acid synthesis by the interacting AccBC and AccAD complexes and is essential for cell growth. Therefore, it has been strongly proposed as a possible target. The pseudopeptide pyrrolidinedione antibiotics, such as moramide B, have recently been discovered to be specifically targeting the AccAD complex (Freiberg et al. 2004). This natural product together with synthetic analogues shows broad-spectrum antibacterial activity. Further studies of structural variants of the natural product have demonstrated significant *in vivo* effect in a murine model of *Staph. aureus* sepsis, (Freiberg et al. 2006). Nonetheless, *in silico* screening using the AccD5 [(5th subunit of acyl-CoA carboxylases (ACCase)] structure of *Myco. Tuberculosis*, identified one inhibitor (Lin et al. 2006). Such inhibitor may serve as drug lead for the development of new tuberculosis.

An initiation condensing enzyme, FabH, and elongation condensing enzymes, FabF/B, are also essential enzymes in fatty acid (type II pathway) synthesis. They are considered highly attractive new targets for the development of antibacterial and antiparasitic compounds. Two natural products namely cerulenin and thiolactomycin inhibit the condensation enzymes FabH and FabF/B, with cerulenin showing selectivity for FabF/B, whereas thiolactomycin (TLM) and its analogs inhibit FabH and FabF/B (Dolak et al. 1986). TLM analogs with its increased potency and better pharmacokinetic proper-ties have been sought by a number of groups, but more work is needed ((Douglas et al. 2002; Kim et al. 2006; Kremer et al. 2000; Sakya et al. 2001; Senior et al. 2003; Senior et al. 2004). Many studies were carried out to identify inhibitors of these enzymes, identified compounds with antibacterial activity against MRSA, *B. subtilis, H. influenzae, E. coli* and *Plasm. falciparum* (He et al. 2004; He and Reynolds 2002; Prigge et al. 2003; Young et al. 2006).

The reduction enzyme FabG, is widely expressed and only a single isoform is known in bacteria. These features suggest that drugs targeting this enzyme would have broad spectrum antibacterial activity, but more work is needed (Zhang and Rock 2004).

FabI, is another essential enzyme which is responsible for performing the last step in the fatty acid synthesis, type II pathway. Triclosan (antiseptic) and isoniazid (the anti-*Myco. tuberculosis* agent) are two marketed antibac-terial agents that target FabI enzyme and its mode of action is well documented as antibacterial target (Parikh et al. 2002). Moreover, optimization studies in the imida-zole series of synthetic 1,4- disubstituted imidazoles reported to be low-micromolar inhibitors of FabI, led to a 16-fold improvement in antibacterial activity and a five fold improvement in potency against the enzyme (Heerding et al. 2001). These clear successes in target-ing the FabI component and the determinant role of FabI in completing each cycle of elongation make this enzyme as one of the most attractive antibacterial targets.

It is well known that FabI is not the only enoyl-ACP reductase in bacteria. Access to and analysis of key bacterial genomes demonstrated that an alternative triclosan-resistant enoyl-ACP reductase, FabK, having no similarity to the prototype FabI; is present in several important clinical pathogens. Both FabI and FabK have been found in pathogens such as *Ent. faecalis* and *Ps. aeruginosa* (Heath and Rock 2000). Accordingly, FabI represents a selective antibacterial target for those pathogens in which its essential such as *Staph. aureus* and *E. coli*. It is worth mentioning that, FabI inhibitors are extremely potent against multidrug-resistant *Staph. aureus* and clinically useful drugs are most likely to come up from this line of development in the near future (Ling etal., 2004; Moir, 2005; Payne et al., 2002; Seefeld et al., 2003). Alternatively, a compound that possesses inhibittory potency against both FabK and FabI would be expected to possess a far broader spectrum of antibac-terial activity.

The *Myco. tuberculosis* homolog of the *E. coli* enoyl-ACP reductase gene (*fabI* in *E. coli*) is *inhA,* with 36% identity. The enzyme InhA has been validated as an anti-mycobacterial target. High-throughput screening of a structurally diverse library of compounds reported indole-5-amides, 4-aryl-substituted piperazines, and various pyrazole derivatives to provide useful core templates dis-playing good InhA inhibition. A second more focused library yielded more potent inhibitors with observed good activities versus *Myco. tuberculosis* and *Plasm. Falcipa-rum,* while having no effect against six other common infectious agents (Ballell et al. 2005).

Acyl carrier protein synthase (AcpS) catalyzes the transfer of the 4'-phosphopantetheinyl group from the coenzyme A, to a serine residue in acyl carrier protein (ACP), thereby activating ACP, an important step in cell wall biosynthesis (Joseph-McCarthy et al. 2005). ACP, the acyl group carrier in type II fatty acid synthesis has attracted attention as a target for drug development.

Inhibitors strategies can target AcpS, inactivate ACP, or block CoA biosynthesis.

AcpS is in many cases is essential for bacteria. One of the encouraging findings is the isolation of a natural pro-duct that inhibits AcpS with antibacterial activity against *Staph. aureus* (Chu et al. 2003). The high resolution structures of several AcpS proteins are known and structure-based design has been employed to identify a class of anthranilic acid inhibitors. Opportunities for syn-thetic modification of this group of inhibitors have also been identified (Joseph-McCarthy et al. 2005).

In most bacteria, CoA is synthesized from pantothenic acid (vitamin B_5) in series of 5 steps and phosphorylation of pantothenate by pantothenate kinase (PanK, CoaA) is the first reaction. This pathway also exists in eukaryotes. However, in most cases there is no sequence homology between the prokaryotic and eukaryotic CoA biosynthetic enzymes (Liu et al. 2006). Moreover, these enzymes are essentially required for bacterial survival and/or virulence. Therefore, there is a potential for having highly specific inhibitors of bacterial CoA enzymes. Development of these inhibitors targeting these enzymes is being actively pursued. Among these, the (N-substituted pantothen-amides) have shown the greatest promise as growth inhibitors of both *E. coli* and *Staph. aureus*. This is most likely because of their incorporation into multiple acyl carrier proteins (Choudhry et al. 2003; Leonardi et al. 2005; Strauss and Begley 2002; Virga et al. 2006; Zhang et al. 2004; Zhao et al. 2003). Although some problems are encountered regarding the activity of pantothen-amides as antibacterial agents, the pantothenamides cha-racterized to date are very effective against drug-resistant *Staph. aureus* and have the potential to be developed further (Zhang et al. 2006).

Furthermore, CoaD, has recently been widely accepted as a unique target for drug discovery. Detailed bioinfor-matic analysis of the CoA pathway shows that only a sin-gle CoaD isozyme is present in bacteria, and its se-quence predicts that it is distinctly different from the multifunctional mammalian protein that carries out the same reaction (Gerdes et al. 2002). Up to date, there is only a single publication reporting CoaD inhibitors (Zhao et al. 2003), but the attractiveness of the target and the availability of high quality x-ray structures (Izard 2002; 2003) suggest that CoaD inhibitors should receive greater attention.

Conclusions

The rising tide of multidrug-resistant pathogens ensures that the search for new antibiotics showing no co-resistance to existing drugs that target well-known pathways, will continue. Many bacterial components appear as promising new targets for progression through the antibiotic drug discovery. Coupled with the increased diversity and number of compounds that have been screened against these validated targets, new antibiotic leads are being identified. Antibacterial activities of such leads include both broad and narrow spectrum mechanisms. The current challenge is to optimize these early-stage discovery leads to make them suitable for clinical evaluation. However, it should be emphasized that the potential of broad spectrum mode of action still carries the risk of inhibiting "beneficial" bacteria and normal flora, which may create some clinical problems. Therefore, it is worth mentioning that all efforts should be directed towards appropriate use of such antibacterial agents.

ACKNOWLEDGEMENT

Authors thank Dr. M Elsheemy, Lincoln, UK and Dr. S Louis, san Diego, USA for editing the manuscript.

REFERENCES

Allsop A, Illingworth R (2002). The impact of genomics and related technologies on the search for new antibiotics. J. App. Microbiol. 92: 7-11.

Ballell L, Field RA, Duncan K, Young RG (2005). New small-molecule synthetic antimycobacterials. Antimicrob. Agents Chemother. 6: 2153-2163.

Calfee MW, Coleman JP, Pesci EC (2001). Interference with pseudomonas quinolone signal synthesis inhibits virulence factor expression by Pseudomonas aeruginosa. Proc. Natl. Acad. Sci. U. S. A. 98: 11633- 11637.

Camilli A, Mekalanos JJ (1995). Use of recombinase gene fusion to identify V. cholerae genes induced during infection. Mol. Microbiol. 18: 671-683.

Cheng W, Chang Y, Wang W (2005). Structural basis for shikimate-binding specificity of Helicobacter pylori shikimate kinase. Bacteriol.23: 8156–8163.

Chhabra SR, Stead P, Bainton NJ, Salmond GP, Stewart GS, Williams P, Bycroft B W (1993). Autoregulation of carbapenem biosynthesis in Erwinia carotovora by analogues of N-(3-oxohexanoyl)-L-homoserine lactone. J. Antibiot. 46: 441–454.

Choudhry AE, Mandichak TL, Broskey JP, Egolf RW, Kinsland C, Begley TP, Seefeld MA, Ku TW, Brown JR, Zalacain M, Ratnam K (2003). Inhibitors of pantothenate kinase: novel antibiotics for staphylococcal infections. Antimicrob. Agents Chemother. 6: 2051-2055.

Chu M, Mierzwa R, Xu L, Yang SW, He L, Patel M, Stafford J, Macinga D, Black T, Chan TM, Gullo V (2003). Structure elucidation of Sch 538415, a novel acyl carrier protein synthase inhibitor from a microorganism. Bioorg. Med. Chem. Lett. 13: 3827–3829.

Daughery M, Vonstein V, Overbeek R, Oserman A (2001). Archaeal shikimate kinase, a new member of the GHMP-kinase family. J. Bacteriol. 183: 292-300

Dolak L, Castle T, Truesdell S, Sebek O (1986). Isolation and structure of antibiotic U68,204, a new thiolactone. J. Antibiot. Tokyo 39: 26-31.

Donlan RM (2001). Biofilms and device-associated infections. Emerg. Infect. Dis. 7: 277-281.

Douglas JD, Senior SJ, Morehouse C, Phetsukiri B, Campbell IB, Besra GS, Minnikin DE (2002). Analogues of thiolactomycin: potential drugs with enhanced anti-mycobacterial activity. Microbiol. 148: 3101–3109.

Eisen MB, Spellman PT, Brown PO, Botstein D (1998). Cluster analysis and display of genome-wide expression patterns. Proc. Natl. Acad. Sci. 95: 14863-14868.

Errington J, Daniel RA, Scheffers D (2003). Cytokinesis in bacteria. Microbiol. Mol. Biol. Rev. 67: 52-65.

Fernandes CL, Santos DS, Basso LA, de Souza ON (2005). Structure prediction and docking studies of chorismate synthase from Mycobacterium tuberculosis. L. N. C. S. 3594: 118- 127.

Fetrow JS, Godzick A, Skolnick J (1998). Functional analysis of the Escherichia coli genome using the sequence to function paradigm: Identification of proteins exhibiting the flutaredoxin/thioredoxin disulphide oxidoreductase activity. J. Mol. Biol. 282: 703-711.

Freiberg C, Brunner NA, Schiffer G, Lampe T, Pohlmann J, Brands M, Raabe M, Häbich D, Ziegelbauer K (2004). Identification and characterization of the first class of potent bacterial acetyl-CoA carboxylase inhibitors with antibacterial activity . J. Biol. Chem. 25: 26066-26073.

Freiberg C, Pohlmann J, Nell PG, Endermann R, Schuhmacher J, Newton B, Otteneder M, Lampe T, Häbich D, Ziegelbauer K (2006). Novel bacterial acetyl coenzyme A carboxylase inhibitors with antibiotic efficacy in vivo. Antimicrob. Agents Chemother. 8: 2707-2712.

Gerdes SY, Scholle MD, D'Souza M, Bernal A, Baev MV, Farrell M, Kurnasov OV, Daugherty MD, Mseeh F, Polanuyer BM, Campbell JW, Anantha S, Shatalin KY, Chowdhury SA, Fonstein MY, Osterman AL (2002). From genetic footprinting to antimicrobial drug targets: examples in cofactors biosynthetic pathways. J. Bacteriol. 184: 4555-4572.

Gerstein M (2000). Integrative database analysis in structural genomics. Nature, Struct. Biol. Sup: pp. 960-963.

Gerstein M, Jansen R (2000). The current excitement in bioinformatics—analysis of whole-genome expression data: how does it relate to protein structure and function?. Cur. Opin. Struct. Biol. 10: 574-584.

Gilmour R, Foster JE, Sheng Q, McClain JR, Riley A, Sun PM, Ng WL, Yan D, Nicas TI, Henry K, Winkler ME (2005). New class of competitive inhibitor of bacterial histidine kinases. J. Bacteriol. 23: 8196–8200.

Givskov M, de Nys R, Manefield M, Gram L, Maximilien R, Eberl L, Molin S, Steinberg PD, Kjelleberg S (1996). Eukaryotic interference with homoserine lactone-mediated prokaryotic signalling. J. Bacteriol. 178:6618-6622.

González JE, Keshavan ND (2006). Messing with Bacterial Quorum Sensing. Microbiol. Mol. Biol. Rev. 70: 859-875.

He X, Reeve AM, Desai UR, Kellogg GE, Reynolds KA (2004). 1,2-dithiole-3-ones as potent inhibitors of the bacterial 3-ketoacyl acyl carrier protein synthase III (FabH). Antimicrob. Agents Chemother. 8: 3093-3102.

He X, Reynolds KA (2002). Purification, characterization, and identifica-tion of novel inhibitors of the beta-ketoacyl-acyl carrier protein synthase III (FabH) from Staphylococcus aureus. Antimicrob. Agents Chemother. 46: 1310-1318.

Heath RJ, Rock CO (2000). A triclosan-resistant bacterial enzyme. Nature 406:145.

Hedi M, Rodwell VW (2004). Enterococcus faecalis mevalonate kinase. Protein Sci. 13: 687-693.

Heerding DA, Chan G, DeWolf WE, Fosberry AP, Janson CA, Jaworski DD, McManus E, Miller WH, Moore TD, Payne DJ, Qiu X, Rittenhouse SF, Slater-Radosti C, Smith W, Takata DT, Vaidya KS, Yuan CC, Huffman HF (2001). 1,4-Disubstituted imidazoles are potential antibacterial agents functioning as inhibitors of enoyl acyl carrier protein reductase (FabI). Bioorg. Med. Chem. Lett. 11: 2061-2065.

Heithoff DM, Conner CP, Mahan M (1997). Dissection the biology of a pathogen during infection. Trends Microbiol. 5: 509-513.

Hentzer M, Wu H, Andersen JB, Riedel K, Rasmussen TB, Bagge N, Kumar N, Schembri MA, Song Z, Kristoffersen P, Manefield M, Costerton JW, Molin S, Eberl L, Steinberg P, Kjelleberg S, Hoiby N, Givskov M (2003). Attenuation of *Pseudomonas aeruginosa* virulence by quorum sensing inhibitors. E. M. B. O. J. 22: 3803–3815.

Hilliard JJ, Goldschmidt RM, Licata L, Baum EZ, Bush K (1999). Multiple mechanisms of action for inhibitors of histidine protein kinases from bacterial two-component systems. Antimicrob. Agents Chemother. 43: 1693-1699.

Hood DW (1999). The utility of complete genome sequences in the study of pathogenic bacteria. Parasitol. 118: S3-S9.

Isomerase, and phylogenetic distribution of isoprenoid biosynthesis pathways. Eur. J. Biochem. 271: 2658–2669.

Izard TA (2002). The crystal structures of phosphopantetheine adenylyltransferase with bound substrates reveal the enzyme's catalytic mechanism. J. Mol. Biol. 315: 487–495.

Izard TA (2003). Novel adenylate binding site confers phosphopantetheine adenylyltransferase interactions with Coenzyme A. J. Bacteriol. 185: 4074–4080.

Izhar M, DeSilva L, Joysey HS, Hormaeche CH (1990). Moderate immunodeficiency does not increase susceptibility to *Salmonella typhimurium aroA* live vaccines in mice. Infect. Immun. 58: 2258-2261.

Jennings LD, Foreman KW, Rush TS, Tsao DHH, Mosyak L, Kincaid SL, Sukhdeo MN, Sutherland AG, Ding W, Kenny CH, Sabus CL, Liu H, Dushin EG, Moghazeh SL, Labthavikul P, Petersen PJ, Tuckman M, Ruzin AV (2004). Combinatorial synthesis of substituted 3-(2-indolyl) piperidines and 2-phenyl indoles as inhibitors of ZipA–FtsZ interaction. Bioorg. Med. Chem. 19: 5115-5131.

Joseph-McCarthy D, Parris K, Huang A, Failli A, Quagliato D, Dushin EG, Novikova E, Severina E, Tuckman M, Petersen PJ, Dean C, Fritz CC, Meshulam T, DeCenzo M, Dick L, McFadyen IJ, Somers WS, Lovering F, Gilbert AM (2005). Use of structure-based drug design approaches to obtain novel anthranilic acid acyl carrier protein synthase inhibitors. J. Med. Chem. 25: 7960-7969.

Keshavan ND, Chowdhary PK, Haines DC, González JE (2005). L-Canavanine made by *Medicago sativa* interferes with quorum sensing in *Sinorhizobium meliloti*. J. Bacteriol. 187: 8427-8436.

Kim P, Zhang YM, Shenoy G, Nguyen QA, Boshoff HI, Manjunatha UH, Goodwin M, Lonsdale JT, Price AC, Miller DJ, Duncan K, White SW, Rock CO, Barry CE, Dowd CS (2006). Structure-activity relationships at the 5-position of thiolactomycin: an intact (5R)-isoprene unit is required for activity against the condensing enzymes from *Mycobacterium tuberculosis* and *Escherichia coli*. J. Med. Chem. 49: 159–171.

Kitayama T, Iwabuchi R, Minagawa S, Shiomi F, Cappiello J, Sawada S, Utsumi R, Okamoto T (2004). Unprecedented olefin-dependent histidine-kinase inhibitory of zerumbone ring-opening material. Bioorg. Med. Chem. Lett. 14:5943-5946.

Krell T, Maclean J, Boam DJ, Cooper A, Resmini M, Brocklehurst K, Kelly SM, Price NC, Lapthorn AJ, Coggins JR (2001). Biochemical and X-ray crystallographic studies on shikimate kinase: the important structural role of the P-loop lysine. Protein Sci. 10: 1137-1149.

Kremer L, Douglas JD, Baulard AR, Morehouse C, Guy MR, Alland D, Dover LG, Lakey JH, Jr. Jacobs WR, Brennan PJ, Minnikin DE, Besra GS (2000). Thiolactomycin and related analogues as novel anti-mycobacterial agents targeting KasA and KasB Condensing Enzymes in *Mycobacterium tuberculosis*. J. Biol. Chem. 275: 16857–16864.

Kuntz L, Tritsch D, Grosdemange-Billiard C, Hemmerlin A, Willem A, Bach TJ, Rohmer M (2005). Isoprenoid biosynthesis as a target for antibacterial and antiparasitic drugs: phosphonohydroxamic acids as inhibitors of deoxyxylulose phosphate reducto-isomerase. Biochem. J. 386: 127–135.

Laupitz R, Hecht S, Amslinger S, Zepeck F, Kaiser J, Richter G, Schramek N, Steinbacher S, Huber R, Arigoni D, Bacher A, Eisenreich W, Rohdich F (2004). Biochemical characterization of *Bacillus subtilis* type II isopentenyl diphosphate

Leonardi R, Chohnan S, Zhang YM, Virga KG, Lee RE, Rock CO, Jackowski C (2005). A pantothenate kinase from *Staphylococcus*

aureus refractory to feedback regulation by coenzyme A. J. Biol. Chem. 280: 3313-3322.

Lin TW, Melgar MM, Kurth D, Swamidass SJ, Purdon J, Tseng T, Gago G, Baldi P, Gramajo H, Tsai SC (2006). Structure-based inhibitor design of AccD5, an essential acyl-CoA carboxylase carboxyltransferase domain of *Mycobacterium tuberculosis*. Proc. Natl. Acad. Sci. U. S. A. 103: 3072–3077.

Ling LL, Xian J, Ali S, Geng B, Fan J, Mills DM, Arvanites AC, Orgueira H, Ashwell MA, Carmel G, Xiang Y, Moir DT (2004). Identification and characterization of inhibitors of bacterial enoyl-acyl carrier protein reductase. Antimicrob. Agents Chemother. 48: 1541–1547.

Liu W, Han C, Hu L, Chen K, Shen X, Jiang H (2006). Characterization and inhibitor discovery of one novel malonyl-CoA: acyl carrier protein transacylase (MCAT) from *Helicobacter pylori*. F. E. B. S. Lett. 580: 697–702.

Margalit DN, Romberg L, Mets RB, Hebert AM, Mitchison TJ, Kirschner MW, Chaudhuri DR (2004). Targeting cell division: small-molecule inhibitors of FtsZ GTPase perturb cytokinetic ring assembly and induce bacterial lethality. P. N. A. S. 32: 11821-11826.

McConkey GA (1999). Targeting the shikimate pathway in the malaria parasite *Plasmodium falciparum*. Antimicrob. Agents Chemother. 43: 175-177.

McDevitt D, Payne DJ, Holmes DJ, Rosenberg M (2002). Novel targets for the future development of antibacterial agents. J. App. Microbiol. 92: 28S–34S.

Moir DT (2005). Identification of inhibitors of bacterial enoyl-acyl carrier protein reductase. Curr. Drug Targets Infect. Disord. 5: 297–305.

Oger P, Farrand SK (2001). Co-evolution of the agrocinopine opines and the agrocinopine-mediated control of TraR, the quorum-sensing activator of the Ti plasmid conjugation system. Mol. Microbiol. 41:1173-1185.

Ohashi Y, Chijiiwa Y, Suzuki K, Takahashi K, Nanamiya H, Sato T, Hosoya Y, Ochi K, Kawamura F (1999). The lethal effect of a benzamide derivative, 3-methoxybenzamide, can be suppressed by mutations within a cell division gene, *ftsZ*, in *Bacillus subtilis*. J. Bacteriol. 4: 1348-1351.

Olsen JA, Severinsen R, Rasmussen TB, Hentzer M, Givskov M, Nielsen J (2002). Synthesis of new 3- and 4-substituted analogues of acyl homoserine lactone quorum sensing autoinducers. Bioorg. Med. Chem. Lett. 12: 325–328.

Paradis-Bleau C, Sanschagrin F, Levesque RC (2005). Peptide inhibitors of the essential cell division protein FtsA. Protein Eng. Des. Sel. 18: 85-91.

Parikh SL, Xiao G, Tonge PJ (2002). Inhibition of InhA, the enoyl reductase from *Mycobacterium tuberculosis*, by triclosan and isoniazid. Biochem. 39: 7645–7650.

Passador L, Tucker KD, Guertin KR, Journet MP, Kende AS, Iglewski BH (1996). Functional analysis of the *Pseudomonas aeruginosa* autoinducer PAI. J. Bacteriol. 178: 5995–6000.

Payne DJ, Miller WH, Berry V, Brosky J, Burgess WJ, Chen E, Jr. DeWolf JW, Fosberry AP, Greenwood R, Head MS, Heerding DA, Janson CA, Jaworski DD, Keller PM, Manley PJ, Moore TD, Newlander KA, Pearson S, Polizzi BJ, Qiu X, Rittenhouse SF, Slater-Radosti C, Salyers KL, Seefeld MA, Smyth MG, Takata DT, Uzinskas IN, Vaidya K, Wallis NG, Winram SB, Yuan CC, Huffman WF (2002). Discovery of a novel and potent class of FabI-directed antibacterial agents. Antimicrob. Agents Chemother. 46: 3118–3124.

Pereira JH, de Oliveira JS, Canduri F, Dias MV, Palma MS, Basso LA, Santos DS, Jr. de Azevedo WF (2004). Structure of shikimate kinase from *Mycobacterium tuberculosis* reveals the binding of shikimic acid. Acta Crystallogr. Sect. 60: 2310-2319.

Persson T, Hansen TH, Rasmussen TB, Skinderso ME, Givskov M, Nielsen J (2005). Rational design and synthesis of new quorum-sensing inhibitors derived from acylated homoserine lactones and natural products from garlic. Org. Biomol. Chem. 3: 253–262.

Pinho MG, Errington J (2003). Dispersed mode of *Staphylococcus aureus* cell wall synthesis in the absence of the division machinery. Mol. Microbiol. 50: 871–881.

Prigge ST, He X, Gerena L, Waters NC, Reynolds KA (2003). The initiating steps of a type II fatty acid synthase in *Plasmodium falciparum* are catalyzed by pfACP, pfMCAT, and pfKASIII. Biochem. 42: 1160-1169.

Raffa RB, Lannuzo JR, Levine DR, Saeid KK, Schwartz RC, Sucic NT, Terleckyj OD, Young JM (2005). Bacterial communication ("Quorum sensing") via ligands and receptors: a novel pharmacological target for the design of antibiotic drugs. J. P. E. T. 312: 417-423.

Rasmussen TB, Bjarnsholt T, Skindersoe ME, Hentzer M, Kristoffersen P, Kote M, Nielsen J, Eberl L, Givskov M (2005a). Screening for quorum-sensing inhibitors (QSI) by use of a novel genetic system, the QSI selector. J. Bacteriol. 187: 1799–1814.

Rasmussen TB, Givskov M (2006). Quorum sensing inhibitors: a bargain of effects. *Microbiol.* 152: 895-904.

Rasmussen TB, Skindersoe ME, Bjarnsholt T, Phipps RK, Christensen KB, Jensen PO, Andersen JB, Koch B, Larsen TO, Hentzer M, Eberl L, Hoiby N, Givskov M (2005b). Identity and effects of quorum-sensing inhibitors produced by *Penicillium* species. Microbiol. 151: 1325-1340.

Reverchon S, Chantegrel B, Deshayes C, Doutheau A, Cotte-Pattat N (2002). New synthetic analogues of N-acyl homoserine lactones as agonists or antagonists of transcriptional regulators involved in bacterial quorum sensing. Bioorg. Med. Chem. Lett. 12: 1153–1157.

Reynolds RC, Srivastava S, Ross LJ, Suling WJ, White EL (2004). A new 2-carbamoyl pteridine that inhibits mycobacterial FtsZ. Bioorg. Med. Chem. Lett. 12: 3161-3164.

Rohmer M, Grosdemange-Billiard C, Seemann M, Tritsch D (2004). Isoprenoid biosynthesis as a novel target for antibacterial and antiparasitic drugs. Curr. Opin. Invest. Drugs. 5:154–162.

Romanowski MJ, Burley SK (2002). Crystal structure of the *Escherichia coli* shikimate kinase I (AroK) that confers sensitivity to mecillinam. Proteins 47: 558-562.

Sakya SM, Suarez-Contreras M, Dirlam JP, O'Connell TN, Hayashi SF, Santoro SL, Kamicker BJ, George DM, Ziegler CB (2001). Synthesis and structure-activity relationships of thiotetronic acid analogues of thiolactomycin. Bioorg. Med. Chem. Lett. 11: 2751–2754.

Schaefer AL, Hanzelka BL, Eberhard A, Greenberg EP (1996). Quorum sensing in *Vibrio fischeri*: probing autoinducer-LuxR interactions with autoinducer analogs. J. Bacteriol. 178: 2897–2901.

Schuster M, Lostroh CP, Ogi T, Greenberg EP (2003). Identification, timing, and signal specificity of *Pseudomonas aeruginosa* quorum-controlled genes: a transcriptome analysis. J. Bacteriol. 185: 2066–2079.

Seefeld MA, Miller WH, Newlander KA, Burgess WJ, Jr. DeWolf WE, Elkins PA, Head MS, Jakas DR, Janson CA, Keller PM, Manley PJ, Moore TD, Payne DJ, Pearson S, Polizzi BJ, Qiu X, Rittenhouse SF, Uzinskas IN, Wallis NG, Huffman WF (2003). Indole naphthyridinones as inhibitors of bacterial enoyl-ACP reductases FabI and FabK. J. Med. Chem. 46: 1627–1635.

Senior SJ, Illarionov PA, Gurcha SS, Campbell IB, Schaeffer ML, Minnikin DE, Besra GS (2003). Biophenyl-based analogues of thiolactonamycin active against *Mycobacterium tuberculosis* mtFabH fatty acid condensing enzyme. Bioorg. Med. Chem. Lett. 13: 3685–3688.

Senior SJ, Illarionov PA, Gurcha SS, Campbell IB, Schaeffer ML, Minnikin DE, Besra GS (2004). Acetylene-based analogues of thiolactomycin, active against *Mycobacterium tuberculosis* mtFabH fatty acid condensing enzyme. Bioorg. Med. Chem. Lett. 14: 373–376.

Smith KM, Bu Y, Suga H (2003a). Induction and inhibition of *Pseudomonas aeruginosa* quorum sensing by synthetic autoinducer analogs. Chem. Biol. 10: 81–89.

Smith KM, Bu Y, Suga H (2003b). Library screening for synthetic agonists and antagonists of a *Pseudomonas aeruginosa* autoinducer. Chem. Biol. 10: 563–571.

Sperandio V, Torres AG, Jarvis B, Nataro JP, Kaper JB (2003). Bacteria-host communication: the language of hormones. Proc. Natl. Acad. Sci. USA 100: 8951-8956.

Steinbacher S, Kaiser J, Gerhardt S, Eisenreich W, Huber R, Bacher A, Rohdich F (2003). Crystal structure of the type II isopentenyl diphosphate:dimethylallyl diphosphate isomerase from *Bacillus subtilis*. J. Mol. Biol. 329: 973–982.

Stephenson K, Hoch JA (2002). Two-component and phosphorelay signal-transduction systems as therapeutic targets. Curr. Opin. Pharmacol. 2: 507-512.

Stephenson K, Hoch JA (2004). Developing inhibitors to selectively target two-component and phosphorelay signal transduction systems of pathogenic microorganisms. Curr. Med. Chem. 11: 765-773.

Stokes NR, Sievers J, Barker S, Bennett JM, Brown DR, Collins I, Errington VM, Foulger D, Hall M, Halsey R, Johnson H, Rose V, Thomaides HB, Haydon DJ, Czaplewski LG, Errington J (2005). Novel inhibitors of bacterial cytokinesis identified by a cell-based antibiotic screening assay . J. Biol. Chem. 48: 39709-39715.

Strauss E, Begley TP (2002). The antibiotic activity of N-pentylpantothenamide results from its conversion to ethyldethia-coenzyme A, a coenzyme A antimetabolite. J. Biol. Chem. 277: 48205-48209.

Strauss E, Falkow S (1977). Microbial pathogenesis: genomics and beyond. Science 276: 707-712.

Sutherland AG, Alvarez J, Ding W, Foreman KW, Kenny CH, Labthavikul P, Mosyak L, Petersen PJ, Rush TS, A. Ruzin A, Tsao DH, Wheless KL (2003). Structure- based design of carboxybiphenylindole inhibitors of the ZipA-FtsZ interaction. Org. Biomol. Chem. 23: 4138-4140.

Tabernero L, Rodwell VW, Stauffacher CV (2003). Crystal structure of a statin bound to a class II hydroxymethylglutaryl-CoA reductase. J. Biol. Chem. 278:19933–19938.

Virga KG, Zhang YM, Leonardi R, Ivey RA, Hevener K, Park HW, Jackowski S, Rock CO, Lee RE (2006). Structure-activity relationships and enzyme inhibition of pantothenamide-type pantothenate kinase inhibitors. Bioorg. Med. Chem. Lett. 14: 1007-1020.

Voynova NE, Rios SE, Miziorko HM (2004). *Staphylococcus aureus* mevalonate kinase: isolation and characterization of an enzyme of the isoprenoid biosynthetic pathway. J. Bacteriol. 1: 61-67.

Vuong C, Gerke C, Somerville GA, Fischer ER, Otto M (2003). Quorum-sensing control of biofilm factors in *Staphylococcus epidermidis*. J. Infect. Dis. 188: 70618.

Walsh C (2003). Antibiotics: actions, origins, resistance, ASM Press, Washington DC.

Wang J, Galgoci A, Kodali S, Herath KB, Jayasuriya H, Dorso K, Vicente F, González A, Cully D, Bramhill D, Singh S (2003). Discovery of a small molecule that inhibits cell division by blocking FtsZ, a novel therapeutic target of antibiotics. J. Biol. Chem. 45: 44424-44428.

Watson WT, Minouge TD, Val DL, von Bodman SB, Churchill ME (2002). Structural basis and specificity of acyl-homoserine lactone signal production in bacterial quorum sensing. Mol. Cell. 9: 685-694.

White EL, Suling WJ, Ross LJ, Seitz LE, Robert C (2002). **2-**Alkoxycarbonylaminopyridines: inhibitors of *Mycobacterium tuberculosis* FtsZ. J. Antimicrob. Chemother. 50: 111-114.

Young K, Jayasuriya H, Ondeyka JG, Herath K, Zhang C, Kodali S, Galgoci A, Painter R, Brown-Driver V, Yamamoto R, Silver LL, Zheng Y, Ventura JI, Sigmund J, Ha S, Basilio A, Vicente F, Tormo JR, Pelaez F, Youngman P, Cully D, Barrett JF, Schmatz D, Singh SB, Wang J (2006). Discovery of FabH/FabF inhibitors from natural products. Antimicrob. Agents Chemother. 2: 519-526.

Zhang YM, Frank MW, Virga KG, Lee RE, Rock CO, Jackowski S (2004). Acyl carrier protein is a cellular target for the antibacterial action of the pantothenamide class of pantothenate antimetabolites. J. Biol. Chem. 279: 50969-50975.

Zhang YM, Rock CO (2004). Evaluation of epigallocatechin gallate and related plant polyphenols as inhibitors of the FabG and FabI reductases of bacterial type II fatty-acid synthase. J. Biol. Chem. 279: 30994–31001.

Zhang YM, White SW, Rock CO (2006). Inhibiting bacterial fatty acid synthesis. J. Biol. Chem. 26: 17541-17544.

Zhao L, Allanson NM, Thomson SP, Maclean JKF, Barker JJ, Primrose WU, Tyler PD, Lewendon A (2003). Inhibitors of phosphopantetheine adenylyltransferase. Eur. J. Med. Chem. 38: 345-349.

Zhu J, Beaber JW, More MI, Fuqua C, Eberhard A, Winans SC (1998). Analogues of the autoinducer 3-oxooctanoyl-homoserine lactone strongly inhibit activity of the TraR protein of *Agrobacterium tumefaciens*. J. Bacteriol. 180: 5398–5405.

5

Mechanistic links between maternal bacterial infection and cerebral palsy

Heping Zhou

Department of Biological Sciences, Seton Hall University, 400 South Orange Avenue, South Orange, New Jersey 07079, USA. E-mail: zhouhepi@shu.edu.

Maternal bacterial infection is known as a causal factor for preterm labor and neonatal morbidity. In recent years, both epidemiological and experimental studies have identified maternal bacterial infection as one of the causal factors for the development of cerebral palsy (CP) in the offspring. This review examines accumulating evidence that as critical mediators of the host's response to fighting the infecting bacteria, inflammatory cytokines and oxidative stress also play important roles in maternal bacterial infection-induced white matter damage and ultimately the development of CP in the offspring. Understanding the actions of cytokines and oxidative stress in CP development could potentially lead to novel and effective therapeutic strategies.

Key words: maternal infection; cerebral palsy; lipopolysaccharide; cytokine; oxidative stress

TABLE OF CONTENT

INTRODUCTION

Cerebral palsy (CP) is defined as a group of non progressive, but often changing, motor impairment syndromes secondary to lesions or anomalies of the brain arising in the early stages of development (Mutch et al., 1992). Cerebral palsy is usually clinically defined at 4 - 5 years of age (Neufeld et al., 2005), and it is the most common childhood disability in neuromotor development, affecting as many as every 2 - 3 out of 1000 children in the U.S. each year (Schendel et al., 2002). The signs include motor deficits (Rosenbaum et al., 2007). CP patients may also exhibit psychoemotional retardation, seizure, mental retardation, speech delay, perceptual and sensory deficits, and many other problems (Rahman et al., 2004; Rosenbaum et al., 2007; Singhi et al., 2003). The most prevalent pathological lesion seen in CP subjects is white matter damage, varying from periventricular leukomalacia (PVL) to diffuse non cystic myeli-

nation disturbances within the white matter, characterized by hypomyelination and loss of oligodendrocytes (OLGs), the myelin-forming cells of the central nervous system (Back et al., 1998; Johnston and Hoon, 2006). Both prenatal and postnatal factors have been found to contribute to the development of CP in infants (Jacobsson and Hagberg, 2004), and maternal bacterial infection is one of the prenatal factors that have been found to be associated with fetal brain damage and increase the risk of developing various neurological disorders including CP in the offspring (Babulas et al., 2006; Neufeld et al., 2005).

Bacterial infection has been found to increase the levels of inflammatory cytokines, such as tumor necrosis factor (TNF)-α, interleukin (IL)-1β, and IL-6 in the peripheral system and in the brain of rodents (Koedel et al., 2004; Stoycheva and Murdjeva, 2005; Turrin et al.,

2001). In the meantime, bacterial infection has also been found to increases the formation of reactive oxygen species (ROS), such as oxygen ions, free radicals, and peroxides, in the liver, heart, lung, blood, and brain in experimental animal models, leading to oxidative stress (Sakaguchi and Furusawa, 2006; Victor et al., 2005). The inflammatory cytokines and ROS act as critical mediators of the host's immune response to fight infecting bacteria (Knight, 2000; Sikora, 2000; Slifka and Whitton, 2000). There is also accumulating evidence that the inflammatory cytokines and ROS may also mediate the development of CP. Several reviews focused on the incidence and prevalence, classification, diagnosis, and prognosis of CP have been published recently (Green and Hurvitz, 2007; Jones et al., 2007b). The objective of this review is to examine the relationship between maternal bacterial infection and the development of CP in the offspring, and the role of inflammatory cytokines and oxidative stress in the induction of white matter damage in the offspring following maternal bacterial infection.

Maternal bacterial infection as a causal factor for CP

There is accumulating evidence that maternal infection not only contributes to pre-term labor, but also increases the risk for neurological pathologies in the offspring (Romero et al., 2007). Several population-based case control studies have found that maternal infection is correlated with the occurrence of CP in preterm and term infants (Matsuda et al., 2000; Neufeld et al., 2005). Furthermore, Wu et al., (2000) conducted a meta-analysis on studies examining the association between CP and clinical chorioamnionitis, an infection of the mem-branes (chorion and amnion) and amniotic fluid, and found that clinical chorioamnionitis is strongly associated with the development of PVL and CP in both preterm and term infants (Wu and Colford, 2000). These studies suggest that maternal infection is a significant risk factor for the development of CP in infants.

Animal model systems using prenatal administration of live bacteria or bacterial products to mimic maternal bacterial infection have been employed to directly examine the linkage between maternal bacterial infection and white matter damage in the offspring (for details, see Table 1). Intracervical administration of 100 µg/kg lipopolysaccharide (LPS, endotoxin), a major component of the bacterial cell wall, on day 15 of pregnancy, or 150 µg/kg LPS on days 15, 17, and 19 of pregnancy leads to decreased immunostaining for oligodendrocyte (OLG)-specific proteins within the white matter of rat pups, suggesting OLG loss (Bell and Hallenbeck, 2002; Toso et al., 2005). Administration of 300 µg/kg LPS to pregnant rats on days 19 and 20 of pregnancy via intraperitoneal injection also induces significant white matter injury in the neonatal rats (Rousset et al., 2006). Pang et al., 2005 inoculated live E. coli to the uteri of pregnant rats on day 17 of pregnancy, and found significant apoptosis and

hypomyelination in periventricular white matter of pups born to E. coli-treated dams as compared to those born to vehicle-treated dams (Pang et al., 2005). Furthermore, administration of LPS into pregnant sheep at mid or late gestation leads to white matter damage in fetal sheep (Mallard et al., 2003; Svedin et al., 2005). Collectively, experimental studies using different animal model systems have provided direct evidence that maternal bacterial infection significantly influences fetal brain development and may lead to the development of CP in the offspring.

Cytokines as mediators for the development of CP

Besides their immunological roles, cytokines have been recognized as important regulators of brain development (Mehler and Kessler, 1997). For example, IL-1β has a significant influence on neuronal survival (Marx et al., 2001), neuronal differentiation (Ling et al., 1998), dendrite development (Gilmore, 2004), and synaptic plasticity (Schneider et al., 1998). IL-6 has been found to affect neuronal survival (Marx et al., 2001) and dentrite development (Gilmore, 2004). There is also evidence that TNF-α is involved in the regulation of neurite growth (Neumann et al., 2002), neuronal survival (Yang et al., 2002), and hippocampal morphogenesis (Golan et al., 2004b). Furthermore, treatment with TNF-α, IL-1β, or IFN-γ induces death of both oligodendrocyte precursor cells (OPCs) and mature oligodendrocytes (OLGs) (Cai et al., 2004; Feldhaus et al., 2004; Nakazawa et al., 2006; Takahashi et al., 2003), and inhibits the differentiation of OPCs (Feldhaus et al., 2004; Vela et al., 2002). Taken together, the inflammatory cytokines play important roles during brain development. In agreement, transgenic mice over expressing IL-6 or TNF-α in the central nervous system develop a spectrum of cellular alterations resulting in pronounced neurological diseases (Wang et al., 2002). Therefore, the inflammatory cytokines have been hypothesized to mediate the development of CP following maternal bacterial infection.

Supporting evidence for the neuroinflammatory hypothesis comes from several areas of research as described below. Firstly, the neonates born with funisitis, a histological marker for fetal inflammatory systemic response, have a higher risk to develop neurological defects includeing cerebral palsy (Romero et al., 2007). Furthermore, white matter damage is frequently associated with activation of microglia, the resident immune cells in the brain (Deguchi et al., 1996). At the molecular level, an increased level of IL-18 in umbilical blood has also been found to correlate with white matter damage in preterm infants (Hagberg et al., 2005), and increased levels of TNF-α, IL-1β, and IL-6 has been found to be associated with white matter damage in cerebral palsy patients (Deguchi et al., 1996; Yoon et al., 2003). Additionally, a single nucleotide G – A base pair substitution at nucleotide -308 relative to the transcriptional start site, a polymorphism associated with higher level of TNF-α expression, has been reported to be associated with an increaseed risk of cerebral palsy in a recent case-control study using DNA samples collected from newborn infants with cerebral palsy as compared to control infants (Gibson et al., 2006). Secondly, experimental studies using animal model systems have found that maternal exposure to bacteria or bacterial products induces inflammation in maternal and fetal tissues (for details, see Table 2). Maternal exposure to LPS has been shown to activate microglial cells in the periventricular region of the offspring rabbits (Kannan et al., 2007), induce the production of TNF-α, IL-1β, and IL-6 in the placenta (Ashdown et al., 2006; Beloosesky et al., 2006; Urakubo et al., 2001), increase the level of IL-6 in the amniotic fluid (Beloosesky et al., 2006; Urakubo et al., 2001), and elevate the level of IL-1β in the fetal plasma of rodents (Ashdown et al., 2006). There is evidence that the cytokines produced by maternal tissues may cross placenta, enter fetal tissue (Dahlgren et al., 2006; Kent et al., 1994), and stimulate fetal immune cells to produce more cytokines (Hebra et al., 2001). After crossing blood brain barrier (Banks, 2005), these cytokines may activate astrocytes and microglia cells, which, in turn, produce more inflammatory cytokines in the fetal brain (McLaurin et al., 1995). Therefore, it is very likely that the increased levels of inflammatory cytokines in the maternal tissue could transfer maternal inflammation to the fetal brain. Furthermore, increased levels of TNF-□ and IL-1β in the fetal brain have been found to be associated with prenatal LPS-induced white matter damage (Bell and Hallenbeck, 2002; Rousset et al., 2006) while microglial activation is associated with intrauterine E. coli-induced PVL and OLG loss in the offspring rodents (Pang et al., 2005). Lastly, treatment with IL-10, an anti-inflammatory cytokine, suppresses intrauterine E. coli-induced microglial activation in the fetal brain, reduces apoptosis in the white matter, and protects the fetal brain from white matter damage in rodents (Pang et al., 2005; Rodts-Palenik et al., 2004) (for details, see Table 2). These findings provide strong support that maternal bacterial infection induced inflammatory cytokines contribute to the development of CP in the offspring.

Oxidative stress as a mediator for the development of CP

The reactive oxygen species (ROS), such as oxygen ions, free radicals and peroxides, induced by bacteria infection plays an important protective role in destroying the infecting microorganisms (Lin et al., 2005; Sakaguchi and Furusawa, 2006; Santos et al., 2007). However, oxidative stress results when the ability of the endogenous antioxidant system is overcome by the generation of ROS, and ROS will modify the nucleic acids, lipids, and proteins, which ultimately leads to cellular damage and cell death (Hald and Lotharius, 2005; Mehlhase et al., 2000). There is increasing evidence that oxidative stress is involved in the development of maternal bacterial infection-induced white matter damage and CP in the offspring. Firstly, examinations of the autopsy brain tissues have found that PVL brains exhibit increased lipid peroxidation in OPCs than non-PVL control brains (Haynes et al., 2003). Clinical studies have also found that CP patients exhibit increased lipid peroxidation and decreased antioxidant capacity as compared to healthy controls (Aycicek and Iscan, 2006). Secondly, there is accumulating evidence that oxidative stress regulates the survival and differentiation of both neuronal and glial cells (Loh et al., 2006; Mahadik et al., 2006; Wang et al., 2004). For example, intracellular depletion of GSH or cysteine leads to ROS accumulation and death of both OPCs and mature OLGs (Wang et al., 2004). Furthermore, OPCs have been found to be more susceptible to oxidative stress induced by intracellular depletion of GSH or cysteine than mature OLGs due to the presence of stronger antioxidant defense mechanisms in mature OLGs (Back et al., 1998; Wang et al., 2004). Finally, experimental animal studies have shown that maternal exposure to LPS potentiates the release of hydroxyl radicals in the brain of pups (Cambonie et al., 2004), increases protein carbonylation, and decreases the ratio of reduced/oxidized forms of glutathione (GSH/GSSG) in the hippocampus of the offspring, suggesting the presence of oxidative stress (Lante et al., 2007) (for details, see Table 3). Furthermore, pre-treatment with antioxidants, such as ascorbic acid and melatonin, significantly decreases prenatal LPS-induced fetal mortality (Chen et al., 2006a; Chen et al., 2006b), diminishes prenatal LPS-induced damages in the hippocampus of the offspring (Lante et al., 2007), and prevents prenatal LPS-induced degeneration of OPCs and hypomyelination in the fetal rat brains (Paintlia et al., 2004). More recently, it has been reported that NAC treatment prevents prenatal LPS-induced decrease of GSH content in the hippocampus, and protects prenatal LPS-induced deficits in long-term potentiation and spatial learning in the offspring (Lante et al., 2008) (for details, see Table 3). Additionally, the ROS pathway may cross-talk with the inflammatory cytokine cascade. For example, NAC has been found to attenuate prenatal LPS-induced expression of inflammatory cytokines, such as TNF-α and IL-1β, in fetal brains, and block the induction of IL-6 in maternal serum and amniotic fluid of rodents following prenatal exposure to LPS (Beloosesky et al., 2006; Paintlia et al., 2004). Taken together, these data suggest that maternal bacterial infection-induced oxidative stress contribute to the genesis of CP in the offspring.

Management of CP

Cerebral palsy patients may exhibit a spectrum of pathologies and clinical phenotypes depending on the extent and location of brain damage. Current strategies for the health care of children with CP include promoting optimal function, fostering the acquisition of new skills, and pre-

Table 1. Reported experimental animal studies to demonstrate the association between prenatal bacterial infection and brain injury.

Bacteria or bacterial products used	Host	Route of treatment	Dose of treatment	Time of treatment	Major Findings	Reference
E. coli LPS serotype O111:B4	Lewis and Fischer 344 rats	i.c.	0.1, 0.2, 0.5, 1, 3 mg/kg maternal weight	day 15 of pregnancy	Intracervical LPS treatment induces dose-dependent fetal mortality. Treatment with 0.1 mg/kg LPS decreases immunohistochemical staining of oligodendrocyte markers within the corpus callosum in the P21 offspring.	(Bell and Hallenbeck 2002)
E. coli LPS serotype O55:B5	Sprague-Dawley rats	i.p.	4 mg/kg or 0.5 mg/kg maternal weight	day 18 of pregnancy	Maternal treatment with 4 mg/kg LPS increases the expression of TNF-α and IL-1β mRNA in the fetal brain; Treatment with 0.5 mg/kg LPS on days 18 and 19 of pregnancy increases glial fibrillary acidic protein-positive astrocytes and decreases myelin basic protein staining in the brain of neonatal rats.	(Cai et al. 2000)
E. coli K1	New Zealand White rabbits	i.u.	$5 \times 10^3 \sim 5 \times 10^5$ CFU	between 24 and 30 d of gestation	E. coli intrauterine inoculation increase cell death and white matter damage in the periventricular region of surviving fetuses.	(Debillon et al. 2000)
E. coli LPS serotype O55:B55	Sheep fetus	Fetal i.v.	1 μg/kg estimated fetal weight	day 91 of pregnancy, (3-5 injections over 5days)	Neuroinjury is found 10-11 days following initial LPS injection	(Duncan et al. 2002)
Gardnerella vaginalis	New Zealand White rabbits	i.u.	$2 \times 10^4 \sim 2 \times 10^6$ CFU	day 20 or 21 of pregnancy	G. vaginalis intrauterine inoculation results in amnionitis and deciduitis, which is associated with increased fetal mortality, reduced birth weight, and increased brain injury in the offspring.	(Field et al. 1993)
E. coli LPS serotype O55:B5	Dunkin-Hartley guinea pig	i.p.	1, 5, 25, 50, 100, 200, 300 μg/kg maternal weight	70% gestation	Maternal LPS treatment elicits a dose-dependent cell death in the brain of the fetus 7 days following injection.	(Harnett et al. 2007)
E. coli LPS serotype O111:B4	Sheep fetus	Fetal i.v.	100 ng/kg estimated fetal weight	midgestation	Fetal LPS treatment results in focal inflammatory infiltrates and cystic lesions in periventricular white matter in 40% of the offspring.	(Mallard et al. 2003)
E. coli	Sprague-Dawley rats	i.u.	1×10^6 CFU	Day 17 of pregnancy	E. coli treatment leads to significant apoptosis in periventricular white matter of P0 pups. Treatment with IL-10 reduces E. coli-induced white matter damage.	(Pang et al. 2005)
E. coli LPS serotype -O55:B5	Wistar rats	i.p.	300 μg/kg maternal weight	days 19 and 20 of pregnancy	Maternal LPS treatment elevates IL-1β mRNA level in the brain of P1 offspring, and increases cell death in the brain of P1 and P7 offspring.	(Rousset et al. 2006)

Table1. Continued

Bacteria or bacterial products used	Host	Route of treatment	Dose of treatment	Time of treatment	Major Findings	Reference
E. coli LPS serotype O55:B4	Sheep fetus	Fetal i.v.	88.7 ng/kg estimated fetal weight	65% or 85% of gestation	Fetal LPS treatment leads to white matter damage, increased microglia activation, and loss of neurofilament staining in the brain of the infants.	(Svedin et al. 2005)
E. coli LPS serotype O111:B4	Fischer 344 rats	i.c.	150 µg/kg LPS maternal weight	days 15, 17, and 19 of pregnancy	Maternal LPS treatment decreases the staining of myelin proteolipid protein, a maker for oligodendrocytes, and causes sensory-motor delays in the offspring.	(Toso et al. 2005)
E. coli	New Zealand White rabbits	i.u.	$10^3 \sim 10^4$ CFU	days 20 or 21 of pregnancy	E. coli intrauterine inoculation results in fetal white matter damage.	(Yoon et al. 1997)

Abbreviations: IL: interleukin; TNF: tumor necrosis factor; i.p.: intraperitoneal; i.u.: intrauterine; i.v.: intravenous; P1: postnatal day 1.

Table 2. Reported experimental animal studies to demonstrate the association between prenatal bacterial infection-induced inflammation and brain injury.

Bacteria or bacterial products used	Host	Route of treatment	Dose of treatment	Time of treatment	Major Findings	Reference
E. coli LPS serotype O111:B4	Sprague-Dawley rats	i.p.	50 µg/kg maternal weight	day 18 of pregnancy	Maternal LPS treatment increases the levels of TNF-α, IL-1β, and IL-6 in maternal plasma and placenta, and elevates the level of IL-1β in fetal plasma.	(Ashdown et al. 2006)
E. coli LPS serotype O55:B5	Sprague-Dawley rats	i.p.	4 mg/kg Or 0.5 mg/kg maternal weight	day 18 of pregnancy	Maternal treatment with 4 mg/kg LPS increases the expression of TNF-α and IL-1β mRNA in the fetal brain; Treatment with 0.5 mg/kg LPS on days 18 and 19 of pregnancy increases glial fibrillary acidic protein-positive astrocytes and decreases myelin basic protein staining in the brain of neonatal rats.	(Cai et al. 2000)
E. coli LPS serotype O55:B5	C3H/HeN mice	i.p.	50 µg/kg maternal weight	day 15 of pregnancy	Maternal LPS treatment increases the levels of TNF-α, IL-6, and IL-1α in the maternal serum, and elevates the levels of IL-6 and IL-1α in the amniotic fluid.	(Fidel et al. 1994)
Gardnerella vaginalis	New Zealand White rabbits	i.u.	$2 \times 10^4 \sim 2 \times 10^6$ CFU	days 20 or 21 of pregnancy	G. vaginalis intrauterine inoculation results in amnionitis and deciduitis, which is associated with increased fetal mortality, reduced birth weight, and increased brain injury in the offspring.	(Field et al. 1993)
E. coli LPS serotype O111:B4	Sprague-Dawley rats	i.p.	0.1 mg/kg maternal weight	day 18 of pregnancy	Maternal LPS treatment increases corticotropin-releasing factor in the fetal brain, elevates TNF-α, IL-6, and IL-10 in the chorioamnion, and upregulates TNF-α, IL-1β, and IL-6 in the placenta.	(Gayle et al. 2004)

Table 2. Continued.

E. coli LPS	Black c-57 mouse	i.p.	0.12 µg/g maternal weight	day 17 of pregnancy	Maternal LPS increases expression of IL-6 in the fetal brain; and impairs distinct aspects of learning and memory in the adult offspring.	(Golan et al. 2005)
Salmonella enteritidis LPS, or Rc mutant E. coli LPS	Sprague-Dawley rats	i.p.	50 µg/kg maternal weight	day 19 of pregnancy	Maternal LPS treatment increases plasma TNF-α concentration in both dams and their fetuses, and induces endotoxin tolerance in P0 offspring.	(Goto et al. 1997)
E. coli LPS serotype O127:B8	New Zealand White rabbits	i.u	20 and 30 µg/kg maternal weight	day 28 of pregnancy	Maternal LPS treatment leads to dose-dependent activation of microglia in the offspring at postnatal day (P)1.	(Kannan et al. 2007)
E. coli LPS serotype O127:B8	C57BL6/J mice	i.p.	50 µg per mouse	day 18 of pregnancy	Maternal LPS treatment increases the expression of MCP-1, IL-6, IL-1β, and VEGF in fetal brain.	(Liverman et al. 2006)
E. coli LPS serotype O111:B4	Sheep fetus	Fetal i.v.	100 ng/kg estimated fetal weight	midgestation	Fetal LPS treatment results in focal inflammatory infiltrates and cystic lesions in periventricular white matter in 40% of the offspring.	(Mallard et al. 2003)
E. coli	Sprague-Dawley rats	i.u.	1x10^6 CFU	day 17 of pregnancy	E. coli treatment leads to microglial activation, astrogliosis, and apoptosis in periventricular white matter of the offspring. Treatment with IL-10 reduces E. coli-induced white matter damage.	(Pang et al. 2005)
E. coli	Sprague-Dawley rats	i.u.	1 x 10^7 CFU	day 17 of pregnancy	IL-10 treatment decreases LPS-induced oligodendrocyte loss, white matter damage, and apoptosis in the neonatal brain.	(Rodts-Palenik et al. 2004)
E. coli LPS serotype -O55:B5	Wistar rats	i.p.	300 µg/kg maternal weight	days 19 and 20 of pregnancy	Maternal LPS treatment elevates IL-1β mRNA level in the brain of P1 offspring, and increases cell death in the brain of P1 and P7 offspring.	(Rousset et al. 2006)
E. coli LPS serotype O127:B8	CD-1 mice	i.u.	10 µg per mouse	day 15 of pregnancy	Maternal LPS increases the levels of TNF-α, IL-1β, IL-6, and IL-10 in the amniotic fluid. Inhibition of PDE4 by rolipram prevents LPS-induced rise of inflammatory cytokines, preterm delivery, and fetal demise.	(Schmitz et al. 2007)
E. coli LPS serotype O55:B4	Sheep fetus	Fetal i.v.	88.7 ng/kg estimated fetal weight	65% or 85% of gestation	Fetal LPS treatment leads to white matter damage, increased microglia activation, and loss of neurofilament staining in the brain of the infants.	(Svedin et al. 2005)
E. coli LPS serotype O55: B5	Sprague-Dawley rats	i.p.	0.5 or 2.5 mg/kg maternal weight	day 18 of pregnancy	Maternal treatment with 0.5 mg/kg LPS increases the levels of TNF-α, IL-1β and IL-6 in the placenta. Maternal treatment with 2.5 mg/kg LPS increases TNF-α, IL-1β, and IL-6 in the placenta.	(Urakubo et al. 2001)

Abbreviations: IL: interleukin; TNF: tumor necrosis factor; i.p.: intraperitoneal; i.u.: intrauterine; i.v.: intravenous; MCP: monocyte chemoattractant protein; P1: postnatal day 1; PDE: phosphodiesterase.

Table 3. Reported experimental animal studies to demonstrate the association between prenatal bacterial infection induced-oxidative stress and brain injury.

Bacteria or bacterial products used	Host	Route of treatment	Dose of treatment	Time of treatment	Major Findings	Reference
E. coli LPS serotype O111:B4	Sprague-Dawley rats	i.p.	100 µg/kg maternal weight	day 18 of pregnancy	Maternal LPS treatment increases the levels of IL-6 and IL-10 in maternal serum, and elevates the level of IL-6 in the amniotic fluid and placenta, which could all be attenuated by N-acetyl cysteine (NAC) treatment.	(Beloosesky et al. 2006)
E. coli LPS serotype O55:B5	Sprague–Dawley rats	i.p.	500 µg/kg maternal weight	day 19 of pregnancy	Maternal LPS treatment increases protein carbonylation in male fetal brain, and leads to impaired spatial recognition in the offspring.	(Lante et al. 2007)
E. coli LPS serotype O55:B5	Sprague–Dawley rats	i.p.	500 µg/kg maternal weight	day 19 of pregnancy	NAC treatment prevents LPS-induced decrease of GSH content in the hippocampus, and protects LPS-induced deficits in long-term potentiation and spatial learning.	(Lante et al. 2008)
E. coli LPS serotype O55:B5	Sprague-Dawley rats	i.p.	1, 2, or 4 mg/kg maternal weight	day 18 of pregnancy	NAC protects LPS-induced fetal mortality, attenuates LPS-induced expression of TNF-α, IL-1β, and iNOS in fetal rat brains, and inhibits LPS-induced hypomyelination.	(Paintlia et al. 2004)

Abbreviations: i.p.: intraperitoneal; NAC: N-acetyl cysteine.

preventing and/or treating the complications of CP (Gibson et al., 2007; Jones et al., 2007a). Several regimens to manage the musculoskeletal motor deficit, a key sign of CP, have been reported. Orthopedic surgery is sometimes used to correct and/or prevent musculoskeletal deformities (Jones et al., 2007a). As an alternative or supplement to orthopedic surgery, botulinum toxin type A (BTX-A), a neurotoxin produced by the bacterium Clostridium botulinum that blocks the release of acetyl cholinesterase, has been reported to decrease spasticity and improve conscious movements such as crawling, standing, walking at 3 – 24 weeks following BTX-A injection (Bjornson et al., 2007; Gibson et al., 2007; Meholjic-Fetahovic, 2007). Hyperbaric oxy-gen treatment (HBOT) and pressurized room air have been shown to improve motor function, yet adverse effects including seizure has also been reported during HBOT treatment (McDonagh et al., 2007).

Medications, such as Baclofen, have also been shown to improve motor function and reduce spasm-related pain (Jones et al., 2007a). In gene-ral, no single treatment method has proven to be sufficient alone and a combination of medi-cal treatments and physical therapy are often carefully designed to manage the motor deficits in CP (Jones et al., 2007a).

DISCUSSION

Normal brain development involves precisely tim-ed cellular and molecular events including cell proliferation, migration, differentiation, myelina-tion, and synaptogenesis (Holmes and McCabe, 2001). For example, OPCs primarily arise from the neuroepithelium of the subventricular zone in mid-late to late gestational and early postnatal mammalian brain, and migrate to the developing white matter. After they reach appropriate axons,

OPCs exit the cell cycle and differentiate into myelin-forming OLGs (Baumann and Pham-Dinh, 2001; Simons and Trajkovic, 2006). All these processes could potentially be affected by envi-ronmental and genetic factors. In order to ensure full and timely myelination of all axonal tracts, the process and timing of oligodendrogenesis, migra-tion, and differentiation must be tightly controlled and coordinated with neurogenesis and neuronal differentiation (Simons and Trajkovic 2006). How-ever, bacterial infection during pregnancy elevates the expression of inflammatory cytokines and increases the formation of ROS in the fetus, which significantly affects the survival, differentiation, maturation trajectory of OPCs, as well as the survival of mature OLGs (Back et al., 1998; Bell and Hallenbeck, 2002; Feldhaus et al., 2004; Wang et al., 2004), leading to OLG loss and subsequently hypomyelination, one of the hall-marks of CP. Consistently, many experimental studies have reported white matter damage induced by maternal bacterial infection occurring at mid-late gestation, late gestation, and early postnatal periods when most of the oligo-dendrogenesis occurs (Back et al., 1998; Levine et al., 1993; Thomas et al., 2000).

It should be emphasized that besides the effects of inflammatory cytokines and oxidative stress, other mechanisms such as increased maternal temperature and fetal hypoxia during maternal bacterial infection may also contribute to fetal brain damage and the development of CP (Coumans et al., 2005; Dalitz et al., 2003). For example, heat stress during gestation has been reported to produce seizures, abnormal writhing movements of the limbs and body, and other gross abnormalities in the central nervous system in experimental animal models (Edwards, 2006; Mottola et al., 1993) while maternal hypoxia has been found to cause white matter damage in the offspring rabbits (Derrick et al., 2007), and reduce cerebral cortex cell density and cell size in mouse (Golan et al., 2004a). Understanding the mechanistic links between maternal bacterial infection and the development of CP in the offs-pring could potentially lead to novel and effective therapeutic strategies.

Conditions such as periodontal disease (Felice et al., 2005), bacterial vaginosis (Oakeshott et al., 2004; Ugwumadu, 2002), and chlamydia (Hou et al., 2006) are some possible scenarios for pregnant women to contract bacterial infection. Overall, the susceptibility and extent of white matter damage during maternal bacterial infection is affected by a combination of genetic factors, severity of the infection, strain and virulence of the infecting microorganism, developmental period of the offspring when bacterial infection occurs, and the existence of any additional environmental factors during prenatal and early postnatal period (Back, 2006; Levine et al., 1993; Meyer et al., 2006; Thomas et al., 2000). Experimental animal models will be useful to thoroughly dissect the mechanisms underlying maternal bacterial infection-induced neurodevelopmental deficits and to examine the neurological defects following maternal exposure to multiple insults.

Conclusion

Maternal bacterial infection is one important prenatal factor that could lead to white matter damage and utimattely CP in the offspring. Inflammatory cytokines and ROS induced during maternal bacterial infection significantly influence fetal brain development and contribute to the development of CP in the offspring.

REFERENCES

Ashdown H, Dumont Y, Ng M, Poole S, Boksa P, Luheshi GN (2006). The role of cytokines in mediating effects of prenatal infection on the fetus: implications for schizophrenia.Mol. Psychiatry. 11:47-55.

Aycicek A, Iscan A (2006) Oxidative and antioxidative capacity in children with cerebral palsy. Brain. Res. Bull 69:666-668.

Babulas V, Factor-Litvak P, Goetz R, Schaefer CA, Brown AS (2006). Prenatal exposure to maternal genital and reproductive infections and adult schizophrenia. Am. J. Psychiatry. 163:927-929

Back SA (2006) Perinatal white matter injury: the changing spectrum of pathology and emerging insights into pathogenetic mechanisms. Ment. Retard. Dev. Disabil. Res. Rev. 12:129-140

Back SA, Gan X, Li Y, Rosenberg PA, Volpe JJ (1998). Maturation-dependent vulnerability of oligodendrocytes to oxidative stress-induced death caused by glutathione depletion. J. Neurosci. 18:6241-6253

Banks WA (2005). Blood-brain barrier transport of cytokines: a mechanism for neuropathology. Curr. Pharm. Des. 11:973-984.

Baumann N, Pham-Dinh D (2001) Biology of oligodendrocyte and myelin in the mammalian central
nervous system. Physiol. Rev. 81:871-927.

Bell MJ, Hallenbeck JM (2002). Effects of intrauterine inflammation on developing rat brain. J Neurosci Res 70:570-579.

Beloosesky R, Gayle DA, Amidi F, Nunez SE, Babu J, Desai M, Ross MG (2006). N-acetyl-cysteine suppresses amniotic fluid and placenta inflammatory cytokine responses to lipopolysaccharide in rats. Am. J. Obstet. Gynecol. 194:268-273.

Bjornson K, Hays R, Graubert C, Price R, Won F, McLaughlin JF, Cohen M (2007). Botulinum toxin for spasticity in children with cerebral palsy: a comprehensive evaluation. Pediatrics 120:49-58.

Cai Z, Lin S, Pang Y, Rhodes PG (2004). Brain injury induced by intracerebral injection of interleukin-1beta and tumor necrosis factor-alpha in the neonatal rat. Pediatr. Res. 56:377-384.

Cai Z, Pan ZL, Pang Y, Evans OB, Rhodes PG (2000). Cytokine induction in fetal rat brains and brain injury in neonatal rats after maternal lipopolysaccharide administration. Pediatr. Res. 47:64-72.

Cambonie G, Hirbec H, Michaud M, Kamenka JM, Barbanel G (2004). Prenatal infection obliterates glutamate-related protection against free hydroxyl radicals in neonatal rat brain. J. Neurosci. Res. 75:125-32.

Chen YH, Xu DX, Wang JP, Wang H, Wei LZ, Sun MF, Wei W (2006a). Melatonin protects against lipopolysaccharide-induced intra-uterine fetal death and growth retardation in mice. J. Pineal. Res. 40:40-47.

Chen YH, Xu DX, Zhao L, Wang H, Wang JP, Wei W (2006b). Ascorbic acid protects against lipopolysaccharide-induced intra-uterine fetal death and intra-uterine growth retardation in mice. Toxicology. 217:39-45.

Coumans AB, Garnier Y, Supcun S, Jensen A, Berger R, Hasaart TH (2005) Nitric oxide and fetal organ blood flow during normoxia and hypoxemia in endotoxin-treated fetal sheep. Obstet. Gynecol. 105:145-155.

Dahlgren J, Samuelsson AM, Jansson T, Holmang A (2006) Interleukin-6 in the maternal circulation reaches the rat fetus in mid-gestation. Pediatr. Res. 60:147-151.

Dalitz P, Harding R, Rees SM, Cock ML (2003). Prolonged reductions in placental blood flow and cerebral oxygen delivery in preterm fetal sheep exposed to endotoxin: possible factors in white matter injury after acute infection. J. Soc. Gynecol. Investig. 10:283-290.

Debillon T, Gras-Leguen C, Verielle V, Winer N, Caillon J, Roze JC, Gressens P (2000). Intrauterine infection induces programmed cell death in rabbit periventricular white matter. Pediatr. Res. 47:736-742.

Deguchi K, Mizuguchi M, Takashima S (1996). Immunohistochemical expression of tumor necrosis factor alpha in neonatal leukomalacia. Pediatr. Neurol. 14:13-16

Derrick M, Drobyshevsky A, Ji X, Tan S (2007). A model of cerebral palsy from fetal hypoxia-ischemia. Stroke 38:731-735.

Duncan JR, Cock ML, Scheerlinck JP, Westcott KT, McLean C, Harding R, Rees SM (2002). White matter injury after repeated endotoxin exposure in the preterm ovine fetus. Pediatr. Res. 52:941-949.

Edwards MJ (2006) Review: Hyperthermia and fever during pregnancy. Birth Defects Res A Clin. Mol. Teratol. 76:507-516.

Feldhaus B, Dietzel ID, Heumann R, Berger R (2004). Effects of interferon-gamma and tumor necrosis factor-alpha on survival and differentiation of oligodendrocyte progenitors. J. Soc. Gynecol. Investig. 11:89-96.

Felice P, Pelliccioni GA, Checchi L (2005). Periodontal disease as a risk factor in pregnancy. Minerva. Stomatol. 54:255-264.

Fidel PL, Jr., Romero R, Wolf N, Cutright J, Ramirez M, Araneda H, Cotton DB (1994). Systemic and local cytokine profiles in endotoxin-induced preterm parturition in mice. Am. J. Obstet. Gynecol. 170:1467-1475.

Field NT, Newton ER, Kagan-Hallet K, Peairs WA (1993). Perinatal effects of Gardnerella vaginalis deciduitis in the rabbit. Am. J. Obstet. Gynecol. 168:988-994.

Gayle DA, Beloosesky R, Desai M, Amidi F, Nunez SE, Ross MG (2004). Maternal LPS induces cytokines in the amniotic fluid and corticotropin releasing hormone in the fetal rat brain. Am. J. Physiol. Regul. Integr. Comp. Physiol. 286:R1024-1029.

Gibson CS, MacLennan AH, Goldwater PN, Haan EA, Priest K, Dekker GA (2006). The association between inherited cytokine polymorphisms and cerebral palsy. Am J. Obstet. Gynecol. 194:674 e1-11.

Gibson N, Graham HK, Love S (2007). Botulinum toxin A in the management of focal muscle overactivity in children with cerebral palsy. Disabil Rehabil 29:1813-22.

Golan H, Kashtuzki I, Hallak M, Sorokin Y, Huleihel M (2004a) Maternal hypoxia during pregnancy induces fetal neurodevelopmental brain damage: partial protection by magnesium sulfate. J. Neurosci. Res. 78:430-41.

Golan H, Levav T, Mendelsohn A, Huleihel M (2004b). Involvement of tumor necrosis factor alpha in hippocampal development and function. Cereb Cortex 14:97-105.

Golan HM, Lev V, Hallak M, Sorokin Y, Huleihel M (2005). Specific neurodevelopmental damage in mice offspring following maternal inflammation during pregnancy. Neuropharmacology 48:903-917.

Goto M, Yoshioka T, Young RI, Battelino T, Anderson CL, Zeller WP (1997). A sublethal dose of LPS to pregnant rats induces TNF-alpha tolerance in their 0-day-old offspring. Am. J. Physiol. 273:R1158-1162.

Green LB, Hurvitz EA (2007). Cerebral palsy. Phys Med Rehabil Clin. N. Am. 18:859-882.

Hagberg H, Mallard C, Jacobsson B (2005). Role of cytokines in preterm labour and brain injury. Bjog 112 Suppl 1:16-18.

Hald A, Lotharius J (2005). Oxidative stress and inflammation in Parkinson's disease: is there a causal link? Exp. Neurol. 193:279-290.

Harnett EL, Dickinson MA, Smith GN (2007) Dose-dependent lipopolysaccharide-induced fetal brain injury in the guinea pig. Am J. Obstet. Gynecol. 197:179 e1-7.

Haynes RL, Folkerth RD, Keefe RJ, Sung I, Swzeda LI, Rosenberg PA, Volpe JJ, Kinney HC (2003) Nitrosative and oxidative injury to premyelinating oligodendrocytes in periventricular leukomalacia. J. Neuropathol. Exp. Neurol. 62:441-450.

Hebra A, Strange P, Egbert JM, Ali M, Mullinax A, Buchanan E (2001). Intracellular cytokine production by fetal and adult monocytes. J. Pediatr. Surg. 36:1321-1326.

Holmes GL, McCabe B (2001). Brain development and generation of brain pathologies. Int. Rev. Neurobiol. 45:17-41.

Hou GQ, Chen SS, Lee CP (2006) Pathogens in maternal blood and fetal cord blood using Q-PCR assay. Taiwan J. Obstet. Gynecol. 45:114-119.

Gilmore JH, Vadlamudi S, Lauder JM (2004) Prenatal Infection and Risk for Schizophrenia: IL-1beta, IL-6, and TNF-alpha Inhibit Cortical Neuron Dendrite Development. Neuropsychopharmacology 29:1221-1229.

Jacobsson B, Hagberg G (2004) Antenatal risk factors for cerebral palsy. Best Pract. Res. Clin. Obstet. Gynaecol. 18:425-436.

Johnston MV, Hoon AH (2006) Cerebral palsy. Neuromolecular Med. 8:435-450.

Jones MW, Morgan E, Shelton JE (2007a). Primary care of the child with cerebral palsy: a review of systems (part II). J. Pediatr. Health Care 21:226-237.

Jones MW, Morgan E, Shelton JE, Thorogood C (2007b) Cerebral palsy: Introduction and diagnosis (part I). J. Pediatr. Health Care 21:146-152.

Kannan S, Saadani-Makki F, Muzik O, Chakraborty P, Mangner TJ, Janisse J, Romero R, Chugani DC (2007). Microglial Activation in Perinatal Rabbit Brain Induced by Intrauterine Inflammation: Detection with 11C-(R)-PK11195 and Small-Animal PET. J. Nucl. Med.

48:946-954.

Kent AS, Sullivan MH, Elder MG (1994) Transfer of cytokines through human fetal membranes. J. Reprod. Fertil. 100:81-84.

Knight JA (2000). Review: Free radicals, antioxidants, and the immune system. Ann Clin. Lab. Sci. 30:145-158.

Koedel U, Rupprecht T, Angele B, Heesemann J, Wagner H, Pfister HW, Kirschning CJ (2004). MyD88 is required for mounting a robust host immune response to Streptococcus pneumoniae in the CNS. Brain 127:1437-1445.

Lante F, Meunier J, Guiramand J, De Jesus Ferreira MC, Cambonie G, Aimar R, Cohen-Solal C, Maurice T, Vignes M, Barbanel G (2008). Late N-acetylcysteine treatment prevents the deficits induced in the offspring of dams exposed to an immune stress during gestation. Hippocampus.Feb. 27, pp. 1-8.

Lante F, Meunier J, Guiramand J, Maurice T, Cavalier M, de Jesus Ferreira MC, Aimar R, Cohen-Solal C, Vignes M, Barbanel G (2007). Neurodevelopmental damage after prenatal infection: role of oxidative stress in the fetal brain. Free Radic. Biol. Med. 42:1231-1245.

Levine JM, Stincone F, Lee YS (1993). Development and differentiation of glial precursor cells in the rat cerebellum. Glia. 7:307-321.

Lin HC, Wan FJ, Wu CC, Tung CS, Wu TH (2005). Hyperbaric oxygen protects against lipopolysaccharide-stimulated oxidative stress and mortality in rats. Eur. J. Pharmacol. 508:249-254.

Ling ZD, Potter ED, Lipton JW, Carvey PM (1998). Differentiation of mesencephalic progenitor cells into dopaminergic neurons by cytokines. Exp. Neurol 149:411-423.

Liverman CS, Kaftan HA, Cui L, Hersperger SG, Taboada E, Klein RM, Berman NE (2006). Altered expression of pro-inflammatory and developmental genes in the fetal brain in a mouse model of maternal infection. Neurosci. Lett. 399:220-225.

Loh KP, Huang SH, De Silva R, Tan BK, Zhu YZ (2006). Oxidative stress: apoptosis in neuronal injury. Curr Alzheimer Res. 3:327-33.7

Mahadik SP, Pillai A, Joshi S, Foster A (2006). Prevention of oxidative stress-mediated neuropathology and improved clinical outcome by adjunctive use of a combination of antioxidants and omega-3 fatty acids in schizophrenia. Int. Rev. Psychiatry. 18:119-131.

Mallard C, Welin AK, Peebles D, Hagberg H, Kjellmer I (2003). White matter injury following systemic endotoxemia or asphyxia in the fetal sheep. Neurochem. Res. 28:215-223.

Marx CE, Jarskog LF, Lauder JM, Lieberman JA, Gilmore JH (2001). Cytokine effects on cortical neuron MAP-2 immunoreactivity: implications for schizophrenia. Biol. Psychiatry 50:743-749.

Matsuda Y, Kouno S, Hiroyama Y, Kuraya K, Kamitomo M, Ibara S, Hatae M (2000). Intrauterine infection, magnesium sulfate exposure and cerebral palsy in infants born between 26 and 30 weeks of gestation. Eur. J. Obstet Gynecol Reprod. Biol. 91:159-164

McDonagh MS, Morgan D, Carson S, Russman BS (2007). Systematic review of hyperbaric oxygen therapy for cerebral palsy: the state of the evidence. Dev. Med. Child Neurol 49:942-947

McLaurin J, D'Souza S, Stewart J, Blain M, Beaudet A, Nalbantoglu J, Antel JP (1995). Effect of tumor necrosis factor alpha and beta on human oligodendrocytes and neurons in culture. Int. J. De. Neurosci. 13:369-381.

Mehler MF, Kessler JA (1997). Hematolymphopoietic and inflammatory cytokines in neural development. Trends Neurosci. 20:357-365.

Mehlhase J, Gieche J, Ullrich O, Sitte N, Grune T (2000). LPS-induced protein oxidation and proteolysis in BV-2 microglial cells. IUBMB Life. 50:331-335.

Meholjic-Fetahovic A (2007).Treatment of the spasticity in children with cerebral palsy. Bosn. J. Basic Med. Sci. 7:363-367.

Meyer U, Nyffeler M, Engler A, Urwyler A, Schedlowski M, Knuesel I, Yee BK, Feldon J (2006). The time of prenatal immune challenge determines the specificity of inflammation-mediated brain and behavioral pathology. J. Neurosci. 26:4752-4762.

Mottola MF, Fitzgerald HM, Wilson NC, Taylor AW (1993). Effect of water temperature on exercise-induced maternal hyperthermia on fetal development in rats. Int. J. Sports Med. 14:248-251.

Mutch L, Alberman E, Hagberg B, Kodama K, Perat MV (1992). Cerebral palsy epidemiology: where are we now and where are we going? Dev. Med. Child Neurol. 34:547-551.

Nakazawa T, Nakazawa C, Matsubara A, Noda K, Hisatomi T, She H,

Michaud N, Hafezi-Moghadam A, Miller JW, Benowitz LI (2006). Tumor necrosis factor-alpha mediates oligodendrocyte death and delayed retinal ganglion cell loss in a mouse model of glaucoma. J. Neurosci. 26:12633-12641.

Neufeld MD, Frigon C, Graham AS, Mueller BA (2005). Maternal infection and risk of cerebral palsy in term and preterm infants. J. Perinatol 25:108-113.

Neumann H, Schweigreiter R, Yamashita T, Rosenkranz K, Wekerle H, Barde YA (2002). Tumor necrosis factor inhibits neurite outgrowth and branching of hippocampal neurons by a rho-dependent mechanism. J. Neurosci. 22:854-862.

Oakeshott P, Kerry S, Hay S, Hay P (2004). Bacterial vaginosis and preterm birth: a prospective community-based cohort study. Br. J. Gen. Pract. 54:119-122.

Paintlia MK, Paintlia AS, Barbosa E, Singh I, Singh AK (2004). N-acetylcysteine prevents endotoxin-induced degeneration of oligodendrocyte progenitors and hypomyelination in developing rat brain. J. Neurosci. Res. 78:347-361.

Pang Y, Rodts-Palenik S, Cai Z, Bennett WA, Rhodes PG (2005). Suppression of glial activation is involved in the protection of IL-10 on maternal E. coli induced neonatal white matter injury. Brain Res. Dev. Brain Res. 157:141-149.

Rahman MM, Akhter S, Karim BA (2004) Epilepsy in children with cerebral palsy. Mymensingh Med. J. 13:67-70.

Rodts-Palenik S, Wyatt-Ashmead J, Pang Y, Thigpen B, Cai Z, Rhodes P, Martin JN, Granger J, Bennett WA (2004) Maternal infection-induced white matter injury is reduced by treatment with interleukin-10. Am J Obstet Gynecol 191:1387-1392.

Romero R, Gotsch F, Pineles B, Kusanovic JP (2007) Inflammation in pregnancy: its roles in reproductive physiology, obstetrical complications, and fetal injury. Nutr. Rev. 65:S194-202.

Rosenbaum P, Paneth N, Leviton A, Goldstein M, Bax M, Damiano D, Dan B, Jacobsson B (2007) A report: the definition and classification of cerebral palsy April 2006. Dev Med Child Neurol. Suppl. 109:8-14.

Rousset CI, Chalon S, Cantagrel S, Bodard S, Andres C, Gressens P, Saliba E (2006). Maternal exposure to LPS induces hypomyelination in the internal capsule and programmed cell death in the deep gray matter in newborn rats. Pediatr. Res. 59:428-433.

Sakaguchi S, Furusawa S (2006). Oxidative stress and septic shock: metabolic aspects of oxygen-derived free radicals generated in the liver during endotoxemia. FEMS Immunol. Med. Microbiol. 47:167-177.

Santos SG, Diniz CG, Silva VL, Martins WA, Cara DC, Souza NC, Serufo JC, Nicoli JR, Carvalho MA, Farias LM (2007). Effects of oxidative stress on the virulence profile of Prevotella intermedia during experimental infection in gnotobiotic mice. J. Med. Microbiol. 56:289-297.

Schendel DE, Schuchat A, Thorsen P (2002) Public health issues related to infection in pregnancy and cerebral palsy. Ment Retard Dev Disabil Res Rev 8:39-45.

Schmitz T, Souil E, Herve R, Nicco C, Batteux F, Germain G, Cabrol D, Evain-Brion D, Leroy MJ, Mehats C (2007). PDE4 inhibition prevents preterm delivery induced by an intrauterine inflammation. J. Immunol. 178:1115-1121.

Schneider H, Pitossi F, Balschun D, Wagner A, del Rey A, Besedovsky HO (1998) A neuromodulatory role of interleukin-1beta in the hippocampus. Proc. Natl. Acad. Sci U S A 95:7778-7783.

Sikora JP (2000). [The role of cytokines and reactive oxygen species in the pathogenesis of sepsis]. Pol. Merkur Lekarski 7:47-50.

Simons M, Trajkovic K (2006). Neuron-glia communication in the control of oligodendrocyte function and myelin biogenesis. J. Cell Sci. 119:4381-9.

Singhi P, Jagirdar S, Khandelwal N, Malhi P (2003). Epilepsy in children with cerebral palsy. J. Child Neurol. 18:174-179.

Slifka MK, Whitton JL (2000). Clinical implications of dysregulated cytokine production. J. Mol. Med. 78:74-80.

Stoycheva M, Murdjeva M (2005) Serum levels of interferon-gamma, interleukin-12, tumour necrosis factor-alpha, and interleukin-10, and bacterial clearance in patients with gastroenteric Salmonella infection. Scand J. Infect. Dis. 37:11-14 .

Svedin P, Kjellmer I, Welin AK, Blad S, Mallard C (2005) Maturational effects of lipopolysaccharide on white-matter injury in fetal sheep. J Child Neurol 20:960-964.

Takahashi JL, Giuliani F, Power C, Imai Y, Yong VW (2003). Interleukin-1beta promotes oligodendrocyte death through glutamate excitotoxicity. Ann Neurol 53:588-95.

Thomas JL, Spassky N, Perez Villegas EM, Olivier C, Cobos I, Goujet-Zalc C, Martinez S, Zalc B (2000) Spatiotemporal development of oligodendrocytes in the embryonic brain. J. Neurosci. Res. 59:471-476.

Toso L, Poggi S, Park J, Einat H, Roberson R, Dunlap V, Woodard J, Abebe D, Spong CY (2005) Inflammatory-mediated model of cerebral palsy with developmental sequelae. Am. J. Obstet. Gynecol. 193:933-941.

Turrin NP, Gayle D, Ilyin SE, Flynn MC, Langhans W, Schwartz GJ, Plata-Salaman CR (2001). Pro-inflammatory and anti-inflammatory cytokine mRNA induction in the periphery and brain following intraperitoneal administration of bacterial lipopolysaccharide. Brain Res. Bull 54:443-453.

Ugwumadu AH (2002). Bacterial vaginosis in pregnancy. Curr. Opin. Obstet. Gynecol. 14:115-118.

Urakubo A, Jarskog LF, Lieberman JA, Gilmore JH (2001). Prenatal exposure to maternal infection alters cytokine expression in the placenta, amniotic fluid, and fetal brain. Schizophr Res. 47:27-36.

Vela JM, Molina-Holgado E, Arevalo-Martin A, Almazan G, Guaza C (2002). Interleukin-1 regulates proliferation and differentiation of oligodendrocyte progenitor cells. Mol. Cell Neurosci. 20:489-502.

Victor VM, Rocha M, Esplugues JV, De la Fuente M (2005) Role of free radicals in sepsis: antioxidant therapy. Curr. Pharm. Des. 11:3141-3158.

Wang H, Li J, Follett PL, Zhang Y, Cotanche DA, Jensen FE, Volpe JJ, Rosenberg PA (2004). 12-Lipoxygenase plays a key role in cell death caused by glutathione depletion and arachidonic acid in rat oligodendrocytes. Eur. J. Neurosci. 20:2049-2058.

Wang J, Asensio VC, Campbell IL (2002) Cytokines and chemokines as mediators of protection and injury in the central nervous system assessed in transgenic mice. Curr Top Microbiol Immunol 265:23-48

Wu YW, Colford JM, Jr. (2000). Chorioamnionitis as a risk factor for cerebral palsy: A meta-analysis. Jama 284:1417-1424.

Yang L, Lindholm K, Konishi Y, Li R, Shen Y (2002). Target depletion of distinct tumor necrosis factor receptor subtypes reveals hippocampal neuron death and survival through different signal transduction pathways. J. Neurosci. 22:3025-3032.

Yoon BH, Kim CJ, Romero R, Jun JK, Park KH, Choi ST, Chi JG (1997) Experimentally induced intrauterine infection causes fetal brain white matter lesions in rabbits. Am. J. Obstet. Gynecol. 177:797-802.

Yoon BH, Park CW, Chaiworapongsa T (2003). Intrauterine infection and the development of cerebral palsy. Bjog 110 Suppl. 20:124-127.

6

Advancements in the diagnosis of bacterial plant pathogens: An overview

Kalyan K. Mondal[1]* and V. Shanmugam[2]

[1]Division of Plant Pathology, Indian Agricultural Research Institute, New Delhi 110 012, India.
[2]Institute of Himalayan Bioresource Technology, Council of Scientific and Industrial Research, Palampur 176 062, Himachal Pradesh, India.

The timely detection and appropriate identification of causal agents associated with disease of crop plants or seeds are considered to be the most important issue in formulating the management strategies for plant diseases. This is particularly important for plant diseases of a bacterial nature, where disease-free planting materials is the only effective way to restrict the disease. The detection of bacterial pathogens still largely depends on cultural, morphological and biochemical properties. The protocol requires skilled taxonomical expertise and is also time and labor intensive. Moreover, it cannot discriminate between closely related strains of same bacterial pathogens. With the advent of polymerase chain reaction (PCR), nucleic acid based techniques have made the diagnostic procedures for plant pathogens, including bacteria, easier than the conventional approaches. The wide acceptability of nucleic acid based techniques is due to them being more sensitive, more accurate, more specific, and much faster than conventional techniques. The serology-based diagnoses are very often preferred over nucleo-based techniques as they are more user-friendly and less cumbersome, besides being sensitive, accurate, specific, and much faster than conventional techniques. This review critically analyzes the recent developments and scope of various nucleic acid- and serology-based techniques for the diagnosis of bacterial plant pathogens.

Key words: Bacterial pathogens, diagnosis, detection.

INTRODUCTION

Bacterial pathogens cause substantial loss to the productivity of major crop plants. Unlike fungal pathogens, bacterial pathogens cannot be contained effectively through chemical methods. Early detection in seeds, planting materials, or ensuring disease-free planting materials through rapid diagnostics are likely the effective means of reducing bacterial disease incidence. The traditional detection protocol, based on cultural, morphological and biochemical properties, requires skilled taxonomical expertise to confirm the identity of the causal bacterium. Though recent automation in the field of conventional biochemical approaches like Biolog-phenotyping, fatty acid methyl esterase (FAME) analyser

etc. have reduced data analysis and interpretation part to a great extent for diagnostician but these approaches are still time and labor intensive requiring incubation and processing time of 2-7 days before being subjected for automation. Also, they cannot discriminate closely related races within pathovar-populations of affecting bacterial pathogens. This review focuses on scope and status of different molecular diagnostics including nucleic acid and serology-based techniques for bacterial plant pathogens.

NUCLEIC ACID-BASED TECHNIQUES

Nucleic acid (NA) based techniques are widely recognized and powerful plant pathogen detection techniques. The target region mostly exploited for bacterial diagnostic is ribosomal DNA, which are present

*Corresponding author. E-mail: mondal_kk@rediffmail.com.

Table 1. Important housekeeping marker genes for the detection of bacterial pathogens.

Gene	Encoded proteins	References
16S rRNA	16S Ribosomal protein subunit	(Lee et al., 1997a; Alvarez, 2004)
23S rRNA	23S Ribosomal protein subunit	(Maes et al., 1996a)
16S-23S rRNA	Internal transcribed spacer region between 16S and 23S ribosomal subunit	(Maes et al.1996b; Song et al., 2004)
rpoB	β subunit of RNA polymerase	(Hocquellet et al., 1999)
groEL	Heat-shock protein	(Yushan et al., 2010)
gyrB	β subunit of DNA gyrase	(Mondal et al., 2012)
recA	Recombinase A protein	(Eisen, 1995; Waleron et al., 2002; Young and Park, 2007)
atpD	ATP synthesis β chain	(Young and Park, 2007)
dnaK	Heat shock protein 70, molecular chaperone DnaK	(Young et al., 2008)
rpoD	Sigma -70 factor of RNA polymerase	(Young et al., 2008)
fyuA	transmembrane protein, TonB-dependent receptor	(Young et al., 2008)
efP	Eleongation factor P protein	(Bui et al., 2010)
glnA	Glutamine synthetase I	(Takle et al., 2007)

in all bacteria at high copy number per genome with highly conserved regions, allowing for very sensitive detection (Mondal et al., 2004). Since the specificity of DNA-based techniques only relies on primer and probe sequences, such assays are also easy to develop and can be transposed into virtually every pathosystem. Molecular markers like restriction fragment length polymorphism (RFLP), random amplified polymorphic DNA (RAPD), amplified fragment length polymorphism (AFLP), as well as *hrp* genes, *pth* gene based markers have been extensively employed to detect and identify pathogenic bacteria affecting different crop plants (Williams et al., 1990; Leite et al., 1994; Opio et al., 1996; Kerkoud et al., 2002, Berg et al., 2005). Besides the aforementioned biomarkers, the important housekeeping marker genes used for multi locus sequence typing (MLST) of bacterial pathogens are listed in Table 1.

Restriction fragment length polymorphism (RFLP)

RFLP analysis has been extensively used in detection and identification of plant pathogens (Mondal et al., 2004). A small DNA segment from a known bacterium, pathogenic to the host plant in question, is used as a probe. The DNA from both the known (as positive control) and suspected bacterial pathogen (isolated from infected plant samples) are digested with the same restriction enzyme(s). If the bacteria in question contain DNA with slightly different base sequences, some of the assayed restriction sites will be missing or in different locations. Therefore, the restriction enzyme-digested DNA will produce unique fragment numbers and sizes. The samples of both control as well as unknown digested bacterial DNA are placed side by side in an agarose gel, and are then separated by size using electrophoresis. The double-stranded DNA fragments are then chemically denatured into single-strands. Then the DNA fragments in gel are transferred to a nylon membrane which fixes their positions and maintains them as single-stranded DNA. The nylon membrane is washed with a solution containing many copies of a radioactive DNA probe. The probe is a very short, single-stranded DNA that will hybridize with its complementary sequence wherever the sequence is found among the DNA fragments on the nylon membrane. This further increases the specificity of this technique. Finally, an x-ray film is exposed to the nylon membrane. When this film is developed, a dark band will appear at each location where the probe hybridized to

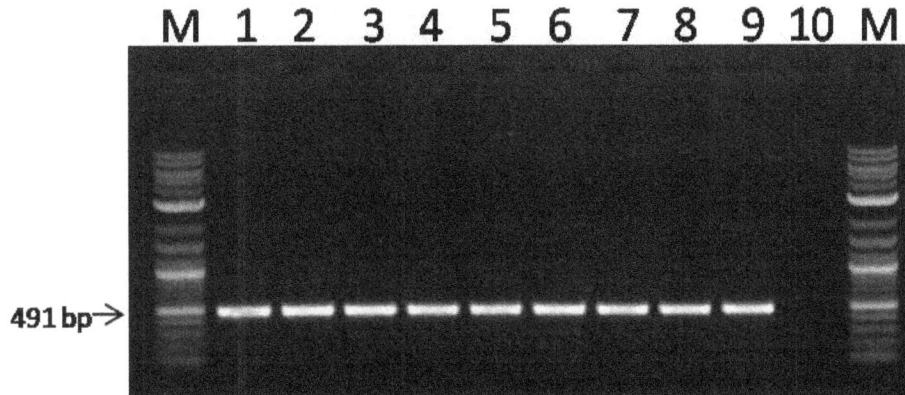

Figure 1. PCR detection of *Xanthomonas axonopodis* pv. *punicae* in pomegranate plant. M 100 kb ladder, Lanes 1 to 8 indicate pomegranate leaf samples, 9 = positive control (purified bacterial DNA) and 10= negative control (*Pantoea agglomerans,* out group bacteria).

complementary DNA. These banding patterns will determine whether the unknown specimen is same to the bacterium that was used as probe.

PCR-aided nucleic acid based diagnosis

The polymerase chain reaction (PCR) is one of the greatest inventions in science by Kary Mullis in 1984 (Mullis, 1987). With the advent of PCR, DNA based techniques have rapidly become the gold standard for detection, and identification of plant pathogens, including bacteria (Jensen et al., 1993; Bereswill et al., 1994; Alexander et al., 2004). This is due to the fact that these techniques overcome many of the shortcomings due to their sensitivity, greater accuracy, specificity, and more rapid results than conventional techniques (Schaad *et al., 2001*). The PCR technique primarily involves three steps in sequential events, namely denaturation, primer annealing and DNA synthesis or chain extension with the help of a thermal cycler. The specificity of the technique depends upon designing of primers that are unique to target pathogens. With the introduction of PCR, the nucleic acid mediated detection of plant pathogens has become easier and more sensitive (Janse, 1988; Henson and French, 1993; Cubero and Graham, 2002). Recently, Mondal et al. (2012) demonstrated a rapid and reliable PCR based detection of *Xanthomonas axonopodis* pv. *punicae* causing bacterial blight of pomegranate. The primer set (KKM5 and KKM6) was synthesized based on sequence alignment of 530 nucleotides of *C*-terminal region in the *gyrB* genes from 15 different bacterial strains. The primer set was validated for amplification of 491 bp of *gyrB* gene. The developed technique could detect the pathogen from infected pomegranate plant samples including leaf (Figure 1), fruit and stem within 3 h, at a detection limit of 0.1 ng of template DNA μl^{-1}.

Random amplified polymorphic DNA (RAPD)

The RAPD markers are demonstrated to be useful for determining polymorphisms among phytopathogenic bacteria (Rafalski et al., 1994; Chen et al., 2003; Grover et al., 2006; Mondal et al., 2008). The utility of random primers for differentiating bacterial blight pathogens in beans, *Xanthomonas campestris* pv. *phaseoli* and *X. campestris* pv. *phaseoli* var fuscans has been well documented (Birch et al., 1997). RAPD based polymorphism comprising Indian isolates of three bacterial pathogens causing common blight, fuscous blight and halo blight in grain legumes was studied (Mondal, 2009). Among the random primers used, SBSB02 yielded polymorphic amplicons (the sequence of the primer SBSB02 is 5'-TGATCCCTGG-3'). The primer specifically amplified two polymorphic fragments of ~300 bp and ~1600 bp in fuscous blight bacterium, but not in common and halo blight bacterium, suggesting that the amplified regions are conserved within the fuscous blight isolates. Thus one of these amplicons could be exploited as sequence characterized amplified regions (SCAR) marker after sequencing and designing primer specific to the region.

Amplified fragment length polymorphism (AFLP)

AFLP involves amplification of specific region of genomic DNA through PCR (using a single primer) followed by the cleaving of the amplified fragments using restriction endonuclease (Figure 2). The technique is thus a combination of RFLP and PCR techniques and is extremely useful in detection of polymorphism between closely related bacterial pathogens. The genomic fingerprints are produced without any prior knowledge of

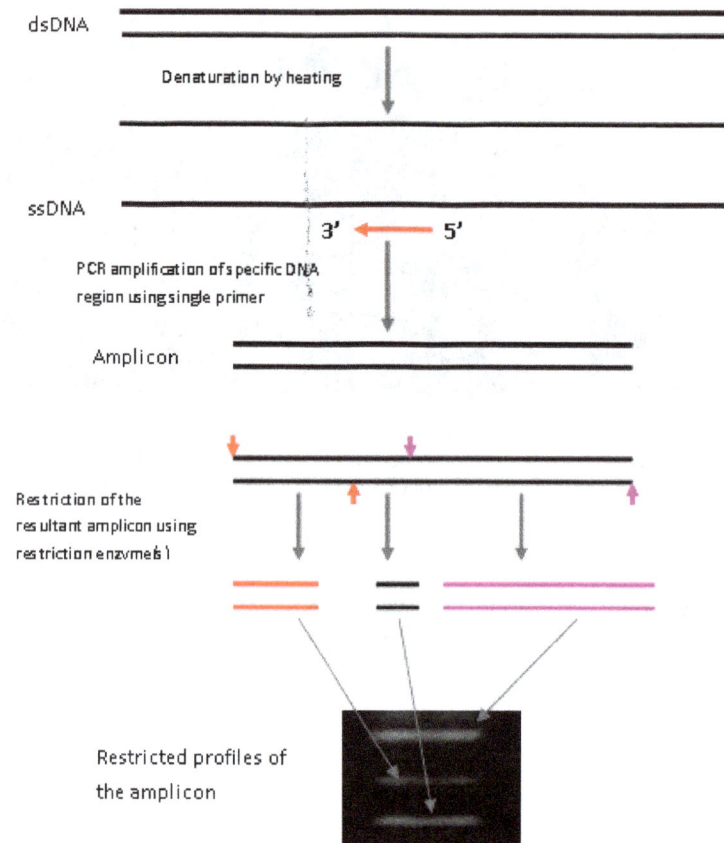

Figure 2. Schematic steps involved in AFLP.

sequences using a limited set of generic primers. The number of fragments detected in a single reaction can be 'tuned' by selection of specific primer set. The AFLP technique is reliable, since stringent reaction conditions are used for primer annealing. AFLP is now preferred over RFLP as it is PCR linked and is very often employed in the detection and differentiation of several bacterial plant pathogens including *Xanthomonas axonopodis* pv. *phaseoli*, *Xanthomonas axonopodis* pv. *phaseoli* var. *fuscans, Pseudomonas syringae* pv. tomato, using primers specific to ribosomal genes or rRNA operons (Manceau and Horvais,1997; Mahuku et al., 2006). A comparative analysis of RAPD and AFLP techniques was undertaken to assess genetic diversity and genetic relatedness within genospecies III of *Pseudomonas syringae* (Clerc et al., 1998). Both techniques showed high discriminating power because strains of *P. syringae* pv. tomato and *P. syringae* pv. Maculicola, which were indistinguishable by other techniques including pathogenicity tests on tomato, were separated into two groups by both RAPD and AFLP analyses. However, the AFLP method was more efficient for assessing intrapathovar diversity than RAPD analysis and allowed clear delineation between intraspecific and interspecific

genetic distances, suggesting that it could be an alternative to DNA pairing studies (Clerc et al., 1998).

Real time PCR

Real time PCR is the most advanced version of PCR wherein the amplification of target sequence can be quantified after each PCR cycle and amplification is pictographically displayed through an attached monitor (Espy et al., 2006). Real time PCR has gained popularity as it not only eliminates the post-PCR processing steps including gel electrophoresis, but also eliminates the exposure to carcinogenic chemicals, ethidium bromide, radioactive isotopes, UV-radiation. It also poses a reduced risk of sample contamination with amplicon and a greater amenity to multiplexing, simultaneous testing for multiple pathogens or samples. The technique can also be implemented in the field by using portable real time PCR machines. There are currently four different fluorescence-detection techniques that are used to detect amplicons, include SYBR green dye based detection, TaqMan probes, fluorescent resonance energy transfer (FRET) probes, and molecular beacons (Schaad and

Frederick, 2002; Mackey et al., 2002).

SYBR green dye based detection

The use of fluorescent dyes such as SYBR Green I (SG) in real time PCR have become increasingly important for diagnostic applications of plant pathogens (Kiltie and Ryan, 1997; Vitzthum and Bernhagen, 2002). The SG binds to the minor groove of the DNA double helix. In solution, the unbound dye exhibits very little fluorescence, however, fluorescence is greatly enhanced upon DNA-binding. At the beginning of amplification, the reaction mixture contains the denatured DNA, the primers, and the dye. The unbound dye molecules weakly fluoresce, producing a minimal background fluorescence signal which is subtracted out during computer analysis. During primer annealing steps, a small amount of dye molecules bind to the double strand. The dye binding results in a dramatic increase of the SG molecules to emit light upon excitation. Subsequently, during elongation, more and more dye molecules bind to the newly synthesized DNA. If the reaction is monitored continuously, an increase in fluorescence is viewed in real-time. Upon denaturation of the DNA for the next heating cycle, the dye molecules are released and the fluorescence signal falls. Fluorescence measured at the end of the elongation step of every PCR cycle indicates the increasing amount of amplified DNA. A real-time SYBR Green I assay was developed for the detection of X. arboricola pv. pruni, the causal agent of bacterial spot disease of stone fruit (Palacio-Bielsa et al., 2011).

TaqMan probes

The TaqMan assay is based on the 5'-3' exonuclease activity of Taq polymerase. In this assay, an oligonucleotide probe is labeled with two dyes; a reporter fluorescent dye attached to its 5' end and a quencher dye attached to its 3' end. Usually, 6-carboxyfluorescein (6FAM) is used as reporter dye at 5' and 6-carboxytetramethyl-rhodamine (TAMRA) is used at 3' end as quencher dye (Holland et al., 1991). When the probe hybridizes to its target template DNA, the reporter dye is cleaved by the 5' nuclease activity of Taq polymerase. As a result of cleavage, the reporter dye emits fluorescent signal, since it is no longer suppressed by the quencher dye. TaqMan probes overcome the error of reducing signals due to mispriming or primer-dimer formation – superior over SYBR green chemistry with DNA binding dyes (more prone to error since any non-specific PCR products, and primer-dimer artifacts, can generate a signal). TaqMan amplicon is generally 60-70 bp, thus reaction is more efficient (contrary to standard PCR, where an amplicon of at least 200 bp is required to detect

efficiently by electrophoretic separation). TaqMan PCR was demonstrated to be useful in detecting R. solanacearum strains (Weller, 2000).

Fluorescent resonance energy transfer (FRET) probes

There are two FRET probes used in this method (Didenko, 2001). Probe 1 contains a fluorescent label at its 3' end, and probe 2 is labeled at its 5' end with a different dye such as light cycler red 640. The two probes are designed so that when they hybridize to the amplified PCR product they are aligned head to tail to bring the two fluorescent dyes in close proximity to each other. The dye attached to the first probe is excited by the light source of the light cycler unit and it emits a green fluorescent light at a slightly longer wavelength. When 2nd probe is in close proximity, the energy emitted by the 1st probe excites the light cycler red 640 dye attached to the 2nd probe and the red fluoresces at a longer wavelength, which is now detected at 640 nm. Fluorescence is measured during annealing step of each cycle.

Molecular beacons

Molecular beacons are short fluorescent oligonucleotide probes that are designed to form stem-loop folding (Didenko, 2001). The probes contain a fluorescent chromophore at the 5' end and a quencher molecule at the 3' end. The probes at resting stage (after end each amplification cycle) form a stem-loop structure due to annealing of the complementary arm sequences that are added on both sides of the probe. The probes at stem-loop form do not emit fluorescence as the energy from fluorophore molecule get transferred to the quencher molecule. However, the probes emit fluorescence when the fluorophore and quencher become separated from each other that is, at the time of annealing to the target template DNA during PCR amplification (Cockerill and Smith, 2002).

Nested-PCR

In nested PCR two rounds of PCR are performed, the second using a primer set internal to those used in the first round. Sometimes immunocapture is used to concentrate pathogenic cells prior to PCR, thus increasing detection sensitivity. It is a technique with increased detection sensitivity through multiple displacement amplification (MDA). MDA exponential amplification of DNA is carried out using random hexamer primers resulting in more or less complete genome amplification for essentially all DNA present in sample (Dean et al., 2002). By performing MDA prior to PCR on a sample containing trace amounts of pathogen genomic DNA one can exponentially increase the quantity and thereby help in pathogen detection. Furthermore, MDA can be used to generate large quantities of genomic DNA from a very limited sample (few bacterial cells in the infected samples), thus providing a supply of genomic DNA that can be used for multiple tests and stored for future use. The nested-PCR has been demonstrated to be useful in ultrasensitive detection of various plant pathogenic bacteria including *Clavibacter michiganensis* subsp. s*epedonicus* (Lee et al., 1997b), *Erwinia amylovora* (Llop et al., 2000) and *Pseudomonas savastanoi* pv. *savastanoi* (Bertolini, 2003b).

Bio-PCR

The sensitivity of a PCR based detection assay can be improved using a short culture enrichment step preceding PCR-amplification and is referred as Bio-PCR (Schaad et al., 1995). Increased sensitivity of the detection assay using culture enrichment step prior to PCR was observed during detection of many bacterial plant pathogens including *Pseudomonas syringae* pv. *phaseolicola* in bean seed extracts, *Acidovorax avenae* ssp. *avenae* in rice seeds (Schaad et al., 1995; López et al., 2003; Song et al., 2004).

Multiplex PCR

Multiplex PCR is useful for the simultaneous detection of multiple pathogens in a single reaction. In multiplex reaction more than one set of primers are used. Multiplex reaction can be run either in a conventional or real time PCR machine. Using multiplex reaction we can reduce the total detection time required for individual pathogens (viral, fungal, bacterial) associated with a given crop samples. A multiplex nested reverse-transcription PCR detection protocol was developed for four RNA viruses and *Pseudomonas savastanoi* pv. *savastanoi* affecting olive trees (Bertolini, 2003a).

Loop-mediated isothermal amplification (LAMP)

The LAMP technique has been shown to be good approach for amplifying nucleic acid with high specificity, efficiency, and rapidity without the need for thermal cycler (Kubota et al., 2007). It is a method for the detection of specific nucleic acid sequences and has the potential to overcome many of the limitations of PCR-based methods. The ability of LAMP to amplify a target nucleic acid sequence under isothermal conditions eliminates the need for thermal cycling equipment, allowing testing to be carried out with minimal equipment (a water bath or heated block). Furthermore, this simplicity of LAMP-based technique facilitates its use in the field or in less well-resourced settings. LAMP method has been successfully used for detection of many plant pathogenic bacteria including epiphytic *E. amylovora* in pear and apple (Temple et al., 2008).

Repetitive sequence based PCR (Rep-PCR)

Diagnosis of bacterial plant pathogen using primers corresponding to specific repetitive sequences like enterobacterial repetitive intergenic consensus (ERIC), repetitive extragenic palindromic (Rep) and repetitive BOX elements (BOX), which are dispersed throughout the bacterial genome, has well been documented (Martin et al., 1992; Louws et al., 1994, 1998; Barak and Gilbertson, 2003). The distribution pattern of these repetitive sequences varies from one bacterium to other, and this comprises the basis of differentiation in a bacterial population. PCR based on these repetitive sequences (often termed as rep-PCR) were found to be effective in identification of bacterial plant pathogens even at race level (Mahuku et al., 2006). The primers corresponding to Rep region are rep-1 R-1[5'-IIIICGICGICATCIGGC-3'] and rep-2 [5'-ICGICTTATCIGGCCTAC-3']; and to ERIC region (Enterobacterial repetitive intergenic consensus sequences) are ERIC1R [5'-ATGT AAGCTCCTGGGGATTCAC-3'] and ERIC2 [5'-AAGTAAGTGACTGGGTGAGCG-3']; to BOXA region (a subunit of the BOX element) is BOX-A1R primer [5'-CTACGGCAAGGCGACGCTGACG-3']. The ERIC-PCR has been demonstrated to be an effective method in determining the genetic diversity among population of many bacterial plant pathogenic genera, including *Xanthomonas* and *Pseudomonas* (Weingrat et al., 1997; Mondal and Mani, 2009). Recently, the use of BOX-PCR in determining the genetic variability among the bacterial flora associated with pomegranate leaf (Figure 3) was documented (Mondal, 2011).

Microarrays

Identification of bacterial plant pathogens by conventional PCR is prone to potentially reporting false-positives. Further, scoring of a gel band continues to be problematical, especially with smeared backgrounds or low and high intensity bands. Criteria for inclusion or exclusion of bands above or below a given size are arbitrary. Eliminating these problems, microarray-based assays has come up as an effective method for the identification as well as differentiation of pathogenic

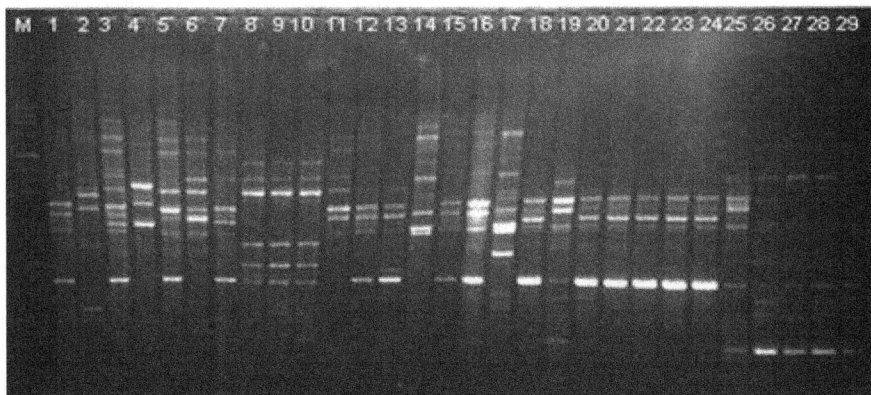

Figure 3. Genomic fingerprints of major bacterial flora associated with pomegranate leaf generated by BOX-PCR. M=1 kb plus ladder (Fermentous Co.) 1-29 bacterial isolates.

bacteria from mixed-culture environmental samples without the problem of false-positivity (Willse et al., 2002; Burton et al., 2006). The method is alternatively known as oligonucleotide arrays (Call et al., 2001). This technique has a greater multiplexing capability compared to conventional PCR, real time PCR or other NA-based techniques. After amplification by PCR, the resultant amplicon is hybridized with species/pathovar-specific DNA probes. The microarray-based techniques are applicable to any microorganism, without requiring a priori knowledge of specific nucleic acid signatures (Beattie, 1997; Loy et al., 2002). Thus dozen or hundreds of plant pathogens can be detected at a time, and this method is most useful in cases when the species being detected is unknown.

Genome-wide comparison between pathogenic and nonpathogenic strains within a species is a useful strategy for identifying candidate genes important for virulence. DNA microarray-based genome composition analysis is a good alternative to full genome sequencing and has been used in comparative studies to analyze various bacterial pathogens including Mycobacterium tuberculosis (Behr et al., 1999), Helicobacter pylori (Salama et al., 2000), Pseudomonas aeruginosa (Wolfgang et al., 2003), Yersinia pestis, and Y. pseudotuberculosis (Hinchliffe et al., 2003). A nonpathogenic strain J1a12 of Xylella fastidiosa associated with citrus could easily be differentiated from the pathogenic strain 9a5c causing variegated chlorosis, through DNA microarray-based comparison (Koide et al., 2004). It was revealed that 14 coding sequences of strain 9a5c are absent or highly divergent in strain J1a12. An arginase and a fimbrial adhesin precursor of type III pilus were confirmed to be absent in the nonpathogenic strain by PCR and DNA sequencing. Thus, the absence of both genes can be associated with the failure of the J1a12 strain to establish and spread in citrus and tobacco plants.

The gene distribution among strains of Ralstonia solanacearum, a highly polymorphic plant pathogenic bacterium, has been a priority area to study the status of known or candidate pathogenicity genes. Based on the use of comparative genomic hybridization on a pangenomic microarray for the GMI1000 reference pathogenic strain, researchers could compare the repertoires of genes among a collection of 18 different strains representative of the biodiversity of the R. solanacearum species (Guidot et al., 2007). Presently, a list of 2,690 core genes has been identified in all tested strains. As a corollary, a list of 2,338 variable genes within the R. solanacearum species has been defined. The hierarchical clustering based on the distribution of variable genes is fully consistent with the phylotype classification, which was previously defined from the nucleotide sequence analysis of four different genes. The presence of numerous pathogenicity-related genes in the core genome indicates that R. solanacearum is an ancestral pathogen.

Advantages of NA-based detection techniques

There is no debate about the value of NA-based techniques to an applied plant pathology programme. The main advantages are high sensitivity (Janse, 1988), rapidity, specificity and the quantifiability. Of further note are the following advantages:

(i) Detection and identification of bacterial pathogens in seed and plant samples for post- and pre-entry quarantine check.
(ii) Quantification of bacterial biomass in host tissue or in environmental samples.
(iii) Identification of bacteria that grow slowly or difficult to grow (fastidious bacteria).
(iv) Genomic characterization of genus, species, pathovars,

of bacterial pathogens.

(v) To provide a more dependable diagnosis in the specific cases where the routine diagnostics through cultural, morphological and biochemical are not conclusive enough.

Disadvantages of NA-based detection techniques

(i) NA-based tests very often yield false positives and false negatives, therefore, plant pathologist should not rely exclusively on NA-based test as the sole evidences for new reports, nor for other important samples.

(ii) Depending on the complexity of the diseased sample, a positive PCR test may not be sufficient evidence for a high-confidence diagnosis.

(iii) Further verification of the identity of PCR amplicons generated using species-specific primers is sometimes required to confirm results. These include analysis of RFLP, sequencing or hybridization to a specie-specific DNA probe. This makes the diagnostic protocol more complex, cost and time intensive.

SEROLOGY BASED DIAGNOSTIC TECHNIQUES

The basic principle of sero-diagnostic lies on the fact that the antigen (target pathogens) is detected using antibodies (specific to the target pathogens concerned); and this reaction is visualized through an enzyme-substrate hybridization. The detection of bacterial plant pathogens with antisera is still the method of choice for many plant diagnostician because of the relative low costs and presence of technical infrastructure based on automated enzyme liked immunosorbent assay (ELISA).

Preparation of antigen and production of antisera

The antigen preparation is emulsified with an equal volume of Freund's complete adjuvant. The emulsified antigen (0.5 ml/animal) is administrated to rabbits through sub-cutaneal mode and foot pad at multiple sites. The antigens (100-150 μg), emulsified in an equal volume of Freund's incomplete adjuvant, are administered to rabbits to produce hyper immune sera. Rabbits are bled through the ear vein 7 days after the last booster dose and sera are separated, and then stored at -20°C.

Mono- and poly-clonal antibodies are required to develop for specific immunological diagnostic assays. Polyclonal antibodies (pAb) are typically raised in rabbits, goats or sheep, and their popularity is evident by the fact that they are frequently used in immunosensor-based assays for bacterial detection. It should be noted that the different epitopes of pAbs may often be recognized on a single cell. Under specific cases where this is undesirable, such as in the case where high specificity is a requirement, monoclonal or recombinant antibodies may be more applicable. Monoclonal antibodies are generated using the hybridoma technology. The bone marrow, primary lymph nodes and the spleen are selected as a source of antibody-producing B cells which are harvested and fused to immortal myeloma cells. The resulting hybrid cells (referred to as hybridomas) subsequently secrete full-length antibodies that are directed towards a single epitope. Suitable candidates, identified by ELISA-based analysis, are then 'cloned out' to ensure that a single cell, producing antibody specific for an individual epitope, is present and the antibody generated can be used for assay development.

Cell parts of different pathogens possess antigenic properties that are employed for antibody production. Like soluble cytoplasmic antigen, insoluble cell wall antigen and lyophilized mycelial antigen in fungi. In bacterial plant pathogens, lipopolysaccharides, exopolysaccharide, cell wall protein, flagellar protein, and outer membrane proteins are used as antigens. The monoclonal antibody when reacted with bacterial antigen (like lipopolysaccharides) yields a low mobility ladder pattern on immuno-blotting experiment.

Enzyme linked immunosorbent assay (ELISA)

Through this technique, the bacterial pathogens can be detected directly in the infected plant materials without culturing the bacteria. Owing to its simplicity, sensitivity and adaptability, this technique has gained popularity in different diagnostic laboratories. Both poly and monoclonal antibodies are used for the detection of phytopathogenic bacteria using ELISA assay. The sensitivity of ELISA assay ranges between 10^5–10^6 CFU ml^{-1}. There are several reports where monoclonal antibody have been employed to detect bacterial plant pathogens including *Xanthomonas campestris* pv. *begoniae*, *X. campestris* pv. *pelargonii* (Benedict et al., 1990), *X. oryzae* pv. *oryzae* and *X. oryzae* pv. *oryzicola* (Benedict et al., 1989), *Corynebacterium sepedonicum* (DeBoer and Wieczorek, 1984), and *Erwinia amylovora* (Gorris et al., 1996). Shanmugam et al. (2002) used polyclonal antibody developed against whole bacterial cells (Courtesy: International Potato Research Centre, Lima, Peru) of *R. solanacearum* infecting potato to detect the survival of *R. solanacearum* in ginger rhizomes stored at different temperatures.

The sensitivity of ELISA can be increased to many folds by adding ethylene diamine tetraacetic acid (EDTA) and lysozyme in the extraction buffer (which helps to release lipopolysaccharide into solution). Pre-heating of bacterial samples to destroy proteins also improves the sensitivity by increasing the signal: noise ratio (Jones et al., 1997).

Lateral flow devices

The lateral flow devices are principally based on ELISA techniques where different types of filters are used as solid support for capturing the antigen-antibody binding reaction (Alvarez, 2004). Lateral flow devices are now available for rapid detection of bacterial pathogens including *Ralstonia solanacearum* (Danks and Barker, 2000), *Clavibacter michiganensis* subsp. *michiganensis* and *X. hortorum* pv. *pelargonii* (Alvarez, 2004).

CONCLUSION

The detection of pathogenic bacteria through conventional strategies involves the aseptic transfer of inoculum from an infected source (soil, plant parts etc.) to a suitable growth medium and subsequent transfer to selective or differential media. The colonies that appear on such media can be distinguished based on their distinctive colony morphologies by ocular inspection and their identification confirmed by rigorous biochemical (sugar utilization, etc.), physiological assays. Though the conventional colony based approach provides an inexpensive and straightforward protocol for quantitative and qualitative bacterial pathogen detection, a major disadvantage of this approach is the lengthy times required to obtain visible and identifiable bacterial colonies. The sero-diagnostics for bacterial pathogens are often preferred being more user-friendly, higher sensitivity, and much faster than conventional techniques. However, the success of any antibody-mediated test depends on the specificity of the antibody. Adequate care should be taken to minimize the cross-reactivity with related antigens that should be checked while evaluating important antigens. The nucleic acid based detection of bacterial pathogens has emerged as a supplement to overcome these bottlenecks. The recent advancement in the area of PCR based approaches further extended its versatility. Assays like real-time PCR, multiplex PCR, nested PCR, Bio-PCR, repetitive PCR, LAMP are among the detection options that provides rapid data analysis with specificity. However, one has to choose the best or combination of options depending upon the needs. For example, when multiple pathogens are to be detected in a minimum time multiplex-PCR would be the best options. While, the pathogen detection limit in a sample is at zero tolerance level, nested PCR, bio-PCR should be carried out to detect even the lower number of bacterial cells in the tested samples and this would also help to differentiate between viable and non-viable cells. And for routine diagnosis of the bacterial pathogens, integrated approaches including 16S rDNA sequencing, MLST, Biolog-phenotyping, fatty acid methyl esterase (FAME) profiling and pathogenicity assay following Kotch's postulate are to be preferred.

REFERENCES

Alexander BCM, Robert BM, Pablo G, Paul G, Robert LG (2004). Genetic diversity and pathogenic variation of common blight bacteria (*Xanthomonas campestris* pv. *phaseoli* and *X. campestris* pv. *phaseoli* var. *fuscans*) suggests pathogen coevolution with the common bean. Phytopathology 94:593-603.

Alvarez, AM (2004). Integrated approaches for detection of plant pathogenic bacteria and diagnosis of bacterial diseases. Annu Rev Phytopathol 42:339-366.

Barak, JD, Gilbertson RL (2003). Genetic diversity of *Xanthomonas campestris* pv. *vitians*, the causal agent of bacterial leafspot of lettuce. Phytopathology 93:596-603.

Beatti KL (1997). Genomic fingerprinting using oligonucleotide arrays. In Caetano-Anolles,G. and Gresshoff,P.M. (eds) *DNA Markers: Protocols, Applications and Overviews*. John Wiley & Sons, New York, NY, pp. 213-224.

Behr MA, Wilson MA, Gill WP, Salamon H, Schoolnik GK, Rane S, Small PM (1999). Comparative genomics of BCG vaccines by whole-genome DNA microarray. Science 284:1520-1523.

Benedict AA, Alvarez AM, Berestecky J, Imanaka W, Mizumoto CY, Pollard LW, Mew TW, Gonzalez CF (1989). Pathovar-specific monoclonal antibodies for *Xanthomonas campestris* pv. *oryzae* and for *Xanthomonas campestris* pv. *oryzicola*. Phytopathology 79:322-328.

Benedict AA, Alvarez AM, Pollard LW (1990). Pathovar-specific antigens of *Xanthomonas campestris* pv. *begoniae* and *X. campestris* pv. *pelargonii* detected with monoclonal antibodies. Appl Environ. Microbiol 56:572-574.

Bereswill, S, Bugert P, Volksch B, Ullrich M, Bender CL, Geider K (1994). Identification and relatedness of coronating producing *Pseudomonas syringae* pathovars by PCR analysis and sequence determination of amplification products. Appl Eniveron Microb. 60:2924-2930.

Berg T, Tesoriero L, Hailstone DL (2005). PCR-based detection of *Xanthomonas campestrie* pathovers in Brassica seed. Plant Pathol. 54:416-427.

Bertolini E, Olmos A, López MM, Cambra M (2003a). Multiplex nested reverse-transcription polymerase chain reaction in a single tube for sensitive detection of four RNA viruses and *Pseudomonas savastanoi* pv. *savastanoi* in olive trees. Phytopathology 93:286-292.

Bertolini E, Penyalver R, García A, Olmos A, Quesada JM, Cambra M, López MM (2003b). Highly sensitive detection of *Pseudomonas savastanoi* pv. *savastanoi* in asymptomatic olive plants by nested PCR in a single closed tube. J. Microbiol. Methods 52:261-266.

Birch PRJ, Hyman LJ, Taylor R, Opio AF, Bragardm C, Toth IK (1997). The RAPD PCR-based differentiation of *Xanthomonas campestris* pv. *phaseoli* and *Xanthomonas campestris* pv. *phaseoli* var. fuscans. Eur. J. Plant Pathol. 103:809-814.

Bui TNL, Vernière C, Jouen E, Ah-You, N, Lefeuvre P, Chiroleu F, Gagnevin L, Pruvost O (2010). Amplified fragment length polymorphism and multilocus sequence analysis-based genotypic relatedness among pathogenic variants of *Xanthomonas citri* pv. *citri* and *Xanthomonas campestris* pv. *bilvae*. Int. J. Syst. Evol. Microbiol. 60:515-525.

Burton JE, Oshota OJ, Silman NJ (2006). Differential identification of *Bacillus anthracis* from environmental *Bacillus* species using microarray analysis. J. Appl. Microbiol. 101:754.

Call DR, Chandler DP, Brockman FJ (2001). Fabrication of DNA microarrays using unmodified oligomer probes. Biotechniques 30:368-379.

Chen YF, He LY, Xu J (2003). RAPD analysis and group division of *Ralstonia solanacearum* strains in China. Acta Phytopathol Sin 33:503-508.

Clerc A, Manceau C, Nesme X (1998). Comparison of randomly amplified polymorphic DNA with amplified fragment length polymorphism to assess genetic diversity and genetic relatedness within genospecies III of *Pseudomonas syringae*. Appl. Environ. Microbiol. 64:1180-1187.

Cockerill FR, Smith TF (2002). Rapid-cycle real-time PCR: a revolution for clinical microbiology. Am. Soc. Microbiol. News 68:77-83.

Cubero J, Graham JH (2002). Genetic relationship among worldwide strains of *Xanthomonas* causing canker in citrus species and design of new primers for their identification by PCR. Appl. Environ. Microbiol. 68:1257-1264.

Danks C, Barker I (2000). On-site detection of plant pathogens using lateral flow devices. *Bull. OEPP/EPPO* 30:421-426.

DeBoer SH, Wieczorek A (1984). Production of monoclonal antibodies to *Corynebacterium sepedonicum*. Phtopathology 74:1431-1434.

Dean FB, Hosono S, Fang L, Wu X, Faruqi AF, Bray-Ward P, Sun Z, Zong Q, Du Y, Du J, Driscoll M, Song W, Kingsmore SF, Egholm M, Lasken RS (2002). Comprehensive Human Genome Amplification using multiple displacement amplification. Proc. Natl. Acta. Sci. USA 99:5261-5266.

Didenko VD (2001). DNA probes using fluorescence resonance energy transfer (FRET): designs and applications. Biotechniques 31:1106-1120.

Eisen JA (1995). The RecA protein as a model molecule for molecular systematic studies of bacteria: comparison of trees of RecAs and 16S rRNAs from the same species. J. Mol. Evol. 41:1105-1123.

Espy MJ, Uhl JR, Sloan LM, Buck-walter SP, Jones MF, Vetter EA, Yao JDC, Wengenack NL, Rosenblatt JE, Cockerill, FR, Smith TF (2006). *Real-time* PCR in clinical microbiology application testing. Clin. Microbiol. Rev. 19:165-256.

Gorris, MT, Cambra M, Llop P, Lopez MM, Lecomte P, Chartier R, Paulin JP (1996). A sensitive and specific detection of *Erwinia amylovora* based on the ELISA DASI enrichment method with monoclonal antibodies. Acta Hortic. 411:41-45.

Grover A, Azmi W, Gadewar AV, Pattanayak D, Naik, PS, Shekhawat GS, Chakrabarti SK (2006). Genotypic diversity in a localized population of *Ralstonia solanacearum* as revealed by random amplified polymorphic DNA markers. J. Appl. Microbiol. 101:1364-5072.

Guidot A, Prior P, Schoenfeld J, Carrere S, Genin S, Boucher C (2007). Genomic structure and phylogeny of the plant pathogen *Ralstonia solanacearum* inferred from gene distribution analysis. J. Bacteriol. 189:377-387.

Henson JM, French R (1993). The polymerase chain reaction and plant disease reaction. Annu. Rev. Phytopathol. 31:81-109.

Hinchliffe SJ, Isherwood KE, Stabler RA, Prentice MB, Rakin A, Nichols RA, Oyston PCF, Hinds J, Titball RW, Wren BW (2003). Application of DNA microarrays to study the evolutionary genomics of *Yersinia pestis* and *Yersinia pseudotuberculosis*. Genome Res. 13:2018-2029.

Hocquellet A, Toorawa P, Bové JM, Garnier M (1999). Detection and identification of the two *Candidatus* Liberobacter species associated with citrus huanglongbing by PCR amplification of ribosomal protein genes of the β operon. Mole Cell Probes 13:373-379.

Holland PM, Abramson RD, Watson R, Gelfand DH (1991). Detection of specific polymerase chain reaction product by utilizing the 5'→3' exonuclease activity of *Thermus aquaticus* DNA polymerase. Proc. Natl. Acad. Sci. USA 88:7276-7280.

Janse JD (1988). A detection method for *Pseudomonas solanacearum* in symptomless potato tubers and some data on its sensitivity and specificity. EPPO Bull. 18:343-351.

Jensen MA, Webster JA, Straus N (1993). Rapid identification of bacteria on the basis of polymerase chain reaction-amplified ribosomal DNA spacer polymorphism. Appl. Environ. Microbiol. 59:945-952.

Jones JB, Somodi GC, Scott JW (1997). Increased ELISA sensitivity using a modified extraction buffer for detection of *Xanthomonas campestris* pv. *vesicatoria* in leaf tissue. J Appl Microbiol 83:397-401.

Kerkoud M, Manceau C, Paulin JP (2002). Rapid diagnosis of *Pseudomonas syringae* pv. *papulans*, the causal agent of blister spot of apple, by polymerase chain reaction using specifically designed *hrpL* gene primers. Phytopathology 92:1077-1083.

Kiltie AE, Ryan AJ (1997). SYBR Green I staining of pulsed field agarose gels is a sensitive and inexpensive way of quantitating DNA double-strand breaks in mammalian cells. Nucleic Acids Res. 25:2945-2946.

Koide T, Paulo AZ, Leandro MM, Ricardo ZN, Veˆncio AY, Matsukuma AMD, Diva CT, Hamza ED, Patrícia BM, Ana Claudia RS, Sergio VA, Aline MS, Suely LG (2004). DNA Microarray-Based Genome Comparison of a Pathogenic and a Nonpathogenic Strain of *Xylella fastidiosa* Delineates Genes Important for Bacterial Virulence. J Bacteriol. 186:5442-5449.

Kubota R, Alvarez AM, Vine BG, Jenkins DM (2007). Development of a loop mediated isothermal amplification method (LAMP) for detection of the bacterial wilt pathogen. *Ralstonia solanacearum*. (Abstr) Phytopathology 97:S60.

Lee IM, Bartoszyk IM, Gundersen-Rindal DE, Davis RE (1997a). Phylogeny and classification of bacteria in the genera*Clavibacter* and *Rathayibacter* on the basis of 16S rRNA sequence analysis. Appl. Environ. Microbiol. 63:2631-2636.

Lee IM, Bartoszyk IM, Gundersen DE, Mogen B, Davis RE (1997b). Nested PCR for ultrasensitive detection of the potato ring rot bacteria *Clavibacter michiganensis* subsp. *sepedonicus*. Appl. Environ. Microbiol. 63:2625-2630.

Leite RP, Minsavage GV, Bonas U, Stall REM (1994). Detection and identification of pathogenic *Xanthomonas* strains by amplification of DNA sequences of *hrp* genes of *Xanthomonas compestris* pv. *vesicatoria*. Appl. Environ. Microbiol. 60:1068-1077.

Llop P, Bonaterra A, Peñalver J, López MM (2000). Development of a highly sensitive nested-PCR procedure using a single closed tube for detection of *Erwinia amylovora* in asymptomatic plant material. Appl. Environ. Microbiol. 66:2071-2078.

López MM, Bertolini E, Olmos A, Caruso P, Gorris MT, Llop P, Penyalver R, Cambra M (2003). Innovative tools for detection of plant pathogenic viruses and bacteria. Int. Microbiol. 6:233-243.

Louws FJ, Bell J, Medina-Mora CM, Smart CD, Opgenorth D, Ishimaru CA, Hausbeck MK, de Bruijn FJ, Fulbright DW (1998). Rep-PCR–mediated genomic fingerprinting: A rapid and effective method to identify *Clavibacter michiganensis*. Phytopathology 88:862-868.

Louws FJ, Fulbright DW, Stephens CT, de Bruijn FJ (1994). Specific genomic fingerprints of phytopathogenic *Xanthomonas* and *Pseudomonas* pathovars and strains generated with repetitive sequences and PCR. Appl Environ Microbiol 60:2286-2295.

Loy A, Lehner A, Lee N, Adamczyk J, Meier H, Ernst J, Schleifer KH, Wagner M (2002). Oligonucleotide microarray for 16S rRNA gene-based detection of all recognized lineages of sulfate-reducing prokaryotes in the environment. Appl. Environ. Microbiol. 68:5064-5081.

Mackay LM, Arden KE, Nitsche A (2002). Survey and summary of real time PCR in virology. Nucleic Acids Res. 30:1292-1305.

Maes M, Garbeva P, Crepel C (1996a). Identification andsensitive endophytic detection of the fire blight pathogen *Erwinia amylovora* with 23S ribosomal DNA sequences and the polymerase chain reaction. Plant Pathol. 45:1139-1149.

Maes M, Garbeva P, Kamoen O (1996b). Recognition and detection in seed of the *Xanthomonas* pathogens that cause cereal leaf streak using rDNA spacer sequences and polymerase chain reaction. Phytopathology 86:63-69.

Mahuku GS, Jara C, Henriquez MA, Castellanos G, Cuasquer J (2006). Genotypic characterization of the common bean bacterial blight pathogens, *Xanthomonas axonopodis* pv. *phaseoli* and *Xanthomonas axonopodis* pv. *phaseoli* var. *fuscans* by rep-PCR and PCR-RFLP of the ribosomal genes. J. Phytopathol. 154:1-35.

Martin B, Humbert O, Camara M, Guenzi E, Walker J, Mitchell T, Andrew P, Rudhomme M, Alloing G, Hakenbeck R, Morrison DA, Boulnois GJ, Claverys JP (1992). A highly conserved repeated DNA element located in the chromosome of *Streptococcus pneumoniae*. Nucleic Acids Res. 20:3479-3483.

Manceau C, Horvais A (1997). Assessment of genetic diversity among strains of *Pseudomonas syringae* by PCR-restriction fragment length polymorphism analysis of rRNA operons with special emphasis on *P. syringae* pv. tomato. Appl. Environ. Micriobiol. 63:498-505.

Mondal KK (2009). RAPD-based polymorphism specific to fuscous blight bacterium in grain legume. Indian Phytopathol. 62:72-74.

Mondal KK (2011). Plant Bacteriology, Kalyani Publishers, New Delhi, (ISBN: 978-93-272-1631-8) pp. 190.

Mondal KK, Bhattacharya RC, Kaundal KR (2004). Biotechnological strategies in the detection, characterization and management of fungal diseases in plant. Botanica 54:1-20.

Mondal KK, Rajendran TP, Phaneendra C, Mani C, Joystna S, Verma G, Kumar R, Kumar A, Singh D, Saxena AK, Pooja, S, Richa, JRK (2012). The reliable and rapid PCR diagnosis for *Xanthomonas axonopodis* pv. *punicae* in pomegranate. Afr. J. Microbiol. Res. 6(30): 5950-5956.

Mondal KK, Mani C (2009). ERIC-PCR generated genomic fingerprints and their correlation with pathogenic variability of *Xanthomonas campestris* pv. *punicae*, the incitant of bacterial blight of pomegranate. Curr. Microbiol. 59:616-620.

Mondal KK, Singh D, Prasad R, Mani C (2008). Differential characterization of three bacterial blights in legumes based on symptomatological, cultural, pigmentation and molecular profiles. In Proc Natl Sym Pl Disease Scenario on Organic Agriculture for Eco-friendly Sustainability, Indian Phytopathological Society, 10-12th Jan '08, Mahabaleshwar, Satara, MS. 69 pp.

Mullis K (1987). Process of amplifying nucleic acid sequences. United states Patent No. 4: 683-202.

Opio AE, Allen DJ, Teri JM (1996). Pathogen variation in *Xanthomonas campestris* pv. *phaseoli*, the causal agent of common bacterial blight in *Phaseolus* beans. Plant Pathol. 45:1126-1133.

Palacio-Bielsa, Ana, Cubero J, Cambra MA, Collados R, Berruete IM, López María M (2011). Development of an efficient real-time quantitative PCR protocol for detection of *Xanthomonas arboricola* pv. *pruni* in Prunus species. Appl. Environ. Microbiol. 7:89-97.

Rafalski A, Tingey S, Williams JGK (1994). Random amplified polymorphic DNA (RAPD) markers. In Plant Molecular Biology (Eds. Gelvin, S.B. and Schilperoot, R.A.), H4:1-8, Kluwer Belgium.

Salama N, Guillemin K, McDaniel TK, Sherlock G, Tompkins L, Falkow S (2000). A whole-genome microarray reveals genetic diversity among *Helicobacter pylori* strains. Proc Natl Acad Sci USA 97:14668-14673.

Shanmugam V, Kumar A, Sarma YR (2004). Survival of *Ralstonia solanacearum* in ginger rhizomes stored at different temperatures. Plant Dis. Res. 19:40-43.

Schaad NW, Frederick R (2002). Real-time PCR and its application for rapid plant disease diagnostics. Can. J. Plant Pathol. 24:250-258.

Schaad NW, Gaush P, Postnikova E, Frederick R (2001). On-site one hour PCR diagnosis of bacterial diseases Phytopathology 91:S79-S89 (Abstract).

Schaad NW, Cheong SS, Tamaki S, Hatziloukas E, Panopoulos NJ (1995). A combined biological and enzymatic amplification (BIO-PCR) technique to detect *Pseudomonas syringae* pv. *phaseolicola* in bean seed extracts. Phytopathology 85:243-248.

Song WY, Kim HM, Hwang CY, Schaad NW (2004). Detection of *Acidovorax avenae* ssp. *avenae* in rice seeds using BIO-PCR. J. Phytopathol. 152:667-676.

Takle WG, Toth IK, Brurberg MB (2007). Evaluation of reference genes for real-time RT-PCR expression studies in the plant pathogen Pectobacterium atrosepticum BMC Plant Biology 7:50 doi:10.1186/1471-2229-7-50.

Temple TN, Stockwell VO, Johnson KB (2008). Development of a rapid detection method for *Erwinia Amylovora* by loop-mediated isothermal amplification (Lamp). Acta Horticulturae (ISHS) 793:497-503.

Vitzthum F, Bernhagen J (2002). SYBR Green I: an ultrasensitive fluorescent dye for double-standed DNA quantification in solution and other applications. Recent Res. Dev. Anal. Biochem. 2:65-93.

Waleron M. Waleron K, Podhajska AJ, Łojkowska E (2002). Genotyping of bacteria belonging to the former *Erwinia* genus by PCR-RFLP analysis of a *recA* gene fragment. Microbiology 148:583-595.

Weingrat H, Volksch B (1997). Genetic fingerprinting of *Pseudomonas syringae* pv using ERIC, REP and 1S50 PCR. J. Phytopathol. 145:339-345.

Weller SA, Elphinstone JG, Smith NC, Boonham N, Stead DE (2000). Detection of Ralstonia solanacearum strains with a quantitative, multiplex, real-time, fluorogenic PCR (TaqMan) assay. Appl. Environ. Microbiol. 66:2853-2858.

Williams JGK, Kubelik AR, Livak KJ, Rafalski JA, Tingey SV (1990). DNA polymorphisms amplified by arbitrary primers are useful as genetic markers. Nucleic Acids Res. 18:6531-6535.

Willse A, Straub TM, Wunschel SC, Small JA, Call DR, Daly DS, Chandler DP (2002). Quantitative oligonucleotide microarray fingerprinting of *Salmonella enterica* isolates. Appl. Environ. Microbiol. 68:6361-6370.

Wolfgang MC, Kulasekara BR, Liang XY, Boyd D, Wu K, Yang Q, Miyada CG, Lory S (2003). Conservation of genome content and virulence determinants among clinical and environmental isolates of *Pseudomonas aeruginosa*. Proc. Natl. Acad. Sci. USA 100:8484-8489.

Young JM, Park DC (2007). Relationships of plant pathogenic enterobacteria based on partial atpD, carA, and recA as individual and concatenated nucleotide and peptide sequences. Syst Appl. Microbiol. 30:343-354.

Young JM, Park DS, Shearman HM, Fargier E (2008). A multilocus sequence analysis of the genus *Xanthomonas*. Syst. Appl. Microbiol. 31:366-377.

Yushan H, Lei L, Weijia L, Xiaoguang C (2010). Sequence analysis of the *groEL* gene and its potential application in identification of pathogenic bacteria. Afr. J. Microbiol. Res. 4:733-1741.

7

Uses of mushrooms in bioremediation: A review

Adenipekun C. O.* and Lawal R.

Department of Botany, University of Ibadan, Ibadan, Nigeria.

One of the major environmental problems facing the world today is the contamination of soil, water and air by toxic chemicals as a result of industrialization and extensive use of pesticides in agriculture. Incineration is currently the most effective and common remediation practice but is costly in terms of money and energy used. A rapid cost effective and ecologically responsible method of clean-up is "bioremediation" which utilizes micro-organisms to degrade toxic pollutants in an efficient economical approach. Toxic chemicals are degraded to less harmful forms. Although, bioremediation by bacteria agents has received attention of workers, the role of fungi has been inadequately studied. The ability of fungi to transform a wide variety of hazardous chemicals has aroused interest in using them for bioremediation. Mushroom forming fungi (mostly basidiomycetes), are amongst nature's most powerful decomposers, secreting strong extra cellular enzymes due to their aggressive growth and biomass production. These enzymes include lignin peroxidases (LiP), manganese peroxidase (MnP) and laccase, etc. Thus, carbon sources such as sawdust, straw and corn cob can be used to enhance degradation rates by these organisms at polluted sites. White rot fungi have been used for biotransformation of pesticides, degradation of petroleum hydrocarbons and lignocellulolytic wastes in the pulp and paper industry. *Phanerochaete chrysosporium*, *Agaricus bisporus*, *Trametes versicolor* and *Pleurotus ostreatus* amongst many mushrooms have been reported in the decontamination of polluted sites. In Nigeria, *Lentinus squarrosulus*, *Pleurotus tuber-regium*, *P. ostreatus* and *P. pulmonarius* have been employed in bioremediation of contaminated soils both *in-situ* and *ex-situ*. This paper highlights the use of fungal mycelia in bioremediation (myco-remediation) and studies on the uses of mushrooms for bioremediation.

Key words: Bioremediation, mushrooms, polluted soils.

INTRODUCTION

The pollution of the environment with synthetic organic compounds has become a major problem world wide. Many of these novel compounds introduced to the nature are called xenobiotics being materials that do not occur naturally in the biosphere and many of which are not easily degraded by the indigenous microflora and fauna (Sullia, 2004).

There are several classes of chemicals that have been targeted by United States Environmental Agency (USEPA) as priority pollutants due to their toxic effects on the environment and human health. These chemicals include polycyclic aromatic hydrocarbons, pentachloro-phenols, polychlorinated biphenyls, 1,1,1- trichloro – 2,2-

bis (4-chlorophenyl) ethane, benzene, toluene, ethyl-benzene xylene and trinitrotoluene. Polycyclic aromatic hydrocarbons (PAH) are recalcitrant environmental contaminants that are generated from the burning of fossil fuels, coal mining, oil drilling and wood burning (Lau et al., 2003; Verdin et al., 2004).

Currently, incineration is the most effective and common remediation practice, but this is extremely costly, in terms of money and energy used. All of these chemical compounds pose a significant threat to the health and vitality of the earth system (Hamman, 2004). The elimination of wide ranges of pollutants and wastes from the environment is therefore an absolute require-ment to promote a sustainable development of our society with low environmental impact. Due to the magnitude of this problem and the lack of a reasonable solution, a rapid cost-effective ecologically responsible method of clean-up is greatly needed (Hamman, 2004).

*Corresponding author. E-mail: oyinpek@yahoo.com.

According to Atlas and Bartha (1992), bioremediation is the onsite enhancement of live soil organisms such as fungi, bacteria and green plants to breakdown hydrocarbon and organic contaminants. It involves the application of organisms and nutrients to contaminated soil to enhance biodegradation. Micro-organisms used in bioremediation must have been tested and proven to be successful in laboratory studies. Bioremediation may be employed to attack specific soil contaminants such as degradation of chlorinated hydrocarbons by bacteria. A mere general approach is the cleanup of oil spills by the addition of nitrate and phosphate fertilizers to facilitate decomposition of crude oil by indigenous or exogenous bacteria.

Loske et al. (1990) reported that the main contaminants in polluted soils are:

1. Polycyclic Aromatic Hydrocarbons (PAH's) viz. residues from the processing of oil, tar, coal and comparable substances.
2. Polychlorinated Biphenyls (PCB's) used as cooling agents in transformers.
3. Dioxines: These are by-products of chemical manufacturing and are found in fly-ashes from combustion processes.

The general approaches to bioremediation are to enhance natural biodegradation by natural organisms (intrinsic bioaugmentation). Unlike conventional technologies, bioremediation can be carried out *in-situ* and *ex-situ*. *In-situ* bioremediation involves treating contaminated materials at the site while *ex-situ* involves the removal of the contaminated material to be treated elsewhere (http//en.wikipedia.org.wiki/bioremediation).

The literature concerning organic chemical wastes (xenobiotics) dealt almost exclusively with bacteria. Whenever bioremediation is discussed, bacteria agents come into focus and fungi are much less studied. It is now becoming apparent that fungi also play an important role in degrading organic materials in the ecosystem and that they have potential for remediating contaminated soils and water.

Alexander (1994) reported that the ability of fungi to transform a wide variety of hazardous chemicals has aroused interest in using them in bioremediation. Fungi are among nature's most powerful decomposers, secreting strong enzymes. The great potential of fungi in bioremediation is by virtue of their aggressive growth, great biomass production and extensive hyphae reach in the environment (Ashoka et al., 2002).

Fungi require substrates such as cellulose or other carbon source as a source of energy. Thus, carbon sources such as corn cobs, straw and sawdust can be easily used to enhance degradation rates by these organisms at polluted sites. Also, the branching filamentous mode of fungal growth allows for more efficient colonization and exploration of contaminated soil (Hamman, 2004). Fungi are unique among microorganisms in that they secrete a variety of extracellular enzymes involved in pollutant degradation. They use a variety of mechanisms to accomplish the degradation of lignin and a wide variety of other environmental pollutants (Asamudo et al., 2005).

The easiest method of treating contaminated soil is to simply add organic matter to as many toxic metals will readily form compounds with the organic materials found in compost (Kellogg and Pettigrew, 2005). The concept of composting originally applicable to waste conversion into mulching and soil conditioner is now being applied to the hazardous waste treatment.

There is a lot of descriptive bioremediation concerning the ability of various fungi and their enzymes to biotransform pesticides (Raj et al., 1992). To date, the most sophisticated fungal approach to environmental clean-up has grown out of prior research on degradation of petroleum hydrocarbons (Atlas and Bartha, 1992; Cerniglia et al., 1992).

MUSHROOMS AND BIOREMEDIATION

White-rot fungi are so called because their degradation processes result in a bleaching of wood substrates (Kirk et al., 1992). They digest lignin in wood by the secretion of enzymes giving wood a bleached appearance.

The white-rot fungi technology is quite different from other well-established methods of bioremediation (for example, bacteria systems). The differences are primarily due to the unusual mechanisms which nature has provided them with several advantages for pollutant degradation (Asamudo et al., 2005).

One distinct advantage these fungi have over bacterial systems is that they do not require preconditioning to the particular pollutant. Bacteria usually must be pre-exposed to a pollutant to allow the enzymes that degrade the pollutant to be induced. The pollutant also must be in a significant concentration, otherwise induction of enzyme synthesis cannot occur. Thus, there is a finite level to which bacteria can degrade pollutants (Asamudo et al., 2005). Various strains of white-rot fungi capable of degrading aromatic compounds were reported by Barr and Aust (1994).

Lang et al. (1995) reported that lignin decomposing white-rot fungi show extraordinary abilities to transform recalcitrant pollutants like polycyclic aromatic hydrocarbons (PAH's). They added that this unique capability may be used for decontamination of oil-polluted soils although lignocellulosic substrates must be supplied for the survival of fungal species in the soil. White-rot fungi have been used in bioremediation of polluted soils and accumulation of heavy metals. They have also been found to be involved in mineralization, bio-deterioration, biodegradation, transformation and co-metabolism (Bennet et al., 2002).

Isikhuemhen et al. (2003) reported that white-rot fungi are increasingly being investigated and used in bioremediation because of their ability to degrade an extremely diverse range of very persistent or toxic environmental pollutants.

WHITE-ROT FUNGI DEGRADATION SYSTEM

The main mechanism of biodegradation employed by this group of fungi is the lignin degradation system of enzymes. Extra-cellular lignin modifying enzymes (LME's) have very low substrates-specificity so they are able to mineralize a wide range of highly recalcitrant organopollutants that is structurally similar to lignin (Mansur et al., 2003; Pointing, 2001). The major components of lignin degradation system include lignin-peroxidase (LiP), manganese peroxidase (MnP), H_2O_2 producing enzymes (Kirk and Farrell, 1987) and laccase, although not all lignolytic fungi show the three types of enzymatic activity.

It has been demonstrated that a lot of species belonging to the group of white-rot fungi are able to degrade lignin, which is a naturally occurring polymer (Hattaka, 1994). This capacity is assumed to result from the activities of extracellular oxidases and laccases (Glenn and Gold, 1983). These enzymes are non-specific; they oxidize a wide range of xenobiotics (Barr and Aust, 1994; Martens et al., 1996). White-rot fungi have been proposed for the biodegradation of polluted sites containing complex mixtures of PAH's such as occurring in creosote, coal tar and crude oil (Loske et al., 1990). Isikhuemhen et al. (2011) carried out a solid state fermentation (SSF) experiment with *Lentinus squarrosulus* (strain MBFBL 201) on cornstalks and evaluated lignocellulolytic enzymes activity. The results showed that *L. squarrosulus* was able to degrade cornstalks significantly after 30 days. Maximum lignocellulolytic enzyme activities were obtained on day 6 of cultivation and are a good producer of exopolysaccharides. The very fast good rate of *L. squarrosulus* makes it an ideal candidate for application in industrial pretreatment and biodelignification of lignocellulosic biomass.

POTENTIALS OF MUSHROOMS IN BIOREMEDIATION

Phanerochaete chrysosporium

Among the fungal systems, *Phanerochaete chrysosporium* is emerging as the model system for bioremediation. *P. chrysosporium has* been known *to* degrade besides lignin macro molecules, many types of organopollutants such as polycyclic aromatic hydro-carbons (PAH's), polychlorinated biphenyls and dioxines, chlorophenols, chlorolignins, nitrocranditics, synthetic dyes and different pesticides. Several powerful degraders for example, *Phanerochaete sordida* (P. Karst. Y). Erikss,

Pleurotus ostreatus (Jacq. Fr.) P. kumm, *Trametes versicolor* (L. Fr.) Lloyd, *Nematolana frowardii* (Speg.) e. Herak, and *Irpex lacteus* (Fr.) Fr. were selected for the study (Sasek and Cajthaml, 2005).

P. chrysosporium has been shown to affect the bioleaching of organic dyes (Nigam et al., 1995). The first extracellular enzyme (ligninase) discovered to deploy-merize lignin, and lignin sub-structured compounds *in vitro* were produced by this organism (Aitken and Irvine, 1989). *P. chrysosporum* has degraded toxic xenobiotics such as aromatic hydrocarbons, chlorination organics, insecticides, pesticides, nitrogen aromatics and laccases; polyphenol oxidases and lignin peroxidases being involved in the degradation process (Barr and Aust, 1994).

Phanerochaete flavido-alba

Phanerochaete flavido-alba has been able to decolorize Olive oil mill wastewater (OMW), a major waste product of olive oil extraction for subsequent use in bioreme-diation assays. Of several media tested, nitrogen-limited *P. flavido-alba* cultures containing 40 mg/L Mn (II) were the most efficient at decolorizing OMW. Decolorization was accompanied by a 90% decrease in the OMW phenolic content. Concentrated extracellular fluids alone (showing manganese peroxidase, but not lignin peroxidase activity) did not decolorize the major OMW pigment, suggesting that mycelium binding forms part of the decolorization process.

Trametes versicolor

Trametes versicolor produced three lignolytic enzymes with efficient degradation capacity on lignin, polycyclic aromatic hydrocarbons, polychlorinated biphenyl mixture and a number of synthetic dyes (Tanaka et al., 1999; Novotny et al., 2004). *T. versicolor* and its enzymes have been reported to delignify and to bleach kraft pulp (Gamelas et al., 2005) and also efficiently dechlorinate and decolorize bleach kraft pulp effluents (Selvam et al., 2002). This presents a good potential to be the base of new environmental friendly technologies for pulp and paper industry. Amaral et al. (2004) also reported the use of *T. versicolor* as biocatalysts for decolorization of different industrial dyes and waste water treatments.

Pleurotus ostreatus (Jacq. Fr.) P. Kumm

Recent studies have shown that *P. ostreatus* is able to degrade a variety of polycyclic aromatic hydrocarbons (PAH) (Sack and Gunthen, 1993). It has the ability to degrade PAH in non-sterile soil both in the presence and in the absence of cadmium and mercury. It has been reported to catalyze humification of anthracene, benzo (a) pyrene and flora in two PAH – contaminated soils from a manufactured gas facility and an abandoned electric cooping plant (Bojan et al., 1999).

Pleurotus tuber-regium (Fries) Singer

The white-rot fungus, *P. tuber-regium* is another fungus examined for its ability to ameliorate crude oil polluted soil. Isikhuemhen et al. (2003) reported that the fungus had the ability to ameliorate crude oil polluted soil and the resulting soil sample supported seed germination and seedling growth of *Vigna unguiculata*. They reported a significant improvement in percentage germination, plant height and root elongation.

In another investigation, observed an increase in nutrient contents (organic matter, carbon, available potassium) in soil polluted with 1 to 40% engine oil concentration after six months of incubation with *P. tuber-regium*. The fungus also brought about an increase in copper content in engine oil polluted soils at 10% concentration followed by a decrease at 20 and 40% concentrations. Bioaccumulation of zinc and nickel was recorded at 20% enzyme oil concentration.

Ogbo et al. (2006) investigated the effect of different levels of spent lubricating oil on the growth of *P. tuber-regium* at different levels (5, 8, 16, 30, 65, 98, 130 and 160%). The fungus grew optimally at 98% level of contamination. Average yield showed that 98% level of contamination gave the highest yield (79.56 g), although inhibition was noticed at 130 and 160%. The shortest and tallest sporophores were recorded in control and 5% level of contamination.

Ogbo and Okhuoya (2009) investigated the effect of crude oil on the yield and chemical composition of *P. tuber-regium* (Fr.) Singer on soils to which sawdust, shredded banana leaves, NPK fertilizer and poultry litter were added. The study showed that crude oil contamination improves the overall well being of the fungus. Substrate composition affects the nutrient status of the mushroom causing variation in the carbohydrate, protein, fat, crude fibre and ash content. Sawdust and poultry litter enhances crude oil fertilizer effect on the fungus while shredded banana leaf blades and NPK fertilizer reduces the fertilizer effect. The quality of the mushroom fruit body from the crude oil contaminated site was comparable to that cultivated under the best substrate composition for the mushroom".

Bioremediation of cutting fluids contaminated soil by *P. tuber-regium* Singer by Adenipekun et al. (2011a) reported that there was an improvement in the nutrient status of the soil and an increase in enzyme activity. A reduction in the pH and heavy metal contents of the soil at the levels of cutting fluids concentrations was detected. The lignin in the rice straw decreased from 34.50% in the control to 8.06 at 30% cutting fluids concentration after 3 months of incubation. The highest TPH loss of 30.84%

was recorded at 20% cutting fluids contamination after 3 months compared to 13.75% at the onset of the experiment. The improvement of the nutrients contents of the soil, bioaccumulation of heavy metals, degradation of TPH, lignin, and increased activity of polyphenol oxidase and peroxidase was due to biodegradation of the cutting fluids.

Lentinus squarrosulus (Mont.) Singer

Lentinus squarrosulus has been found to mineralize soil contaminated with various concentrations of crude oil resulting in increased nutrient contents in treated soil. Adenipekun and Fasidi (2005) reported the ability of *L. squarrosulus* to mineralize soil contaminated with various concentrations of crude oil (1 to 40%). They found that nutrient contents were generally higher after 6 months of incubation except potassium levels which were not increased. The highest rate of biodegradation was at 20% concentration after 3 months and 40% after 6 months of incubation (Adenipekun and Fasidi, 2005).

Adenipekun and Isikhuemhen (2008) investigated the bioremediation of engine oil polluted soil by *L. squarrosulus*. Results indicated that contaminated soils had increased organic matter, carbon, available phosphorus while nitrogen and available potassium reduced. A relative high percentage degradation of total petroleum hydrocarbon (TPH) was observed at 1% engine oil concentration (94.46%) which decreased to 64.05% TPH degradation at 40% engine oil concentration after 90 days of incubation. The concentrations of Fe, Cu, Zn and Ni recovered from the fungal biomass complex increased with increase of engine oil contamination. The improvement of nutrient content values as well as the bioaccumulation of heavy metals at all levels of engine oil concentrations tested through inoculation with *L. squarrosulus* is of importance for the bioremediation of engine polluted soils.

Isikhuemhen et al. (2010) carried out the preliminary studies on Mating and Improved Strain Selection in the Tropical Culinary-Medicinal Mushroom *L. squarrosulus* Mont. (Agaricomycetideae). The rapid mycelia growth and enhanced enzyme production by *L. squarrosulus* have biotechnological applications for wood and pulp, textile, and tanning, as well as in the bioremediation of oil spills. Furthermore, the results showed that intrastock mating resulted in some dikaryons that outperform their parent strains in growth rate and speed to primordial initiation. This reinforces the importance of breeding and selection during the domestication of wild edible and medicinal fungi of interest.

Pleurotus pulmonarius

Adenipekun et al. (2011b) also worked on the management of cement and battery polluted soils using

Pleurotus pulmonarius. A general increase was observed in the carbon content, organic matter, phosphorus and potassium and a decrease in percentage nitrogen, calcium, and pH after 10 weeks of incubation. The lead content was constant in both polluted soils while a significant decrease was observed in the copper, manganese and nickel contents of the soils. The polyaromatic hydrocarbons (PAH) content also decreased from 6.86 in control to 0.56 after 10 weeks of incubation.

OTHER WHITE-ROT FUNGI

The vegetative growth responses of three local edible mushrooms, *P. pulmonarius, L. squarrosulus* and *P. tuber-regium* on different concentrations of crude oil, automotive gasoline oil (AGO), fresh engine oil and spent engine oil were investigated. *P. tuber-regium* grew fastest among the three organisms on all pollutants and radial growth was observed at all concentrations. Almost the same pattern of growth was observed for *P. pulmonarius* and *L. squarrosulus*. Radial growth for both was supported by crude oil and AGO at all concentrations whereas growth on engine oil (fresh and spent) was not observed beyond 10% concentration. The ability of these mushrooms to tolerate the pollutants and grow on them suggests they could be employed as bioremediation agents on sites contaminated by these pollutants (Adedokun and Ataga, 2006).

Olive oil mill wastewater (OMWW) was used as a substrate for the culture of a mixture of edible fungi in order to obtain a potentially useful microbial biomass and to induce a partial bioremediation of this fastidious waste (Laconi et al., 2007). The fungal mixture grew fairly well in the treated OMWW and reached a maximum of biomass production within about 14 days of fermentation at room temperature. Up to 150 to 160 g of wet biomass was obtained per liter of OMWW. Analysis of the partially dehydrated biomass revealed a protein content of about 13 and 6 g% of row fiber. A relevant presence of unsaturated fatty acids was found, as well as the presence of significant amounts of vitamins A and E, nicotinic acid, calcium, potassium and iron in the *Pleurotus* species.

Emuh (2010) reported that the crude oil and heavy metals present in polluted soil are broken down and absorbed by mushroom hypha and mycelia through the secretion of enzymes into environmentally safe levels resulting in carbon (IV) oxide, water and perhaps biomass. Zebulun et al. (2011) worked on the decontamination of anthracene-polluted soil through a white rot fungus (*P. ostreatus*) induced biodegradation. They reported that time, level of contamination and fungal treatment affected the rate of degradation of all levels of anthracene degradation (76 to 89%) compared to control soil (33 to 51%). It was reported that the release of lignolytic enzymes such as lignin peroxidase, laccase and manganese peroxidase by *P. ostreatus* are associated with the degradation of anthracene. Increase in the concentration of anthracene at different sampling dates in some of the soil samples was also observed.

Mycoremediation (fungal treatment or fungal-based technology) is the application of fungi in the remediation of polluted soil and aqueous effluents. It is an aspect of bioremediation which is defined as the biological process that involves the filtration of agricultural/industrial water runoff, decomposition of hydrocarbon based contaminants, concentration/removal of heavy metal from soil and other substrates by a biological organism (Stamets, 1999).

Mycoremediation of crude oil and palm kernel contaminated soils by *P. pulmonarius* was tested by Adenipekun and Lawal (2011). They reported an increase in the nutrient contents of palm kernel sludge contaminated soil. A decrease in the heavy metal contents was observed at all level of crude oil except lead Pb which increased at 5 and 20% crude oil contamination. An increase was observed at all levels of palm kernel sludge contamination except Zn which decreased from 2.97 to 2.75 mg/kg, Lead Pb also decreased at 2.5, 20 and 40%, and Cu at 1 and 40% in palm kernel sludge contamination after 2 months of incubation. The Total Petroleum Hydrocarbon showed a percentage loss of 40.80% at 1% crude oil concentration and 9.28% at 40% crude oil contaminated soil after 2 months. The lignin content of the rice straw reduced from 13.56% in the control to 7.71% and organic matter content decreased from 37.96 to 17.90% after 2 months. The improvement of nutrient content value as well as the bioaccumulation of heavy metals at all levels of crude oil concentrations tested through inoculations with *P. pulmonarius* is of importance for the myco-remediation of crude oil and palm kernel sludge polluted soil.

Okparanma et al. (2011) reported that spent white-rot fungi (*P. ostreatus*) substrate can be used to remediate Nigerian oil-based drill cuttings containing poly-aromatic hydrocarbons and also serve the dual purpose of reducing the bulk volume of the spent fungal substrate as a waste and the incidence of occurrence of PAH's in the environment. After 56 days of composting, the total amount of residual PAH's in the residual cuttings decreased from 19.75 to 7.62%, while the overall degradation of PAH's increased to between 80.25 and 92.38% with increasing substrate addition.

The use of higher fungi like mushrooms has been known in the remediation of polluted soil for some years now. Research has shown that mushroom species like *P. ostreatus* and *P. chrysosoporium* have emerged as model systems for studying xenobiotics degradation. A great deal still remains to be learned about the fundamentals of how this white-rot fungus mineralizes pollutants; not surprisingly, even less is known about the degradative mechanisms used by fungi in general. Though oxidative enzymes play a major role, organic acids and chelators excreted by the mushroom also contribute to the process and these still need to be studied. Also, most of the work on mycoremediation that has been carried out in Nigeria especially is under laboratory conditions. It is recommended that *in situ* applications of white-rot fungus on polluted soil should be carried out especially in the Niger Delta area of Nigeria where the problem of oil spillage is a daily occurrence (Fasidi et al., 2008).

CONCLUSION

The application of white-rot fungi in bioremediation is expected to be relatively economical because the fungi can be grown on a number of inexpensive agricultural or forest wastes such as rice straw, corn cobs and sawdust. The fungal inoculum can also be mass-produced by current simple techniques used to produce fungal spawn.

In the quest, for economical and ecologically sound methods for environmental remediation, the use of mushrooms is a very good approach and solution. More intensive research needs to be carried out on the potentials of bioremediation and ecology of a large number of edible mushrooms effective for bioremediation. The challenges faced in the field application such as contamination by other fungi especially *Penicillium* sp., *Aspergillus* spp. needs to be also looked into and solutions recommended.

REFERENCES

Adedokun OM, Ataga AE (2006). Effect of crude oil and oil product or growth of some edible mushrooms. J. Appl. Sci. Environ. Manag. 10(2):91-93.

Adenipekun CO, Fasidi IO (2005). Bioremediation of oil polluted soil by *Lentinus subnudus*, a Nigerian white rot fungus. Afr. J. Biotechnol. 4(8):796-798.

Adenipekun CO, Isikhuemhen OS (2008). Bioremediation of engine oil polluted soil by the tropical white-rot fungus, Lentinus squarrosulus Mont. (Singer). Pakistan J. Biol. Sci. 11(12):1634-1637.

Adenipekun CO, Lawal Y (2011). Mycoremediation of crude oil and palm kernel contaminated soils by *Pleurotus pulmonarius* Fries (Quelet). Nature. Sci. 9(9):125-131.

Adenipekun CO, Ejoh EO, Ogunjobi AA (2011a). Bioremediation of cutting fluids contaminated soil by *Pleurotus tuber-regium* Singer. Environmentalist 32:11-18.

Adenipekun CO, Ogunjobi AA, Ogunseye OA (2011b). Management of polluted soils by a white-rot fungus, *Pleurotus pulmonarius*. Assumption University of Technol J. 15(1):57-61.

Aitken MB, Irvine RL (1989). Stability testing of ligninase and Mn-peroxidase from *Phanerochaete chrysosporium*. Biotech. Bioengr. 34:1251-1260.

Alexander M (1994). Biodegradation and Bioremediation. 2nd Edn.Academic Press,San Diego.

Amaral PFF, Fernandes LAFD, Tavares APM, Xavier AMR, Cammarota MC, Coutinho JAP, Coelho MAZ (2004). Decolorization of dyes from textile wastewater by *Trametes versicolor*. Environ. Technol. 25(11):1313-1320.

Asamudo NU, Dada AS, Ezeronye OU (2005). Bioremememediation of textile effluent using *Phanerochaete chrysosporium*. Afr. J. Biotechnol. 4(13):1548-1553.

Ashoka G, Geetha MS, Sullia SB (2002). Bioleaching of composite

textile dye effluent using bacterial consortia. Asian J. Microb. Biotechnol. Environ. Sci. 4:65-68.

Atlas RM, Bartha R (1992). Hydrocarbon biodegradation and oil spill bioremediation. Adv. Microbiol. Ecol. 12:287-338.

Barr BP, Aust D (1994). Mechanisms of white-rot fungi use to degrade pollutant. Environ. Sci. Technol. 28:78-87.

Bennet JW, Wunch KG, Faison BD (2002). Use of fungi in biodegradation: of fungi in bioremediation pg. 960 -971 In: Manual of Environmental Microbiology Washington D.C.: ASM Press.

Bojan BW, Lamar RT, Burjus WD, Tien M (1999). Extent of humification of anthrecene, fluoranthene adbenzo (a) pyrene by *Pleurotus ostreatus* during growth in PAH-contaminated soils. Lett. Appl. Microbial. 28:250-254.

Cerniglia CE, Sutherland JB, Crow SA (1992). Fungal metabolism of aromatic hydrocarbons. In: G.Winkelmann (ed.) Microbial Degradation of Natural products. VCH press. Weinheim, Germany pp.193-217.

Emuh FN (2010). Mushroom as a purifier of crude oil polluted soil. Inter. J. Sci. Nat. 1(2):127-132.

Fasidi IO, Kadiri M, Jonathan SG, Adenipekun CO, Kuforiji OO (2008). Cultivation of Edible Tropical Mushrooms by University Press Publishing House, University of Ibadan. ISBN 978-121-438-4. p. 81.

Gamelas JAF, Tavares APM, Evtuguin DV, Xavier MRB (2005). Oxygen bleaching of Kraft pulp with polyoxometaltes and laccase applying a novel multi-stage process. J. Mol. CatallYsis B: Enzymatic 33(3-6):57-64.

Glenn JK, Gold MH (1983). Decolorization of several polymeric dyes by the lignin degrading basidiomycetes *Phanerochaete chrysosporium*. Appl. Environ. Microbiol. 45:1741-1747.

Hamman S (2004). Bioremediation capabilities of white- rot fungi. Biodegradation 52:1-5.

Hattaka A (1994). Lignin–modifying enzymes for selected white rot fungi production and mode in lignin degradation. Microbiol. Rev. 13:125-135.

Isikhuemhen OS, Anoliefo G, Oghale O (2003). Bioremediation of crude oil polluted soil by the white rot fungus, *Pleurotus tuber-regium* (Fr) Sing. Environ. Sci. Pollut. Res. 10:108-112.

Isikhuemhen OS, Mikiashvilii NA, Adenipekun CO, Ohimain EI, Shahbazi G (2011). The tropical white rot fungus, *Lentinus squarrosulus* Mont: lignocellolytic enzymes activities and sugar release from cornstalks under solid state fermentation. World J. Microbiol. Biotechnol. 28(5):1961-1966.

Isikhuemhen OS, Adenipekun CO, Ohimain EI (2010). Preliminary studies on mating and improved selection in the tropical culinary medicinal mushroom *Lentinus squarrosulus* Mont. (Agaricomycetideae) Int. J. Med. Mushrooms 12(2):177-183.

Kellogg S, Pettigrew S (2005). Low- Tech Bioremediation: Microbiology for the myriad. http//yeoileconsciousnessshoppe. com/art177.html.

Kirk TK, Farrell RL (1987). Enzymatic combination – microbial degradation of lignin. Annu. Rev. Microbiol. 41:465-505.

Kirk TK, Lamar RT, Glaser JA (1992). Potential of white-rot fungi in bioremediation. Biotechnol. Environ. Sci. Mol. Appl. pp. 131-138.

Laconi S, Giovanni M, Cabiddu A, Pompei R (2007). Bioremediation of olive oil mill waste water and production of microbial biomass. Biodegradation 18(5):559-566.

Lang E, Eller I, Kleeberg R, Martens R, Zadrazil F (1995). Interaction of white rot fungi and micro-organisms leading to biodegradation of soil pollutants. In: Proceedings of the 5th International FZK/ TNo Conference on contaminated soil. 30th Oct- 5Nov 1995, Maustrient. The Netherlands by Van de Brink WJ, Bosman R and Arend F. contaminated soils 95:1277-1278.

Lau KL, Tsang YY, Chiu SW (2003). Use of spent mushroom compost to bioremediate PAH- contaminated samples. Chemosphere 52:1539-46.

Loske D, Huttermann A, Majerczk A, Zadrazil F, Lorsen H, Waldinger P (1990). Use of white rot fungi for the clean-up of contaminated sites. In: Coughlan MP, Collaco (eds.) Advances in biological treatment of lignocellulosic materials. Elsevier, London pp. 311-321.

Mansur M, Arias ME, Copa-Patino JL, Flardh M, Gonzalez AE (2003). The white–rot fungus *Pleurotus ostreatus* secretes laccase isozymes with different substrate specificites. Mycologia 95(6):1013-20.

Martens R, Wetzstein, Zadrazil F, Capelari M, Hoffmann P, Schnieer

N (1996). Degradation of the fluoroquinolone enrofloxacin by wood-rotting fungi. Appl. Environ. Microbiol. 62:4206-4209.

Nigam P, Banat IM, McMullan G, Dalel S, Marchant R (1995). Microbial degradation of textile effluent containing Azo, Diazo and reactive dyes by aerobic and anaerobic bacterial and fungal cultures. 36th Annu. Conf. AMI, Hisar pp.37-38.

Novotny C, Svobodova K, Erbanova P, Cajthaml T, Kasinath A, Lange E, Sasek V (2004). Ligninolytic fungi in bioremediation: extacellular enzyme production and degradation rate. Soil Biol. Biochem. 36(10):1545-1551.

Ogbo ME, Okhuoya JA, Anaziah OC (2006). Effect of different levels of spent lubricating oil on the growth of Pleurotus tuber-regium (Fries) Singer. Nig. J. Bot. 19(2):266-270.

Ogbo EM, Okhuoya JA (2009). Effect of crude oil contamination on the yield and chemical composition of Pleurotus tuber-regium (Fr.) Singer. Afr. J. Fd. Sci. 3(11):323-327.

Okparanma RN, Ayotamuno JM, Davies DD, Allagoa M (2011). Mycoremediation of polycyclic aromatic hydrocarbons (PAH)-contaminated oil-based drill cuttings. Afr. J. Biotechnol. 10(26):5149-5156.

Pointing SB (2001). Feasibility of Bioremediation by White-rot fungi. Appl. Microbiol. Biotechnol. 51:20-33.

Raj HG, Saxena M, Allameh A, Mukerji KG (1992). Metabolism of foreign compounds by fungi: In: Arora KK, Elander RP and Kukerji KG (ed.), Handbook of Applied Mycology, 4:881-904. Fungal Biotechnology Marcel Dekker, Inc; New York, N.K.

Sack U, Gunther T (1993). Metabolism of PAH by fungi and correction with extracellular enzymatic activities. J. Basic Microbiol. 33:269-277.

Sasek V, Cajthaml T (2005). Mycoremediation. Current state and perspectives. Int. J. Med. Mushrooms 7(3):360-361.

Selvam K, Swaminathan K, Song MH, Chae KS (2002). Biological treatment of a pulp and paper industry effluent by Fomes lividus and Trametes versicolor. World J. Microbiol. Biotechnol. 18(6):523-526.

Sullia SB (2004). Environmental Applications of Biotechnology. Asian J. Microbiol. Biotechnol. Environ. Sci. 4:65-68.

Stamets P (1999). Helping the Ecosystem through Mushroom cultivation. Fungi Perfecti. Mushrooms and the Ecosystem Ecosystem. http://www.fungi.com/mycotech/mycov.html.

Tanaka H, Itakura S, Enoki A (1999). Hydroxyl radical generation by an extracellular low- molecular–weight substance and phenol oxidase activities during wood degradation by the white–rot basidiomycetes Trametes versicolor. J. Biotechnol. 75(1):57-70.

Verdin AA, Sahraoui L-HR, Durand R (2004). Degradation of benzo(a)pyrene by mitosporic fungi and extracellular oxidative enzymes. Int. Biodeterior. Biodegr. 53:65-70.

Zebulun OH, Isikhuemhen OS, Iyang H (2011). Decontamination of anthracene-polluted soil through white rot fungus-induced biodegradation. Environmentalist 31:11-19.

8

Therapeutic potential of antimicrobial peptides from insects

Rodney Hull, Rodrick Katete and Monde Ntwasa[*]

School of Molecular and Cell Biology, University of the Witwatersrand, Wits, 2050, South Africa.

The first antimicrobial peptides were isolated from the cecropia moth *Hyalophora cecropia* in 1980. Since then a plethora of antimicrobial peptides have been isolated from other arthropods, invertebrates and chordates. With the emergence of antibiotic resistant bacterial pathogens and the promising activity of these peptides, attempts are being made to use these peptides as new antimicrobial agents. Other researchers are interested in using these peptides to improve the resistance of crops and livestock to infections, while another line of research is interested in using these peptides to control vector borne diseases. Despite the promising antibacterial, antiviral, anti-protozoan and anti-tumor activity of these peptides, relatively few peptides have made it to clinical trials. Problems associated with the development of these peptides into effective antimicrobial agents include their higher cost, proteolysis or decreased activity in physiological environments and mass production. This review will focus specifically on the development of insect antimicrobial peptides into useful chemotherapeutic agents.

Key words: Insect, antimicrobial peptide, drug discovery.

INTRODUCTION

Interest in antimicrobial peptides began in 1980 when cecropin was isolated from the cecropia moth *Hyalophora cecropia*. This was 50 years after the initial observations in the 1920s that insects released a bacteriolytic substance into their blood when challenged with bacteria. Since the isolation of cecropin there has been interest in the use of these antimicrobial peptides in therapeutic, biocontrol and agricultural applications. In total there are about 559 antimicrobial peptides identified to date, isolated from plants, vertebrates and invertebrates. Of these peptides the majority are bactericidal, a significant amount are antifungal while only a small number are anti-viral or act against cancer cells (Wang and Wang, 2004).

Most antimicrobial peptides tend to be highly basic, which facilitates their interaction with the microbial cell membrane (Lauth et al., 1998). All of these antimicrobial peptides are synthesised as precursor peptides that can

based on their amino acid sequence and structural characteristics, as follows:

Antimicrobial peptides can be divided into classes be up to five times larger than the active peptide (Trenczek, 1997).

1) The linear amphipathic α helix forming peptides (Bulet et al., 2004) (Table 1, Figure 1A).
2) The cystine rich or cyclic antimicrobial peptides (Table 2, Figure 1 B-F).
3) The lysozymes.
4) The proline rich peptides (Otvos, 2002) (Table 3).
5) The glycine rich peptides (Rees et al., 1997) (Table 3).

Classes 4 and 5 apply specifically to insect antimicrobial peptides and may be combined into a class of peptides rich in one or more amino acids.

This review will focus on the development of insect antimicrobial peptides into useful therapeutic products and their application to agriculture and disease prevention. This will be followed by a description of the

*Corresponding author. E-mail monde@biology.wits.ac.za.

Table 1. Amphipathic linear antimicrobial peptides (AMPs) from different organisms. Linear amphipathic antimicrobial peptides isolated from plants and animals. The table includes mainly arthropod antimicrobial peptides, and some vertebrate and plant peptides.

Sources	Name	Size (amino acids)	Structure	Activity	Commercialised or not	References
Arthropods						
Lepidoptera	Cecropin	31-39	α helix hinge α helix	Gram negative, Gram positive	No	Steiner et al. (1981)
Diptera	Cecropin	31-39	α helix hinge α helix	Gram negative, Gram positive	No	Okada and Natori (1983)
Dipteran (mosquito)	Cecropin	31-39	α helix hinge α helix	Gram negative, Gram positive, Protozoa	No	Lowenberger et al. (1999)
Synthetic (cecropin analogue)	Shiva-3	35	α helix hinge α helix	Gram negative, Gram positive, Protozoa	No	Jaynes et al. (1988)
Synthetic cecropin analogue	D2A21	23	α helix hinge α helix	Bacteria, Virus, Tumor	Yes	Chalekson et al. (2003)
Oxyopes kitabensis	Oxypinin	48	α helix	Bacteria, Hemolytic, Fungi	No	Corzo et al. (2002)
Parabuthus schlechteri	Parabutoporin	40-50	α helix	Bacteria, Hemolytic, Fungi	No	Moerman et al. (2002)
Opistophthalmus carinatus	Opistoporin	40-50	α helix	Bacteria, Hemolytic, Fungi	No	Moerman et al. (2002)
Lycosa carolinensis	Lycotoxins	25-27	α helix	Gram negative bacteria, Yeast	No	Yan and Adams (1998)
Cupiennius salei	Cupiennin	35	α helix hinge, α helix	Bacteria	No	Kuhn-Nentwig et al. (2002) and Pukala et al. (2007)
Pachycondyla goeldii	Ponericin G	30	α helix	Bacteria, Insecticidal, Hemolytic	No	Bulet et al. (2004) and Orivel et al. (2001)
Stomoxys calcitrans	Stomoxyn	42	α hinge α hinge α	Bacteria, Fungi, Yeast, Protozoa	No	Boulanger et al. (2002b)
Psuedocanthotermes spiniger	Spinigerin	25	α helix hinge α helix	Gram negative, Gram positive, Fungi	No	Lamberty et al. (2001a)
Bombyx mori	Moricin	42	α helix	Gram positive, Gram negative	No	Hara and Yamakawa (1995)
D. melanogaster	Andropin	34	α helix hinge α helix	Gram positive	No	Samakovlis et al. (1991)
Ceratitis capitata	Cerratotoxin	29	α helix	Gram positive, Gram negative, Hemolytic	No	Marchini et al. (1993)
Apis mellifera	Mellitin	26	αα	Gram positive, Gram negative, Hemolytic	No	Anderson et al. (1980)
Synthetic (mellitin analogue)	Hecate	23	α helix hinge α helix	Bacteria, Virus, Tumor	No	Arrowood et al. (1991)
Agelaia pallipes pallipes	Protonectin	12	α helix	Bacteria	No	Mendes et al. (2004)
Vertebrates						
Sus scrofa	Cecropin P1	31	α helix hinge α helix	Bacteria	No	Lee et al. (1989)
Styela clava	Styelin	32	α helix	Bacteria, Heolytic	No	Zhao et al. (1997)
	Clavanin	23	α helix histidine rich	Bacteria	No	Lee et al. (1997)
Xenopus laevis	Magainin-2	23	α helix	Bacteria, Fungi	No	Zasloff (1987)
Sythetic Magainin	Pexiganan	22-	Magainin analogue	Bacteria	Passed Phase III trial	Jacob and Zasloff (1994)
Rana temporaria	Temporin	10-13	α helix	Bacteria, Fungi, Hemolytic	No	Simmaco et al. (1996)
Litoria species	Caerin	25	αα	Bacteria	No	Steinborner et al. (1997)
Rana rugosa	Gaegurin	37	Helix kink helix	Bacteria	No	Park et al. (1995)
Homo sapiens	Ovispirin-1	18	α helix	Bacteria, Fungi, Hemolytic	No	Sawai et al. (2002)
Synthetic ovispirin	Novispirin G-10		α helix	Bacteria, Fungi	No	Sawai et al. (2002)
Other						
Entamoeba histolytica	Amoebapore	77	αααα	Bacteria, Fungi	No	Lynch et al. (1982)

problems associated with the development of antimicrobial peptides as therapeutic treatments.

MODES OF ACTION

Antimicrobial peptides generally act in two ways. Firstly, they act by destabilizing the cell membrane and secondly by entering the cell and interacting with specific targets. However, before these peptides can act on the bacterial membrane, they must first pass the bacterial cell wall. A possible way this could be accomplished is the self promoted uptake model. This model relies on the displacement of polyanionic cations that bridge the LPS molecules. This disrupts the outer wall through which peptides pass (Shai et al., 2001).

Those peptides which destabilize the cell membrane are thought to act by one of three models. The first model is the barrel stave model, where peptide monomers associate and form a pore in the membrane made up of bundles of helices (Shai et al., 2001). The second model, the carpet model, involves peptides gathering on the membrane surface and forming toroidal pores which disrupt membrane structure (Huang, 2000). The final model is the toroidal pore model which involves the peptides binding to the surface of the membrane and causing a thinning of the membrane as the peptides associate with lipid headgroups (Huang, 2000).

APPLICATIONS OF ANTIMICROBIAL PEPTIDES

Since the discovery of penicillin antibiotics have been widely used to combat many previously fatal infectious diseases. However, their continued widespread and excessive use has led to the emergence of many antibiotic resistant strains. Much of the current interest in antimicrobial peptides is a direct result of these antibiotic resistant pathogens and the consequent need for new antibiotics. Many think that antimicrobial peptides could be the new generation of antibiotics.

There are many ways in which antimicrobial peptides may play a role as antimicrobials or therapeutic agents. Firstly, they can be used as stand alone antimicrobial agents, such as conventional antibiotics. Secondly, they can be used in conjunction with other antimicrobial agents to increase the efficiency of antimicrobial activity. Thirdly, they can be used to enhance the patient's own innate immune system. Finally, they can be used to neutralize endotoxins resulting from septic shock (Gordon et al., 2005).

Stand alone antimicrobial agents

Many bacterial pathogens that are becoming increasingly resistant to current antibiotics have been found to be sensitive to antimicrobial peptides isolated from insects.

For example it has been found that strains of the pathogenic *Staphylococcus aureus* that are resistant to antibiotics such as methicillin, are sensitive to the defensin isolated from the beetle *Allomyrina dichotoma* (Yamada et al., 2005). A cecropin analogue peptide, D2A21, was shown to be more successful in the treatment of infected wounds than standard treatments, with 100% of rats with infected wounds surviving compared to a 50% survival rate in the control (Chalekson et al., 2003). There are also peptides isolated from chordates with similar activity. Halocidin, a peptide isolated from a tunicate, was active against resistant strains of *S. aureus* and *Pseudomonas aerginosa*. Halocidin is a heterodimer consisting of an 18 residue domain linked to a 15 residue domain via a disulfide bond. It was found that a synthetic 18 residue heterodimer was more active than the natural peptide (Jang et al., 2002).

Antimicrobial peptides have been extensively investigated for their use as ophthalmic and dental antimicrobials. In ophthalmic applications the peptide can be applied directly to the infected site in the form of eye drops and the amount of active peptide can easily be increased by additional dosing (Gordon et al., 2005). Despite these advantages the only sign of antimicrobial peptides having a promising application in ophthalmology applications, is in the disinfection of contact lenses. The increased use of contact lenses has also led to an increase in contact lens associated cornea infections (Sousa et al., 1996). Cecropin analogues Shiva-11, D_5C as well as hecate were all tested for their abilities to disinfect contact lenses as well as contact lens solutions. D_5C was able to exponentially increase the ability of existing contact lens sterilizing solutions to sterilise contact lenses (Sousa et al., 1996). Shiva 11 and hecate were both able to kill bacterial isolates from infected contact lenses (Gordon et al., 2005).

Antimicrobial peptides isolated from insects have been shown to have the ability to prevent mortality in mice that have been infected by influenza virus. These peptides, named alloferons, were isolated from the blow fly *Calliphora vicina*. Alloferon shares some structural elements to the influenza virus hemaglutinin protein. This may mean that alloferon is able to either interfere with viral assembly, or viral attachment to the cell (Chernysh et al., 2002). A highly successful mellitin derivative named hecate was produced by altering the charge distribution of mellitin, while at the same time retaining its three dimensional structure (Baghain et al., 1997). Hecate demonstrated antiviral activity against *herpes simplex virus-1*. At relatively low concentrations, hecate reduced plaque formation. However, it did not interfere with the virus's ability to synthesize proteins. Furthermore, it was observed that Hecate was able to prevent (HSV-1)-induced cell fusion and virus spread, with no cytotoxic effects (Baghain et al., 1997).

Both cecropin and mellitin were able to inhibit the production of HIV-1 in infected cells. These peptides

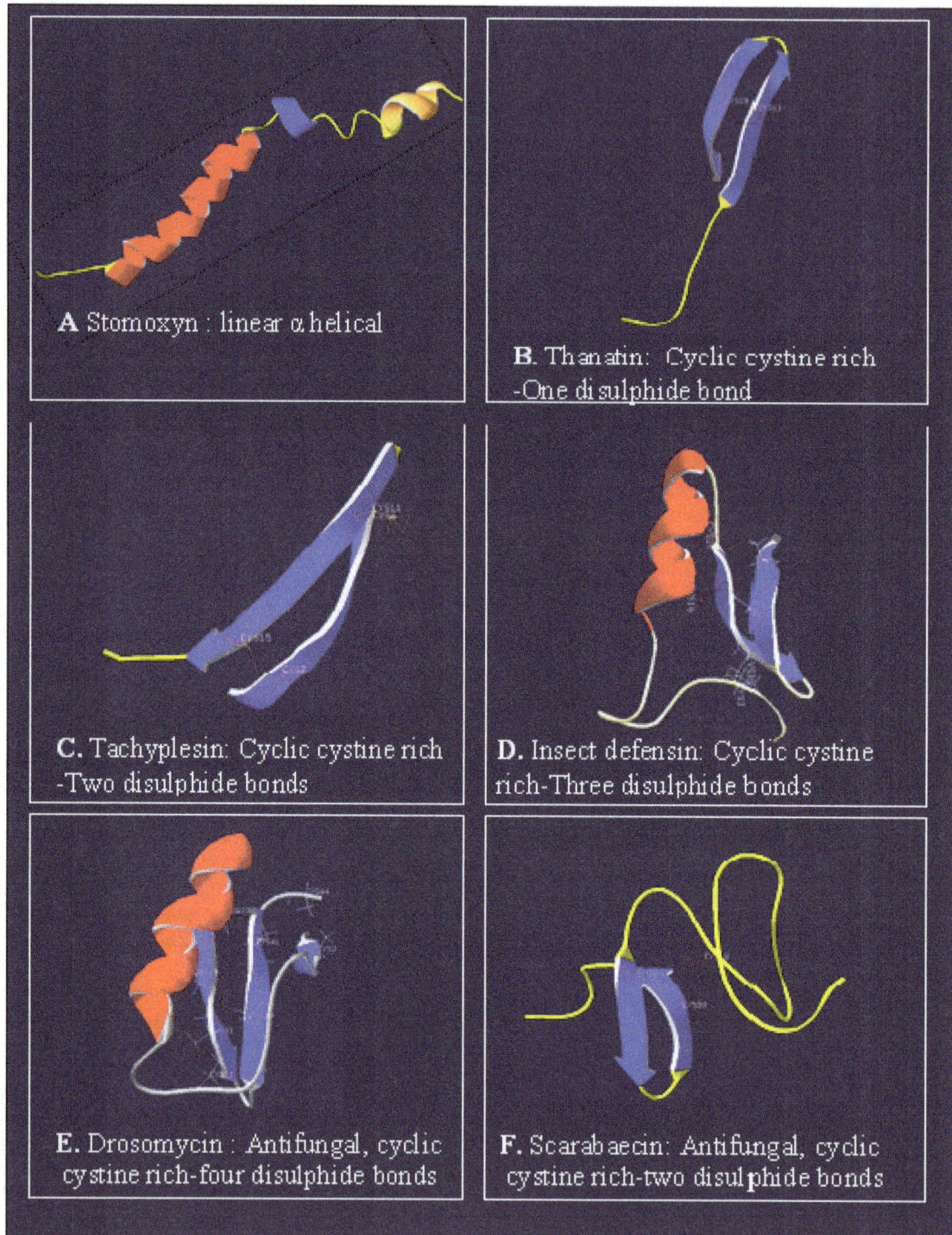

Figure 1. Structural representations of some arthropod anti-microbial peptides representing the different classes. Structural information was obtained from the Protein Data Bank (Berman et al., 2000). (A) The linear α helical peptide stomoxyn (Landon et al., 2006) isolated from the dipteran *Stomoxys calcitrans*. (B) Thanatin (Mandard et al., 1998), isolated from the hemipteran *Podisus maculiventris*, is the only insect cyclic cystine rich peptide with only one disulphide bond. (C) Cyclic cysteine rich peptide containing two disulfide bonds tachyplesin (Laederach et al., 2002), isolated from the horseshoe crab *Tachyplesus tridentatus*. (D) The insect defensin isolated from *Phormia terranovae* (Cornet et al., 1995) (E) The antifungal peptide drosomycin (Landon et al., 1997) is a cyclic cystine rich peptide containing 4 disulphide bonds. (F) Another antifungal peptide scarabeacin (Hemmi et al., 2003) was isolated from the rhinoceros beetle *Orychtes rhinoceros*. Although this peptide is a cyclic cystine rich peptide with two disulfide bonds, its structure is very different to that of tachyplesin.

Table 2. Cyclic cystine rich antimicrobial peptides (AMPs) from different organisms (plants and animals).

Sources	Name	Size (amino acids)	Structure	Activity	Commercialised or not	References
One disulphide bond				Invertebrates		
Podisus maculiventris	Thanatin	21	Anti parallel β sheet with a tail	Gram positive, Gram negative Fungus	No	Fehlbaum et al. (1996)
Myrmecia pilosula	Pilosulin	24-36	β sheet	Bacteria	No	Inagaki et al. (2004)
Two disulfide bonds						
Oryctes rhinoceros	Scarabeacin	38-40	Two stranded antiparrallel β-sheet after a helical turn	Fungi	No	Tomie et al. (2003)
Acanthoscurria gomesiana	Gomesin	18	Anti parallel β sheet	Bacteria, Fungi, Yeast, Hemolytic, Eukaryotic, Parasites	No	Mandard et al. (2002)
Androctonus australis	Androctonin	25	Anti parallel β sheet	Yeasts, Bacteria Fungi	No	Bulet et al. (2004)
Limulus polyphemus	polyphemusin	18	Anti parallel β sheet	Yeasts, Bacteria, Fungi	No	Bulet (2004: 162)
Tachyplesus tridentatus	tachyplesin	17	Anti parallel β sheet	Yeasts, Bacteria, Fungi, Hemolytic, Anti viral	No	Kawano (1990: 2)
Three disulfide bonds						
Acrocinus longimanus	Alo-3	36	Knottin type fold	Yeast	No	Barbault et al. (2003)
Tachypleus tridentatus	tachystatin					Osaki et al. (1999)
Mytillus edulis	Defensin	38-43	Disulfide-stabilised αβmotif	Gram positive Gram negative	No	Charlet et al. (1996) and Mitta et al. (1999)
Dipteran	Defensin	38-43	Disulfide-stabilised αβmotif	Gram positive	No	Chalk et al. (1995), Matsuyama and Natori (1988a), Lehane et al. (1997), Boulanger et al. (2002a) and Lauth et al. (1998)
Hemipteran	Defensin	42-43	Disulfide-stabilised αβmotif	Gram positive	No	Cociancich et al. (1994), Chernysh et al., 1996) Lopez et al., 2003)

Table 2. Contd.

Hymenoptera	Defensin	38-43	Disulfide-stabilised αβmotif	Gram positive	No	Rees et al. (1997)
Arachnida	Defensin	38-43	Disulfide-stabilised αβmotif	Gram Positive, Gram negative	No	Ehret-Sabatier et al. (1996), Johns and Sonenshine (2001) and Cocianich et al. (1993)
Coleopteran	Defensin	43	Disulfide-stabilised αβmotif	Gram Positive, Gram negative	No	Yamaauchi (2001) and Moon et al. (1994)
Aeschna cyanea	Defensin	38		Gram positive, Gram negative	No	Bulet et al. (1992)
A. gambiae	Defensin	38-43	Disulfide-stabilised αβmotif	Gram positive	No	Richman et al. (1997)
Apis mellifera	Royalisin	51	Disulfide-stabilised αβmotif	Gram positive	No	Fujiwara et al. (1990)
Lepidoptera	Defensin	32-36		Gram positive	No	Seitz et al. (2003), Volkoff et al. (2002) and Mandrioli et al. (2003)
Pseudacanthotermes spniger	Termicin	36	ááá cysteine stabilized motif	Fungi	No	Lamberty et al. (2001a)
Heliothis virescens	Heliomicin	44	βαββ	Fungi, Yeast	No	Lamberty et al. (1999)
Synthetic Heliomicin analogue				Fungi	Pre-clinical trials	
Ascaris suum	ASABF	71		Bacteria	No	Pillai et al. (2003)
Drosophila melanogaster	Drosomycin	44	βαββ	Fungi	No	Fehlbaum et al. (1994)
Mytillus edulis	Mytillin			Gram positive, Gram negative	No	Charlet et al. (1996) and Mitta et al. (1999)
5 disulfide bonds						
Tachypleus tridentatus	Tachycitin	73	β sheet	Gram-negative, Gram positive Fungi	No	Suetake et al. (2000)
Vertebrate						
One disulphide bond						
Halocynthia aurantium	Halocidin	32	Two subunits connected by a single disuphide	Bacteria	No	Jang et al. (2002)
Rana brevipoda	Brevinin	24	Hydrophobic region, a proline hinge C-terminal loop region	Bacteria	No	Morikawa et al. (1992)
Rana esculenta	Esculentin	46		Bacteria	No	Simmaco et al. (1993)
Rana catesbeiana	Ranalexin	20	α-helical	Bacteria	No	Clark et al. (1994)

Table 2. Contd.

Two disulfide bonds						
Sus scrofa	Protegrin I	17	β sheet	Bacteria, Viruses, Fungi	No	Storici and Zanetti (1993)
Synthetic protegrin	Iseganan			Bacteria, Fungi	Failed Phase III trial	Toney (2002)
Three disulfide bonds						
Gallus gallus	Gallinacin-1	36-39			No	Harwig et al. (1994)
Mus musculus	Cryptidin	70		Bacteria	No	Ouellette et al. (1989)
Plants						
Phytolacca americana	PAFP-S	38	Knottin-type fold	Fungi	No	Shao et al. (1999)
Triticum turgidum	γ-1-P thionin			Fungi, Bacteria, Tumor cells	No	Carrasco et al. (1981)
Chassalia parviflora	Circulin A	30	Knottin-type fold	HIV	No	Fujikawa et al. (1965) and Daly et al. (1999)
Fungi						
Pseudoplectania nigrella	Plectasin	40	3 disulfide Coil αhelix coil	Bacteria	Yes	Fungi

achieve this by decreasing both HIV-1 transcription and the number of viral gene products (Wachinger et al., 1998)from the beetle *Allomyrina dichotoma* were found to protect mice from endotoxic shock by acting as anti-inflammatory agents. The peptides were found to accomplish this by inhibiting the production of tumor necrosis factor α (Koyama et al., 2006). The authors suggested a possible mechanism underlying the peptides' ability to block TNF-α production. This involves the peptide preventing LPS from binding to LPS receptors on the surface of macrophages (Koyama et al., 2006). Moreover, the small alloferon peptides isolated from C. vicina were able to stimulate interferon production in mice as well as stimulating mouse spleen lymphocyte cytotoxicity (Chernysh et al., 2002). The horseshoe crab peptide Tachyplesin III was also able to efficiently kill *P. aeruginosa*– a multidrug resistant pathogen. This effect was significantly enhanced when conventional antibiotics were used in conjunction with tachyplesin (Cirioni et al., 2007). Like the beetle defensins, tachyplesin was also able to protect the mice from endotoxic shock following bacterial lysis (Cirioni et al., 2007). The highly effective hemipteran peptide thanatin was found to be effective against multidrug resistant isolates of *Enterobacter aerogenes and Klebsiella pneumoniae*. These strains show increased resistance to antibiotics due to increased altered membrane permeability, which allows them to expel antibiotics regardless of structure. Like tachyplesin III, thanatin also restored the antibiotic susceptibility of these resistant isolates. This must be achieved by making the membrane of the bacteria more porous to antibiotics or interfering with the bacteria's ability to expel specific character of the cell membrane. Comparison of bacteria and tumor cell membranes point to the common characteristic that both are negatively charged (Leushner and Hansel, 2004). This is due to tumor cell membranes containing a small amount of negative phosphatidylserine, making them 3-9% more negative than normal cells (Zwaal and Schroit, 1997).

Mellitin derivatives have been produced by changing a few L-amino acids with D-enantiomers.

Table 3. AMPs rich in particular amino acids from different organisms.

Sources	Name	Size (amino acids)	Structure	Activity	Commercialised or not	References
			Invertebrates			
Drosophila melanogaster	Drosocin	19	Proline rich O-glycosylated	Gram negative	No	Bulet et al. (1993)
Phormia terranovae and other diptera	Diptericin	100-110	O glycosylated glycine rich domain and a proline rich domain	Gram-negative bacteria	No	Dimarcq et al. (1988)
Palomena prasina	Metalnikowin	15-16	Proline rich non-glycosylated	Fungi	No	Chernysh et al. (1996)
Hymenoptera	Apideacin	19	Proline rich non-glycosylated	Gram negative	No	Casteels et al. (1989)
Hymenoptera	Abeacin	39	Proline rich non-glycosylated	Bacteria	No	Casteels et al. (1990)
Bombyx mori Trichoplusia ni	Lebocin	179 133	Proline rich O-glycosylated	Bacteria	No	Hara and Yamakawa (1995), and Liu et al. (2000)
Drosophila melanogaster	Metchnikowin	28	Proline rich O-glycosylated		No	Levashina (1998: 75)
Pyrrhocoris apterus	Pyrochoricin	20	Proline rich O-glycosylated reverse turns at both the C and N termini	Gram-positive bacteria	No	Cociancich et al. (1994)
Myrmecia gulosa	Formeacin	16	Proline rich O-glycosylated	Gram negative	No	Mackintosh et al. (1998b)
Hyalophora cecropia and other lepidoptera	Attacin	214-224	Random loops and B sheets	Gram-negative	No	Hultmark et al. (1983)
Sarcophaga peregrina and other diptera	Sarcotoxin	63	Random loops and B sheets	Gram-negative	No	Ando et al. (1987)
Pyrrhocoris apterus	Hemiptericin	133	Glycine rich Equally large number of positively and negatively charged residues	Gram negative bacteria	No	Cociancich et al. (1994)
Zophopas atratus	Coleoptericin	74	Glycine rich	Gram-negative bacteria	No	Bulet et al. (1991)
Holotrichia diomphalia..	Holotricin	72	High glycine and proline	Gram-negative bacteria	No	Lee et al. (1994)
Apis mellifera	Hymenopteacin	93		bacteria	No	Casteels et al. (1993)
Hyalophora.cecropia. Helicoverpa armigera. Trichoplusiani	Gloverin		α helical	Gram-negative bacteria	No	Axén et al. (1997), Mackintosh et al. (1998a) and Lundström et al. (2002)
Acolaepta luxuriosa.	Acaloleptin	71		Gram negative bacteri	No	Imamura et al. (1999)
			Vertebrates			
Bos Taurus	Indolicidin	13		Bacteria, Viruses, Fungi	No	Selsted et al. (1992)
Synthetic Indolicidin	Omiganan			Bacteria, Fungi	Failed phase II trials	Isaacson (2003)
Homo sapiens	Lactoferricin-B		Tryptophan rich	Bacteria, Fungi	No	Bellamy et al. (1992)

This resulted in peptides that were no longer hemolytic, but were very effective against tumor cells (Papo and Shai, 2003). Another mellitin analogue, hecate, was found to be toxic to breast cancer cells (Leuschner et al., 2003). Hecate's effectiveness has been increased by creating hecate hormone conjugates. By conjugating hecate to hormones, whose receptors are found on the surface of cancer cells e.g. luteinizing hormone, the cell selectivity of hecate can be increased (Hansel et al., 2007).

The naturally occurring alloferon, as well as synthetic alloferon from the blow fly *C. vaicin,* was able to slow tumor growth. It was, however, unable to eliminate cancer cells at high concentrations (Chernysh et al., 2002).

Application to transgenics

The tobacco budworm *H. virescens,* and the fall webworm *H. cunea* are both agricultural pests (Ourth et al., 1994; Park et al., 1997). Research into the responses of these insects to microbial infections could be important in the future control of their populations. Methods could be designed to inhibit insects defense, and increase the rates of mortality in the field (Ourth et al., 1994; Park et al., 1997).

Conversely, the cultivation and conservation of useful insect species could be boosted by increasing the defenses these insects have against pathogenic microorganisms (Destoumieux et al., 1997). Research into the structure and genetics of immune peptides may ensure the long-term survival and continued use of these insects (Destoumieux et al., 1997).

Concerns have been expressed about the levels of antimicrobial peptides present in domesticated poultry, due to intense breeding and in some cases in-breeding (Joerger, 2003). A decrease in the levels of antimicrobial peptides may leave breeding and domestic populations at the risk of microbial infections. Fisheries and fish breeding projects are also significantly adversely affected by infectious diseases caused by pathogens such as *Vibrio anguillarum.* Consequently, pilot studies have been performed on the use of cecropin-mellitin hybrids (CEME) to combat infections in Coho Salmon (Jia et al., 2000). This synthetic peptide caused no adverse effects when it was injected into young salmon. When fish were infected with lethal doses of *V. anguilarum* the mortality rate in the fish that received CEME treatment was 4 to 5 times lower than that in the control group (Jia et al., 2000).

This approach could also be applied in useful crops via insertion of genes encoding antimicrobial peptides into their genomes, thus providing the plants with defense mechanisms against various fungal and bacterial pathogens. Indeed transgenic tobacco expressing heliomicin and drosomycin was found to have higher resistance to fungal infections (Banzet et al., 2002).

Antimicrobial peptides as drug delivery systems

Drug delivery molecules need to be able to penetrate thetarget cells, while displaying low levels of toxicity. The hemipteran peptide pyrrhocoricin has been investigated as a potential drug delivery tool due to its low toxicity to eukaryotic cells, as well as the ease with which it penetrates bacterial as well as human cells. It was shown that the peptide and a synthetic pyrrhocoricin analogue could penetrate human fibroblasts and dendritic cells (Otvos et al., 2004). This feature of the peptide to enter but not kill eukaryotic cells is due to its mode of action, which is described below.

The mode of entry for the short glycine rich peptides apidaecin and drosocin is expected to be the same for pyrrhocoricin (Kragol et al., 2001). All three of these peptides have been found to enter the target cell and become rapidly distributed in all cellular compartments (Kragol et al., 2001). The initial step of the entry of these peptides into bacterial cells is expected to involve their association with some component of the outer cell membrane. After successful invasion of the periplasmic space the peptides then bind to some inner membrane component (Castle et al., 1999). However, this association does not destabilize the plasma membrane or result in the permeabilisation of the target cells (Otvos et al., 2004). The peptide is then translocated into the cell where it interacts with specific targets. Pyrrhocoricin interacts with the bacterial chaperonins and heat shock proteins GroEl and DnaK. The pyrrhocoricin binding site on DnaK has been identified to be the hinge region of the protein's multihelical lid that covers the peptide-binding site (Kragol et al., 2001). Upon binding pyrrhocoricin prevents the helical lid from opening and this leads to inhibition of protein folding. Pyrrhocoricin also inhibits DnaK's ATPase activity, which further narrows the probable binding site to one of the helices between helix D and E (Kragol et al., 2001).

Control of insect borne diseases

There are many insects that act as vectors of parasites that have a major effect on human health, crop plants and domestic animals. The historical method of disease control has been through the control of the insect vector (Durvasula et al., 1997). These methods have become less effective due to an increasing emergence of insecticide and drug resistant insects. Even biological methods of control, such as sterile male release are less effective than they once were (Mogi et al as cited by (Durvasula et al., 1997). One of the strategies to control these diseases is to develop methods that reduce ability of the parasite or pathogen to infect the insect host. This can be achieved by genetically modifying the immune response of the vector, or through the use of transgenic symbiotic bacteria (Boulanger et al., 2001) (Durvasula et al., 1997).

The mosquito, *A. gambiae* is the principal vector of human malaria in Africa. The number of cases of mosquito-transmitted malaria is increasing resulting in about 3 million deaths per year. Mosquitoes also spread lymphatic filariasis and arboviruses such as Dengue fever and yellow fever (Lowenberger, 2001). Despite the high transmission rate in the field, the vast majority of parasite and mosquito contact do not result in infection. This is also true in the tsetse fly *G. moristans* where infection rates are low (Boulanger et al., 2001). *Plasmodium bergei* elicits an immune response in the form of antimicrobial peptides in *A. gambiae*, and *A aegypti*, while *Trypanasoma brucei brucei* elicits a similar response in *G. moristans moristans* (Boulanger et al., 2001). The development of *P. bergei* in Anopheline mosquitoes can be stopped by administering exogenous cecropin to the mosquito (Gwadz et al., 1989). Another strategy being investigated is the use of the scorpion venom peptide scorpine. Scorpine was found to block the development of malaria parasites in tissue culture , as well as in the gut of transgenic fruit flies (Possani et al., 2002).

Further evidence for this line of thought is demonstrated by the peptide stomoxyn. Despite living in the same area, being exposed to the same pathogens and having similar physiology, *S. calcitrans* is not a vector for trypanosome parasites, while the tsetse fly *G. moristans* is (Boulanger et al., 2002). The immune response of these insects was compared. Both insects release defensins into the anterior region of their gut. However, stomoxyn is unique to *S. calcitrans,* and could not be identified in the tsetse fly (Boulanger et al., 2002). By creating a strain of tsetse fly that can express this peptide within its gut, the number of effective vectors for the trypanosome parasite could be decreased.

It is due to these facts that the strategies of transgenic insect vectors expressing enhanced and more effective forms of cecropins (Boulanger et al., 2001), or the use of genetically transformed bacteria expressing antimicrobial peptides (Durvasula et al., 1997) are being investigated. The generally accepted strategy to accomplish this would be to place the transcription of the antimicrobial peptide under the control of the promoters for the transcription of proteolytic gut enzymes in the transgenic mosquitoes (Possani et al., 2002). However, ecologists have pointed out problems with using transgenic mosquitoes. These include doubts as to whether the transgenic mosquitoes could effectively compete with wild type mosquitoes for mates or resources (Enserink, 2002). However, male crickets are able to advertise their pathogen resistance ability to females using their calling song. Males with a higher chirp rate in their song were found to possess increased pathogen resistance. Therefore any males with a more effective immune response should be able to out-compete any other males (Ryder and Siva-Jothy, 2000). Unfortunately, this may become a problem in an area with limited resources, as it is thought that the mounting of an immune response by the host insect is costly and must be compensated for by an increased resource intake

(Moret and Schmid-Hempel, 2000).

A solution to this problem could be to imitate the *D. melanogaster* attacin, where the processing of the immature peptide gave rise to the mature attacin, and a short proline rich peptide (Asling et al., 1995). This means that the insect is getting two peptides for the price of one, and both peptides are still being properly secreted. Another problem ecologists have pointed out is the ability of *P. falciparum* to develop resistance to the host resistance transgenes (Enserink, 2002). However, if the resistance gene is an enhanced antimicrobial peptide that is still under the control of the recognition pathways of the innate immune system, it may not be that easy for *P. falciparum* to develop resistance to these genes. Finally, there is the problem of multiple vectors in a given area. In these areas transgenic members of each of these species would have to be released into the area (Enserink, 2002). Bacterial symbionts of insect vectors of diseases have been engineered to express antimicrobial peptides. The assassin bug *R. prolixus* is the main vector of Chagas disease, which is spread by the parasite *Trypanosoma cruzi*. Transgenic bacterial symbionts of *R. prolixus* have been transformed to express cecropin (Durvasula et al., 1997). The expression of this peptide resulted in the total elimination of the parasite, or at least in reduction of its number. Any negative effects such as toxicity to gut flora or insect cells do not appear to be a problem (Durvasula et al., 1997).

OBSTACLES TO THERAPEUTIC USE

Obstacles to the commercial use of antimicrobial peptides include high cost, difficulties in mass production, loss of activity under physiological conditions, potential toxicity, peptide aggregation, peptide stability (Marr et al., 2006) and newer, effective conventional antibiotics (Gordon et al., 2005). These conventional antibiotics represent a safer more trustworthy drug development option for pharmaceutical companies. In addition to these problems very few antimicrobial peptides have received FDA approval. This has resulted in many pharmaceutical companies abandoning antimicrobial peptides (Gordon et al., 2005).

Cost

One of the many strategies used to lower the cost of production is to use smaller peptides (Marr et al., 2006). This has led to the design of synthetic and model antimicrobial peptides. Model amphipathic peptides have been constructed to optimise the effect of the a-helices involved in the disruption and channel formation within bacterial cell membranes (Blondelle and Houghten, 1992). Cecropin-mellitin hybrids have been designed in order to maximise the active range of peptides, while minimising its effect on the host's cells. The hybrids are small molecules that minimize immunogenicity and

Table 4. Unusual insect antimicrobial peptides that do not fall into any category. Three unusual peptides have been isolated from insects, which cannot be grouped into any class or group. The jelleines appear to be secondary metabolites from mastoporan processing. They and the alloferons are two of the smallest antimicrobial peptides that have been isolated.

Sources	Name	Size amino acid	Activity	Commercialised or not	References
Calliphora viciana	Alloferon	12-13	Cancer Virus	No	Chernysh et al. (2002)
H. virescens	Viresin		Gram negative bacteria	No	Chung and Ourth (2000)
Apis mellifera	Jelleines	8-9	Bacteria and yeast	No	Fontana et al. (2004)

expense of synthesis (Boman et al., 1989). Additionally synthetic oligopeptides based on the structure of the *A. dichotoma* defensin have been synthesised and some of these analogues have greater activity against *S. aureus* than the original peptide (Yamada et al., 2005). These hybrid peptides were found to be more effective antibacterial and antimalarial peptides, and they did not lyse mammalian red blood cells (Boman et al., 1989).

In order to lower the costs of production, smaller synthetic peptides are being created. One strategy being used is to extract active linear fragments from insect defensins (Lee, 2002). Tenecin 1 is a defensin isolated from *Tenebrio molitor*. An active fragment corresponding to the β sheet region was successfully purified. In addition to this the α helical region of tenecin 1, which was originally inactive, could be transformed into an active peptide by single and double amino acid substitutions to increase the positive charge of the peptide (Ahn et al., 2006). In addition to these shortened peptides being easier to produce, they may also present lower levels of antigenicity (Lee, 2002), and higher levels of activity (Boman et al., 1989) than the natural parent peptide.

Insects possess many antimicrobial peptides that are naturally small and should be cost effective to synthesize. These include the alloferons as well as the jelleines (Table 4).

Production

Difficulties around mass production of antimicrobial peptides are a major obstacle in the use of these peptides as therapeutic agents. Biological expression of antimicrobial peptides in bacteria is difficult and normally, failure can be attributed to the recombinant protein being inactive or the protein being toxic to the bacterial cell. Attempts to produce pilosulin in *Escherichia coli* resulted in an inactive protein being produced (Inagaki et al., 2004). However, active human β defensin has been successfully produced in *E. coli* as a GST-Fusion protein (Si et al., 2007). This is surprising as the defensins require the formation of disulfide bridges. The insect peptide moricin from the silkworm *Bombyx mori* was a perfect candidate for production in bacteria. Firstly, it contains unique structural elements at its C-terminus, making chemical synthesis difficult (Hara and Yamakawa, 1995). Secondly, it contains no disulfide bonds and

should therefore fold properly in bacteria. This led to moricin being successfully produced using two separate protein fusion systems (Hara and Yamakawa, 1996).

A possible solution to the difficulties experienced in the mass production of antimicrobial peptides may be the fungal defensin plectasin. Plectasin was isolated from the saprophytic fungus *Psuedoplectania nigrella*. Plectasin resembles arachnid and primitive insect defensins in structure; it shows high activity against bacteria and low levels of cytotoxicity. Being produced in a fungus meansthat plectasin could be produced in high amounts (Mygind et al., 2005a).

Proteolysis, inactivation and toxicity

Due to the high content of basic residues found in most antimicrobial peptides they would be prone to degradation by trypsin-like proteases. Insect pathogenic strains of *P. aeruginosa* are able to avoid antimicrobial activity by degrading cecropin using a cecropin specific protease activity (Jarosz, 1997). Additionally, cecropin degrading enzymes from *Bacillus larvae* were able to effectively remove cecropin activity from the hemolymph of honeybees (Jarosz and Gliński, 1990).

Interest has been shown in the use ofpyrrhocoricin to treat infections because of its low toxic effects on mammalian cells. Furthermore, pyrrhocoricin is more resistant to degradation in mammalian sera than many other peptides (Otvos et al., 2000). Studies on the *in vivo* effectiveness of pyrrhocoricin demonstrated that it was able to protect mice from an *E.coli* infection. At high doses, however, the peptide killed the mice (Otvos et al., 2000). Derivatives were designed to decrease toxicity and increase antimicrobial activity. The most effective of these pyrrhocoricin derivatives contained unnatural amino acids at both termini, and was more resistant to the proteolytic actions of the mouse sera (Otvos et al.,2000).

Another strategy to improve the resistance of antimicrobial peptides to the action of proteases is the use of D enantiomers (Papo and Shai, 2003). This strategy can only be used with membrane lytic peptides that do not interact with specific targets. Good initial results have already been demonstrated using this strategy with mellitin D, L derivatives (Papo and Shai, 2003).

Serum, culture media as well as the physiological

Table 5. Antimicrobial peptides derived from bacteria. Summary of some of the antimicrobial peptides isolated from bacteria and fungi. The table gives details of the peptides structure, range of activity and the current status of any attempts to commercialise the peptide or its derivatives.

Sources	Name	Structure	Activity	Commercialised or not	Reference
Actinoplanes sp.	Ramoplanin	Cyclic lipopeptide	Gram-positive	No	Cavalleri et al. (1984)
Streptomyces roseosporus	Daptomycin	Cyclic lipopeptide	Gram-positive	Yes	Eliopoulos et al. (1986)
Bacillius polymyxa	Polymixin B		Gram-positive	Yes	Watt and Vanderift (1950)
Lactococcus lactis	Nisin	Cyclic lipopeptide	Gram-positive	Yes	Berridge (1949)
Leuconostoc gelidum	Leucocin	Cyclic lipopeptide	Gram-positive		Hastings et al. (1991)

concentrations of salts have been reported to inhibit the activity of antimicrobial peptides (Johansson et al., 1998). This is probably due to the presence of NaCl and divalent cations. Once again this problem may be overcome through peptide engineering. A synthetic version of the human antimicrobial cathelicidin peptide LL37 named WLBU2 was able to kill *S. aureus* in an isotonic saline environment and continued to kill the bacteria in environments with increasing salt concentrations. LL37 was inactive at all tested salt concentrations (Deslouches et al., 2005).

RESISTANCE TO ANTIMICROBIAL PEPTIDES

However, there are bacteria defense mechanisms against antimicrobial peptides. For instance attacins and cecropin are degraded by a protease from *Bacillus thuringeiensis* (Dalhammar and Steiner, 1984). This was further demonstrated by the finding that when this protease is injected into the hemolymph of the tsetse fly *G. moristans.* The flies become more sensitive to bacterial infections (Kaaya et al., 1987).

The dangerous pathogenic *S. aureus* is able to resist high concentrations of many membrane permeabilising peptides by incorporating a higher degree of positive charges into its bacterial membrane. Since these peptides operate by using their positive charge to associate with the negatively charged membrane, an increase in the positive charge of the membrane makes it difficult for the peptides to associate with the membrane (Peschel et al., 1999). The bacteria are able to do this via the *dlt* gene products, which allows the incorporation of D-alanine into the cell wall teichoic acids (Peschel et al., 1999). This means that *S. aureus* strains carrying a mutation in the *dlt* operon are more sensitive to the positively charged antimicrobial peptides.

Another strategy adopted by bacteria to change their outer membrane, is the incorporation of a large concentration of lysine residues in the outer membrane. This would also change the charge of the membrane to a more positive charge (Wang et al., 2002). Fears have been raised that the use of antimicrobial peptides based on human natural defense peptides may result in increased resistance of pathogens to our own immune system (Bell and Gouyon, 2003). There has been no indication that this has been the case with peptides that have been on the market for many years such as nisin (Hancock, 2003).

PEPTIDES UNDER CLINICAL DEVELOPMENT

Many peptides are already present in over the counter medicines and in foods. Polymoxin B is present in topical creams and eye ointment while Nisin has been used as a preservative in food since the late 1960s (Table 5). Additional peptides in clinical development are listed in Table 6.

CONCLUSION

Despite the problems associated with the production and cost of antimicrobial peptides, we now know enough concerning structure and design to create novel synthetic peptides that show little or no cytotoxicity to non-cancerous human cells. These synthetic peptides may also overcome the inactivation of antimicrobial peptides *in vivo*. Shortened synthetic peptides have the added advantage of being cheap to produce.

It is also interesting to note that only Iseganan failed clinical trials due to lack of activity. Pexiganan was refused approval by drug agencies, not due to lack of performance, but because it was no more effective than present treatments. The decision to refuse approval of pexiganan reflects continued faith in older traditional antibiotics, more than it does the failure of the drug as an effective antimicrobial.

Insects represent a rich resource of antimicrobial peptides, many of which are extremely small and therefore cheap to synthesise. Others, such as moricin, can be mass produced in bacteria. The success of the synthetic mellitin analogue hecate and the cecropin derivative D2A21 in the treatment of bacterial infections, breast cancer, and viral inhibition, shows that these peptides are a promising avenue for drug development.

Table 6. Antimicrobial peptides undergoing clinical development.

Peptide	Origin	Stage of development	Activity (target)	Results of trials	References
D2A21	Cecropin analogue	Phase I trial testing	Treatment of infected wounds.	Found to be efficient in preventing burn infections	Schroder and Harder (2006)
Insect defensins (ETD151)	Heliomycin derivative	Pre-clinical trial testing	Treatment of fungal infections.	Promising results	Andres and Dimarq (2005)
Insect defensin	E.g dragonfly defensins	Pre-clinical trial testing	Gram positive bacterial infections.		Andres and Dimarq (2005), Bulet et al. (1992)
Pexiganan	Synthetic derivative of magainin	Phase III trial testing	Pathogenic infections of diabetic foot ulcers	Drug was refused FDA approval, on the basis that is was no more effective than conventional antibiotics	Moore (2003)
Iseganan	Protegrins.	Abandoned	Bacterial infections	Failed to show any promising result	Gordon et al. (2005)
Omiganan	Synthetic analogue of indolicidin	Phase III trial testing	Active against bacteria and fungi	Failed to prevent infections, but it was able to decrease the instances of catheter tube colonization by bacteria.	Gordon et al. (2005)
MBI 594AN	Cathelicidin, based peptide	Treatment of acne	Phase II trials	Highly effective in reducing inflammation and the formation of lesions	Gordon et al. (2005)

Additionally the ability to isolate smaller active components of insect antimicrobial peptides would lead to lower production costs.

Finally, developing drugs based upon arthropod antimicrobial peptides would partially remove the concern for pathogens developing increased resistance to these peptides. If we base antimicrobial peptide drugs upon mammalian peptides we run the risk of giving pathogens an increased opportunity to develop resistance against our own innate defense peptides (Bell and Gouyon, 2003).

As antibiotic resistance becomes an increasing threat the value and potential of antimicrobial peptides as a new source of antimicrobial compounds increases. This combined with other potential uses such as their ability to act synergistically with conventional antibiotics, their ability to prevent endotoxic shock and the difficulty pathogens have in developing resistance, makes antimicrobial peptides an important source of future chemotherapeutic agents.

REFERENCES

Ahn HS, Cho W, Kang SH, Ko SS, Park MS, Cho H, Lee KH (2006). Design and synthesis of novel antimicrobial peptides on the basis of alpha helical domain of Tenecin 1, an insect defensin protein, and structure-activity relationship study. Peptides, 4: 640-648.

Anderson D, Terwilliger TC, Wickner W, Eisenberg D (1980). Melittin forms crystals which are suitable for high resolution X-ray structural analysis and which reveal a molecular 2-fold axis of symmetry. J. Biol. Chem., 255: 2578-2582.

Ando K, Okada M, Natori S (1987). Purification of sarcotoxin II, antibacterial proteins of *Sarcophaga peregrina* (flesh fly) larvae. Biochemistry 13: 226-230.

Andres E, Dimarq JL (2005). Clinical development of antimicrobial peptides. Int. J. of Antimicrob. Agents 25: 448-452.

Arrowood MJ, Jaynes JM, Healey MC (1991). *In vitro* activities of lytic peptides against the sporozoites of *Cryptosporidium parvum*. Antimicrob. Agents Chemother., 35: 224-227.

Asling T, Dushay MS, Hultmark D (1995). Identification of early genes in the Drosophila immune response by PCR-based differential display: Attacin A gene and the evolution of attacin like proteins. Insect Biochem. Mol. Biol. 25: 511-518.

Axén A, Carlsson A, Engström A, Bennich H (1997). Gloverin, an antibacterial protein from the immune hemolymph of Hyalophora pupae. Eur. J. Biochem. 247: 614-619.

Baghain A, Jaynes J, Enright F, Kousoulas K (1997). An Amphipathic a-Helical Synthetic Peptide Analogue of Melittin Inhibits Herpes Simplex Virus-1 (HSV-1)-Induced Cell Fusion and Virus Spread. Peptides, 18: 177-183.

Banzet N, Latorse M-P, Bulet P, Francois C, Derpierre C, Dubad M (2002). Expression of insect cysteine rich antifungal peptides in transgenic tobacco enhances resistance to a fungal disease. Plant Sci., 162: 995-1006.

Barbault F, Landon C, Guenneugues M, Meyer JP, Schott V, Dimarcq JL, Vovelle F (2003). Solution structure of Alo-3: A new knottin-type antifungal peptide from the insect *Acrocinus longimanus*. Biochem., 42: 14434-14442.

Bell G, Gouyon P-H (2003). Arming the enemy: The evolution of resistance to self-proteins. Microbiology, 149: 1367-1375.

Bellamy W, Takase M, Wakabayashi H, Kawase K, Tomita M (1992). Antibacterial spectrum of lactoferricin B, a potent bactericidal peptide derived from the N-terminal region of bovine lactoferrin. J. Appl. Bacteriol., 73: 472-479.

Berman HM, Westbrook J, Feng Z, Gilliland G, Bhat TN, Weissig H, Shindyalov IN, Bourne PE (2000). The Protein

Data Bank. Nucleic Acids Res., 28: 235-242.

Berridge NJ (1949). Preparation of the antibiotic nisin. Biochem. J., 45: 486-493.

Blondelle SE, Houghten R (1992). Design of model amphipathic peptides having potent antimicrobial activities. Biochem., 31: 12688-12694.

Boman HG, Wade D, Boman IA, Wahlin B, Merrifield RB (1989). Antibacterial and antimalarial properties of peptides that are cecropin melittin hybrids. FEBS Lett., 259: 103-106.

Boulanger N, Ehret-Sabatier L, Brun R, Zachary D, Bulet P, Imler JL (2001). Immune response of Drosophila melanogaster to infection with the flagellate parasite Crithidia sp. Insect Biochem. Mol. Biol., 31: 129-137.

Boulanger N, Munks RJ, Hamilton JV, Vovelle F, Brun R, Lehane MJ, Bulet P (2002). Epithelial innate immunity. A novel antimicrobial peptide with antiparasitic activity in the blood-sucking insect Stomoxys calcitrans. J. Biol. Chem. 277: 49921-49926.

Bulet P, Cociancich S, Dimarq J, Lambert J, Reichart JM, Hoffmann D, Hetru C, Hoffmann JA (1991). Isolation of a coleopteran insect of a novel inducible antibacterial peptide and of new members of the insect defensin family. J. Biol. Chem., 266: 24520-24525.

Bulet P, Cociancich S, Reuland M, Sauber F, Bischoff R, Hegy G, Dorsselaer V, Hetru C, Hoffmann JA (1992). Novel insect defensin mediates the inducible antibacterial activity in larvae of the dragonfly Aeschna cyanea (Paleoptera, Odonata). Eur. J. Biochem., 209: 977-984.

Bulet P, Stöcklin R, Menin L (2004). Anti-microbial peptides: From invertebrates to vertebrates. Immunol. Rev., 198: 169-184.

Carrasco L, Vázquez D, Hernández-Lucas C, Carbonero P, García-Olmedo F (1981). Thionins: Plant peptides that modify membrane permeability in cultured mammalian cells. Eur. J. Biochem., 116: 185-189.

Casteels P, Ampe C, Jacobs F, Tempst P (1993). Functional and chemical characterization of Hymenoptaecin, an antibacterial polypeptide that is infection-inducible in the honeybee (Apis mellifera). J. Biol. Chem., 268: 7044-7054.

Castle M, Nazarian A, Yi SS, Tempst P (1999). Lethal effects of apidaecin on Escherichia coli involve sequential molecular interactions with diverse targets. J. Biol. Chem., 12: 32555-32564.

Cavalleri B, Pagani H, Volpe G, Selva E, Parenti F (1984). A-16686, a new antibiotic from Actinoplanes. I. Fermentation, isolation and preliminary physico-chemical characteristics. J. Antibiot. (Tokyo), 37: 309-317.

Chalekson CP, Neumiester MW, Jaynes J (2003). Treatment of infected wounds with the antimicrobial peptide D2A21. Trauma, 54: 770-774.

Charlet M, Chernyshi S, Phillippe H, Hetru C, Hoffman JA, Bulet P (1996). Isolation of several cysteine rich antimicrobial peptides from the blood of a mollusc, Mytilus edulis. J. Biol. Chem., 271: 21808-21813.

Chernysh S, Cociancich S, Briand JP, Hetru C, Bulet P (1996). The inducible antibacterial peptides of the Hemipteran insect Palomena prasina: Identification of a unique family of proline rich peptides and of a novel insect defensin. J. Insect Physiol., 42: 81-89.

Chernysh S, Kim SI, Bekker G, Pleskach VA, Filatova NA, Anikin VB, Platanov VG, Bulet P (2002). Antiviral and Antitiumor peptides from insects. Proc. Natl. Acad. Sci. USA., 99: 12628-12632.

Chung KT, Ourth DD (2000). Viresin. A novel antibacterial protein from immune hemolymph of Heliothis virescens pupae. Eur. J. Biochem., 267: 677-683.

Cirioni O, Ghiseli R, Silvestri C, Kamsay W, Orlando F, Mocchegiani F, Matteo FD, Riva A, Lukasiak J, Scalise G, Saba V, Giacometti A (2007). Efficacy of Tachyplesin III, Colistin, and imipenem against a multiresistant Psuedomonas aerginosa strain. Antimicrob. Agents Chemother., 51: 2005-2010.

Clark DP, Durell S, Maloy WL, Zasloff M (1994). Ranalexin. A novel antimicrobial peptide from bullfrog (Rana catesbeiana) skin, structurally related to the bacterial antibiotic, polymyxin. J. Biol. Chem., 269: 10849-10855.

Cociancich S, Dupont A, Hegy G, Lanot R, Holder F, Hetru C, Bulet JAHP (1994). Novel inducible antibacterial peptides from hemipteran insect,the sap sucking bug Pyrrhocoris apterus. Biochem. J., 300: 567-575.

Cociancich S, Ghazi A, Hetru C, Hoffmann JA, Letellier L (1993). Insect defensin, an inducible antimicrobial peptide, forms insect dependant channels in Micrococcus luteus. J. Biol. Chem., 268: 19239-19245.

Cornet B, Bonmatin JM, Hetru C, Hoffmann JA, Ptak M, Vovelle F (1995). Refined three-dimensional solution structure of insect defensin A. Structure, 3: 435-448.

Corzo G, Villegas E, Go´mez-Lagunas F, Possani DL, Belokoneva SO, Nakajima T (2002). Oxyopinins, Large Amphipathic Peptides Isolated from the Venom of the Wolf Spider Oxyopes kitabensis with Cytolytic Properties and Positive Insecticidal Cooperativity with Spider Neurotoxins. J. Biol. Chem., 277: 23627-23637.

Dalhammar G, Steiner H (1984). Characterization of inhibitor A, a protease from Bacillus thuringiensis which degrades attacins and cecropins, two classes of antibacterial proteins in insects. Eur. J. Biochem., 139: 247-252.

Daly NL, Koltay A, Gustafson KR, Boyd MR, Casas-Finet JR, Craik DJ (1999). Solution structure by NMR of circulin A: A macrocyclic knotted peptide having anti-HIV activity. J. Mol. Biol., 285: 333-345.

Deslouches B, Islam K, Craigo JK, Paranjape SM, Montelaro RC, Mietzner TA (2005). Activity of the de novo engineered antimicrobial peptide WLBU2 against Pseudomonas aeruginosa in human serum and whole blood: Implications for systemic applications. Antimicrob. Agents Chemother., 49: 3208-3216.

Destoumieux D, Bulet P, Loew D, Dorsselaer AV, Rodriguez J, Bachere E (1997). Penaeidins, a new family of antimicrobial peptides isolated from the shrimp Penaeus vannamei (Decapoda). J. Biol. Chem., 272: 28398-28406.

Dimarcq JL, Keppi E, Dunbar B, Lambert J, Reichhart JM, Hoffmann D, Rankine SM, Fothergill JE, Hoffmann JA (1988). Insect immunity. Purification and characterization of a family of novel inducible antibacterial proteins from immunized larvae of the dipteran Phormia terranovae and complete amino-acid sequence of the predominant member, diptericin A. Eur. J. Biochem., 171: 17-22.

Durvasula RV, Gumbs A, Panakal A, Kruglov O, Askoy S, Merrifield RB, Richards FF, Beard CB (1997). Prevention of insect borne disease: An approach using transgenic symbiotic bacteria. Proc. Natl. Acad. Sci. USA., 94: 3274-3278.

Ehret-Sabatier L, Loew D, Goyffon M, Fehlbaum P, Hoffmann JA, Bulet P (1996). Characterization of novel cysteine rich antimicrobial peptides from scorpion blood. J. Biol. Chem., 271: 29537-29544.

Eliopoulos G, Willey S, Reiszner E, Spitzer PG, Caputo G, Jr RM (1986). In vitro and in vivo activity of LY 146032, a new cyclic lipopeptide antibiotic. Antimicrob. Agents Chemother., 30: 532-535.

Enserink M (2002). Ecologists see flaws in transgenic mosquito. Science, 297: 30-31.

Fehlbaum P, Bulet P, Chernysh S, Briand JP, Roussel JP, Letellier L, Hetru C, Hoffmann JA (1996). Structure-activity analysis of thanatin, a 21-residue inducible insect defense peptide with sequence homology to frog skin antimicrobial peptides. Proc. Natl. Acad. Sci.USA., 93: 1221-1225.

Fehlbaum P, Bulet P, Michaut L, Laguerx M, Broekaerts WF, Hetru C, Hoffmann JA (1994). Septic injury of Drosophila induces the synthesis of a potent antifungal peptide with sequence homology to plant antifungal peptides. J. Biol. Chem., 269: 33159-33163.

Fontana R, Mendes MA, de Souza BM, Konno K, César LM, Malaspina O, Palma MS (2004). Jelleines: A family of antimicrobial peptides from the Royal Jelly of honeybees (Apis mellifera). Peptides, 25: 919-928.

Fujikawa K, Suketa Y, Hayashi K, Suzuki T (1965). Chemical structure of circulin A. Experientia., 15: 307-308.

Fujiwara S, Imai J, Fujiwara M, Yaeshima T, Kawashima T, Kobayashi K (1990). A potent antibacterial protein from royal jelly. J. Biol. Chem., 265: 11333-11337.

Ge Y, MacDonald D, Henry MM, Hait HI, Nelson KA, Lipsky BA, Zasloff MA, Holroyd KJ (1999). In vitro susceptibility to pexiganan of bacteria isolated from infected diabetic foot ulcers. Diagn. Microbiol. Infect. Dis., 35: 45-53.

Gordon YJ, Romanowski EG, McDermott AM (2005). A review of Antimicrobial Peptides and Their therapeutic potential as Anti-

Infective Drugs. Curr. Eye Res., 30: 505-515.

Gwadz RW, Kaslow D, Lee JY, Maloy WL, Zasloff M, Miller LH (1989). Effects of magainins and cecropins on the sporogonic development of malaria parasites in mosquitoes. Infect. Immun., 57: 2628-2633.

Hancock REW (2003). Concerns regarding resistance to self-proteins. Microbiology, 149: 3343.

Hansel W, Enright F, Leuschner C (2007). Destruction of breast cancers and their metastases by lytic peptide conjugates in vitro and in vivo.
Mol. Cell Endocrin., 260-262: 183-189.

Hara S, Yamakawa M (1995). A novel peptide family isolated from the silkworm Bombyx mori. Biochem. J., 310: 651-656.

Hara S, Yamakawa M (1996). Production in Eschericia coli of Moricin, a novel Type Antibacterial protein from the silkworm Bombyx mori. Biochem. Biophys. Res. Comm., 220: 664-669.

Harwig S, Swiderekk K, Kokryakova VN, Tana L, Leeb TD, Panyutich EA, Aleshinac GM, Shamovac OV, Lehrera RI (1994). Gallinacins: Cysteine-rich antimicrobial peptides of chicken leukocytes. FEBS Lett., 342: 281-285.

Hastings JW, Sailer M, Johnson K, Roy KL, Vederas JC, Stiles ME (1991). Characterization of leucocin A-UAL 187 and cloning of the bacteriocin gene from Leuconostoc gelidum. J. Bacteriol., 173: 7491-7500.

Hemmi H, Ishibashi J, Tomie T, Yamakawa M (2003). Structural Basis for New Pattern of Conserved Amino Acid Residues Related to Chitin-binding in the Antifungal Peptide from the Coconut Rhinoceros Beetle Oryctes rhinoceros. J. Biol. Chem., 278: 22820-22827.

Huang HW (2000). Action of antimicrobial peptides: Two state model. Biochem., 39: 8347-8502.

Hultmark D, Engström A, Andersson K, Steiner H, Bennich H, Boman HG (1983). Insect immunity. Attacins, a family of antibacterial proteins from Hyalophora cecropia. EMBO J., 2: 571-576.

Imamura M, Wada S, Koizumi N, Kadotani T, Yaoi K, Sato R, Iwahana H (1999). Acaloleptins A: inducible antibacterial peptides from larvae of the beetle, Acalolepta luxuriosa. Arch. Insect Biochem. Physiol., 40: 88-98.

Inagaki H, Akagi M, Imai HT, Taylor RW, Kubo T (2004). Molecular cloning and biological characterisation of Novel Antimicrobial peptides, pilosulin 3 and pilosulin 4 from a species of the Austrailian ant genus Myrmecia. Arch. Biochem. Biophys., 428: 170-178.

Isaacson RE (2003). MBI-226. Micrologix/Fujisawa. Curr. Opin. Investig. Drugs, 4: 999-1003.

Jacob L, Zasloff M (1994). Potential therapeutic applications of magainins and other antimicrobial agents of animal origin. Ciba Found. Symp., 186: 197-216.

Jang WS, Kim HN, Lee YS, Nam MH, Lee IH (2002). Halocidin: A new antimicrobial peptide from hemocytes of the solitary tunicat Halocynthia aurantium. FEBS Lett., 521: 81-86.

Jarosz J (1997). Identification of immune inhibitor from Pseudomonas aeruginosa of inducible cell-free antibacterial activity in insects. Cytobios, 89: 73-80.

Jarosz J, Gliński Z (1990). Selective inhibition of cecropin-like activity of insect immune blood by protease from American foulbrood scales. J. Invertebr. Pathol., 56: 143-149.

Jaynes JM, Burton CA, Barr SB, Jeffers GW, Julian GR, White KL, Enright FM, Klei TR, Laine RA (1988). In vitro cytocidal effect of novel lytic peptides on Plasmodium falciparum and Trypanosoma cruzi. FASEB J., 13: 2878-2883.

Jia X, Patrzykat A, Devlin RH, Ackerman PA, Iwama GK, Hancock RE (2000). Antimicrobial peptides protect coho salmon from Vibrio anguillarum infections. Appl. Environ. Microbiol., 66: 1928-1932.

Joerger RD (2003). Alternatives to Antibiotics: Bacteriocins, Antimicrobial Peptides and Bacteriophages. Poult. Sci., 82: 640-647.

Johansson J, Gudmundsson GH, Rottenberg ME, Berndt KD, Agerberth B (1998). Conformation-dependent Antibacterial Activity of the Naturally Occurring Human Peptide LL-37. J. Biol. Chem., 273: 3718-3724.

Johns R, Sonenshine DE (2001). Identification of a defensin from the hemolymph of the American dog tick, Dermacentor variabilis. Insect Biochem. Mol. Biol., 31: 857-865.

Kaaya GP, Flyg C, Boman HG (1987). Induction of cecropin and attacin like antibacterial factors in the heamolymph of Glossina moristans moristans. Insect Biochem., 17: 309-315.

Koyama Y, Motobu M, Hikosaka K, Yamada M, Nakamura K, Saido-Sakanaka H, Asaoka A, Yamakawa M, Sekikawa K, Kitani H, Shimura K, Nakai Y, Hirota Y (2006). Protective effects of antimicrobial peptides derived from the beetle Allomyrina dichotoma defensin on endotoxic shock in mice. Int. Immunopharmacol., 6: 234-240.

Kragol G, Lovas S, Varadi G, Condie BA, Hoffma R, Otvos L (2001). The antibacterial peptide pyrrhocoricin inhibits the ATPase actions of DnaK and prevents chaperone -assisted protein folding. Biochem., 40: 3016-3026.

Kuhn-Nentwig L, Iler JrM, Schaller J, Walz A, Dathe M, Nentwig W (2002). Cupiennin 1, a New Family of Highly Basic Antimicrobial Peptides in the Venom of the Spider Cupiennius salei (Ctenidae). J. Biol. Chem., 277: 11208-11216.

Laederach A, Andreotti AH, Fulton DB (2002). Solution and micelle-bound structures of tachyplesin I and its active aromatic linear derivatives Biochem., 41: 12359-12368.

Lamberty M, Ades S, Uttenweiler-Joseph S, Brookhart G, Bushey D, Hoffmann JA, Bulet P (1999). Isolation from the lepidopteran Heliothis virescens of a novel insect defensin with potent antifungal activity. J. Biol. Chem., 274: 9320-9326.

Landon C, Meudal H, Boulanger N, Bulet P, Vovelle F (2006). Solution structures of stomoxyn and spinigerin, two insect antimicrobial peptides with an alpha-helical conformation. Biopolymers, 81: 92-103.

Landon C, Sodano P, Hetru C, Hoffmann J, Ptak M (1997). Solution structure of drosomycin, the first inducible antifungal protein from insects. Protein Sci., 6: 1878-1884.

Lauth X, Nesin A, Briand JP, Roussel JP, Hetru C (1998). Isolation, characterisation and chemical synthesis of a new insect defensin from Chironomus plusmosus (Diptera). Insect Biochem. Mol. Biol., 28: 1059-1066.

Lee I, Zhao C, Cho Y, Harwig SS, Cooper EL, Lehrer RI (1997). Clavanins, alpha-helical antimicrobial peptides from tunicate hemocytes. FEBS Lett., 400: 158-162.

Lee JY, Boman A, Sun CX, Andersson M, Jörnvall H, Mutt V, Boman H (1989). Antibacterial peptides from pig intestine: Isolation of a mammalian cecropin. Proc. Natl. Acad. Sci. USA, 86: 9159-9162.

Lee KH (2002). Development of short antimicrobial peptides derived from host defense peptides or by combinatorial libraries. Curr. Pharm. Des., 8: 795-813.

Lee SY, Moon HJ, Kurata S, Kurama T, Natori S, Lee BL (1994). Purification and molecular cloning of cDNA for an inducible antibacterial protein of larvae of a coleopteran insect, Holotrichia diomphalia. J. Biochem. (Tokyo), 115: 82-86.

Leuschner CM, Enright F, Gawronska B, Hansel W (2003). Membrane disrupting lytic peptide conjugates destroy hormone dependent and independent breast cancer cells in vitro and in vivo. Breast Cancer Res.Treat., 78: 17-27.

Leushner C, Hansel W (2004). Membrane Disrupting Lytic Peptides for Cancer Treatments. Curr. Pharmacol. Des., 10: 2299-2310.

Liu G, Kang D, Steiner H (2000). Trichoplusia ni lebocin, an inducible immune gene with a downstream insertion element. Biochem. Biophys. Res. Commun., 269: 803-807.

Lowenberger C (2001). Innate immune response of Aedes aegypti. Insect Biochem. Mol. Biol., 31: 219-229.

Lundström A, Liu G, Kang D, Berzins K, Steiner H (2002). Trichoplusia ni gloverin, an inducible immune gene encoding an antibacterial insect protein. Insect Biochem. Mol. Biol., 32: 795-801.

Lynch EC, Rosenberg IM, Gitler C (1982). An ion-channel forming protein produced by Entamoeba histolytica. EMBO J., 1: 801-804.

Mackintosh JA, Veal DA, Beattie AJ, Gooley AA (1998b). Isolation from an ant Myrmecia gulosa of two inducible O-glycosylated proline-rich antibacterial peptides. J. Biol. Chem., 273: 6139-6143.

Mackintosh J, Gooley A, Karuso P, Beattie A, Jardine D, Veal D (1998a). A gloverin-like antibacterial protein is synthesized in Helicoverpa armigera following bacterial challenge. Dev. Comp. Immunol., 22: 387-399.

Mandard N, Bulet P, Caille A, Daffre S, Vovelle F (2002). The solution structure of gomesin, an antimicrobial cysteine-rich peptide from the spider. Eur. J. Biochem., 269: 1190-1198.

Mandard N, Podano S, Habbe L, Bonmatin JM, Bulet P, Hetru C, Ptak M, Vovelle F (1998). Solution structure of thanatin, a potent bactericidal and fungicidal insect peptide, determined from proton two-dimensional nuclear magnetic resonance data. Eur. J. Biochem., 256: 404-410.

Marchini D, Giordano P, Amons R, Bernini LF, Dallai R (1993). Purification and primary structure of ceratotoxin A and B, two antibacterial peptides from the female reproductive accessory glands of the medfly *Ceratitis capitata* (Insecta:Diptera). Insect Biochem. Mol. Biol., 23: 591-598.

Marr AK, Gooderham WJ, Hancock RE (2006). Antibacterial peptides for therapeutic use: obstacles and realistic outlook. Curr. Opin. Pharmacol., 6: 468-472.

Mendes MA, Bibiana M, Souza DM, Riberio M, Plama MS (2004). Structural and biological characterisation of two novel peptides from the venom of the neotropical social wasp *Agelaia pallipes pallipes*. Toxiconomy, 44: 67-74.

Mitta G, Vandenbulckle F, Hubert F, Roch P (1999). Mussel defensins are synthesized and processed in granulocytes then released into the plasma after bacterial challenge. J. Cell Sci., 112: 4233-4242.

Moerman L, Bosteels S, Noppe W, Willems J, Clynen E, Schoofs L, Thevissen K, Tytgat J, Van Eldere J, Van Der Walt J, Verdonck F (2002). Antibacterial and antifungal properties of alpha-helical, cationic peptides in the venom of scorpions from southern Africa. Eur. J. Biochem., 269: 4799-4810.

Moore A (2003). The big and small of drug discovery. Biotech versus pharma: Advantages and drawbacks in drug development. EMBO Rep., 4: 114-117.

Moret Y, Schmid-Hempel P (2000). Survival for immunity: The price of imune system activation for bumblebee workers. Science, 290: 1166-1168.

Morikawa N, Hagiwara K, Nakajima T (1992). Brevinin-1 and -2, unique antimicrobial peptides from the skin of the frog, *Rana brevipoda porsa*. Biochem. Biophys. Res. Commun., 189: 184-190.

Mygind PH, Fischer RL, Schnorr KM, Hansen MT, Sönksen CP,Ludvigsen S, Raventós D, Buskov S, Christensen B, De Maria L,Taboureau O, Yaver D, Elvig-Jørgensen SG, Sørensen MV, Christensen B, Frimodt-Moller N, Lehrer RI, Zasloff M, Kristensen HH (2005a). Plectasin is a peptide antibiotic with therapeutic potential from a saprophytic fungus. Nature, 437: 975-980.

Osaki T, Omotezako M, Nagayama R, Hirata M, Iwanaga S, Kasahara J, Hattori J, Ito I, Sugiyama H, Kawabata S (1999). Horseshoe crab hemocyte-derived antimicrobial polypeptides, tachystatins, with sequence similarity to spider neurotoxins. J. Biol. Chem., 274: 26172-26178.

Otvos L (2002). The short proline-rich antibacterial peptide family. Cell Mol. Life Sci., 59: 1138-1150.

Otvos L, Cudic MY, Chua B, Deliyannis GC, Jackson D (2004). An Insect Antibacterial Peptide-Based Drug Delivery System. Mol. Pharm., 1: 220-232.

Otvos L, insug O, Rogers ME, Consolvo PJ, Condie BA, Lovas S, Bulet P, Blaszcyk-Thurin M (2000). Interaction between heat shock protiens and antimicrobial peptides. Biochem., 39: 14150-14159.

Ouellette AJ, Greco RM, James M, Frederick D, Naftilan J, Fallon JT (1989). Developmental regulation of cryptdin, a corticostatin/defensin precursor mRNA in mouse small intestinal crypt epithelium. J. Cell Biol.,108: 1687-1695.

Ourth DD, Lockey TD, Renis HE (1994). Induction of cecropin-like and attacin-like antibacterial but not antiviral activity in Heliothis virescens larvae. Biochem. Biophys. Res. Comm., 200: 35-44.

Page J-M, Dimarcq J-L, Quenin S, Hetru C (2003). Thanatin activity on multidrug resistant clinical isolates of *Enterobacter aerogenes* and *Klebsiella pneumoniae*. Int. J. Antimicrob. Agents, 22: 265-269.

Papo N, Shai Y (2003). New Lytic Peptides Based on the D,L Amphipathic Helix Motif Preferentially Kill Tumor Cells Compared to Normal Cells. Biochem., 42: 9346-9354.

Park JM, Lee JY, Moon HM, Lee BJ (1995). Molecular cloning of cDNAs encoding precursors of frog skin antimicrobial peptides from *Rana rugosa*. Biochim. Biophys. Acta, 1264: 23-25.

Park SS, Shin SW, Park DS, Oh HW, Boo KS, Park HY (1997). Protein purification and cDNA cloning of a Cecropin like peptide from the larvae of fall webworm *Hyphantria cunea*. Insect Biochem. Mol. Biol., 27: 711-720.

Peschel A, Otto M, Jack RW, Kalbacher H, Jung G, Gotz F (1999). Inactivation of the dlt operon in *Staphylococcus aureus* confers sensitivity to defensins, protegrins and other antimicrobial peptides. J. Biol. Chem., 274: 8405-8410.

Pillai AS Ueno H, Zhang Y, Kato Y (2003). Induction of ASABF (*Ascaris suum* antibacterial factor)-type antimicrobial peptides by bacterial injection: Novel members of ASABF in the nematode *Ascaris suum*. Biochem. J., 371: 663-668.

Possani LD, Corona M, Zurita M, Rodríguez MH (2002). From noxiustoxin to scorpine and possible transgenic mosquitoes resistant to malaria. Arch. Med. Res., 33: 398-404.

Pukala TL, Boland MP, Gehman JD, Kuhn-Nentwig L, Separovic F, Bowie JH (2007). Solution Structure and Interaction of Cupiennin 1a, a Spider Venom Peptide, with Phospholipid Bilayers. Biochem., 46: 3576-3585.

Rees JA, Moniatte M, Bulet P (1997). Novel Antibacterial peptides isolated from a European Bumblebee, *Bombus pascuorum* (Hymenoptera, Apoidea). Insect Biochem. Mol. Biol., 27: 413-422.

Ryder JJ, Siva-Jothy MT (2000). Male calling song provides a relliable signal of immune function in a cricket. Proc. R. Soc. Lond. B., 267: 1171-1175.

Sawai MV, Waring AJ, Kearney WR, McCray PB, Forsyth WR, Lehrer RI, Tack BF (2002). Impact of single-residue mutations on the structure and function of ovispirin/novispirin antimicrobial peptides. Protein Eng., 15: 225-232.

Schroder JM, Harder J (2006). Antimicrobial Peptides in Skin disease. Drug Discov. Today, 3: 93-100.

Selsted ME, Novotny MJ, Morris WL, Tang YQ, Smith W, Cullor JS (1992). Indolicidin, a novel bactericidal tridecapeptide amide from neutrophils. J. Biol. Chem., 267: 4292-4295.

Shai Y, Oren Z (2001). From "carpet" mechanism to de-novo designed diastereomeric cell-selective antimicrobial peptides. Peptides, 22: 1629-1641.

Shao F, Hu Z, Xiong YM, Huang QZ, Wang CG, Zhu RH, Wang DC (1999). A new antifungal peptide from the seeds of *Phytolacca americana*: Characterization, amino acid sequence and cDNA cloning. Biochem. Biophys. Acta, 1430: 262-268.

Si L-g, Liu X-C, Lu Y-Y, Wang G-Y, Li W-M (2007). Soluble expression of active human β-defensin-3 in *Escherichia coli* and its effects on the growth of host cells. Chin. Med. J., 120: 708-713.

Simmaco M, Mignogna G, Barra D, Bossa F (1993). Novel antimicrobial peptides from skin secretion of the European frog *Rana esculenta*. FEBS Lett., 324: 159-161.

Simmaco M, Mignogna G, Canofeni S, Miele R, Mangoni ML, Barra D (1996). Temporins, antimicrobial peptides from the European red frog *Rana temporaria*. Eur. J. Biochem., 242: 788-792.

Sousa LB, Mannis MJ, Schwab IR, Cullor J, Houstani H, Smith W, Jaynes J (1996). The use of synthetic cecropin (D5C) in disinfecting contact lens solutions. CLAO, 22: 114-115.

Steinborner ST, Waugh RJ, Bowie JH, Wallace JC, Tyler MJ, Ramsay SL (1997). New caerin antibacterial peptides from the skin glands of the Australian tree frog *Litoria xanthomera*. J. Pept. Sci., 3: 181-185.

Storici P, Zanetti M (1993). A novel cDNA sequence encoding a pig leukocyte antimicrobial peptide with a cathelin-like pro-sequence. Biochem. Biophys. Res. Commun., 196: 1363-1368.

Suetake T, Tsuda S, Kawabata S, Miura K, Iwanaga S, Hikichi K, Nitta K, Kawano K (2000). Chitin-binding proteins in invertebrates and plants comprise a common chitin-binding structural motif. J. Biol. Chem., 275: 17929-17932.

Tomie T, Ishibashi J, Furukawa S, Kobayashi S, Sawahata R, Asaoka A, Tagawa M, Yamakawa M (2003). Scarabaecin, a novel cysteine-containing antifungal peptide from the rhinoceros beetle, Oryctes\ minoceros. Biochem. Biophys. Res. Commun., 307; 261-266.

Toney JH (2002). Iseganan (IntraBiotics Pharmaceuticals). Curr. Opin. Investig. Drugs, 3: 225-228.

Trenczek T (1997). Biological mediators of insect immunity. Annu. Rev.

Entomol., 42: 611-643.

Wachinger M, Kleinschmidt A, Winder D, Pechmann NV, Ludvigsen A, Neumann M, Holle R, Salmons B, Erfle V, Brack-Werner R (1998). Antimicrobial peptides melittin and cecropin inhibit replication of human immunodeficiency virus 1 by suppressing viral gene expression. J. Gen. Virol., 79: 731-740.

Wang X, Thoma RS, Carroll JA, Duffin KL (2002). Temporal generation of multiple antifungal proteins in primed seeds. Biochem. Biophys. Res. Comm., 292: 236-242.

Wang Z, Wang G (2004). APD: The Antimicrobial Peptide Database. Nucleic Acids Res., 32: D590-592.

Watt JY, Vanderift WB (1950). Laboratory observations on the actions of aureomycin, circulin, polymyxins B, D, and E on *Endamoeba histolytica*. J. Lab. Clin. Med., 36: 741-746.

Yamada M, Nakamura K, Saido-Sakanaka H, Asaoka A, Yamakawa M, Yamamoto Y, Koyama Y, Hikosaka K, Shimizu A, Hirota Y (2005). Therapeutic effect of modified oligopeptides from the beetle *Allomyrina dichotoma* on methicillin-resistant *Staphylococcus aureus* (MRSA) infection in mice. J. Vet. Med. Sci., 67: 1005-1011.

Yan L, Adams ME (1998). Lycotoxins, antimicrobial peptides from venom of the wolf spider *Lycosa carolinensis*. J. Biol. Chem., 273: 2059-2066.

Zasloff M (1987). Magainins, a class of antimicrobial peptides from Xenopus skin: Isolation, characterization of two active forms, and partial cDNA sequence of a precursor. Proc. Natl. Acad. Sci. USA., 84: 5449-5453.

Zhao C, Liaw L, Lee IH, Lehrer RI (1997). cDNA cloning of three cecropin-like antimicrobial peptides (Styelins) from the tunicate, *Styela clava*. FEBS Lett., 412: 144-148.

Zwaal RFA, Schroit AJ (1997). Pathophysiologic implications of membrane phospholipid asymmetry in blood cells. Blood, 89: 1121-1132.

9

Taxoids: Biosynthesis and *in vitro* production

Priti Maheshwari[1], Sarika Garg[2] and Anil Kumar[3*]

[1]Faculty of Arts and Science, Department of Biological Sciences, 4401, University Drive, University of Lethbridge, Lethbridge, Alberta, T1K 3M4, Canada.
[2]Max Planck Unit for Structural Molecular Biology, c/o DESY, Gebaüde 25b, Notkestrasse 85, D- 22607 Hamburg, Germany
[3]School of Biotechnology, Devi Ahilya University, Khandwa Road Campus, Indore – 452001, India.

Taxoids viz. paclitaxel and docetaxel are of commercial importance since these are shown to have anti-cancerous activity. These taxoids have been isolated from the bark of *Taxus* species. There is an important gymnosperm, *Taxus wallichiana* (common name, 'yew') used for the isolation of taxoids. Due to cutting of the trees for its bark, population of the plant species are threatened to be endangered. Therefore, these are required to be protected globally. Plant cell culture techniques have been exploited for the isolation of mutant cell lines, production of secondary metabolites and genetic transformation of the plants. *In vitro*, culture of *Taxus* not only helps in conservation but is also helpful in the production of paclitaxel and other taxoids. Various strategies tested globally for the commercial production of taxoids are discussed. Different *Taxus* species, their origin, diterpenoids obtained from different parts of the tree and their applications are discussed. Although, detailed taxoid biosynthetic pathway is not well known, an overview of the pathway has been described. Micropropagation of *Taxus* and regeneration of transgenic plants has been described. Although, several protocols have been reported for the production of some important taxoids, a rapid, reproducible and economically viable protocol required for the efficient production of taxoids has yet to be established. Supplementation of the biotic and abiotic elicitor(s) to the cell suspension cultures of *Taxus* has been shown to increase the growth of the cell biomass as well as paclitaxel production due to pathway stimulation. Up-scaling of *Taxus* cell lines capable of over-producing taxoids could only make the industrial production of paclitaxel feasible. Here, we have reviewed *Taxus wallichiana* cell cultures in terms of their capabilities of biomass and secondary metabolites production.

Key words: Docetaxel, paclitaxel, taxanes, taxus.

INTRODUCTION

Plant tissue culture techniques have made the isolation and culture of cells, tissues and organs of many plant species possible, and have been exploited for the produ-ction of various secondary metabolites including several economically important metabolites (Wink et al., 2005). Most of the current strategies for the improved secondary metabolites production involve use of the bioreactors and genetically engineered whole plants and / or cultivation of cells derived from such plants or tissues (Rischer and Oksman-Caldentey, 2005; Ho et al., 2005). These techniques allow manipulation of the pathways by engineering rate limiting genes and cell cultures to obtain enhanced production of desired metabolites. The plant tissue culture techniques also involve induction of the hairy roots and development of transgenic plants through crown gall formation. *Taxus* species belong to the family Taxeaceae (common name 'Yew') and are the source of taxoids, a group of potent chemotherapeutic anti-cancerous agents. The most important compounds are paclitaxel (Taxol);

---

*Corresponding author. E-mail: ak_sbt@yahoo.com.

Abbreviations: BA - benzylaminopurine; 10-DAB - 10-deacetyl baccatin III; DR - Dragendorf's reagent; GA_3 - gibberellic acid; GI - growth index; IAA - indole – 3 – acetic acid; JA - Jasmonic acid; MDR - multi-drug resistance; MS - Murashige and Skoog; NAA - ∝-naphthalene acetic acid; NCI - National Cancer Institute; PVP - polyvinyl pyrrolidone; WPM - woody plant medium.

Table 1. Distribution of the different Taxus species

Taxus species	Distribution
T. baccata	Europe and Asia (European yew)
T. brevifolia	Northwest Pacific (Pacific yew)
T. canadensis	Canada
T. celebica	Asia
T. cuspidata	Japan (Japanese yew)
T. floridana	Northwest Florida
T. globosa	Mexico / El Salvador
T. wallichiana	Indian Himalayas (Himalayan yew)

and its biogenetic precursor, 10-deacetyl baccatin III (10-DAB) that is chemically transformed to docetaxel (Lavelle et al., 1995). Taxoids are complex polyoxygenated diterpenoids having phenylisoserine moieties. These compounds are active in various *in vitro* and *in vivo* pre - clinical models viz. cell lines, human tumor stem cells, murine grafted tumors, human xenografts etc. and have been extensively used as drugs against ovarian, breast, lung, head, neck, prostate, cervical cancers and AIDS related Kaposi's sarcoma (Ketchum et al., 2007; Itokawa, 2003).

The toxicity of the yew tree is known since long time and has been explained due to the presence of taxine, a complex mixture isolated from the leaves of the tree. Taxine has been shown to be a mixture of seven alkaloids with taxine A and B as major components (Prasain et al., 2001). Contrary to taxoids, taxines do not exhibit anti-tumor activity and are relatively abundant in plants. These taxines can be used as starting material for the semi-synthetic production of paclitaxel derivatives (Croteau et al., 2006).

The first taxoid was discovered in 1960s by the National Cancer Institute (NCI) during a large scale plant screening program. It was shown that the bark of Pacific Yew, *Taxus brevifolia* was active against many cancers (Baloglu and Kingston, 1999; Itokawa, 2003). The active ingredient of the bark was later identified as paclitaxel. It exhibits anti-tumor activity against murine leukemia cells (Appendino, 1993), ovarian (McGuire et al., 1998) and breast cancers (Gotaskie and Andreassi, 1994). These findings stimulated an interest in the chemistry, biology and pharmacology of different species of *Taxus* (Wani et al., 1971). In 1983, NCI carried out phase I clinical trials. Afterwards, John Hopkins Oncology Center reported striking clinical results for the treatment of breast cancer. In the same year, Bristol Myers Squibb (Princeton, NJ) became the partner of NCI to commercialize the drug and was granted the patent for Taxol®. Yew tree pharmaceuticals (Haarlem, The Netherlands) developed Yew taxan®, a paclitaxel formulation. Rhone Poulenc (Paris, France) was granted approval for Taxotere® (a semi - synthetic taxane with the generic name docetaxel) in more than 33 countries.

After the FDA approval of paclitaxel and docetaxel as anticancer drugs, efforts are made for the search of new anticancer drugs and a number of new analogues are prepared for getting taxoids with better potency and availability to the targeted cells. Besides, efforts are made to search taxoids efficient in targeting multi-drug resistant cancers or capable of acting as modulators in multi-drug resistance (MDR) (Odgen, 1988; Ketchum et al., 2007). Studies have also been carried out for searching improved methods for getting natural or synthetic preparations of these important anticancer drugs (Dubois et al., 2003). Since, there are reports indicating undesired side effects on using paclitaxel, it has been considered important to develop new taxoids with fewer side effects, superior pharmacological properties, and improved activity against various classes of tumors. This approach resulted in the discovery and development of "Second generation taxoids" with various modifications in the baccatin skeleton and the phenylisoserine side chain (Lin and Ojima, 2000).

The contents of paclitaxel have been found to be very low in the bark (0.017% dry weight in *T. brevifolia*). Therefore, globally, efforts are made to overcome the supply problem since yield by isolation from the bark of the slow-growing yew trees is limited (Witherup et al., 1990). Alternative sources of paclitaxel viz. use of needles, stem bark, heart wood, roots of several other species of *Taxus* have been tried (Donovan, 1995). Besides, plant cell cultures as well as chemical and biotechnological semi-synthesis have been intensively investigated for the production of paclitaxel (Taxol) and docetaxel (Taxotere) in last few years (Chattopadhyay et al., 1995; Nicolaou et al., 1994). Besides, production of paclitaxel using fungal cultures has also been reported (Stierle et al., 1993). In this review, an effort has been made to correlate various aspects of *in vitro* production of taxoids.

Origin and distribution of *Taxus* species

The different *Taxus* species are considered as a single species descending from *Paleotaxus rediviva*, a fossil angiosperm that was abounded about 200 million years ago in the Triassic era, and later got confined to the temperate zones of the northern hemisphere during various glaciations (Hartzell, 1991). Although, the members of taxeaceae family are apparently similar to the conifers, but due to the absence of cones and resin ducts, are placed in a separate order named as Taxales. Yews are dioecious evergreen trees or shrubs and produce seeds having an edible fleshy aril. Many scientists believe that eight to ten species of *Taxus* are there (Table 1) (Krusmann, 1972; Appendino et al., 1993) whereas, others consider these species as simply different geo-graphical varieties. Besides the regular species, two hybrids are also known namely *Taxus x media* Rehder which is a cross between *Taxus baccata* and *Taxus cuspidata* and

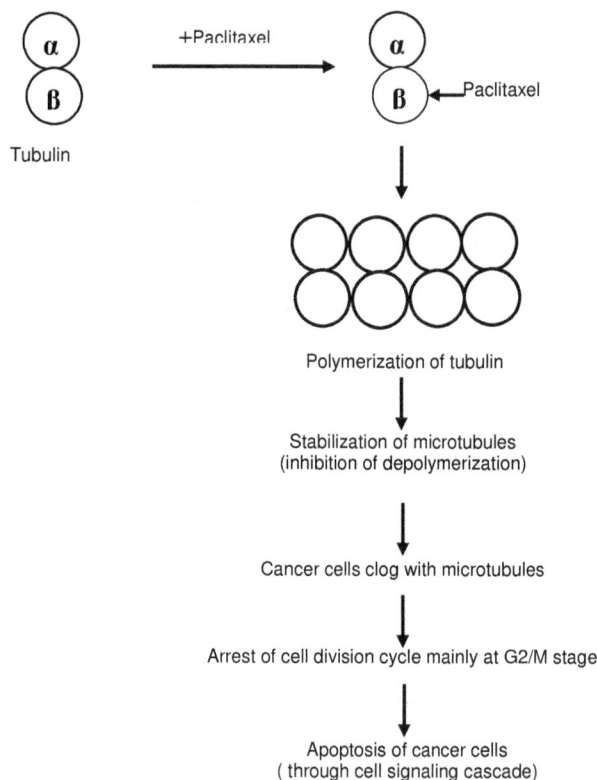

Figure 1. A flow diagram showing mode of action of paclitaxel in the apoptosis of cancer cells.

Taxus x hunnewelliana Rehder, a cross between *T. cuspidata* and *Taxus canadensis*.

Taxus diterpenoids and their uses

Ancient Greeks named the yew as '*Taxus*' due to two important aspects of the tree: Taxon means bow and Toxicon is poison (Hartzell, 1991). The Irish Druids had high regards for the tree due to its poisonous nature and was believed to be efficacious against fairies and witches (Wani et al., 1971). The structure of paclitaxel was first reported in 1971 (Wani et al., 1971). It is a complex and highly oxygenated diterpenoid amide. The compound and its related precursors/ products are commonly called as taxanes or taxoids. Generally, a taxoid compound contains an oxetane ring and is oxygenated at positions C-1, C-2, C-5, C-7, C-9, C-10 and C-13. However, rarely taxoids are oxygenated at C-14 position. Presence of a phenyl isoserine ester group at C-13 position of the diterpenoid taxanes is a critical feature for the anti – cancerous property of taxoids (Ojima et al., 1992). Taxol C is found to be more potent anti-cancerous agent than paclitaxel (Kobayashi et al., 1994). In addition, docetaxel, an analogue of taxol, has almost the same potency against cancers as paclitaxel (Guenard et al., 1993). Shen et al. (1999) isolated many diterpenoids namely

chinentaxunine, 10-deacetyl taxol A, 10-deacetyl-7-*epi*-taxol, 10-deacetyl-10-oxo-7-*epi*-taxol, taxinine M, taxchinin A, 10-deacetyl taxinine B and taxuspine X from the seeds of *Taxus chinensis*. Shi et al (1999) isolated several new taxane diterpenoids from the seeds of *Taxus chinensis* var. *mairei* and *Taxus yunnanensis*. Brubaker et al. (2006) showed that zoledronic acid, an osteolysis inhibitory agent, in combination with docetaxel effectively inhibits growth of the prostate and bone tumors and therefore is a potential treatment option. By the use of fluorescent taxoids, paclitaxel binding site has been shown on the outer side of the microtubules (Diaz et al, 2005). Chen and Hong (2006) studied chemical modify-cation of Taxus diterpenoids isolated from *T. brevifolia* and bisindole alkaloids isolated from *Catha-ranthus roseus* for efficient use as antitumor agents. Shen et al. (2005) isolated two new bicyclic taxoids namely tasumatrols M (1) and N (2) from the leaves and twigs of *Taxus sumatrana* and determined their structures by using 1D and 2D NMR and chemical derivatization. Zhang et al. (2006) isolated a new abeotaxane diterpenoid from the seeds of *Taxus mairei* and identified its structure as 2α, 5α, 13α-trihydroxy-7β, 10β-diacetoxy-2(3----20) abeotaxa-4(20), 11-dien-9-one by using 1D and 2D spectral analysis.

Mode of action

Taxoids inhibit the growth of cancer cells by affecting microtubules. During normal cell growth, microtubules are formed during cell division. However, once the cell stops dividing, the microtubules are broken down or destroyed. The microtubules produced in the presence of taxoids are resistant to disassembly by physiological stimuli, and cells exposed to these agents exhibit an accumulation of disorganized microtubule arrays. It affects the normal mitotic process and eventually results in cell death (Schrijvers and Vermorken, 2000). Paclitaxel binds to the ß-subunit of the tubulin heterodimer. After binding, the complex stimulates the polymerization of tubulin and stabilizes the resultant microtubules. It leads in the inhibition of their de-polymerization. Thereafter, cancer cells get clogged with the microtubules and are unable to divide resulting in the arrest of cell division cycle mainly at the G2 / M stage. This causes apoptosis of the cancer cells through cell signaling cascade. Hence, taxoids may be considered as a new class of anti - mitotic agents different from other plant derived anti-microtubule agents such as colchicine, podophyllotoxin and *Vinca* alkaloids that inhibit microtubule assembly. The mode of action of paclitaxel for the apoptosis of the cancer cells has been shown in Figure 1. Paclitaxel/taxoids are lipophilic in nature and most likely accumulate into the extracyto-plasmic compartments (the cell wall) and not sequestered within the cellular compart-ments viz. vacuoles / plastids. *Taxus* cells avoid its toxic effect on self by excluding it

Table 2. Characteristics of the enzymes involved in taxoid biosynthetic pathway.

Enzyme	Opt. pH	Km	ORF of cDNA	Residues	Mol. wt. (kDa)	Reference
Geranyl geranyl pyrophosphate synthase	--	--	1179	393	42.6	Stierle et al., 1993
Taxadiene synthase	8.5	3µM for prenyl substrate	2586	862	98.3	Nicolaou et al., 1994
Taxadien-5α-yl acetate 10β – hydroxylase (Taxane 10β – hydroxylase)	--	--	1494	498	56.69	Krusmann, 1972
Taxadien-5α-ol-O-acetyltransferase	9.0	4.2µM for taxadienol and 5.5µM for acetyl CoA	1317	439	49.079	Hartzell, 1991
10-Deacetylbaccatin III-10β-O-acetyltransferase	7.5	10 µM for 10-deacetylbaccatin III & 8 µM for acetyl CoA	1320	440	49.052	Grothaus et al., 1993
Taxane 2α-O-benzoyltransferase	8.0	0.64µM for 2-debenzoyl-7,13-diacetylbaccatin III & 0.30µM for benzoyl coenzyme A	1320	440	50.089	Saikia et al., 2000

from the protoplast and excreting it into the cell wall. This system is considered as a natural defense system in cells against toxic effects of their own compounds.

Taxoid biosynthetic pathway

The detailed taxoid biosynthetic pathway is not well known. Floss and Mocek (1995) carried out feeding studies with advanced labeled taxoids, and gave an overview of the pathway. Taxoid biosynthesis occurs in non-photosynthetic stem tissue in vivo, and in non-photosynthetic, undifferentiated cells in culture systems in vitro. It involves several enzymatic steps leading to the formation of a tetracycline skeleton by cyclization of the universal diterpenoid precursor, geranyl geranyl pyrophosphate and afterwards oxygen and acyl groups are added to this taxane core by oxygenation at multiple positions mediated by cytochrome P450 mono-oxygenases. Croteau et al. (2006) showed that taxol biosyn-thetic pathway

involves 19 discrete enzymatic steps from primary metabolism. It is shown that there is precursor flux to the production of taxoids other than taxol (Kikuchi and Yatakai, 2003; Takeya, 2003; Ketchum et al., 2007). Any effort to improve the yield of paclitaxel and its precursors must consider numerous and diversionary taxoid bio-synthetic pathway outlined by Ketchum et al. (2007). A brief summary of the characteristics of the few enzymes studied is given in Table 2. Over the past years, significant advances have been made in the identification of the genes that code the enzymes involved in this metabolic pathway. In a study, cDNAs of five enzymes involved in taxoid biosynthesis have been isolated, cloned, expressed and characterized (Walker and Croteau, 2001). These are the enzymes of the first committed step: taxadiene synthase (which constructs the taxane skeleton), a cytochrome P450 taxane hydroxylase and three taxoid O – acyltransferases. Long and Croteau (2005) sho-wed involvement of a microsomal 2'- hydroxyl-lase in

the cytochrome P450 mediated C-13 side chain hydroxylation of β-phenylalanoyl baccatin III to form phenylisoserinoyl baccatin III. There are several enzymes involved in taxoid biosynthesis whose genes are yet to be identified. A line diagram (Figure 2) describes various known steps involved in the biosynthesis. The entire pathway can be summarized in the following steps.

Taxadiene formation

The first committed step of taxol biosynthesis is catalyzed by taxadiene synthase, an enzyme first isolated from T. brevifolia stem. The enzyme catalyzes the cyclization of the universal diterpenoid precursor, geranylgeranyl pyrophosphate to taxa-4 (5), 11(12)-diene resulting in the formation of a taxane skeleton. This enzyme mediates a unique intramolecular hydrogen migration in the B / C ring closure step. It is a monomer (mol wt ~79 kDa) and requires Mg^{2+} for the enzyme activity (Hezari et al., 1995). Its cDNA has been cloned

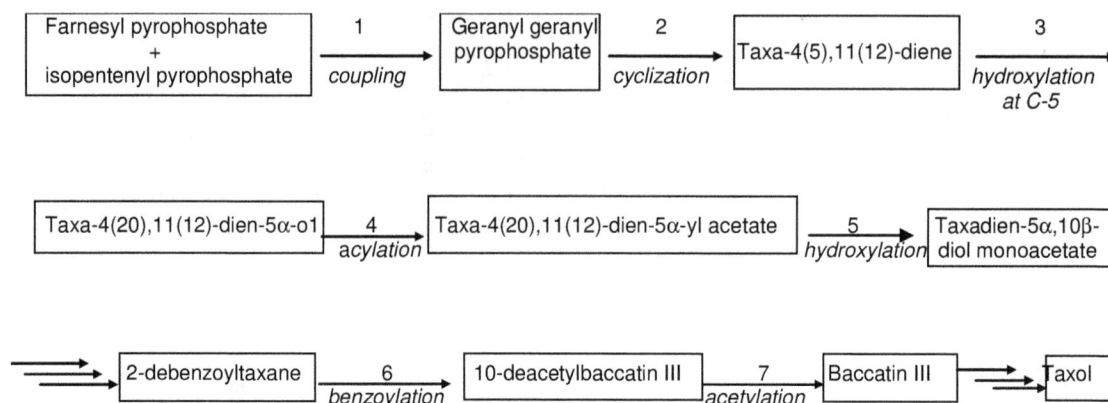

Figure 2: Taxoid biosynthesis pathway. Enzymes involved in various steps are: 1. Geranylgeranyl pyrophosphate synthase; 2. Taxadiene synthase; 3. Cytochrome P450 taxadiene 5α-hydroxylase; 4. Taxa-4(20),11(12)-diene-5α-ol-O-acetytranferase; 5. Cytochrome P450 taxane 10β-hydroxylase, 6. Taxane 2α-O-benzoyltransferase and 7. 10-deacetyl baccatin III-10-O-acetyltransferase. Multiple arrows indicate multiple convergent steps.

and functionally expressed in *E. coli* (Wildung and Croteau, 1996). Geranylgeranyl pyrophosphate is synthesized by a prenyl transferase named geranylgeranyl pyrophosphate synthase. Its gene has been isolated and sequenced. This enzyme is of special interest as it leads to the formation of a branched point progenitor of a variety of diterpenoids and tetraterpenoids (Wildung and Croteau, 1996; Hefner et al., 1998). Methyl jasmonate modulates the production of this enzyme at transcriptional level.

Taxadienol formation

Kingston et al. (1993) suggested hydroxylation of taxa (4)5, 11(12) taxadiene at C-5 position catalyzed by the enzyme, taxadiene 5α-hydroxylase. This hydroxylation results in the formation of taxa-4 (20), 11(12)-dien-5α-ol, the second step in taxoid biosynthesis. The cytochrome P450 monooxygenases bring about hydroxylation of various intermediates of the pathway (Jennewein et al., 2003). Subsequent oxy-genations depend on the relative availability of oxygen functional groups on the taxane ring. The suggested order of oxygenation after C5 is C10, C2, C9, C13 (Hefner et al., 1996).

Acylation of taxadienol

Since naturally occurring taxoids bear an acylation at C-5 position, it is considered that acetylation of taxa-4 (20),11(12)-dien-5α–ol in the presence of Acetyl CoA could be the next step in paclitaxel biosynthesis (Kingston et al., 1993). An another important enzyme, taxane 10β -hydroxylase has also been isolated that catalyzes the conversion of taxadien-5α-yl acetate into

10β -hydroxy taxadien-5α-yl acetate. The third specific intermediate in the pathway is taxa-4(5), 11(12)-diene 5α-yl acetate. Reverse genetic approach led to the isolation of a full length cDNA encoding taxa-4 (20), 11(12)-dien-5α-ol-O-acetyltransferase, an enzyme that catalyzes the first acylation step using taxadienol and acetyl coA as substrates and is highly specific towards the C5 hydroxyl position. The taxa-4 (20), 11(12)-dien-5α-acetate provides a functional group that is further used for the extension of oxetane ring by epoxidation of 4(20)- double bond. Subsequent reactions include additional oxygenations, hydroxy group acylations, oxidation to ketone and ultimately generation of the oxetane ring. The formation of an oxetane ring is a central step in the biosynthesis of all the bio-active taxoids. Another identified transacetylation reaction in the taxol biosynthetic pathway involves hydroxylation at C10 position of the 10-DAB. The reaction is catalyzed by the enzyme, 10-deacetyl-baccatin III–10-O-acetyltransferase. It leads to the formation of the last diterpene intermediate, baccatin III in the pathway using 10- DAB and acetyl coA as substrates. A full length cDNA clone of this enzyme has been isolated from *T. cuspidate* (Walker and Croteau, 2000). The last enzymatic step (acylation) in the taxol biosynthetic pathway is catalyzed by several full length acyl transferases (CoA thioester dependent), amongst which 3'–N-debenzoyl – 2'- deoxytaxol N–benzoyltransferase is identified, cloned and expressed in *E. coli*. The enzyme catalyzes the conversion of 2-debenzoyl-7,13-diacetyl-baccatin III and benzoyl coenzyme A into 7,13- diacetyl-baccatin III. This enzyme can be exploited to improve the production yield of taxol in genetically engineered systems (Walker et al., 2002). Structural studies of these acyltransferases have revealed presence of a conserved sequence HXXXDG. The histidine residue is considered to be essential for the catalytic activity of these enzymes.

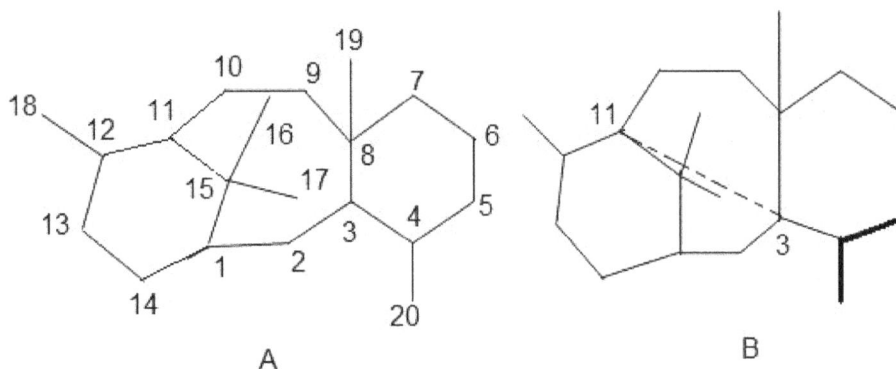

Figure 3. 6/8/6 and 5/7/6 ring systems of taxoids.

TAXOID COMPOUNDS IDENTIFIED IN *TAXUS*

During the search for improved analogues of paclitaxel, a large number of chemical constituents in *Taxus* species are identified (Parmar et al., 1999). There are several reports on the isolation of taxoids, flavonoids, abeotaxanes, lignans and glycosides from *Taxus*. In addition to taxoids, *Taxus wallichiana* has been used for the isolation of a wide range of non - taxoid compounds viz. higher isoprenoids, lignans, flavonoids, sugar derivatives, apocarotenoids and phenolic compounds. Approximately 120 taxoids have been isolated to date from the Japanese yew (Kobayashi and Shigemori, 2004). The taxoid compounds identified in various *Taxus* species have been summarized in Table 3. The normal taxoids contain 6/8/6-membered ring system (Figure 3A) but rearranged taxoids (naturally occurring) possess a 5/7/6/ or 6/10/6 membered ring system (Figure 3B). Taxoids with 5/7/6 type of ring system are known as (15→1) abeotaxoids or nortaxoids whereas those with 6/10/6 type of ring systems as 2 (3 → 20) abeotaxoids.

Normal taxanes

The naturally occurring diterpenoids containing taxane ring are the typical constituents. In addition to paclitaxel, taxine also belongs to this class. The compounds of this group are responsible for the toxicity of leaves and berries.

3, 11-Cyclotaxanes

The 3,11-cyclotaxanes can be produced upon irradiation of corresponding 13-oxo-taxa-11-enes (Appendino et al., 1992b, 1993).

11(15 →1) Abeotaxanes

Taxoids isolated from different *Taxus* species have been found to possess a rearranged nor-taxoid skeleton, 11

(15→1) abeotaxanes. Taxchinin A was the first naturally occurring rearranged taxoid identified as 11 (15→ 1) abeotaxane. Whereas, Taxchinin B isolated from *Taxus chinensis* was the first identified 11 (15→1) obeotaxoid with an oxetane ring. The first natural taxoid, identified to have 11(15→1) abeotaxane ring was brevifoliol (Balza et al., 1991). It was initially assigned a normal taxane skeleton but later corrected to 11(15→1) abeoskeleton (Datta et al., 1994; Georg et al., 1995). Wollifoliol with 11 (15→1) and 11(10→9) are the only bisabeotaxanes isolated from *T. wallichiana*.

Pretaxoids

Epitaxol, acetyltaxol and taxol C can be categorized under this class. These contain ring structure as 6/8/6/4 (A/B/C/D) (Wani et al., 1971).

New taxoids

A series of new taxoids, named taxuspines possessing various skeletons containing 5/7/6, 6/10/6, 6/5/5/6, or 6/8/6, or 6/12-membered ring systems have been isolated from *T. cuspidata* (Kobayashi and Shigemori, 2002).

ISOLATION OF TAXOIDS

In spite of the low yield, until recently, *T. brevifolia* was the sole source of paclitaxel (100 mg/kg of dry bark). Due to short supply compared to increased demand for the drug, investigations started searching for alternate sources viz. other species or tree parts, cell cultures, fungi, synthetic and semisynthetic approaches. Extensive chemical investigations on the Himalayan yew have resulted in the isolation and characterization of several unique taxoids in addition to paclitaxel (Shinozaki et al., 2002; Shi et al., 2001; Cheng et al., 2000; Kiyota et al., 2002; Shen et al., 2002). The contents of paclitaxel and 10-DAB in the shoots of *T .wallichiana* have been repor-

Table 3. Taxoids from different plant parts of *Taxus wallichiana*.

Taxoids	Plant part(s)	Reference (s)
10,13-deacetyl obeobaccatin IV	Stem bark	Wickremesinhe & Arteca, 1993
2-deacetoxydecinnamoyl taxinine J	Stem bark	Navia-Osorio et al., 2002
Taxol and 10-deacetyl baccatin III	Needles	Floss & Mocek, 1995
2-deacetyltaxinineJ, Brevifaliol, 13α-acetoxy brevifoliol, 2α-acetoxy brevifoliol, baccatin IV	Stem bark	Fett-Neto et al., 1992
1 ß-hydroxybaccatin I	Stem bark, needle	Fett-Neto et al., 1992, Kim et al., 1995
Dihydrotaxol, 9-o-benzoyl-9, 10-dideacetyl-11(15→1) abeobaccatinVI and 9-o-benzoyl-9-o-acetyl-10-de-o-acetyl 11(15→1) Abeobaccatin VI	Stem bark	Menhard et al., 1998
19-hydroxybaccatin III	Needles	Seki et al., 1997
10-Deacetyltaxol	Stem bark, needle	Seki et al., 1997
1-hydroxy-2-deacetoxy taxinine J and 7, 2'-bisdeacetoxy austrospicatine	-do-	Yukimune et al., 2000
Pactitaxel	-do-	Kim et al., 1995; Yukimune et al.,, 2000
7-Xylosyl-10-deacetyl taxol C	Needles	Walker et al., 2002; Yukimune et al., 2000
10-deacetyl cephalomannine	-do-	Seki et al., 1997
Taxuspinanane A	-do-	Kobayushi et al., 1994
Brevifoliol	Needles	Furmanowa et al., 1997
7-deacetoxy-10-debenzoyl brevifoliol	-do-	Fett-Neto et al., 1992; Guenard et al., 1993
Taxichinin A (2-acetoxy brevifoliol)	Needles	Yukimune et al., 2000
Taxacuotin & Taxayuntin	Heartwood	Walker et al., 2002
10-debenzoyltax-chinin C and 7, 9, 13-deacetyl abeobaccatin	Stem bark	Furmanowa et al., 1997
1-hydrocybaccatin I and 2-deacetoxy decinnamoyltaxinine J	Needle	Guenard et al., 1993
Wallifoliol	Heart wood	Walker et al., 2002
Yunnaxane and taxusin	Aerial parts	Ketchum et al., 1999
Dantaxusin (II) and dantaxusin (D2)	Needles	Phisalaphong & Linden, 1999
5 alpha O – (3' – dimethyl amino – 3' – phenyl proionyl) taxinine M (1)	Needles	Seki et al., 1997
19-hydroxybaccatin III [3], 10-deacetylcephalomannine [4]) and 10-deacetyltaxol [5].	Leaves	Baebler et al., 2002
5 alpha,13 alpha-diacetoxy-taxa-4[20],11-diene-9 alpha,10 beta-diol [1], 7 beta, 13 alpha-diacetoxy-5 alpha-cinnamyloxy-2[3-->20]-abeo-taxa-4[20],11-diene-2 alpha, 10 beta-diol [2], and 2 alpha,10 beta,13 alpha-triacetoxy-taxa-4[20],11-diene-5 alpha,7 beta,9 alpha-triol [3]	Needles	Linden, & Phisalaphong, 2000
l-beta-hydroxy-7beta-acetoxytaxinine [1] and lbeta,7beta-dihydroxytaxinine [2]	Roots	Wu & Lin, 2003
Taxumairols G (1), H (2), I (3), J [4], and L [5]	Needles	Morrita et al., 2005
Taxezopidines M and N	Seeds	Morrita et al., 2005
2,20-*O*-diacetyltaxumairol N (1) and 14ß-hydroxy-10-deacetyl-2-*O*-debenzoylbaccatin III (2)	Needles and stems	Xia et al., 2005

ted to vary between 18.3-40.6 mg/kg and 247.6-594.9 mg/kg, respectively on dry weight basis (Poupat et al., 2000). Further, the bark and needle leaves have been found to be enriched in taxoids along with other non-taxoids viz. polyphenols, glycosides, lignans and flavor-noids. The needles are renewable sources of both taxoid and non-taxoid compounds. A few taxoids and some sugar derivatives have also been isolated some twigs, seeds, roots and heart wood of *T. wallichiana* (Chattopadhyay et al., 1994, 1997, 1999a, 1999b). The needles also contain a group of 11 toxic alkaloid com-pounds, collectively called as taxines (0.4 – 0.7% of the fresh plant material). The taxines were isolated even before paclitaxel was discovered. The needles of *T. baccata* were the first identified source for 10-DAB (Appendino et al., 1992a). Furthermore, the needles of *T. wallichiana* have also been identified as the only source of a unique taxane, wollifoliol, having 5/6/6/6/4 ring sys-tem (Vander Velde et al., 1994). Various organs including stem bark and leaves of *T. wallichiana* and a few other species, *Taxus yunnanensis* and *T. cuspidata* have been found to accumulate C-14 oxygenated taxoids (Zhang et al., 1995). All C-14 oxygenated taxoids exhibit poor cytotoxicity due to lack of the side chain and C-4 (20), 5-oxetane ring. In 1984, docetaxel was semi - synthetically produced from 10-DAB, by combining it with a fully syn-thetic side chain. This was an efficient process of using a renewable source i.e. needles in place of bark which opened possibilities for the pharmaceutical deve-lopment of taxoids.

Micropropagation of *taxus* species

Several tissue culture techniques used for the micro-propagation of *Taxus* are summarized in Table 4. The *Taxus* species are found recalcitrant towards direct regeneration from axillary buds, nodal, internodal and leaf explants. *Taxus* seeds have a long dormancy period of 1 - 2 years, hence dormancy breaking agents such as gibberellic acid (GA_3) are used for *in vitro* germination. Regeneration through *in vitro* germination of zygotic embryos of several *Taxus* species has been successful (Chee, 1994). Washing of mature zygotic embryos under tap-water for 7 days, followed by culturing for 7 days on modified Murashige and Skoog (MS) or Heller's media is also used for breaking the dormancy. However, embryo derived seedlings showed nutrient deficiency and that could be overcome by the supplementation of additional magnesium sulfate in the media (Flores et al., 1993). Rooting in seedlings is induced by the supplementation of boric acid in the liquid medium. Another approach involves culture of zygotic embryos of *T. baccata* on woody plant medium (WPM) supplemented with 5% activated charcoal, resulting in 65% shoot induction and complete regeneration after 2 months of incubation (Majda et al., 2000). Besides dormancy, light conditions play a critical role in embryo germination. An initial dark incubation of 2 - 4 weeks, followed by subsequent trans-fer of embryos to white cool light conditions proved effi-cient (Majda et al., 2000). Chee (1995) reported conver-sion of adventitious bud primordia into multiple shoots

arising from the zygotic embryos on ½ B5 medium supplemented with 10 µM BA. However, by this method only 58% of the shoot primordia explants could be developed into complete plantlets. Micropropagation of *Taxus* trees is also achieved by culturing microcuttings of *T. x media* in a medium supplemented with 1 mg/lkinetin and 2% sucrose. Induction of somatic embryogenesis has been reported from friable calli and immature zygotic embryos in the pre-sence of 2, 4-D, 1 mg/l glutamine, cytokinin, BA and auxin, NAA. Subsequent transfer of embryo-genic calli onto the WP medium supplemented with 4 µM BA, 1 µM kinetin and 1 µM NAA for 6 to 8 weeks supported maturation of the somatic embryos (Chee, 1996).

Taxus: *In vitro* expression of taxoids

Christen *et al.* (1989) for the first time reported the pro-duction of paclitaxel using *Taxus* cell cultures. After-wards, many scientists attempted to optimize conditions for prolific growth of cells and enhanced taxane pro-duction in several tissue culture systems viz. cell, hairy root, shoot and embryogenic cultures of different *Taxus* species (Mirajalili and Linden, 1996; Ketchum et al., 1995). Amongst various approaches available for *in vitro* production of taxoids, cell suspension cultures have been found to be the most reliable yielding 0.8% taxoids on dry biomass basis (Jaziri et al., 1996; Yukimune et al., 1996; Cusido et al., 1999). Moreover, cell suspension cultures secrete 90% of the total taxoids synthesized into the culture medium (Srinivasan et al., 1995). In addition to paclitaxel and 10-DAB, a range of other regular C-14 taxoids has also been identified in cells and culture media from cell cultures of various *Taxus* species (Banerjee et al., 1996; Vanek et al., 1999; Agarwal et al., 2000).

Establishment of cell cultures of *taxus* species

Rohr was the first scientist who developed callus cultures from different explants of *T. baccata* including micro-spores (Rohr, 1973). Subse-quently, David and Plastira (1976) standardized mineral and phytohormonal com-position of the culture medium required for efficient prolix-feration of the calli generated from mature stem explants in *T. baccata* L. A number of culture parameters affect the callus and cell suspension culture of various *Taxus* species (Table 5). Basal media such as Gamborg's (B5) (Gamborg et al., 1968; Wickremesinhe and Arteca, 1994), MS (Murshige and Skoog, 1962) and Woody Plant Medium (WPM) (Loyd and McCown, 1981) have been used for the initiation, proliferation and maintenance of the callus and cell suspension cultures in *Taxus* species. Fre-quently, supple-mentation of organic substances such as casein hydrolyzate, polyvinyl pyrrolidone (PVP), ascorbic acid and other essential amino acids such as glutamine, aspartic acid, proline, phenyl alanine along with vitamins in the medium enhanced callus/cell growth.

Table 4. Organ cultures in various Taxus species.

Culture type	Taxus sp.	Explants	Media	Hormone(s)	Remark(s)	Reference(s)
Embryo culture	T. brevifolia	Embryos	B5	NS[a]	63% full seedlings	Hezari et al., 1995
	T. baccata	Embryos	MS/ Heller	GA$_3$	100% full seedlings (1 week)	Hefner et al., 1998
	T. brevifolia	Embryos	B5	NS	Mature embryos, 36% full seedlings	Wildung & Croteau, 1996
	T. brevifolia	Embryos	½ B5	BA	Multiple shoots , 50% rooting	Chee, 1995; Chang et al.,2001
	T. baccata	Zygotic embryos	WPM	BA/ Zeatin	65% seedlings development	Kingston et al., 1993
Shoot multiplication	T x media	Shoot tips	Hogland	Kinetin	NS	Walker & Croteau, 2000
	T. floridana	Needles	MS/ B5	NAA	Glutamine used to induce somatic embryos in callus	Wu & Lin, 2003
Microcutting	T. cuspidata	Embryos	WPM	NAA+ 2,4 –D+ Kinetin	Immature embryos, mature in hormone free medium	Appendino et al., 1992b
Somatic embryogenesis	Taxus species	Zygotic embryos	MCM	2,4 –D/ BA	2 stages cultures	Parmar et al., 1999

[a]NS- not specified

Most of the research groups have observed that auxins (IBA, NAA, 2,4-D and Picloram) at the varying level of 1.0-10.0 mg/l used in combination with cytokinins (BAP/Kinetin) at the level of 0.1 - 0.5 mg/l were effective for the initiation of callusing. Inorganic compounds like VOSO$_4$ have also been reported to be instrumental in enhancing the cell growth and multiplication in different Taxus species. The influence of plant growth factors on the rate of cell proliferation is complex and depends on both the basal medium composition and plant genotype. A major problem associated with Taxus cultures is the tendency to secrete phenolics, hampering the growth of calli and after a certain period leading to blackening / browning that results in death of the cells (Gibson et al., 1993). Use of anti-oxidants (phenolic binding compounds) like PVP, citric acid, ascorbic acid, activated charcoal, and fre-quent sub-culturing of cell cultures have been found to prevent ill effects of phenolics leach-outs. Simul-taneous incubation of cultures under dark conditions leads to lesser leaching of phenolics. The inoculum size, period of subculture, light intensity and photoperiod are also the deciding factors for the multiplication, phenotypic appearance and proliferation of Taxus cells (Wickremsinhe and Arteca, 1993; Navia-Osorio et al., 2002).

Synthesis and accumulation of paclitaxel and 10-DAB in the cell cultures of Taxus species

For optimum cell growth and taxoid production, selection of explants, nutritional/ phytohormonal supplementation, pH, use of antioxidants and amino acids have been observed as the deciding factors (Jaziri et al., 1996). Several protocols have been reported for the production of some important taxoids and up-scaling of these cell cultures (Navia-Osorio et al., 2002). Almost every Taxus species has been studied in terms of the optimization of the growth media to yield maxi-mum biomass (Mirajallili and Linden, 1996). Ketchum et al.(1995) formulated a new TMS growth medium for the callus cultures of T. brevifolia. Similarly, Fett- Neto et al. (1992) studied the effect of various nutritional components on paclitaxel production in T. cuspidata cell cultures. Srinivasan et al. (1995) studied the kinetics of biomass accumulation and paclitaxel production in T. baccata cell suspension cultures. Fur-thermore, a strong correlation is found between biomass production and taxol accumulation in callus cultures. Paclitaxel has been found to accumulate in high amounts (1.5 mg/l of culture) in the second phase of the growth curve. A similar level of paclitaxel has also been observed in T. brevifolia suspension cultures (Kim et al., 1995). Seventeen known taxoids and 10 abietanes have been isolated from the dark brown callus culture of T. cuspidata cultivated on a modified Gamborg's B5 medium with 0.5 or 1.0 mg/l NAA (Bai et al., 2005). Production of the

Table 5. Effect of variation in the nutrient and phytohormones used in the media for culturing the explants from *Taxus* species

Taxus sp.	Explants used	Basal medium	Phytohormones (mg/l)	% Carbohydrate	Anti-oxidant (mg/l)	Additive(s)	Type of culture(s)	Reference (s)
T. baccata	Gametophyte	B&N	NS[a]	Sucrose	NS	NS	Callus	Chattopadhyay et al., 1999a
T. baccata	Old stem segments	Heller, MS	2,4-D + kinetin	NS	NS	Thiamine + Nurse culture	Callus	Lucas, 1956
T x media	Stem	B5	NS	NS	NS	Casein hydrolysate	Cell suspension	Mirajalili & Linden, 1996
T. cuspidata	Young needle	B5	2,4-D + GA_3	Sucrose	PVP	Casein hydrolysate	Callus	Wu & Ge, 2004
T. floridana	Young needle	MS	NAA+BA	Sucrose	PVP	Glutamine	Callus, embryo	Wu & Lin, 2003
T. brevifolia	Stem, bark and needle	B5	2,4-D	NS	NS	Casamino acids	Callus	Gibson et al., 1993
T. baccata	Stem	WPM	NAA+BA	Sucrose	NS	NS	Callus	David & Plastira, 1976
		B5	Picloram+ kinetin+ ABA/GA3	NS	NS	Glutamine	Cell suspension cultures	Kobayashi & Shigemori, 2002
		B5	Picloram+ kinetin + ABA	Sucrose + fructose	Ascorbic acid, citric acid	Aspartic acid + arginine+ proline+ KM,N&N vitamins	Cell suspension	Wang et al., 2001
		B5	NS	NS	NS	NS	Cell suspension	Flores et al., 1993
		Eriksson medium	2,4-D + kinetin + BA	Sucrose	NS	Glutamine	Cell suspension	Chattopadhyay et al., 1994
		B5	2,4-D+ Kinetin	Sucrose or fructose	NS	Casein hydrolysate, Kao Michayluk vitamins	Protoplasts, callus, cell suspension	Wu et al., 2001
		B5	2,4-D+ Kinetin	Sucrose	NS	2 x B5 vitamins & $VOSO_4$ as elicitor	Callus, cell suspension	Wickremesinhe & Arteca, 1994
T. baccata	Young stems	B5	NAA+BA	Sucrose	Ascorbic acid	Glutamine	Cell suspension	Christen et al., 1991
T. wallichiana	Leaf needle, stem	B5	2,4-D + kinetin	Sucrose	NS	Casein hydrolysate	Callus, cell suspension	Wang et al., 2000
T. cuspidata	mature and immature seeds	B5	2,4-D+GA	Sucrose	NS	$KCl,KI,NaH_2PO_4,MgSO_4.7H_2O, (NH4)_2 SO_4$	Callus, cell suspension	Xu et al., 1998
T. media	Zygotic embryo	WPM	NAA	Sucrose	NS	Methyl jasmonate	Callus, cell suspension	Agarwal et al., 2000; Chee, 1995

[a]NS- not specified

taxoids does not occur in the stationary growth phase of the cell cultures. Seki et. al. (1997) and Morita et al. (2005) focused their studies on plant cell cultures of T. cuspidata for the continuous production of taxoids. Addition of carbohydrates (sucrose and fructose) in the midway of the growth cycle increased the rate of the cell growth and paclitaxel accumulation. Supplementation of biotic and abiotic elicitor(s) in the cell suspension cultures of Taxus has been shown to affect the growth of cell biomass as well as paclitaxel production by pathway stimulation. Several reports have shown that methyl jasmonate (abiotic chemical elicitor) if added in the second growth phase of suspension cultures, strongly promoted taxane biosynthesis (Menhard et al., 1998; Yukimune et al., 1996, 2000; Ketchum et al., 1999; Phisalaphong and Linden, 1999; Mirajilili and Linden, 1996). Jasmonic acid (JA) in 100 µM concentration if added at the 7th day after subculture also increased taxane content in the culture medium (Baebler et al., 2002). Oligosac-charide addition stimulates the effect of methyl jasmonate in Taxus canadensis culture (Furmanowa et al., 1997; Linden and Phisalaphong, 2000). Paclitaxel and baccatin III production in suspension cultures of T. media can be improved by a two-stage culture method by adding methyl jasmonate (220 µg/l FW) together with mevalonate (0.38 mM) and N-benzoylglycine (0.2 mM). Under these conditions, 21.12 mg/l of paclitaxel and 56.03 mg/l of baccatin III were obtained after 8 days of the culture in the pro-duction medium (Cusido et al., 2002). Ten known taxoids, paclitaxel, 7-epi-taxol, taxol C, baccatin VI, taxayuntin C, taxuyunnanine C, yunnanxane and an abietane, taxa-mairin A, have been produced in the callus culture of T. cuspidata cultivated on a modified Gamborg's B5 medium in the presence of 0.5 mg/l NAA. After stimulation with 100 µM methyl jasmonate, five more taxoids have been found in addition to the above-men-tioned compounds (Bai et al, 2004). Aspergillus niger, an endophytic fungus, isolated from the inner bark of T. chinensis, if added as an elicitor (40 mg/l) in the late exponential-growth phase resulted in more than two-folds increase in the yield of the taxol and about a six-folds increase in the yield of the total taxoids (Wang et al., 2001). Addition of a trivalent ion of a rare earth element, lanthanum (1.15 to 23.0 µM) also promoted taxol production in suspension cultures of Taxus species (Wu et al., 2001). Maximum synthesis and accumulation of taxoids occurred in the period just after subculture. Thereafter a steady decrease in the paclitaxel content occurred (Wickremesinhe and Arteca, 1994). Consistent pro-duction of taxoids in a continuous culture system can be achieved only by their removal just after optimum production. Commercial resins are used for the removal of paclitaxel from T. bre-vifolia suspension cultures (Christen et al., 1991). It is observed that higher initial sucrose concentration in culture medium repressed cell growth leading to a longer lag phase. This could be overcome by a low initial su-

crose concentration (20 g/l) and subsequent sucrose feeding (fed-batch culture) resulting in a high taxane yield of 274.4 mg/l (Wang et al., 2000). It has also been shown that initial addition of 1.0 – 2.0 mmol/l phenylalanine into the medium, followed by addition of 73.0 mmol/l sucrose and 173.3 mmol/l mannitol at the 28th day of culture, strongly promoted cell growth and taxoid production. The variable amount of taxoids in callus lines of different Taxus species and in the callus lines of the same Taxus species could be explained on the basis of the fact that secondary metabolite accumulation occurred due to dynamic equili-brium between the product formation, transport, storage, turnover and degradation of com-pounds. The levels of paclitaxel obtained in cell/organ culture systems of various Taxus species are described in Table 6. Taxol production can be enhanced by the use of self-immo-bilized cell aggregates, free and calcium alginate gel particle-immobilized cells (Xu et al., 1998). Besides, taxol biosynthesis has also been enhanced up to 40 - 70% by the addition of adsorbents like ion ex-changer, XAD-4 in cell cultures. A new method for the production of taxoids involved culturing bacteria isolated from yew (Sphingo-monas sp., Bacillus sp., Pantoea sp. or Curtobacterium sp. isolated from T. canadensis) or its mutated forms. Taxane production has also been reported using bio-transformation of pro–taxanes namely paclitaxel, 10-deacetylcephalomannine, 7-epitaxol, 10-deacetyl-7-epitaxol, 7-epicephalomannine, 7-epibaccatin-III, 7-xylosyltaxol, 7-xylosyl-cephalomannine, taxagifine, delta-benzoyloxy taxagifine, 9-acetyloxy taxusin, 9-hydroxy taxusin, taxane-Ia, taxane-Ib, taxane-Ic or taxane-Id into taxanes. Another method used for the high yield of taxane production from Taxus sp. was fed–batch mode of suspension culture with inclusion of an inhibitor of phenylpropanoid metabolism. This method is high yielding, producing 15 mg taxane/l/day, 10 mg taxol/l/day and / or 15 mg baccatin-III/l/day. The culture medium is exchanged at least once during taxane production, and taxane may be removed from the culture during produc-tion (Roberts et al., 2003).

Production of various other taxoids in cell cultures of Taxus species

Cell cultures have been found to produce a variety of irregular and unique taxoids in addition to regular taxoids. A few species produced detectable amounts of other related taxoids such as baccatin, epitaxol, 10-deacetyl taxol, 7-epi-10-deacetyl taxol etc. (Table 7). Cell cultures of T. baccata produced small quantities (4.4 µg/l FW) of the new bioactive taxoids including 7-o-xylosides of taxol C; 10-deacetyl taxol C; N-methyl taxol and a C-13 oxygenated taxoid taxucultine (Ma et al., 1994a). In a subsequent report, Ma et al. (1994 b) reported four new C-14 oxygenated taxoids, along with yunnaxane. Simi-larly, Cheng et al. (1996) reported production of seven C-14 oxygenated taxoids from the cell suspension cultures

Table 6. Content of paclitaxel in cell lines and organ cultures of different *Taxus* species.

Taxus species	Content of Paclitaxel	Reference
(a) Cell Line		
T. baccata	1.50 mg/l	Chattopadhyay et al., 1999
T. cuspidata	0.30 mg/l	Roberts et al., 2003
T. canadensis	5.95 mg/l	Wang et al., 2001
T. media	110.00 mg/l	Agarwal et al., 2000
T. baccata	48.30 mg/l	Agarwal et al., 2000
T. brevifolia	0.50 mg/l	Agarwal et al., 2000
T. baccata	13.00 mg/l	Christen et al., 1991
T. wallichiana	0.50 mg/l (d.w.)	Wang et al., 2000
T. yunanensis	3.00 mg/l (d.w.)	Loyd & McCown, 1981
Organ culture (culture type)		
T. brevifolia (shoot)	1.0μg /gFW[a]	Wildung & Croteau, 1996
T. brevifolia (Root)	6.2 μg /g FW	Wildung & Croteau, 1996
T. x media(shoot)	2.1 μg /gFW	Wildung & Croteau, 1996
T.x media (Root)	8.1 μg /gFW	Wildung & Croteau, 1996
T. media Hicksii (Root)	619μg /g DW[b]	Wildung & Croteau, 1996
T. cuspidata (Root)	423μg /g DW	Wildung & Croteau, 1996
T. cuspidata var special (Root)	332 μg /gDW	Wildung & Croteau, 1996
T. baccata (seedling)	40,000 μg /gDW	Kingston et al., 1993

[a] = FW, fresh weight; [b] = DW, dry weight

of *T. yunnanensis*. Out of these, three were entirely new compounds characterized as 10-de-acetyl yunnaxane; 2α-hydroxy-5α, 10β, 14β-triacetoxy-4 (20), 11-taxadiene and 2α, 10β, 14β-triacetoxy-5α-acetoxy-4(20), 11-taxadiene. 4(5), 11(12)-taxadiene and their deo-xygenated derivatives are potential biosynthetic precu-rsors of paclitaxel and other taxoids. Microbial mediated hydroxyl-lation of 4(5),11(12)-taxadiene by *Absidia coerula* resulted in a 20-hydroxylated metabolite with a molecular formula of $C_{28}H_{40}O_9$. Menhard et al. (1998) reported prod-uction of a mixture of 16 different C-13 and C-14 oxy-genated taxoids in cell cultures of *T. chinensis*. Twelve of these taxoids were found to be esterified at C-13 position, whereas four of them were oxygenated at C-14 position. A cell suspension line of *T. wallichiana* has been found to produce a group of three C-14 oxygenated taxoids and an epoxide of baccatin.

OTHER TISSUE CULTURE TECHNIQUES USED FOR TAXOIDS PRODUCTION

Several other tissue culture techniques such as organ culture, genetic transformation and protoplast culture have been found to play an important role in taxoids production.

Organ cultures

These cultures possessed a high potential for paclitaxel production. A yield of 40 mg of paclitaxel per g dry weight *Taxus* has been reported by Majda et al. (2000).

Genetic transformation and regeneration of transgenic plants

Successful transformation of *Taxus* cells with *Agro-bacterium rhizogenes* and *Agrobacterium tumefaciens* has been achieved (Han et al., 1994). *Taxus* embryos serve as a good starting material for establishing hairy root cultures (Plaut- CarCasson, 1994). Use of *Nicotiana* feeder cells and acetosyringone further enhanced the transformation rate of the embryos. The highest hairy root biomass has been obtained in B5 liquid medium supple-mented with 0.5 g/l L-glutamine. Transgenic plantlets are regenerated from these hairy roots in 2 steps, by culturing in B5 liquid medium supplemented with 1 mg/l NAA, followed by transfer into a medium containing 5 mg/l BAP. Both hairy roots and the plants possessed detectable amount of paclitaxel. In another report, hairy root cultures of *Taxus x media var. Hic*ksii were estab-lished from shoot tips, young leaves and hypocotyls of 8 weeks old seedlings cultured on plant growth factor free modified DCR medium with 20 g/l sucrose and solidified with 6 g/l phyto-agar. On addition of 100 μM methyl jasmonate, paclitaxel contents in these hairy root cultures increased from 69 to 210 μg/g dry wt. whereas, the 10 DAB contents were not affected (Furmanowa and

Table7. Type and level of various taxoids other than paclitaxel and DAB in cell cultures of *Taxus* species.

Taxus species	Type of culture	Name of the taxoid expressed	Max.level of taxoid detected	Reference(s)
T. baccata	Callus culture	Taxol C	4.4 mg /kg FW	Banerjee et al., 1996
T. baccata	-do-	10-deacetyl taxol C	1.0 mg /kg FW	Banerjee et al., 1996
T. baccata	-do-	N-methyl taxol	2.2 mg /kg FW	Banerjee et al., 1996
T. baccata	-do-	Taxcultine	1.6mg/kg F.W.	Banerjee et al. 1996
T. chinensis and *T. wallichiana*	cell suspension	2α, 5α, 10β, 14β-tetraacetoxy-4[20], 11-taxadiene C and Yunnanaxane	145 mg/ l	Chattopadhyay et al., 1997
-do-	-do-	2α, 5α, 10β-triacetoxy-14β-propionyloxy-4[20], 11-taxadiene	11.2 mg /l	Banerjee et al., 1996
T. chinensis and *T. wallichiana*	-do-	2α, 5α, 10β-triacetoxy-14β-(2-methyl)-butyryloxy-4[20], 11-taxadiene	2.5 mg/ l	Chattopadhyay et al., 1997
T. chinensis	-do-	9-dehydrobaccatin III	0.2 mg /l	Flores et al., 1993
T. media, T. baccata, T. brevifolia,	-do-	Baccatin III	53.6 mg /l	Flores et al., 1993; Agarwal et al., 2000
T. chinensis	-do-	13-dehydroxy-10-deacetyl baccatin III	0.9 mg /l	Flores et al., 1993,
-do-	-do-	9-dihydro-13-acetoxy baccatin III	0.9 mg /l	Flores et al.,1993
-do-	-do-	9-dihydro-13-dehydroxy baccatin III	3.2 mg /l	Flores et al.,1993,
-do-	-do-	13-dihydroxy baccatin III	1.9 mg /l	Flores et al.,1993
-do-	-do-	Baccatin VI	0.5 mg /l	Flores et al.,1993,
T. canadensis	-do-	13-deacetoxy baccatin I	0.75 mg /l	Flores et al.,1993
T. baccata	-do-	2α, 5α, 10β-triacetoxy-14β-propionlofy-4[20], 11-taxadiene	1.4 mg /l	Flores et al.,1993
-do-	-do-	10-deacetyl-7-xylosyl taxol B	0.027 mg /l	Ketchum et al., 1999
-do-	-do-	10-deacetyl-7-xylosyl taxol C	0.087 mg /l	Ketchum et al., 1999
-do-	-do-	7-epitaxol	0.102 mg /l	Ketchum et al., 1999
-do-	-do-	10-deacetyl-7-epitaxol	0.064 mg /l	Ketchum et al., 1999
-do-	-do-	10-deacetyltaxol	0.122 mg /l	Vander Velde et al., 1994

Table 8. Comparison of the paclitaxel recovery from *in vivo* and *in vitro* tissue biomass of *Taxus* species.

Plant material	Average paclitaxel content (% D.W.)	Reference
Bark of mature tree (100 yrs old)	0.017	Yukimune et al., 2000
Taxus needles	0.005	Ma et al., 1994a
Taxus cell cultures	0.800	Agarwal et al., 2000
Taxomyces andreanae	---	Rohr, 1973

Syklowska, 2000).

Production of paclitaxel through protoplast culture

Luo et al. (1999) isolated viable protoplasts from 20 day old friable calli of *T. yunnanensis*. These protoplasts were cultured and maintained on B5 salts having KM-vitamins and supplemented with 0.45 M fructose, 3.0 mg/l 2, 4-D and 0.1 mg/l kinetin (Luo et al., 1999). Protoplast derived colonies varied in terms of the growth and paclitaxel production. Though these colonies were not promising for paclitaxel production, but provided a source for obtaining cell lines with high paclitaxel productivity after muta-genesis. Aoyagi reported 6 times more paclitaxel accum-ulation in protoplast culture of *T. cuspidata* after immobi-lization in agarose gel (in shaking cultures) compared with the cell suspension cultures (Aoyagi et al., 2002). This may also be a promising approach for taxane prod-uction.

COMMERCIAL FEASIBILITY OF *IN VITRO* TAXOID PRODUCTION

Several problems have been observed in the isolation of taxoids from *Taxus*, for instance the growth of yew trees is very slow. Complete chemical synthesis of taxoids on an industrial scale seems to be impractical. A comparison of different sources for paclitaxel isolation (stem bark of mature trees, needles from 4 year old trees, cell sus-pension culture and cultures of *Taxomyces andreaneae*) indicated that cell cultures of *Taxus* are the only potential source for the commercial production of paclitaxel as well as other taxoids (Table 8). The growth rate of callus cultures is slow, hence, repeated use of *Taxus* cells is necessary for economical production of paclitaxel. The highest concentration of paclitaxel reported is 0.8% on dry cell weight basis (Yukimune et al., 1996). This value is approximately 470 times higher than the paclitaxel contents observed in the bark. In a study, out of the 27 different genotypes screened, *T x media* "Sargentii" was found to be enriched in paclitaxel, yielding 0.069 and 0.032% taxoids from the leaves and callus, respectively (Parc et al., 2002). Large scale production of taxoids (3 mg/l taxol and 74 mg/l of total taxanes) was achieved from the suspension cultures of *T. cuspidata* using a bal-loon type bubble bioreactor. In cell suspension cul-tures of *T. wallichiana,* established in shake flasks and in a 20 l airlift bioreactor running for 28 days in a batch mode, the maximum yield of paclitaxel was 20.84 mg/l on the 24th day and of baccatin III was 25.67 mg/l on the 28th day (Yukimune et al., 2000).

Taxol® is manufactured by Bristol-Myers Squibb in a 9-steps semi-synthetic process from 10-DAB originally isolated and purified from yew species such as *T. baccata* in a multi-steps process. Aphios Corporation manufactures Taxol in a 4-steps process that is cost-effective and environmental-friendly. This process pro-duces 10-DAB, a precursor of taxoids and cephalo-mannine as by-product, that may be semi-synthetically converted to Taxol in a 3-steps process. Total synthesis of taxane is costly, low yielding and therefore is not a good alternate for its commercial supply. Therefore, at present, semi-synthesis is the major industrial mode for the production of paclitaxol and other related taxoids (Commercon et al., 1992, 1995). Although semi-synthesis is efficient, but purification of the precursors required from the plant tissues is difficult. On the other hand, isolation of taxol from *Taxus* cell cultures requires fewer steps than purification from intact tissue.

A few other organisms have been found capable of producing taxol. These include a diverse group of endo-phytic fungi of *Taxus* species viz. *Taxomyces andrea-nae*, *Pestalotiopsis*, *Fusarium* and *Alternaria*, but these are yet to be exploited for the commercial production of these drugs (Stierle et al., 1993; Strobel et al., 1996; Kim et al., 1999). Induced *Taxus* cell cultures provide a good model for the study of complex biosynthetic pathway for the synthesis of Taxol. The most promising approach for taxol production is *Taxus* cell cultures elicited with methyl jasmonate for increased taxol production (Yukimune et al., 2000; Baebler et al., 2002). Novel taxane derivatives can also be obtained by genetic manipulation (Han et al., 1994). Hence, in the present scenario, biological method of taxol production is commercially feasible. It requires understanding of the complete taxol biosynthetic pathway and the enzymes involved in it. This will be helpful in improving taxol production by manipulation of several steps. Over-expression of the genes in transgenic plants may also be helpful in increasing the level of the taxoids to commercially significant levels. Once the entire pathway is studied and the genes identified, it might be possible to suppress the undesirable side routes by various sense and anti-sense techniques and also to direct the pathway towards other intermediates possibly some new taxoid derivatives. Furthermore, cell cultures can produce a range of other taxanes that may prove a powerful way in knowing the biochemistry, enzymology and molecular biology of taxoids production. Hence, it may be concluded that up-scaling of *Taxus* cell lines that are capable of overproducing taxoids could only make the industrial production of paclitaxel feasible.

Conclusion

Taxoids viz. paclitaxel and docetaxel are of commercial importance since these are shown to have anti-can-cerous activity. These taxoids have been isolated from the bark of *Taxus* species. Plant cell culture techniques have been exploited for the isolation of mutant cell lines, production of secondary metabolites and genetic transfor-mation of the plants. *In vitro*, culture of *Taxus* not only helps in conservation but is also helpful in the production of paclitaxel and other taxoids. Various stra-tegies tested globally for the commercial production of taxoids are discussed. Different *Taxus* species, their origin, diterpenoids obtained from different parts of the tree and their applications are discussed. Although, detailed taxoid biosynthetic pathway is not well known, an overview of the pathway has been described. Micropro-pagation of *Taxus* and regeneration of transgenic plants has been described. On the basis of comparative studies on pacli-taxel production in different systems, it is concluded that at present *Taxus* cell suspension cultures are relatively more suitable and could be considered as an alternate for the direct production of paclitaxel as well as other related taxanes. Of course, feasibility of a cost effective production of paclitaxel through tissue/cell culture requires more advancement in optimization of the amount and rate of the biomass and taxoid production. Commercial feasibility of *in vitro* taxoid production has also been discussed.

ACKNOWLEDGEMENTS

The authors acknowledge the facilities of the Department of Biotechnology, Ministry of Science and Technology, Government of India, New Delhi (DBT) under the Bioinformatics Sub-center used in the preparation of the manuscript. The authors also acknowledge the efforts of Dr. Shipra Agarwal who initiated this work.

REFERENCES

Agarwal S, Banerjee S, Chattopadhyay SK, Kulshreshtha M, Musudanan KP, Mehta VK, Kumar S (2000). Isolation of taxoids from cell suspension cultures of Taxus wallichiana. Planta Med. 66: 1-3.

Aoyagi H, DiCosmo F, Tanaka H (2002). Efficient paclitaxel production using protoplasts isolated from cultured cells of Taxus cuspidata. Planta Med. 68(5): 420-424.

Appendino G (1993). Taxol (Paclitaxel). Historical and ecological aspects. Fitoterapia 64: 5-25.

Appendino G, Barboni L, Gariboldi P, Bambardelli E, Gabetta B, Viterbo D (1993). Revised structure of brevifoliol and some baccatin VI derivatives. J. Chem. Commun. 20: 1587-1589.

Appendino G, Gariboldi P, Pisetta A, Bombardelli E, Gabetta B (1992a). Taxanes from Taxus baccata. Phytochemistry 31: 4253-4257.

Appendino G, Lusso P, Gariboldi P, Bombardelli E, Gabetta B (1992b). A 3, 11-cyclotaxane from Taxus baccata. Phytochemistry 31: 4259-4262.

Baebler S, Camloh M, Kovac M, Ravnikar M, Zel J (2002). Jasmonic acid stimulates taxane production in cell suspension culture of yew (Taxus x media). Planta Med. 68(5): 475-476.

Bai J, Kitabatake M, Toyoizumi K, Fu L, Zhang S, Dai J, Sakai J, Hirose K, Yamori T, Tomida A, Tsuruo T, Ando M. (2004). Production of biologically active taxoids by a callus culture of Taxus cuspidata. J. Natural Products 67(1): 58-63.

Bai J, Ito N, Sakai J, Kitabatake M, Fujisawa H, Bai L, Dai J, Zhang S, Hirose K, Tomida A, Tsuruo T, Ando M (2005). Taxoids and Abeotaxanes from Callus Cultures of Taxus cuspidata. J. Nat. Prod. 68(4): 497 -501.

Baloglu E, Kingston DGI (1999). The texane diterpenoids. J. Nat. Prod. 62: 1448-1472.

Balza F, Tachibana S, Barnos H, Towers GHN (1991). Brevifoliol, A Taxane from *Taxus brevifolia*. Phytochemistry 30: 1613- 1614.

Banerjee S, Upadhyay N, Kukreja AK, Ahuja PS, Kumar S, Saha GC, Sharma RP, Kulshrestha M, Chattopadhyay SK (1996). Taxanes from the in vitro cultures of the Himalayan Yew-Taxus wallichiana. Planta Med. 62: 333-335.

Brubaker KD, Brown LG, Vessella RL, Corey E (2006). Administration of zoledronic acid enhances the effects of docetaxel on growth of prostate cancer in the bone environment. BMC Cancer 6: 15 doi:10.1186/1471-2407-6-15.

Chang SH, Ho CK, Chen ZZ, Tsay JY (2001). Micropropagation of Taxus mairei from mature trees. Plant Cell Reports 20: 496-502.

Chattopadhyay SK, Kulshrestha M, Saha GC, Sharma RP, Jain SP, Kumar S (1997). The taxoids and the phenolic constituents of the heartwood of the Himalayan yew Taxus wallichiana. JAMAPS. 19: 17-21.

Chattopadhyay SK, Kulshrestha M, Saha GC, Tripathi V, Sharma RO (1999b). Studies on the Himalayan Yew Taxus wallichiana: Part VI-Isolation of non taxoid constituents. Indian J. Chem., 38B: 246-247.

Chattopadhyay SK, Saha GC, Kulshrestha M, Sharma RP, Kumar S (1994). Studies on the Himalayan yew. Part IV. Isolation of

dihydrotaxol and dibenzoylated taxoids. Indian J. Chem. 35B: 754-756.

Chattopadhyay SK, Sharma RP (1995). A taxane from the Himalayan Yew: Taxus wallichiana. Phytochemistry 39: 935-936.

Chattopadhyay SK, Tripathi V, Sharma RP, Shawl AS, Joshi BS, Roy R (1999a). A brevifoliol analogue from the Himalyan yew Taxus wallichiana. Phytochemistry 50(1):131-133.

Chee PP (1994). In Vitro culture of zygotic embryos of Taxus species. Hort. Science 29: 695-697.

Chee PP (1995). Organogenesis in Taxus brevifolia tissue cultures. Plant cell reports 14: 560- 565.

Chee PP (1996). Plant regeneration from somatic embryos of Taxus brevifolia. Plant cell reports 16: 184-187.

Chen SH, Hong J (2006). Novel tubulin interacting agents : A tale of Taxus brevifolia and Catharanthus roseus based dru discovery. Drugs Fut. 31: 123-150.

Cheng K, Fang W, Yang Y, Xu H, Merg C, Kong M, He W, Fang QC (1996). C –14 oxygenated taxanes from Taxus yunnensis cell cultures. Phytochemistry 42: 73-75.

Cheng Q, Oritani T, Horiguchi T (2000). Two novel taxane diterpenoids from the needles of Japanese yew, Taxus cuspidata. Biosci. Biotechnol. Biochem. 64(4): 894-898.

Christen AA, Bland J, Gibson DM (1989). Cell culture as a means to produce taxol. Proc. Am. Asso. Cancer Res. 30: 566.

Christen AA, Gibson DM, Bland J (1991). Production of taxol or taxol-like compounds in cell culture. US Patent No. 5 019, 504.

Commercon A, Bezard D, Bernard F, Bourzat JD (1992). Improved protection and esterification of a precursor of the Taxotere and Taxol side chains. Tetrahedron Letters 33: 5185-5188.

Commercon A, Bourzat JD, Didier E, Lavelle F (1995). Taxane Anticancer Agents: Basic Science and Current Status (Georg GI, Chen TT, Ojima I, Vyas DM eds.). American Chemical Society, Washington DC. pp. 233-246.

Croteau R, Ketchum REB, Long RM, Kaspera R, Wildung MR (2006). Taxol biosynthesis and molecular genetics. Phytochem. Rev. 5: 75-97.

Cusido RM, Palazon J, Bonfill M, Navia-Osorio A, Morales C, Pinol MT (2002). Improved paclitaxel and baccatin III production in suspension cultures of Taxus media. Biotechnol. Prog. 18(3): 418-423.

Cusido RM, Palazon J, Navia-Osoria A, Mallol A, Bonfill M, Morales C, Pinol MT (1999). Production of taxol and baccatin III by a selected Taxus baccata callus line and its derived cell suspension cultures. Plant Sci. 146: 101-107.

Datta A, Aube J, Georg GI, Mitscher LA, Jayasinghe LR (1994). The first Synthesis of a C-9 carbonyl modified baccatin III derivative and Its conversion to novel taxol® and taxotere® analogues. Bioorganic and Medicinal Chemistry Letters. 4: 1831-1834.

David A, Plastira V (1976). Istophysoplogie vegetale compartment en culture in vitro de cellules isolees de deux gymnosperms: Taxus baccata L et Pinus pinaster Sol. C. R. Acad. Sci. Paris. 282:1159-1162.

Diaz JF, Barasoain I, Souto AA, Amat-Guerri F, Andreu JM (2005). Macromolecular accessibility of fluorescent taxoids bound at a paclitaxel binding site in the microtubule surface. J. Biol. Chem. 280: 3928-3937.

Donovan TA (1995). The paclitaxel dilema. Taxane J. 1: 28-41.

Dubois J, Guénard D, Guéritte F (2003). Recent developments in antitumour taxoids. Expert Opinion on Therapeutic Patents. 13(12): 1809-1823.

Fett-Neto AG, DiCosmo F, Reynolds WF, Sakata K (1992). Cell culture of Taxus as a source of the antineoplastic drug taxol and related taxanes. Bio-Technology 10: 1572 –1575.

Flores T, Wagner LJ, Flores HE (1993). Embryo culture and taxane production in Taxus species. In vitro Cell Dev. Biol. 29: 160-165.

Floss HG, Mocek U (1995). Taxol: Science and applications, Suffness, M., ed., CRC press, Boca Raton, FL. pp. 191-208.

Furmanowa M, Glowniak K, Syklowska Baranek K, Zgorka G, Jozefczyk A (1997). Effect of picloram and methyl jasmonate on growth and taxanes accumulation in callus culture of Taxus x media var. Hatfieldii. Plant Cell Tissue Organ Culture. 49: 75-79.

Furmanowa M, Syklowska BK (2000). Hairy root cultures of Taxus x media var. Hicksii Rehd. as a new source of paclitaxel and 10-

deacetylbaccatin III hairy root culture ST: chemotherapeutic drug production from Agrobacterium rhizogenes-infected. Biotechnol. Lett. 22(8): 683-86.

Gamborg OL, Miller RA, Ojima K (1968). Nutrient requirements of suspension cultures of soybean root cells. Exp. Cell Res. 50: 148-151.

Georg GI, Harriman GCB, Velde DGV, Boge TC, Cheruvallath ZS, Datta A, Hepperle M, Park H, Himes RH (1995) Medicinal Chemistry of Taxol: Chemistry, structure-activity relationships and conformational analysis. In: Taxane Anticancer Agents. Basic Science and Current Status (Georg GI, Ojima I, Vyas DM eds.), American Chemical Society Symposium series. pp. 217-232.

Gibson DM, Ketchum REB, Vance NC, Christen AA (1993). Initiation and growth of cell lines of Taxus brevifolia (Pacific yew). Plant Cell Rep.12: 479-482.

Gotaskie GE, Andreassi BF (1994). Paclitaxel: a new antimitotic chemotherapeutic agent. Cancer Pract. 2(1): 27-33.

Grothaus PG, Raybould TJG, Bignami GS, Lazo CB, Byrnes JB(1993). An enzyme immunoassay for the determination of taxol and taxanes in Taxus sp. tissues and human plasma. J. Immunol. Methods 158(1): 5-15.

Guenard D, Gueritte-Voegelein F, Potier P (1993). Taxol and Taxotere: discovery, chemistry and structure-activity relationships. Acc. Chem. Res. 26: 160-167.

Han KH, Fleming P, Walker K, Laper M, Chilton WS, Mocek U, Gordon MP, Floss HG (1994).Genetic transformation of mature Taxus: An approach to genetically control the in vitro production of the anticancer drug, Taxol. Plant Sci. 95: 187-196.

Hartzell H (1991). The yew tree, a thousand wipers. Biography of a species. Europe, OR: Hulogosi.

Hefner J, Ketchum REB, Croteau R (1998). Cloning and functional expression of a cDNA encoding geranylgeranyl pyrophosphate synthase from Taxus canadensis and assessment of the role of this prenyltransferase in cells induced for Taxol production. Arch. Biochem. Biophys. 360: 62-74.

Hefner J, Rubenstein SM, Ketchum REB, Gibson DM, Williams RM, Croteau R (1996) Cytochrome P450 catalyzed hydroxylation of taxa-4(20),11(12)-diene to taxa-4(20),11(12)-dien-5α-ol: the first oxygenation step in taxol biosynthesis. Chem. Biol. 3: 479-489.

Hezari M, Lewis NG, Croteau R (1995). Purification and characterization of taxa-4 (5), 11(12)diene synthase from Pacific Yew (Taxus brevifolia) that catalyzes the first committed step in taxol biosynthesis. Arch. Biochem. Biosyn. 322: 437-444.

Ho CK, Chang SH, Lung H Jr, Tsai CJ, Chen KP (2005) The strategies to increase taxol production by using Taxus mairei cells transformed with TS and DBAT genes. Int. J. Appl. Sci. Eng. 3: 179-185.

Itokawa H (2003). Taxoids occurring in the genus Taxus. In: Taxus-The Genus Taxus (Itokawa, H. and Lee, K.H. eds.). Taylor & Francis, London. pp. 35-78.

Jaziri M, Zhiri A, Guo YW, Dupont JP, Shimomura K, Hamada H, Vanhaelen M, Homes J (1996). Taxus species cell, tissue and organ cultures as alternative sources for taxoid production: A literarure survey. Plant Cell Tissue Organ Culture. 46: 59-75.

Jennewein S, Rithner CD, Williams RM, Croteau R (2003) Taxoid metabolism: Taxoid 14β hydroxylase is a cytochrome P450 dependent monooxygenase. Arch. Biochem. Biophys. 413: 262-270.

Ketchum REB, Gibson DM, Greenspan Gallo L (1995). Media optimization for maximum biomass production in cell cultures of pacific yew (T. Brevifolia). Plant Cell Tiss. and Organ Cult. 42: 185-193.

Ketchum REB, Horiguchi T, Qiu D, Williams RM, Croteau RB (2007). Feeding cultured Taxus cells with early precursors reveals bifurcations in the taxoid biosynthetic pathway. Phytochemistry 68: 335-341.

Ketchum REB, Tandon M, Gibson DM, Begley T, Shuler ML (1999). Isolation of labeled 9-Dihydrobaccatin III and related taxoids from cell cultures of Taxus canadensis elicited with methyl jasmonate. J. Nat. Prod. 62: 1395-1398.

Kikuchi Y, Yatagai M (2003). The commercial cultivation of Taxus species and production of taxoids. In: Itokawa H, Lee KH eds. Taxus-The Genus Taxus Taylor & Francis, London, pp. 151-178.

Kim JH, Yun JH, Hwang YS, Byun SY, Kim DI (1995). Production of

taxol and related taxanes in Taxus brevifolia Cell culture effect of sugar. Biotech. Lett. 17: 101-106.

Kingston DGI, Molinero AA, Rimotdi, JM (1993). The taxane diterpenoids. Prog. Chem. Org. Nat. Prod. 61(206): 1469-1474.

Kiyota H, Shi QW, Oritani T (2002). A new drimane from the heartwood of the Japanese yew, Taxus cuspidata. Nat. Prod. Lett. 16(1): 21-24.

Kobayashi J, Shigemori H (2002). Bioactive taxoids from the Japanese yew Taxus cuspidata. Med. Res. Rev. 22(3): 305-328.

Kobayashi J, Shigemori H (2004). Biological activity and chemistry of taxoids from the Japanese yew, Taxus cuspidata. J. Natural Prod. 67(2): 245-256.

Kobayushi J, Ogiwava A, Hosoyama H, Shigemori H, Yoshida N, Sasaki T, Li Y, Iwasaki S, Naito M, Tsuruo T (1994). Taxuspines A-C, new taxoids from Japanese yew Taxus cuspidata, inhibiting drug transport activity of P-glycoprotein in multidrug-resistant cells. Tetrahedron 50(25): 7401-7416.

Krusmann G (1972). Handbuch der Nadelgehoze. Berlin, Hamburg: Verlag Paul Parey.

Lavelle F, Combeau C, Commercon A (1995). Taxoids: structural and experimental propertoies. Bull. Cancer, 82: 249–264.

Lin S, Ojima I (2000). Recent strategies in the development of taxane anticancer drugs. Expert opinion on Therapeutic Patents 10(6): 869 – 889.

Linden JC, Phisalaphong M (2000). Oligosaccharides potentiate methyl jasmonate induced production of paclitaxel in Taxus candensis. Plant Sci. 158: 41–51.

Long RM, Croteau R (2005). Preliminary assessment of the C-13 side chain 2'- hydroxylase involved in taxol biosynthesis. Biochem. Biophys. Res. Commun. 338: 410-417.

Loyd DG, McCown BH (1981). Commercially feasible micropropagation of mountain laurel (Kalmia latifolai) by use of shoot tip culture. Comb. Proc. Int. Plant Propag. Soc. 30: 421–427.

Lucas H (1956). Ueber ein den blattern von Taxus baccata L. Enthaltenes alkaloid (das taxin). Arch. Pharm. 85: 145-149.

Luo JP, Mu Q, Gu YH (1999). Protoplast culture and paclitaxel production by Taxus yunnensis. Plant Cell Tiss. and Org. Cult. 59: 25-29.

Ma W, Park GL, Gomez GA, Neider MH, Adams TL, Aynsley JS, Sahai OP, Smith RJ, Stahlhat RW, Hylands P (1994). New bioactive taxoids from cell cultures of Taxus baccata. J. Nat. Prod. Lloydia 57:116-122.

Ma W, Stahlhut RW, Adams TL, Park GL, Evans WA, Blumenthal SG, Gomez GH, Neider MH, Hylands PJ (1994). Yunnaxane and its homologous esters from cell cultures of Taxus chinensis Var. Mairei.. J. Nat. Prod.. 57:1320-1324.

Majda JP, Sierra MI, Tames-Sanchez R (2000). One step more towards taxane production through enhanced Taxus Propagation. Plant Cell Reports 19: 825-830.

McGuire WP, Rowinsky EK, Rosenshein NB, Grumbine FC, Ettinger DS, Armstrong DK (1998). Taxol: A unique anti-neoplastic agent significant activity in advanced ovarian epithelial neoplasmas. Ann. Intern. Med. 111: 273-279.

Menhard B, Eiseureich W, Hylands PJ, Bacher A, Zenk MH (1998). Taxoids from cell cultures Taxus chinensis. Phytochemistry 49: 113-125.

Mirajalili N, Linden JC (1996). Methyl jasmonate induced production of taxol in suspension cultures of Taxus cuspidata: ethylene interaction and induction models. Biotechnol. Prog. 12: 110-118.

Morita H, Izumi M, Yusuke H, Jun'ichi K (2005). Taxezopidines M and N, Taxoids from the Japanese Yew, Taxus cuspidata. J. Nat. Prod. 68(6): 935 -937

Murshige T, Skoog F (1962). A revised medium for rapid growth and bioassays with tissue cultures. Plant Physiol. 15: 473-497.

Navia-Osorio A, Garden H, Cusido RM, Palazon J, Alfermann AW, Pinol MT (2002). Production of paclitaxel and baccatin III in a 20-L airlift bioreactor by a cell suspension of Taxus wallichiana. Planta Med. 68(4): 336 - 340.

Nicolaou KC, Dai WM, Guy RK (1994). Chemistry and biology of Taxol. Angew. Chem. Int. Ed. Engl. 33: 15–44.

Odgen L (1988) Taxus (yews)- a highly toxic plant. Vet. Hum. Toxicol. 30: 653-564.

Ojima I, Habas I, Zhao M, Zucco M, Park YH, Sun CM, Brigaud T (1992). New and efficient approaches to the semi-synthesis of taxol and its C-13 side chain analogs by means of β lactam synthon method. Tetrahedron 48: 6985-7012.

Parc G, Canaguier A, Landre P, Hocquemiller R, Chriqui D, Meyer M (2002). Production of taxoids with biological activity by plants and callus culture from selected Taxus genotypes. Phytochemistry 59(7): 725-730.

Parmar VS, Jha A, Bisht KS, Taneja P, Singh SK, Kumar A, Denmark PP, Jain R, Olsen SE (1999). Constituents of the yew tree. Phytochemistry 50: 1267-1304.

Phisalaphong M, Linden JC (1999). Kinetic studies of paclitaxel production by Taxus canadensis cultures in batch and semicontinuous with total cell recycle. Biotechnol. Prog. 15(6):1072-1077.

Plaut-CarCasson Y (1994). Plant tissue culture method for taxol production. WO Patent NO. 94/20606.

Poupat C, Hook I, Gueritte F, Ahond A, Guenard D, Adeline MT, Wang XP, Dempsev D, Breuillet S, Potier P (2000). Neutral and basic taxoid contents in the needles of Taxus species. Planta Med. 66(6): 580 – 584.

Prasain JK, Stefanowicz P, Kiyota T, Habeichi F, Konishi Y (2001). Taxines from the needles of Taxus wallichiana. Phytochemistry 58(8): 1167-1170.

Rischer H, Oksman-Caldentey KM (2005). Biotechnological utilization of plant genetic resources for the production of phytopharmaceuticals. Plant Gen. Resour. 3: 83-89.

Roberts SC, Naill M, Gibson DM, Shuler ML (2003). A simple method for enhancing paclitaxel release from Taxus canadensis cell suspension cultures utilizing cell wall digesting enzymes. Plant Cell Rep. 21(12): 1217-20.

Rohr R (1973). Production de cals par les gemetophyte males de Taxus baccata L. Cultivars sur un milieu artificiel. Etude en microscopie photonique et electronique. Coryologia, 25: 177–189.

Saikia D, Khanuja SPS, Shasany AK, Darokar MP, Kukreja AK, Kumar S (2002). Assessment of diversity among Taxus wallichiana accessions from northeast India using RAPD analysis. Plant Genetic Resources Newsletter 121: 27 –31.

Schrijvers D, Vermorken JB (2000). Role of Taxoids in Head and Neck Cancer. The Oncologist 5(3): 199-208.

Seki M, Ohzora C, Takeda M, Furusaki S (1997). Taxol (paclitaxel) production using free and immobilized cells of Taxus cuspidata. Biotechnol. Bioeng. 53(2): 214-19.

Shen YC, Chang YT, Lin YC, Lin CL, Kuo YH, Chen CY (2002). New Taxane Diterpenoids from the Roots of Taiwanese Taxus mairei. Chem. Pharm. Bull. 50(6): 781-787.

Shen YC, Chen YJ, Chen CY (1999). Taxane diterpenoids from the seeds of Chinese yew Taxus chinensis. Phytochemistry, 52: 1565-1569.

Shen YC, Hsu SM, Lin YS, Cheng KC, Chien CT, Chou CH, Cheng YB (2005). New bicyclic taxane diterpenoids from Taxus sumatrana. Chem. Pharm. Bull. 53: 808-810.

Shi QW, Oritani T, Kiyota H, Murakami R (2001). Three new taxoids from the leaves of the Japanese yew, Taxus cuspidata. Nat. Prod. Lett. 15(1): 55-62.

Shi QW, Oritani T, Sugiyama T, Murakami R, Wei H (1999). Six new taxane diterpenoids from the seeds of Taxus chinensis var. mairei and Taxus yunnanensis. J. Nat. Prod. 62: 1114-1118.

Shinozaki Y, Fukamiya N, Fukushima M, Okano M, Nehira T, Tagahara K, Zhang SX, Zhang DC, Lee KH (2002). Dantaxusins C and D, two novel taxoids from Taxus yunnanensis. J. Nat. Prod. 65(3): 371- 374.

Srinivasan V, Pestchanker S, Moser T, Hirasuna T, Taticek RA, Schuler ML (1995). Taxol production in bioreactors: Kinetics of biomass accumulation, nutrient uptake and taxol production by cell suspensions of Taxus baccata. Biotechnol. Bioeng. 47: 666-676.

Stierle A, Storbel G, Stierle D (1993). Taxol and Taxane production by Taxomyces andreanae, an endophytic fungus of pacific yew. Science 260: 214- 216.

Strobel GA, Hess WM, Ford E, Sidhu RS, Yang X (1996). Taxol from endophytes and the issue of biodiversity. J. Ind. Microbiol. Biotechnol. 17: 417– 423.

Takeya K (2003). Plant tissue culture of taxoids In: Itokawa H, Lee KH eds. Taxus- The Genus Taxus Taylor & Francis, London, pp. 134-150.

Vander Velde DG, Georg GI, Gollapudi SR, Jampani HB, Liang XZ,Mitscher LA, Ye QM (1994). Wallifoliol, a taxol congener with a novel carbon skeleton, from Himalayan Taxus wallichiana. J. Nat. Prod. 57(6): 862-867.

Vanek T, Mala J, Saman D, Silhava I (1999). Production of taxanes in a bio reactor by Taxus baccata cells. Planta Medica 65: 275-277.

Walker K, Croteau R (2000). Molecular cloning of a 10-deacetylbaccatin III-10-O-acetyl transferase cDNA from Taxus and functional expression in Escherichia coli. Proc. Natl. Acad. Sci. U.S.A. 97(2): 583-87.

Walker K, Croteau R (2001). Taxol biosynthetic genes. Phytochemistry, 58: 1-7.

Walker K, Long R, Croteau R (2002). The final acylation step in taxol biosynthesis: cloning of the taxoid C13-side-chain N-benzoyltransferase from Taxus. Proc. Natl. Acad. Sci. U.S.A. 99(14): 9166-9171.

Wang C, Wu J, Mei X (2001). Enhancement of Taxol production and excretion in Taxus chinensis cell culture by fungal elicitation and medium renewal. Appl. Microbiol. Biotechnol. 55(4): 404-410.

Wang HQ, Yu JT, Zhong J (2000). Significant improvement of taxane production in suspension cultures of Taxus chinensis by sucrose feeding strategy. J. Process Biochem. 35(5): 479-483.

Wani MC, Taylor HL, Wall ME, Coggon P, McPhail AT (1971). Plant tumor agent VI. The isolation and structure of taxol, a novel antileukemic and antitumour agent from Taxus brevifolia. J. Am. Chem. Soc. 93: 2325-2327.

Wickremesinhe ERM, Arteca RN (1993). Taxus callus cultures: Initiation, growth optimization, characterization and taxol production. Plant Cell Tiss. Org. Cult. 35: 181-193.

Wickremesinhe ERM, Arteca RN (1994). Taxus cell suspension cultures: optimizing growth and Taxol production. J. Plant Physiol. 144: 183-188.

Wildung MR, Croteau R (1996). A cDNA clone for taxadiene synthase, the diterpene cyclase that catalyses the committed step of taxol biosynthesis. J. Biol. Chem. 271: 9201-9204.

Wink M, Alfermann, AW, Franke R, Wetterauer B, Distl M, Windhovel J, Krohn O, Fuss E, Garden H, Mohagheghzadeh A, Wildi E Ripplinger P (2005). Sustainable bioproduction of phytochemicals by plant in vitro cultures: anticancer agents. Plant Gen. Resour. 3: 90-100.

Witherup KM, Look SA, Stasko MW, Ghiorzi TJ, Cragg GM (1990). Taxus spp. needles contain amounts of taxol comparable to the bark of Taxus brevifolia:analysis and isolation. J. Nat. Prod. 53: 1249-1255.

Wu J, Ge X (2004). Oxidative burst, jasmonic acid biosynthesis and taxol production induced by low energy ultrasound in Taxus chinensis cell suspension cultures. Biotechnol. Bioeng, 85(7): 714-721.

Wu J, Lin L (2003). Enhancement of taxol production and release in Taxus chinensis cell culture by ultrasound, methyl jasmonate and in situ solvent extraction. Appl. Microbiol. Biotechnol. 62(2-3): 151-155.

Wu J, Wang C, Mei X (2001). Stimulation of taxol production and excretion in Taxus spp cell cultures by rare earth chemical lanthanum. J. Biotechnol. 85(1): 67-73.

Xia ZH, Peng LY, Zhao Y, Xu G, Zhao QS, Sun HD (2005). Two new taxoids from Taxus chinensis. Chemistry & Biodiversity 2: 1316-1319.

Xu JF, Yin PQ, Wei XG, Su ZG (1998). Self-immobilized aggregate culture of Taxus cuspidata for improved taxol production. Biotechnol.Tech. 12(3): 241-244.

Yukimune Y, Hara Y, Nomura E, Seto H, Yoshida S (2000). The configuration of methyl jasmonate affects paclitaxel and baccatin III production in Taxus cells. Phytochemistry 54(1):13-17.

Yukimune Y, Tabata H, Higashi Y, Hara Y (1996). Methyl jasmonate induced over production of paclitaxel and baccatin III in Taxus cell suspension cultures. Nature Biotech. 14:1129-1132.

Zhang JH, Fang QC, Liang XT, He CH, Kong M, He WY, Jin XL (1995). Taxoids from the barks of Taxus wallichiana. Phytochemistry 40: 881-884.

Zhang ML, Li LG, Cao CM, Li ZP, Shi QW, Kiyota H (2006) A new abeotaxane diterpenoid from the seeds of Taxus mairei. Chinese Chem. Lett. 17: 27-30.

10

Current trends in molecular epidemiology studies of *Mycobacterium tuberculosis*

Mohammad Asgharzadeh[1]* and Hossein Samadi Kafil[2]

[1]Biotechnology Research Center, Tabriz University of Medical Sciences, Tabriz, Iran.
[2]Department of Bacteriology, School of Medicine, Tarbiat Modares University, P.O. Box: 14115-33, Tehran, Iran.

Mycobacterium tuberculosis **is one of the most harmful human pathogens worldwide, and there are mass efforts for controlling this pathogen. One of the powerful tools to find out and control this pathogen is molecular epidemiology techniques. Currently, wide ranges of techniques are available to type** *mycobacterium tuberculosis,* **and choosing the correct technique as a portable and standard method is difficult. IS6110 restriction fragment length polymorphism (RFLP) remains gold standard method on genotyping to date, but it is labor intensive and inefficient on samples which have fewer than six copy numbers of IS6110. In the recent years, some new methods have been introduced for genotyping of** *mycobacterium tuberculosis* **such as mycobacterial interspersed repetitive unit-variable number tandem repeat (MIRU-VNTR) and Spoligotyping. The present review tries to introduce new approaches in molecular techniques for epidemiological investigation of tuberculosis and to illustrate advantages and problems associated with them.**

Key words: Molecular epidemiology, *Mycobacterium tuberculosis*, fingerprinting, transmission, genotype.

TABLE OF CONTENT

INTRODUCTION

Despite mass BCG vaccination and anti tuberculosis drugs, tuberculosis is one of the major killers among the infectious diseases, causing about eight million cases and two to three million deaths occurs annually (Anh et al., 2000). Current trends suggest that these numbers could rise to 12 million cases and 4 million death, by the year 2010 (Bifani et al., 2002). Tuberculosis is the number one cause of death among individuals infected with human immunodeficiency virus (HIV) (Dye et al., 1999).

Young children are likely to develop this disease after infection and are significantly more likely to develop extrapulmonary and severe disseminated disease than adults (Walls et al., 2004).

Increasing incidence of coinfection of tuberculosis with HIV, especially in developing countries, decrease children burdened of tuberculosis, emergence of multidrug-resistant tuberculosis, and immigrations caused new trends for molecular epidemiology and DNA fingerprinting of *Mycobacterium tuberculosis*. Also it introduced new molecular epidemiology markers for case finding and studying manner of transmission and reactivation of tuberculosis in regional investigations.

*Corresponding author. E– mail: Asgharzadehmo@yahoo.com.

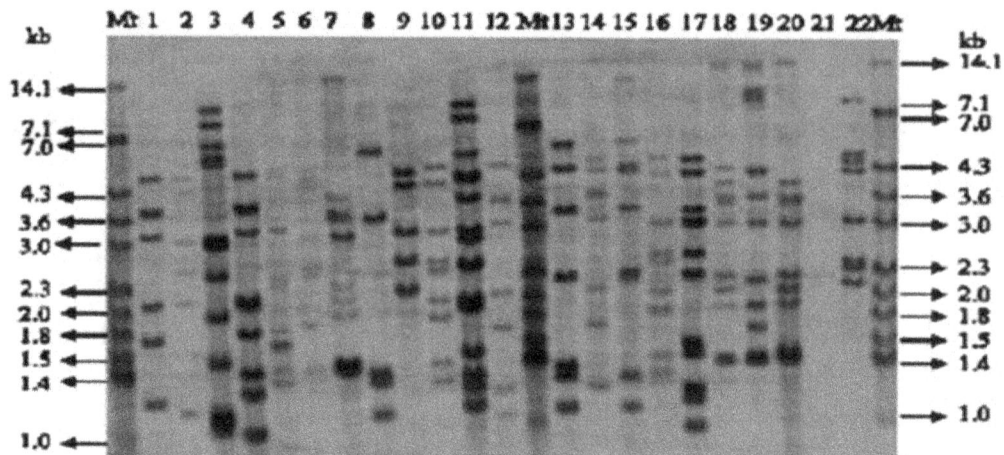

Figure 1. IS6110 fingerprinting in different isolates and reference *Mycobacterium tuberculosis* strain Mt14323. The size of hybridized fragments depends on the distance from this site to the next Pvu II site in DNA.

IS6110

IS6110 was first described by Thierry et al. (1990). Its standard approach to genotyping *M. tuberculosis* isolate was based on restriction-fragment-length polymorphism (RFLP) analysis (van Embden et al., 1993; Asgharzadeh et al., 2006).

IS6110 is a novel Mycobacterial insertion element formed the basis of a reproducible genotyping technique. Mycobacterial DNA is digested with the restriction enzyme PvuII (Asgharzadeh et al., 2007a). The IS6110 DNA probe was prepared by *in vitro* amplification of 245-bp fragment using the polymerase chain reaction and was labeled by digoxgenin. Fragments after digestion with PvuII were transferred from gel to positively charged nylon membrane. And then Hybridization was performed with 245bp probe of insertion sequence and detected by colorimetric system (Asgharzadeh et al., 2007b; van Embden et al., 1993; van Soolingen et al., 1994). The IS6110 probe hybridizes to IS6110 DNA to the right of the Pvu II site in IS6110. The size of hybridized fragments depends on the distance from this site (PvuII digested) to the next PvuII site in DNA (Figure 1). Advantage of this method is by discriminatory power on detecting epidemiological relations, whereas patients with epidemiological links have similar pattern of RFLP and it can reveal source of infection as a reactivated infection (Small et al., 1993a; van Rie et al., 1999) or transmission (Barnes et al., 2003; Bauer et al., 1998; Barnes et al., 1997; Burman et al., 1997; Chin et al., 1998; Dooley et al., 1992; Edlin et al., 1992; Durmaz et al., 2003; Pena et al., 2003; Niobe-Eyangoh et al., 2003). But it has several restrictions like, it is still expensive, requires subculturing the isolates for several weeks for obtaining sufficient DNA, labour intensive, discriminatory and polymorphism of isolates decrease and strains with fewer than six IS6110 insertion sites (Asgharzadeh et al.,

2006; Asgharzadeh et al., 2007c; Bauer at al., 1999; Cowan et al., 2002; Goguet de la Salmoniere et al., 1997; Kremer et al., 1999; Lee et al., 2002; Mazars et al., 2001).

In order to estimate the stability of IS6110 RFLP patterns, studies have examined serial isolates collected from patients with persistent disease (de Boer et al., 1999; Niemann et al., 1999; Warren et al., 2002) and have demonstrated that the IS6110 banding pattern may change over time in a subset of these patients. When survival analysis was applied to the RFLP data collected from these patients, de Boer et al. (1999) calculated that the half life of the IS6110 banding pattern was on the order of 3.2 year. In another study, demonstrated Changes were observed in 4% of strains, and half-life ($t_{1/2}$) of 8.74 year was calculated (Rhee et al., 2000).

The early rate of Change in RFLP pattern probably reflects the change that occurred during active growth prior to therapy, while the low rate may reflect change occurring during or after treatment.

IS6110-based typing is the most widely applied genotyping method in the molecular epidemiology of *M. tuberculosis* and the gold standard to which other currently described method (van Embden et al., 1993; Rhee et al., 2000). Probably if overcome obstacles of IS6110-based genotyping, it can be incorporated into, routine, prospective and population-based method in identifying unsuspected outbreaks.

IS1081

In order to overcome the problem of absence or low copy number, an alternative molecular markers have been identified by van Soolingen et al. (1993). IS1081, first time identified by Collins and Stephens (1991). This insertion element is a 1324-bp sequence found in *M. tuberculosis* complex. It had some restrictions in comparing with IS6110; it has a lower degree of polymorphism than IS6110, and can be due to its low transpositional activity (van Soolingen et al., 1993; van Soolingen et al., 1992; Kanduma et al., 2003). The copy number in IS1081 is lower than that of IS6110, which limits its use in epidemiological studies (Kanduma et al., 2003). Also it cannot be used to differentiate BCG from the other members of *M. tuberculosis* complex (van Soolingen et al., 1993).

PGRS

Ten percent of the *M. tuberculosis* genome consists of the genes that encode *M. tuberculosis* specific PE, PPE and PE-PGRS proteins, the PE family proteins are involved in iron acquisition, which is a critical process for *M. tuberculosis* survival (Ahmed et al., 2004).

PE-PGRS (polymorphic GC-rich sequences family represents an extension of the PE protein family with multiple repeats of glycine–glycine alanine asparagine) motifs (Choudhary et al., 2003; Chakhaiyer et al., 2004). There are numerous copies of PGRS repetitive element in the *M. tuberculosis* complex (De wit et al., 1990; Ross et al., 1992; Poulet sand Cole et al., 1995). It consists of many tandem repeats of 96 bp GC rich consensus sequence. PGRS elements are present in 26 sites of *M. tuberculosis* chromosome (Poulet sand Cole et al., 1995). PGRS – based genotyping involve *sma I* digestion of DNA and probing for a 32bp oligonucleotide (Chaves et al., 1996). PGRS is similar to the standardized IS6110 fingerprinting in which it requires purified DNA for southern blot hybridization and banding pattern analysis (Yang et al., 2000). PGRS fingerprinting has proven to be useful for differentiating *M. tuberculosis* strains with fewer than six copies of IS6110 that could not be readily differentiated by IS6110 fingerprinting (Barnes et al., 1997; Braden et al., 1997; Burman et al., 1997; Chaves et al., 1996). A better correlation between DNA fingerprinting data and the results of conventional epidemiology was found when a combination of the IS6110 and PGRS fingerprinting was applied than when IS6110 was used alone (Burman et al., 1997; Chaves et al., 1996; Yang et al., 2000). However the difficulties in computerizing the analysis of PGRS fingerprints is due to complexity of fingerprint patterns that have limited the wide use of PGRS fingerprinting (Yang et al., 2000).

Spoligotyping

The direct repeat locus in *M. tuberculosis* contains 10 to 50 copies of 36-bp direct repeat, that are separated from one another by spacer which have different sequences.

However, the spacer sequences between any two specific direct repeats are conserved among strains.

Because strains differ in terms of the presence or absence of specific spacer, the pattern of spacers in a strain can be used for genotyping (spacer oligonucleotide

typing) or spoligotyping (Barnes et al., 2003; Filliol et al., 2000; Kamerbeek et al., 1997; Groenen et al., 1993; Hermans et al., 1991).

Spoligotyping is a PCR-based technique (Kamerbeek et al., 1997; Hermans., 1992). With using one set of primers it is possible to simultaneously amplify all the unique nonrepetitive sequences, spacers or between the direct repeats. The presence or absence of spacers is then determined via southern hybridization (Figure 2). Individual strains are distinguished by the number of spacers missing from the complete spacer set that was defined by sequencing this region from a large number of *M. tuberculosis* strains (Burgos et al., 2002).

This method have same advantage and by comparing with IS6110 RFLP method, it can be performed by small amount of DNA and a little time after inoculation of bacteria to liquid culture can be perform(Barnes et al., 2003; Kamerbeek et al., 1997). Also spoligotyping is useful for discrimination between isolates of *M. tuberculosis* with few copy number of IS6110 (Goyal et al., 1999; Goguet et al., 1997). This technique is specific for *M. tuberculosis* and other atypical strains dose not have any signal in analyzing with spoligotyping (Burges et al., 2002).

Usually the results of spoligotyping are expressed as positive or negative for spacers. Therefore, it can be expressed in a digital format (Dale et al., 2001). This method is simple, rapid, robust and economical mean for typing *M. tuberculosis* complex (Kremer et al., 1999; Kanduma et al., 2003; Barnes et al., 2003; Burgos et al., 2002; Goyal et al., 1999). However, the differentiating power of spoligotyping is less than IS6110 typing when high copy number strains are being analyzed (Kremer et al., 1999; Kamerbeek et al., 1997; Goyal et al., 1999; Diaz et al., 1998; Doroudchi et al., 2000). This method is a candidate for use in resource-poor situations (Burgos et al., 2002).

MIRU-VNTR

Variable Number Tandem Repeat (VNTR) typing is an invaluable tool for genotyping and provides data in a simple and format based on the number of repetitive sequences in so called polymorphic micro – and mini satellite regions (Mazars et al., 2001; Asgharzadeh et al., 2007c). VNTR introduced for *M. tuberculosis* which named Mycobacterial Interspersed Repetitive Unit-Variable Number Tandem Repeat (MIRU-VNTR) (Magdalena et al., 1998; Mazars et al., 2001). Out of 41 different loci in the genome of *M. tuberculosis* have been identified by supply et al (Supply et al., 2000).

Twelve loci were identified as hypervariable repetitive units (Magdalena et al., 1998; Supply et al., 2000; Supply et al., 2001). These fragments should be accurately sized for determining the number of repeats at each locus. The repeated units are 52 to 77 nucleotides in length and therefore power of this method may be com-

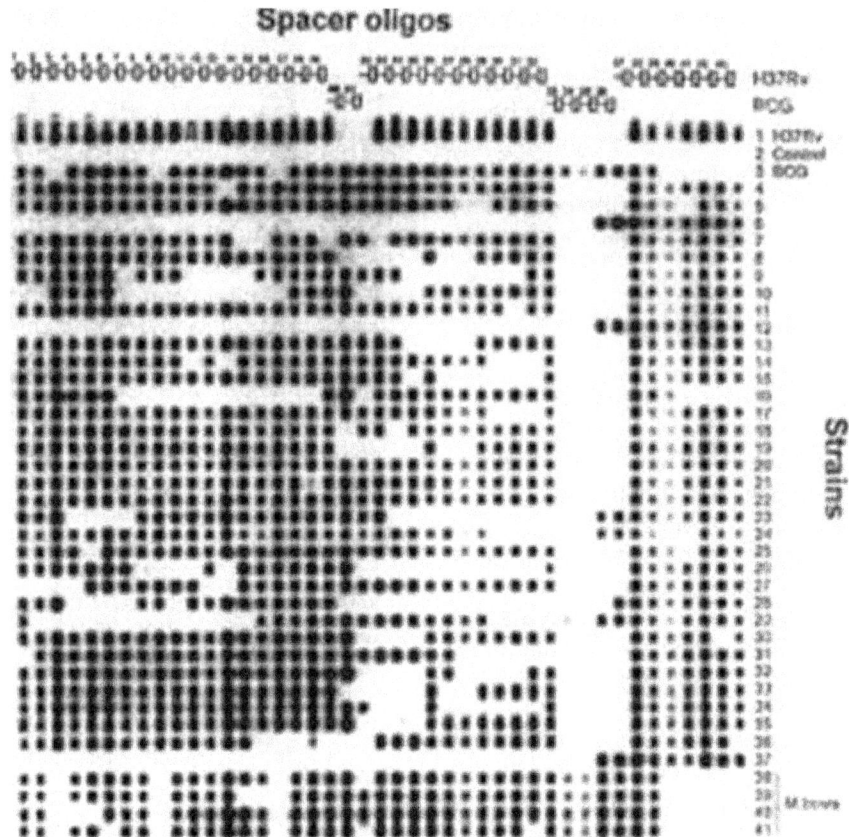

Figure 2. Spoligotyping, strains differ in terms of the presence or absence of specific spacer, the pattern of spacers in a strain can be used for genotyping (spacer oligonucleotide typing).

parable to that of IS6110- RFLP.

Moreover, MIRU typing showed its usefulness in studying the population structure of *M. tuberculosis* (Sola et al., 2003; Supply et al., 2003). MIRU-VNTR can also be used in global databases as each typed strain is assigned a 12-digict number corresponding to the num-ber of repeats at each MIRU-VNTR loci (Mazars et al., 2001). Various studies demonstrated the importance of MIRU-VNTR method for tracking epidemiological key events, such as transmission or relapse and provide non ambiguous data (Mazars et al., 2001), and, which are highly portable between different laboratories (Mazars et al., 2001; Frothingham et al., 1998; Supply et al., 1997).

Approximately 20 million possible combinations of 12 loci alleles are possible (Mazars et al., 2001; Supply et al., 2001). Most important advantage of this method is its simpler performance than IS6110, applying directly to *M. tuberculosis* cultures without DNA purification (Barnes et al., 2003). Based on high discriminatory power (Mazars et al., 2001; Asghaerzadeh et al., 2007c) and when comparing MIRU-VNTR with IS6110-RFLP or spoligotyping, MIRU-VNTR produced suitable discriminatory power (Mazars et al., 2001; Supply et al., 2001; Barlow et al., 2001; Cowan et al., 2002). Hardship of this method is

associated with accurate sizing of multiple small fragments (Burgos et al., 2002). This can be partly overcome by combining multiplex PCR with a fluorescence-based DNA analyzer (Mazars et al., 2001). A population-based study indicated that the use of 12- locus-based MIRU-VNTR typing combined with spoligotyping as a first-line approach provided an adequate discrimination in most cases for large-scale genotyping of M. tuberculosis in the United States (Cowan et al., 2005). In a recent study based on a worldwide collection of tubercle bacillus isolates defined an optimized set of 24 MIRU-VNTR loci, including a subset of 15 discriminatory loci proposed to be used as a first-line typing method. These 15- and 24-locus sets reliably improved the discrimination of *M. tuberculosis* isolates compared to the original 12-locus set (Supply et al., 2006; Oelemann et al., 2007) (Table 1).

RAPD

Random amplification of polymorphic domains (RAPD) is a PCR-based method for genotyping *M. tuberculosis* (Palitta et al., 1993). This method is based on amplifycation of the spacer region between the genes coding for 16S and 23S rRND of M. tuberculosis and digestion by

Table 1. Clustering results by genotyping methods.

Methods	No. of unique isolates	No. of clustered isolates	No. of clusters	No. of distinct types	Clustering rate (%)
Spoligotyping	56	98	16	72	53.2
MIRU-VNTR 12 old	84	70	20	104	32.5
MIRU-VNTR 12 old + Spoligotyping	106	48	14	120	22.1
IS6110-RFLP	115	39	13	128	16.9
MIRU-VNTR 15	115	39	12	127	17.5
MIRU-VNTR 24	117	37	11	128	16.9
MIRU-VNTR 15+ Spoligotyping	120	34	11	131	14.9
MIRU-VNTR 24 + Spoligotyping	120	34	11	131	14.9

Table 1. Brought from Oelemann et al., 2007. 154 isolates investigated for Genotyping by IS6110 DNA fingerprints and MIRU-VNTR typing with 12, 15 and the full set of 24 loci. 24 loci MIRU-VNTR made a slightly better resolution than IS6110. However, both methods yielded identical clustering rates of 16.9% due to minor differences in cluster compositions. The use of the discriminatory subset of 15 MIRU-VNTR loci only marginally affected the resolution, with a clustering rate of 17.5%. Spoligotyping increased the resolution of MIRU-VNTR typing based on the 24 loci and 15 loci to 14.9%.

restriction enzymes (Abed et al., 1995).

Patterns generated this method can be easily analyzed, have discriminatory power, have satisfactory results and may serve as rapid screening test for typing a large number of clinical isolates into clusters for further subtyping by more sophisticated methods (Vrioni et al., 2004). But in some studies reproducibility and the final discriminative power of the RAPD-based method was found to be limited (Frothingham et al., 1995; Glennon et al., 1995).

PFGE

Pulsed–field gel electrophoresis (PFGE) has been widely used to type various microorganisms in both outbreak and population based studies and is available in many clinical laboratories (Singh et al., 1999). But, this method is not commonly used in epidemiological studies of *M. tuberculosis*.

Most published PFGE protocols for *M. tuberculosis* are technically challengeable. Biosafety considerations and the unique cell wall composition of the organism have led to the development of protocols that are high complex and difficult to reproduce (Singh et al., 1999). Little has been done for developing PFGE patterns as a standardized method for analyses of M. tuberculosis, and therefore, it is a limited method for population based molecular epidemiologic studies of *M. tuberculosis*. Also there are contradictory reports on the genetic diversity captured by PFGE and on its utility for molecular epidemiology (D'Amato et al., 1995; Miller et al., 1994; Schirm et al., 1993).

Current application of molecular epidemiology

There are several available applications for molecular epidemiology in control of tuberculosis. The first important role of genotyping studies is on intensive use of these studies on trace outbreaks (Bifani et al., 1996; Frieden et al., 1996; Moss et al., 1997; Valway et al., 1994). Combination of genotyping with epidemiological investigations reveal the occurrence of an outbreak and by controlling the manner of disease transmission, also new molecular methods prepared a rapid typing of isolates where an outbreak is suspected in a hospital ward (Kanduma et al., 2003).

Molecular genotyping methods are important in detecting the dominance of transmission or reinfection in a population. In reinfections, was a documented source should be sought. In anti-tuberculosis – treatment trails, it is important to determine whether recurrent tuberculosis is due to relapse of reinfection when the former represents treatment failure.

The transmission index, defined as the mean number of tuberculosis cases resulting from recent transmission of a potential source case, has been used to qualify transmission between different subpopulations (Borgdorff et al., 2000). Finding the exact transmission of index is important in regional and global controlling programs.

Laboratory cross-contamination is a significant problem in detection of tuberculosis and cause abuse of drugs for treatment. The occurrence of cross-contamination is most likely when acid fact smears are negative and only one specimen is culture –positive (Burmen at al., 2000).

Genotyping can be used to confirm the occurrence of cross contamination in the laboratories and have been identified as a useful measure for avoiding false-positive cultures (Small et al., 1993b; Behr et al., 1997; Asgharzadeh et al., 2007d).

Molecular epidemiology studies have an important role of improving disease control in coinfections, exactly coinfection of tuberculosis and HIV. These studies reveal progress and spread of tuberculosis amongst HIV infected patients (Daley et al., 1992). These studies are useful in evaluating the development of drug resistant in patients with relapse or transmission of tuberculosis and investigation in spread of drug resistant strains among hospitalized patients (Daley et al., 1992; Ritacco et al., 1997; Anastasis et al., 1997; Angarano et al., 1998; Breathnach et al., 1998).

Beijing genotype is a distinct family of tuberculosis which was found first time in Beijing–china. This genotype is prevalent in young individuals and indicates ongoing transmission of tuberculosis (Doroudchi et al., 2000; Anh et al., 2000). These genotypes are more virulent (Anh et al., 2000), and in detection of global epidemiology of this genotype have an important role in controlling tuberculosis.

Conclusion

Several different molecular methods have been made available for epidemiological and evolutionary studies as a result of comparative genomic studies. Now the challenge is to compile standardized molecular fingerprinting patterns originating from highly networked, multi-centric, genotypic analysis in the databases for interlaboratory use and for further references. Rapid genotyping method are needed to overcome low reproducibility, not proven application, less discrimination method such as Mycobacterial interspersed repeat units variable number tandem repeats (MIRU- VNTR). Also A combination of typing methods based on more rapid and slower molecular clacks maybe able to exactly differentiate between the contributions of remote and recent transmission in clustering of infected patient and controlling outbreaks

ACKNOWLEDGMENT

Authors thank Tuberculosis and Lung Disease Research Center and Biotechnology Research Center, Tabriz University of Medical Sciences for granting our studies on molecular epidemiology of tuberculosis.

REFERENCES

Abed Y, Bolted C, Mica P (1995). Identification and strain differentiation of Mycobacterium species on the basis of DNA 16S - 23S spacer region polymorphism. Res. Microbiol. 146: 405-413.

Ahmed N, Hasnain SE (2004). Genomics of Mycobacterium tuberculosis: old threats & new trends. Indian J. Med. Res. 120: 207-212.

Anastasis D, Pillai G, Rambiritch V, Abdool Karim SS (1997). .A retrospective study of human immunodeficiency virus infection and drug resistant tuberculosis in Durban. South Africa. Int J Tuberc lung Dis. 1: 220-224.

Angarano G, Carbohara S, Costa D, Gori A (1998). Drug-resistant tuberculosis in human immunodeficiency virus infected persons in Italy. Int. J. Tuberc. Lung. Dis. 2: 303-311.

Anh DD, Brgdorff MW, Van LN, Lan NTN, Van Grokom T, kremer K, van Soolingen D (2000). Mycobacterium tuberculosis Beijing genotype emerging in Vietnam. Emerg Infect Dis. 6: 302-305.

Asgharzadeh M, Shahbabian K, Majidi J, Aghazadeh AM, Amini C, Jahantabi AR, and Rafi A (2006). IS6110 restriction fragment length polymorphism typing of Mycobacterium tuberculosis isolates from East Azarbaijan province of Iran. Mem inst oswaldo cruz. 101: 517-521.

Asgharzadeh M, Shahbabian K, Kafil HS, and Rafi A (2007a). Use of DNA fingerprinting in identifying the source case of tuberculosis in East Azarbaijan province of Iran. J. Med. Sci. 7: 418-421.

Asgharzadeh M, yousefee S, kafil HS, Nahaei MR, Ansarin K, Akhi MT, (2007b). Comparing transmission of Mycobacterium tuberculosis in East Azarbaijan and west Azarbaijan provinces of Iran by using IS6110 RFLP method. Biotechnol. 6: 273-277.

Asgharzadeh M, Khakpour M, Zahraei T, Kafil HS (2007c). Use of Mycobacterial interspersed repetitive unite–variable number tandem repeat typing to study Mycobacterium tuberculosis isolates from East Azarbaijan province of Iran. Pak. J. Biol. Sci. 10, 3769-3777.

Asgharzadeh M, Yousefee S, Nahaei MR, Akhi MT, Ansarian K, Kafil HS (2007d). IS6110 fingerprinting of Mycobacterium tuberculosis strains Isolated from northwest of Iran. Res. J. Microbiol. 2: 940-946.

Barlow RE, Gascoyne-Binzi DM, Gillespie SH, Dickens A, Shaman Q, Hawkey PM (2001). Comparison of VNTR and IS6110-RFLP analysis for discrimination of high and low copy number IS6110 M. tuberculosis isolates. J. Clin. Microbiol. 39: 2453-2457.

Barnes PF, Yangn ZH, Preston- Martin S, Pogoda M, Jones E, Otaya M, Eisenah D, Knowles L, Harvey S, Cave MD (1997). Patterns of tuberculosis transmission in central Los Angeles. JAMA 278: 1156-1163.

Barnes PF, Cave MD, (2003). Molecular epidemiology of tuberculosis. N. Eng. J. Med. 349: 1149-1156.

Bauer J, Yang Z, Poulsen S, Andersen AB (1998). Results from 5 years of nationwide DNA fingerprinting of mycobacterium tuberculosis complex isolates in country with a low incidence of M. tuberculosis infection. J. Clin. Microbiol. 36: 305-308.

Bauer J, Anderson K, Kremer K, and Miorner H (1999). Usefulness of spoligotyping to discriminate IS6110 low copy number mycobacterium tuberculosis complex strains cultured in Denmark. J. Clin. Microbiol. 37: 2602-2606.

Behr MA, Small PM (1997). Molecular fingerprinting of mycobacterium tuberculosis. how can it help the clinician? Clin. Infec.t Dis. 25: 806-810.

Bifani PJ, Plikaytis BB, Kapuru V, Sock bauerk W, Lutfey ML, Pan X, Lutfey ML, Moghazeh SL, Eisner W, Daniel TM, Kaplan MH, Crowford JD, Musser JM, Kreiswirth BN (1996). Origin and interstate spread of a New York City multidrug resistant M. tuberculosis clone family identification of a variant outbreak of mycobacterium tuberculosis TB via population based molecular epidemiology. J. Am. Med. Assoc. 282: 2321-2327.

Bifani PJ, Mathema B, Kurepina NE, Kreiswirth BN (2002). Global dissemination of the Mycobacterium tuberculosis W-Beijing family strains. Trends Microbiol. 10: 45-52.

Borgdorff MW, Behr MA, Nagelkerke NJ, Hopewell PC, Small PM (2000). Transmission of tuberculosis in San francisae. Int. J. Tuberc. Lung. Dis. 4: 287-294.

Braden CR, Templeton GL, Cave MD, Valway S, Onorato IM, Castro KG, Moers D, Yang Z, Stead WW, Bates JH. (1997). Interpretation of restriction fragment length polymorphism analysis of Mycobacterium tuberculosis isolates from a state with a larg rural population. J. infect. Dis. 175: 1446-1452.

Breathnach AS, de Ruiter A, Holdsworth GM, Bateman NT, O'Sullivan DG, Rees PJ, Snashall D, Milburn HJ, Peters BS, Watson J, Drobniewski FA, French JL (1998). An outbreak of multidrug resistant tuberculosis in a London teaching hospital. J. Hosp. Infect. 39: 111-117.

Burman WJ, Reves RR, Hawkes AP, Rietmeijer CA, Wilson ML, Yang ZH, El-Hajj H, Bates JH, Cave MD (1997). The incidence of false-positive cultures for Mycobacterium tuberculosis. Am. J. Respir. Crit. Care. Med. 155: 1140-1146.

Burmen WJ, Reves RR (2000). Review of false-positice cultures for unnecessary treatment. Clin Infect Dis. 31: 1390-1395.

Burgos MV, Pym AS (2002). Molecular epidemiology of tuberculosis. Eur. Respir. J. 20: 545-655.

Chakhaiyer P, Nagalakshmi Y, Aruna B, Urth KJ, Katoeh VM, Hasnain SE (2004). Regions of high antigenicity within the hypothetical PPE-MPTR open reading frame. RV 2608 show a differential humoral response and a law Tcell response in various categories of patients with tuberculosis. J. Infect. Dis. 190: 1234-1244.

Chaves F, Yang Z, Haji H, Alonso M, Burman WJ, Eisenach KD, Dronda F, Bates JH, Cave MD (1996). Usefulness of the secondary probe pTBN12 in DNA fingerprinting of Mycobacterium tuberculosis. J. Clin. Microbiol. 34: 1118-1123.

Chin DP, Deriemer K, Small PM, Povce de Leon A, Steinhart R, Schecter GF, Delay CL, Moss AR, Paz EA, Paz A, Jasmer AM, Agasino CB, Hopewell PC (1998). Differences in contributing factors to tuberculosis incidence in U.S. born and foreigh-born persons. Am. J. Respire. Crit. Care. Med. 158: 1797-1803.

Choudhary RK, Mulhopadhyay S, Chakhaiyer P, Sharma N, Murthy KY, Katoch VM (2003). PPE antigen RV 2430c of Mycobacterium tubercoulosis induces a strong B-cell response. Infect. Immun. 71: 6338-6343.

Collins DM, Stephens DM, (1991). Identification of an insertion sequence, IS1081, in Mycobacterium bovis. FEMS Microbiol Lett. 83: 11-16.

Cowan LS, Mosher L, Diem L, Massey JP, Crawford JT (2002). Variable number tandem repeat typing of Mycobacterium tuberculosis. Isolates with low copy number of IS6110 by using Mycobacterial interspersed repetitive units. J. Clin. Microbiol. 40: 1592-1602.

Cowan L S, Diem L, Monson T, Wand P, Temporado D, Oemig TV, Crawford JT (2005). Evaluation of a two-step approach for large-scale, prospective genotyping of Mycobacterium tuberculosis isolates in the United States. J. Clin. Microbiol. 43: 688–695.

Dale JW, Brittain D, Cataldi AA, Cousins D, Crawford JD, Driscoll J, Heersma H, Lilebaek T, Quituqua T, Rastogi N, Skuse R, Sola C, van Soolingen D, Vincent V (2001). Spacer oligonucleotide typing of bacteria of the Mycobacterium tuberculosis complex: recommendations for standardized nomenclature. Int. J. turberc. Lung. Dis. 5: 216-219.

Daley CL, Small PM, Scheeter GF, Schoolink GK, McAdam RA, Jacobs WR, Hopewell PC (1992). An outbreak of tuberculosis with accelerated progression among persons infected with the Human immunodeficiency virus. An analysis using restriction – fragment –length polymorphism. N. Eng J. Med. 326: 231-235.

D'Amato RF, Wallman AA, Hochstein LH, Colaninno PM, Scardamaglia M, Ardila E (1995). Rapid diagnosis of pulmonary tuberculosis by using roche AMPLI-COR Mycobacterium tuberculosis PCR test. J. Clin. Microbiol. 33: 1832-1834.

de Boer AS, Bergdorf MW, de Hass PE, Negelkerke NJ, van Embden JD, van Solingen D (1999). Analysis of rate of change of IS6110 RFLP patterns of Mycobacterium tuberculosis based on serial patient isolates. J. Infect. Dis. 180: 1238-1244.

de Wit D, Steyn S, Shoemaker S, and Sagin M (1990). Direct detection of M. tuberculosis in clinical specimens by DNA amplification. J. Clin. Microbiol. 28: 2137-2441.

Diaz, R, Kremer K, de Haas PE, Gomez RI, Marrero A, Valdivia JA, van Embden JDA, van Soolingen D (1998). Molecular epidemiology of tuberculosis in cuba outside of Havana, July 1994- June 1995: utility of spoligotyping versus IS6110 restriction fragment length polymorphism. Int. J. tuberc. Lung. Dis. 2: 743-750.

Dooley SW, Villarino ME, Lawrence M, Salinas L, Amil S, Rullan JV, (1992). Nosocomial transmission of tuberculosis in a hospital unit for HIV infected patients. JAMA. 267: 2632-2635.

Doroudchi M, Kremer K, Basiri EA, Kadivar MR, van Soolingen D, Ghaderi AA (2000). IS6110-RFLP and spoligotyping of Mycobacterium tuberculosis isolates in ran. Scan. J. Infec.t Dis. 32: 663-668.

Durmaz R, Gunal S, Yang Z, Ozerol H, Cave MD (2003). Molecular epidemiology of tuberculosis in turkey. Clin. Microbial. Infect. 9: 873-877.

Dye C, Scheele S, Dolin P, Pathania V, Raviglions MC (1999). Consensus statement. Global burden of tuberculosis: estimated incidence, prevalence, and mortality by country. Who global surveillance and monitoring project. JAMA. 282: 677-686.

Edlin BR, Tokars JI, Grieco MH, Crawford JT, Williams J, Sordillo EM, Ong KR, Kilburn JO, Dudley SW, Castro KG (1992). An outbreak of multidrug resistant tuberculosis among hospitalized patients with a acquired immunodeficiency syndrome. N. Eng.l J. Med. 326: 1514-1521.

Filliol L, Sola C, Rastogi N (2000). Detection of a previously unamplified spacer within the DR Locus of Mycobacterium tuberculosis: epidemiological implication. J. Clin. Microbiol. 38: 1231-1234.

Frothingham R (1995). Discrimination of M. tuberculosis strains by PCR. J. Clin. Microbiol. 33, 2801-2802.

Frothingham R, Meeker- O conell WA, and Forbes KJ (1998). IS6110 transposition and evolutionary scenario of the direct repeat locus in a group of closely related Mycobacterium tuberculosis strains. J. Bacterial. 180: 2102-2109.

Frieden TR, Sherman LF, Maw KL, Fujiwara PI, Crawfold JT, Nivin B, Sharp V, Hewlett D, Brudney K, Alland D, Kreisworth BN (1996). A multi-institution outbreak of highly drug–resistant tuberculosis: epidemiology and clinical outcomes. J. Am. Med. Assoc. 276: 1229-1232.

Glennon M, and Smith T (1995). Can random amplified polymorphic DNA analysis of the 16S-23S spacer region of M. tuberculosis differentiate between isolates. J. Clin. Microbiol. 33: 3359-3360.

Goguet de la Salmoniere YO, Li HM, Tarrea G, Bunschoten A, van Embden J, Gicquel B (1997). Evaluation of spoligotyping in a study of the transmission of Mycobacterium tuberculosis. J. Clin. Microbial. 35: 2210-2214.

Goyal M, Lawn S, Afful B, Acheampong JW, Griffin G, Shaw R (1999). Spoligotyping in molecular epidemiology of tuberculosis in Ghana. J. Infect. 38: 171-175.

Groenen PMA, Bunschoten AE, van Soolingen D, van Embden JDA (1993). Nature of DNA polymorphism in the direct repeat cluster of Mycobacterium tuberculosis application for strain differentiation by a novel typing method. Mol. Microbial. 10, 1057-1065.

Hermans PWM, van Soolingen D, Bik EM, de Hass PEW, Dale JW, van Embden JAD (1991). The insertion element IS987 from M. bovis BCG is located in a hot spot integration region for insertion elemerts in M. tuberculosis complex strains. Infect. Immun. 56: 2695-2705.

Hermans PW, van Soolingen D, van Embden JD (1992). Characterization of a major polymorphic tandem repeat in Mycobacterium tuberculosis and its potential use in the epidemiology of mycobacterium kansasii and mycobacterium gordonae. J. Bacteriol. 174: 4157-4165.

Kamerbeek J, Schouls L, Kolk A, van Agtreveld M, van Soolingen, Kuijper S, Bunschoten A, Molhuizen H, Shaw R, Goyal M, van Embden J (1997). Simultaneous detection and strain differentiation of Mycobacterium tuberculosis for diagnosis and epidemiology. J. Clin. Microbiol. 35: 907-914.

Kanduma E, McHugh TD, Gillespie SH (2003). Molecular methods for Mycobacterium tuberculosis strain typing: a users guide. J. Appl. Microbiol. 94: 781-791.

Kremer K, van Soolingen D, Frothingham R, de Hass WH, Hermans PW, Martin C, Palittapongaarnpim P, Plikaytis BB, Riley LW, Yakrus MA, Musser JM, van Embden JDA, (1999). Comparision of methods based on different molecular epidemiological markers for typing of Mycobacterium tuberculosis complex strains: interlaboratory study of discriminatory power and reproducibility. J. Clin. Microbiol. 37: 2607-2618.

Lee AS, Tang LL, Lim IH, Bellamy R, and Wong SY (2002). Discrimination of single–copy IS6110 DNA Fingerprinting of Mycobacterium tuberculosis isolates by High-resulotion mini satellite–based typing. J. Clin. Microbiol. 40: 657-659.

Magdalena j, Vanchee A, Supply P, Comille L (1998). Identification of a new DNA region specific for members of Mycobacterium tuberculosis complex. J. Clin. Microbiol. 36: 937-943.

Mazars E, Lesjean S, Banuls AL, Gilbert M, Vincent V, Gicquel B, Tibayrenc M, Locht C, Supply P (2001). High–resulotion minisatellite based Typing as a portable approach to global analysis of Mycobacterium tuberculosis molecular epidemiology. Proc. Natl. Acad. Sci. USA. 98: 1901-1906.

Miller N, Hernandez SG, and Cleary TJ (1994). Evaluation of Gen-probe Ampified mycobacterium tuberculosis direct test and PCR for direct detection of Mycobacterium tuberculosis in clinical specimens. J. Clin. Micobiol. 32: 393-397.

Moss AR, Alland D, Telzak E, Hewlett DJR, Sharp V, Chillade P (1997). Acity-Wide outbreak of a multiple–drug resistant strains of Mycobacterium tuberculosis in New York. Int. J. Tuberc. Lung. Dis. 1: 115-121.

Niemann ER, Rusch–Gerdes S (1999). Stability of Mycobacterium tuberculosis IS6110 restriction fragment length polymorphism patterns and spoligotypes determined by analyzing serial isolates from patients with drug resistant tuberculosis. J. Clin. Microbiol. 37: 407-412.

Niobe-Eyangoh SN, Kuaban C, Sarlin P, Cunin P, Thornon J, Sola C (2003). Genetic Biodiversity of Mycobacterium tuberculosis complex strains from patients with pulmonary tuberculosis in Cameroon. J. Clin. Microbiol. 41: 2247-2553.

Oelemann MC, Diel R, Vatin V, de Haas W, Ru¨sch-Gerdes S, Locht C, Niemann S, Supply P (2007). Assessment of an Optimized Mycobacterial Interspersed Repetitive- Unit–Variable-Number Tandem-Repeat Typing System Combined with Spoligotyping for Population-Based Molecular Epidemiology Studies of Tuberculosis. J. Clin. Microbiol. 45: 691-697.

Palitta pongarnim P, Chomyc S, Fanning A, Kunimoto D (1993). DNA fragment length polymorphism analysis of Mycobacterium tuberculosis isolates by arbitrarily primed polymerase chain reaction. J. Infect. Dis. 167: 975-978.

Pena MJ, Caminero JA, Campos-Herrero MZ, Rodrigues-Gullego JC, Garcia-laorden MI (2003). Epidemiology of tuberculosis on Gran Cannria: a 4 year population study using traditional and molecular approaches. Thorax. 58: 618-622.

Poulet sand Cole ST (1995). Characterization of the polymorphic GC-rich repetitive sequence (PGRS) present in M. tuberculosis. Arch. Microbiol. 16, 87-95.

Rhee JT, Tanaka MM, Behr MA, Agasino CB, Paz EA, Hopewell PC, Small PM, (2000). Use of multiple markers in population based molecular epidemiologic studies of tuberculosis. Int. J. Tuberc. Lung. Dis. 4: 1111-1119.

Ritacco V, Di Lonardo M, Reniero A, Ambroggi M, Barrera L, Dambrosi A, López B, Isola N, and Kantor IN (1997). Nasocominal spread of human immunodeficiency virus related multidrug–resistant tuberculosis in Buenos Aires. J. Infect. Dis. 176: 637-642.

Ross BC, Raios K, Jackson K, and Dwyer B (1992). Molecular cloning of a highly repeated DNA element from M. tuberculosis and its use as an epidemiological tool. J. Clin. Microbiol. 30: 942-946.

Schirm J, Oostendorp LAB, Mulder JG (1993). Comparision of Amplicor, in-house PCR, and conventional culture for detection of Mycobacterium tuberculosis in clinical samples. J. Clin. Microbiol. 35: 193-196.

Singh SP, Salamon H, Lahti C, Farid-Moyer M, Small PM (1999). Use of pulsed field gel electrophoresis for molecular epidemiologic and population genetic studies of Mycobacterium tuberculosis. J. Clin. Microbiol. 37: 1927-1931.

Small PM, Shafer RW, Hopewell PC, Singh SP, Desmond E, Sierra MF, Murphy MJ, Desmond E, Sierra MF, Schoolnik GK (1993a). Exogenous reinfection with multidrug-resistant M. tuberculosis in patients with advanced HIV infection. N. Engl. J. Med. 328: 1137-1144.

Small PM, McClenny NB, Singh SP, Schoolnik GK, Tompkins LS, Mickelsen PA (1993b). Molecular strain typing of Mycobacterium tuberculosis to confirm cross contamination in the mycobacteriology laboratory and modification of procedures to minimize occurrence of false-positive cultures. J. Clin. Microbiol. 31: 1677-1682.

Sola C, Filliol L, Legrand E, Lesjean S, Locht C, Supply P, Rastogi N (2003). Genotyping of the Mycobacterium tuberculosis complex using MIRUS Association with VNTR and spoligotyping for molecular epidemiology and evolutionary genetics. Infect. Genet. Evol. 3: 125-133.

Supply P, Magdalena J, Himpens S, Locli C (1997). Identification of novel intergenic repetitive units in a Mycobacterial two- Component system operon. Mol. Microbiol. 26: 991-1003.

Supply P, Mazars E, Lesjean S, Vincent V, Gicquel B, Locht C (2000). Variable human minisatellite-like regions in the Mycobacterium tuberculosis genome. Mol. microbial. 36: 762-771.

Supply P, Lesjean S, Savine E, Kremer K, van Soolingen D, Locht C (2001). Automated high-throughput genotyping for study of global epidemiology of Mycobacterium tuberculosis based on Mycobacterial interspersed repetitive units. J. Clin. Microbiol. 39: 3363-3571.

Supply P, Warren RM, Banuls AL, Lesjean S, Vander Spay GD, Lewis LA, Tibayrenc M, van Helden PD, Locht C (2003). Linkage disequilibrium between minisatellite loci supports clonal evaluation of Mycobacterium tuberculosis in a high tuberculosis incidence area. Mol. Microbiol. 47: 529-538.

Supply P, Allix C, Lesjean S, Cardoso-Oelemann M, Rusch Gerdes S, Willery E, Savine E, de Haas P, van Deutekom H, Roring S, Bifani P, Kurepina N, Kreiswirth B, Sola C, Rastogi N, Vatin V, Gutierrez MC, Fauville M, Niemann S, Skuce R, Kremer K, Locht C, van Soolingen D, (2006). Proposal for standardization of optimized mycobacterial interspersed repetitive unit-variable-number tandem repeat typing of Mycobacterium tuberculosis. J. Clin. Microbiol. 44: 4498–4510.

Thierry D, Brisson- Noel A, Vincent–Levy–Freboult V, Nguyen S, Guesdon KL, Gicquel B (1990). Characterizing of a Mycobacterium

tuberculosis insertion sequence, IS6110, and its application in diagnosis. J. Clin. Microbiol. 28: 2668-2673.

Valway SE, Greifinger RB, Papania M, Kiburn JO, Woodley C, Diferdinando GT, Dooley S (1994). Multidrug resistance tuberculosis in the new york state prison system. 1990-1991, J. infect. Dis. 170: 151-156.

van Embden JDA, Cave MD, Crawford JT, Dale JW, Eisenach KD, Gicquel B, Hermans P, Martin C, McAdam R, Shinnick TM, Small P (1993). Strain identification of Mycobacterium tuberculosis by DNA Fingerprinting recommended for a standardized methodology. J. Clin. Microbiol. 31: 406-409.

van Soolingen D, Hermans PW, de Haas PE. and van Embden JD (1992). Insertion element IS1081 associated restriction fragment length polymorphism in M. tuberculosis complex species: a reliable tool for recognizing mycobacterium bovis BCG. J. Clin. Microbiol. 30: 1772-1777.

van Soolingen D, de Hass PE, Hermans PW, Groenen PM, van Embden JD (1993). Comparision of various repetitive DNA elements as genetic markers for strain differentiation and epidemiology of M. tuberculosis. J. Clin. Microbiol. 29: 2578-2586.

van Soolingen D, de Hass PEW, Hermans PWM, and van Embden JDA, (1994). DNA fingerprinting of Mycobacterium tuberculosis. Methods. Enzymol. 236: 196-205.

Van Rie A, warren R, Richardson M, Victor TC, Gie RP, Enarson DA, Beyers N, van Helden PD (1999). Exogenous reinfection as a cause of recurrent tuberculosis after curative treatment. N. Eng. J. Med. 341: 1174-1179.

Vrioni G, Levidiotou S, Mastsiota-Bernard P, and Marinis E (2004). Molecular characterization of Mycobacterium tuberculosis isolates presenting various drug susceptibility profiles from Greece using three DNA typing methods. J. Infec. 48: 253-262.

Walls T, Shingadia D (2004). Global epidemiology of Pediatric tuberculasis. J. Infect. 48, 13-22.

Warren RM, vander Spay GD, Richardson M, Beyers N, Borgdorff MW, Behr MA, van Helden PD (2002). Calculation of the stability of the IS6110 Banding pattern in patients with persistent Mycobacterium tuberculosis disease. J. Clin. Microbiol. 40: 1705-1708.

Yang ZH, Ijaz K, Bates JH, Eisenach KD, Cave MD (2000). Spoligotyping and polymorphic GC-Rich Repetitive sequence fingerprinting of Mycobacterium tuberculosis strains having few copies of IS6110. J. Clin. Microbiol. 38: 3572-3576.

11

Developments in biochemical aspects and biotechnological applications of microbial phytases

Bijender Singh[1]*, Gotthard Kunze[3] and T. Satyanarayana[2]

[1]Department of Microbiology, Maharshi Dayanand University, Rohtak- 124 001, India.
[2]Department of Microbiology, University of Delhi South Campus, Benito Juarez Road, New Delhi-110 021, India.
[3]Leibniz-Institut für Pflanzengenetik und Kulturpflanzenforschung (IPK), Corrensstr. 3, D-06466 Gatersleben, Germany.

Phytases belong to the class of phosphatases, which catalyze the hydrolysis of phytic acid to inorganic phosphate and *myo*-inositol phosphate derivatives. The enzyme has potential applications in food and feed industries for ameliorating digestibility and assimilation of nutrients of foods and feeds by mitigating the anti-nutritional effects of phytic acid. Phytases have been shown to be useful in improving growth of poultry, pigs and fishes, and they play a role in promoting growth of plants, as well as improve the nutritional quality of bread, soymilk and oil seed cakes by dephytinization. The crystal structures of some phytases have been analyzed for understanding the reaction mechanism. The phytases with desirable properties have been generated through protein engineering approaches, since native phytases do not possess all the properties of an ideal additive feed/food. Recent developments on the characteristics of an ideal phytase, crystal structure, protein engineering, and the potential biotechnological applications of microbial phytases with special reference to their utility in improving growth performance of monogastrics, dephytinization of foods and feeds, plant growth promotion, and combating environmental phosphorus pollution will be discussed in this review.

Key words: Phytic acid, microbial phytase, crystal structure, monogastrics, protein engineering, dephytinization, plant growth promotion.

INTRODUCTION

The hydrolysis of phytic acid to *myo*-inositol and inorganic phosphate by phytases (*myo*-inositol hexakisphosphate phosphohydrolase) is an important reaction for energy metabolism, metabolic regulation and signal transduction pathways in biological systems. Although phytic acid (*myo*-inositolhexakis phosphate), which is an organic form of phosphorus (P), is abundantly present in plants' materials (1 to 5% by weight) such as edible legumes, cereals, oilseeds, pollen and nuts, it is largely unavailable to monogastrics like poultry birds, pigs, fishes and humans, due to the lack of adequate levels of phytases (Wodzinski and Ullah, 1996; Vohra and Satyanarayana, 2003; Vats and Banerjee, 2004; Greiner and Konietzny, 2006; Rao et al., 2009). The

phytic acid present in the plant derived foods acts as an anti-nutritional factor, since it causes mineral deficiency due to efficient chelation of metal ions such as Ca^{2+}, Mg^{2+}, Zn^{2+} and Fe^{2+}, which form complexes with proteins, and thus affect their digestion and also inhibit certain digestive enzymes like α-amylase, trypsin, acid phosphatase and tyrosinase (Harland and Morris, 1995). Due to the lack of adequate levels of phytases in monogastric animals, phytic acid is excreted in faeces, which on degradation by soil microorganisms, release phosphorus in the soil. The phosphorus reaches aquatic bodies that cause eutrophication (Mullaney et al., 2000). Phytic acid can be removed by some physical (autoclaving, cooking and steeping) and chemical (ion exchange and acid hydrolysis) methods, but these methods decrease the nutritional value of foods. The reduction of phytic acid content in foods and feeds by enzymatic hydrolysis using phytase is desirable since it improves their nutritional value. Besides its immense

*Corresponding author. E-mail: ohlanbs@gmail.com.

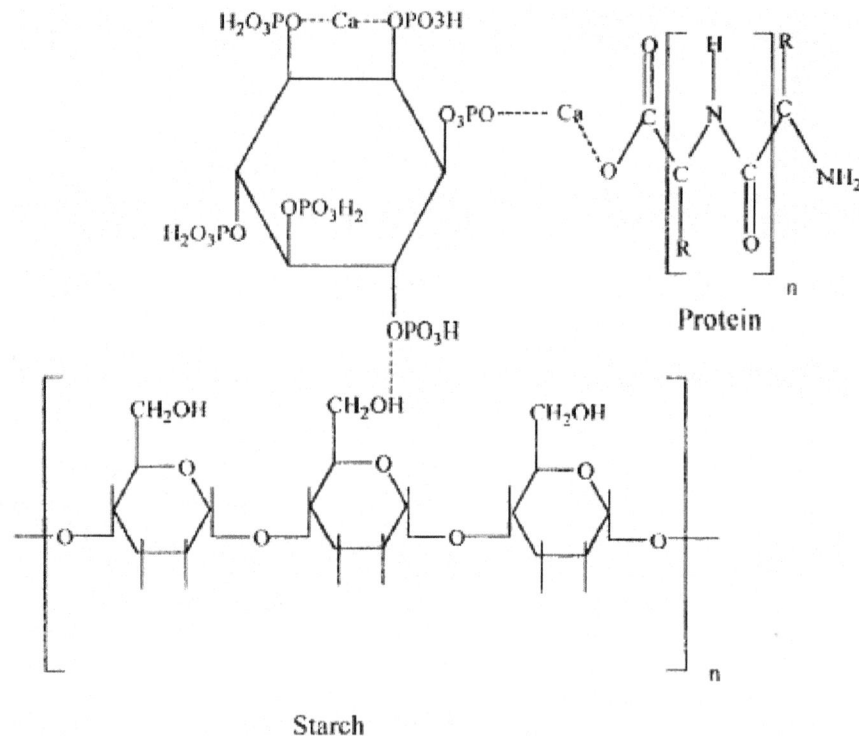

Figure 1. Interaction of phytic acid with metals, proteins and carbohydrate.

commercial value in food and feed industries, the enzyme has potential applications in other fields too. The annual sale of commercial supplemental phytase is estimated at US$ 50 million, which is one-third of the entire feed enzyme market (Sheppy, 2001) and recently increased to 150 million euro (Greiner and Konietzny, 2006). The term phytase has been used in this article to mean microbial phytase.

During the last 15 years, phytases have attracted considerable attention from both scientists and entrepreneurs in the areas of nutrition, environmental protection and biotechnology. Undoubtedly, increasing public concern regarding the environmental impact of high phosphorus levels in animal excreta has driven the biotechnological development of phytase and its application in animal nutrition. The feeding trials have shown the effectiveness of supplemental microbial phytases in improving utilization of phytate-P and the phytate-bound minerals by swine, poultry and fishes (Lei and Stahl, 2001; Singh et al., 2006; Cao et al., 2007; Selle and Ravindran, 2007, 2008; Rao et al., 2009). Inorganic P supplementation of the diets for swine and poultry can be obviated by including adequate amounts of phytase along with an appropriate manipulation of other dietary factors (Han et al., 1997). As a result, the P excretion of these animals may be reduced by about 50% (Lei et al., 1993a, b; Satyanarayana and Vohra, 2003; Vohra et al., 2006). The cost and thermotolerance constraints of the current commercial phytases have,

however, precluded the widespread use of these enzymes in animal feeds.

Several reviews have been published recently on the phytases, which mainly focused on the production, characteristics and their basic applications (Pandey et al., 2001; Vohra and Satyanarayana, 2003; Vats and Banerjee, 2004; Greiner and Konietzny, 2006; Kaur et al., 2007; Fu et al., 2008; Rao et al., 2009). None of these dealt with the ideal and designer phytase, crystal structure and the directed evolution and protein engineering of phytases; and therefore, a comprehensive account is given in this review on the recent developments on all these aspects of microbial phytases and their potential biotechnological applications.

PHYTIC ACID: A FRIEND OR A FOE

Phytic acid is the major storage form of phosphorus in cereals, legumes and oilseeds (Maga, 1982; Tyagi et al., 1998). It has several physiological roles and also affects the functional and nutritional properties of food ingredients. The correct chemical description of phytic acid is *myo*-inositol 1, 2, 3, 4, 5, 6-hexakis dihydrogen phosphate (IUPAC-IUB, 1977). Phytic acid occurs primarily as salts of mono- and divalent cations (for example, potassium-magnesium salt in rice and calcium-magnesium-potassium salt in soybeans) in discrete regions of cereal grains and legumes (Figure 1). It accumulates in seeds and grains during ripening along with other storage substances such as starch and lipids. In cereals and legumes, phytic acid accumulates in the aleurone particles and globoid crystals, respectively (Reddy et al., 1982; Tyagi et al., 1998).

Besides phosphate storage, phytate acts as a strong chelator for divalent metal cations and exists as a stable metal-phytate complex with metal ions in plants (Asada et al., 1969; Reddy et al., 1982). Phytic acid in seeds and grains serves as a phosphorus store, an energy store, a source of cations, a source of myo-inositol, and also helps in initiating dormancy. Phytic acid may also serve several other unknown functions in seeds (Reddy et al., 1982). Graf et al. (1987) suggested that the role of phytic acid in seeds is a natural antioxidant during dormancy. Phytic acid has been shown to exert an antineoplastic effect in animal models of both colon and breast carcinomas. The presence of undigested phytic acid in the colon may protect against the development of colonic carcinoma (Iqbal et al., 1994). The inositol phosphate intermediates play an important role in the transport of materials into the cell, and the role of inositol triphosphates, especially in signal transduction and regulation of cell functions in plant and animal cells, is a very active area of research in order to understand signaling pathways (Wodzinski and Ullah, 1996; Vohra and Satyanarayana, 2003; Greiner and Konietzny, 2006; Rao et al., 2009). Besides these functions, phytic acid also acts as an anti-nutritional factor in several ways due to the interactions with metal ions, proteins and enzymes.

PHYTASES

Phytases myo-inositolhexaphosphate phosphohydrolase) hydrolyze phytic acid to myo-inositol and inorganic phosphates through a series of myo-inositol phosphate intermediates, and eliminate its anti-nutritional characteristics. Phytase is widespread in nature, and it occurs in microorganisms, plants and some animals (Wodzinski and Ullah, 1996; Vohra and Satyanarayana, 2003; Angelis et al., 2003; Vats and Banerjee, 2004; Kaur et al., 2007; Fu et al., 2008; Rao et al., 2009; Raghavendra and Halami, 2009). A large number of bacteria, filamentous fungi and yeasts have been reported to produce phytase extra- and intra-cellularly as well as in the cell-bound form (Shieh and Ware, 1968; Wodzinski and Ullah, 1996; Pandey et al., 2001; Vohra and Satyanarayana, 2003; Vats and Banerjee, 2004; Kaur et al., 2007; Fu et al., 2008; Rao et al., 2009). A list of phytase producing organisms is given in Table 1. There are two types of phytases as classified by Nomenclature Committee of the International Union of Biochemistry and Molecular Biology (NC-IUBMB) in consultation with the IUPAC-IUBMB Joint Commission on Biochemical Nomenclature (JCBN): 3-phytase (EC 3.1.3.8) that first hydrolyses the ester bond at the 3

position of myo-inositol hexakisphosphate, and is mainly reported in microorganisms; and the 6-phytase (EC 3.1.3.26) that first hydrolyses the ester bond at the 6 position of myo-inositol hexakisphosphate, and is mostly reported in plants. This had also been reported in some basidiomycetous fungi (Lassen et al., 2001). An alkaline 5-phytase from lily pollen was found to start phytate hydrolysis at the D-5 position (Barrientos et al., 1994). Phytases can be broadly categorized into two major classes based on the pH for activity: acid phytases and alkaline phytases (Figure 2). More focus has been on acidic phytases because of their applicability in animal feeds and broader substrate specificity than those of alkaline phytases. Recently, phytases have also been classified as HAP (Histidine acid phosphatase), BPP (β-Propeller phytase), CP (cysteine phosphatase) and PAP (purple acid phosphatase) based on their catalytic properties (Mullaney and Ullah, 2003).

AN IDEAL PHYTASE AND ITS DESIGNING

The phytase that has the desirable characteristics for application in animal feed industry can be called an 'ideal phytase', which should be active in the stomach, stable during animal feed processing and storage, and easily processed by the feed manufacturer for its suitability as an animal feed additive.

Phytase should not be detected at the end of the small intestine. This is necessary because in this way the phytase produced by genetically modified organisms should not enter the environment (Jongbloed et al., 1992). Furthermore, it should be effective in releasing phytate-P in the digestive tract and stable to resist proteases (trypsin and pepsin) and inactivation by heat during feed pelleting and storage with low cost of production. The ability of any given phytase to hydrolyze phytate-P in the digestive tract is determined by its properties, such as catalytic efficiency, substrate specificity, temperature and pH optima, which are resistance to proteases. As the stomach is the main functional site of supplemental phytase, a phytase with pH optimum in the acidic range is desirable for improving nutrition. Also, phytase must exhibit resistance to pepsin and trypsin, which are encountered in the intestine. Since the food and feeds are often processed through a pelleting machine at 65 to 80°C with steam to eliminate salmonellae, an ideal phytase must be able to withstand the high temperature and steam encountered during the pelleting process. Similarly, an enzyme that can tolerate long-term storage or transport at ambient temperature is generally preferred for food and feed industry. Finally, a phytase produced in high yield and purity by a relatively inexpensive system is attracting for food industries worldwide. It is now realized that any single phytase may never be 'ideal' for all feeds and foods. For example, the stomach pH in finishing pigs is much more acidic than

Table 1. Optimized culture conditions for the production of phytase by various microorganisms.

Microbial strain	pH_{opt}	T_{opt}	Fermentation	Culture conditions		Reference
				Carbon source	Nitrogen source	
Filamentous fungi						
A. fumigatus SRRC 322	5.0	37	SmF[*]	Hylon starch	NaNO$_3$	Mullaney et al. (2000)
A. niger	5.5	30	SmF	Glucose starch	--	Vats and Banerjee (2005)
A. ficuum	5.0	30	SmF	Corn starch, glucose	NaNO$_3$	Shieh and Ware (1968)
A. oryzae	6.4	37	SmF	Glucose	(NH$_4$)$_2$SO$_4$	Shimizu (1993)
Rhizopus oligosporus	5.5	27	SmF	Corn starch, glucose	NaNO$_3$	Casey and Walsh (2004)
R. oryzae	5.5	30	SSF[#]	Glucose	NH$_4$NO$_3$	Ramachandaran et al. (2005)
Mucor racemosus	5.5	30	SSF	Starch	NaNO$_3$	Roopesh et al. (2005)
Peniophora lycii	5.5	26	SmF	Maltodextrin, soya flour	Peptone	Lassen et al. (2001)
Thermoascus aurantiacus	5.5	45	SmF	Starch, glucose, wheat bran	Peptone	Nampoothiri et al. (2004)
Rhizomucor pusillus	8.0	50	SSF	Wheat bran	Asparagine	Chadha et al. (2004)
Myceliopthora thermophila	5.5	45	SmF	Glucose	NaNO$_3$	Mitchell (1997)
Sporotrichum thermophile	5.0	45	SmF	Starch, glucose	Peptone	Singh and Satyanarayana (2008a)
S. thermophile	5.0	45	SSF	Sesame oil cake, glucose	(NH$_4$)$_2$SO$_4$	Singh and Satyanarayana (2006a)
Yeasts						
Pichia anomala	6.0	25	SmF	Glucose	Beef extract	Vohra and Satyanarayana (2001)
Schwanniomyces castellii	4.4	77	SmF	Galactose	(NH$_4$)$_2$SO$_4$	Segueilha et al. (1992)
Arxula adeninivorans	5.5	28	SmF	Galactose	Yeast extract	Sano et al. (1999)
P. rhodanensis	4.5	70	SmF	Glucose	-	Nakamura et al. (2000)
P. spartinae	4.5	75	SmF	Glucose	-	Nakamura et al. (2000)
Candida krusei	4.6	40	SmF	Glucose	Polypeptone	Quan et al. (2001)
Bacteria						
B. subtilis	7.0	37	SmF	Glucose	NH$_4$NO$_3$	Kerovuo et al. (1998)
B. amyloliquefaciens	6.8	37	SmF	Glucose	Casein, peptone	Idriss et al. (2002)
Escherichia coli	7.0	37	SmF	--	Tryptone	Sunita et al. (2000)
Klebsiella aerogenes	7.0	30	SmF	Sodium phytate	Yeast extract	Tambe et al. (1994)
Lactobacillus sanfranciscensis[*]	5.5	37	SmF	Maltose, glucose	Yeast extract	Angelis et al. (2003)
L. fructivorans[*]	5.5	37	SmF	Maltose, glucose	Yeast extract	Angelis et al. (2003)
L. lactis subsp lactis[*]	5.5	37	SmF	Maltose, glucose	Yeast extract	Angelis et al. (2003)
L. rhamnosus[*]	6.5	37	SmF	Glucose	Yeast extract	Raghavendra and Halami (2009)
L. amylovorus[*]	6.5	37	SmF	Glucose	Yeast extract	Raghavendra and Halami (2009)
Pediococcus pentosaceus[*]	6.5	37	SmF	Glucose	Yeast extract	Raghavendra and Halami (2009)

[*]SmF = Submerged fermentation, [#]SSF = Solid state fermentation.

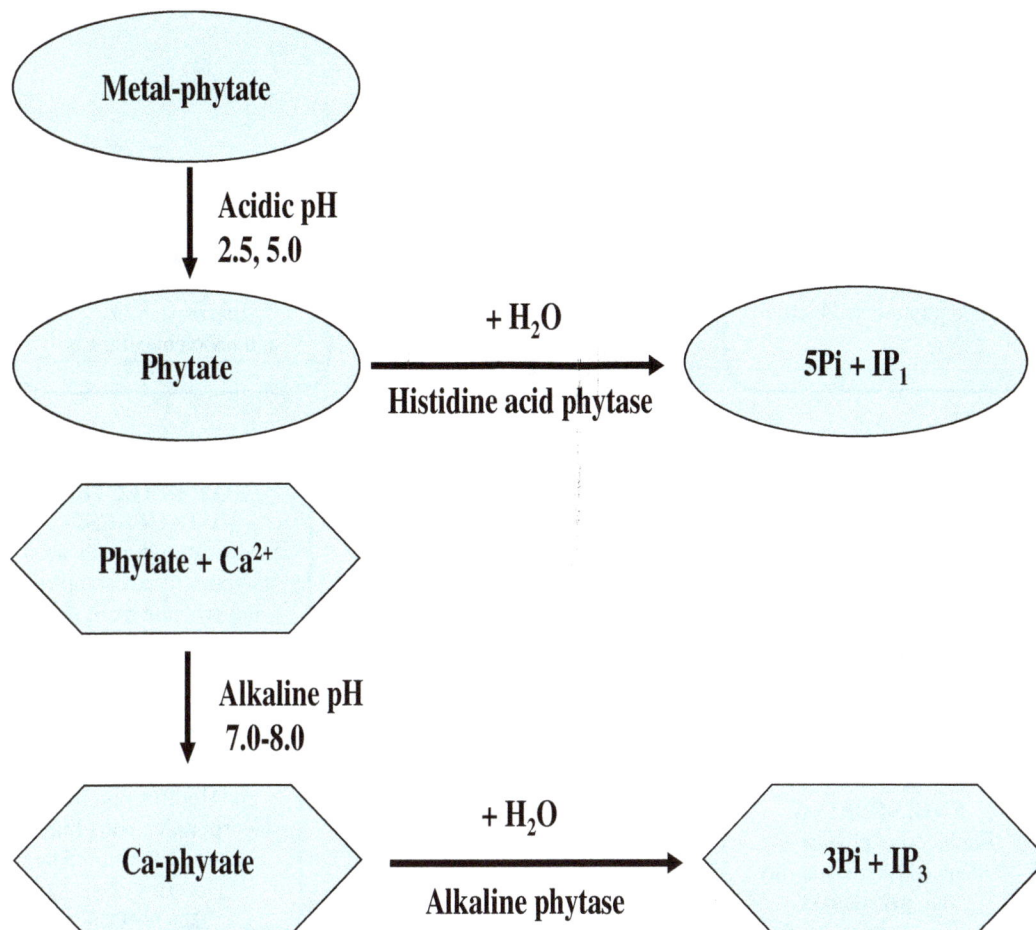

Figure 2. Schematic representation of the hydrolysis of substrate by histidine acid phytase and β-propeller phytase.

that of weanling pigs (Radcliffe et al., 1998). Thus, phytase with optimum pH close to 3.0 will perform better in the former than in the latter. For poultry, an enzyme would be beneficial if it is active over broad pH range, that is, acidic (stomach) to neutral pH (crop) (Riley and Austic, 1984). Phytases used for aquaculture application require a lower temperature that is optimum than the swine or poultry (Ramseyer et al., 1999). The choice of an organism for phytase production is, therefore, dependent upon the target application. Nowadays, there is a great demand for the development of an ideal phytase using directed evolution and protein engineering.

Based on this, the desirable and ideal phytase could be designed as per target application. All these features are not present within a single phytase, and therefore, based on the sequence of the available phytases, a consensus phytase could be designed (Lehman et al., 2000a, b, c). Genetic engineering techniques such as site directed mutagenesis could be employed for further ameliorating the properties. The strategies used for the designing and developing of an ideal phytase are presented in Figure 3.

BIOCHEMICAL AND MOLECULAR CHARACTERISTICS OF PHYTASES

The major properties of enzymes are useful in determining their potential in different industries. The biochemical and molecular properties of some phytases are presented in Table 2.

Phytases with high temperature optima are desirable in the animal feed industry because feed pelleting involves a step of 80 to 85°C for few seconds (Wyss et al., 1999a). Phytase of *A. fumigatus* (Pasamontes et al., 1997b) and *A. niger* NRRL 3135 (Howson and Davis, 1983) exhibited optimum activity at 37°C and at 55°C, respectively. Phytase of *S. castellii* was optimally active at 77°C (Segueilha et al., 1992) and that of *Arxula adeninivorans* at 75°C (Sano et al., 1999). The phytases from *Pichia rhodanensis* and *P. spartinae* showed optimal reaction temperature at 70 to 75°C and 75 to 80°C, respectively (Nakamura et al., 2000), while that of *Pichia anomala* showed optimal activity at 60°C (Vohra and Satyanarayana, 2002). Among the thermophilic fungi,

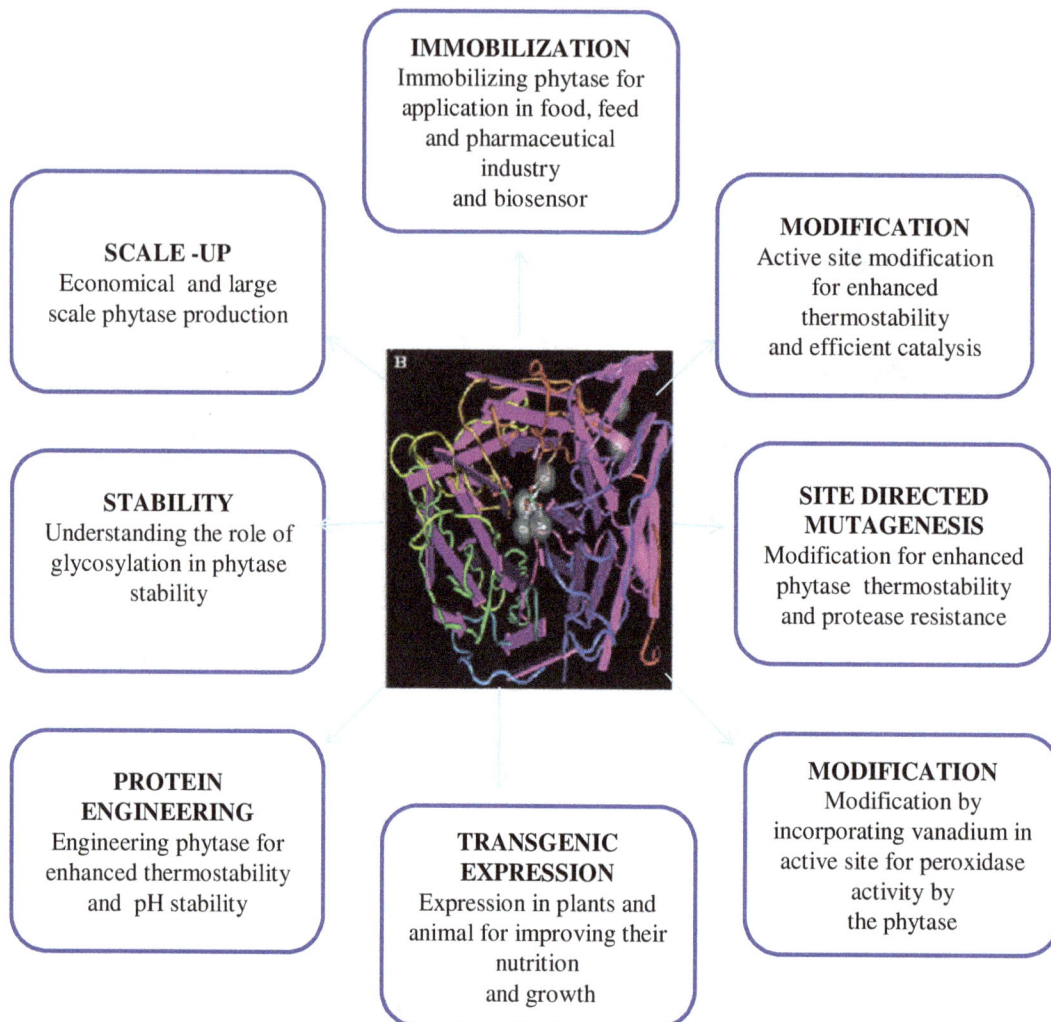

Figure 3. Designing an ideal phytase for biotechnological applications.

Thermomyces lanuginosus phytase exhibited optimum activity at 65°C (Berka et al., 1998), and that of *Rhizomucor pusillus* at 70°C (Chadha et al., 2004). Phytases of *Thermoascus aurantiacus* (Nampoothiri et al., 2004) and *S. thermophile* (Singh and Satyanarayana, 2009) were optimally active at 55°C and 60°C, respectively (Table 3). Phytase from *B. subtilis* (Powar and Jagannathan, 1982), *E. coli* (Greiner et al., 1993), *Klebsiella aerogenes* (Tambe et al., 1994), *Enterobacter* sp.4 (Yoon et al., 1996), *K. oxytoca* MO-3 (Jareonkitmongkol et al., 1997), *Selenomonas ruminantium* (Yanke et al., 1998) were optimally active in the temperature range between 50 and 60°C, while phytase of *Aerobacter aerogenes* had an optima at 25°C (Greaves et al., 1967), and that of *Bacillus* sp. DS11 at 70°C (Kim et al., 1998).

Most microbial phytases studied so far show their optimum activity in the acidic pH range (Pandey et al., 2001; Vohra and Satyanarayana, 2003; Vats and

Banerjee, 2004; Singh and Satyanarayana, 2009; Rao et al., 2009). Phytases from fungal origin exhibit optimal activity at pH 4.5 to 5.5, while some bacterial enzymes at pH 6.5 to 7.5. For the phytase of *Aerobacter aerogenes* (Greaves et al., 1967), *Pseudomonas* sp. (Irving and Cosgrove, 1971), *E. coli* (Greiner et al., 1993), *Selenomonas ruminantium* (Yanke et al., 1998) and *Lactobacillus amylovorus* (Sreeramulu et al., 1996), the pH optimum was between 4.0 and 5.5. The pH optimum for the phytase of *Enterobacter* sp.4 (Yoon et al., 1996) and *Bacillus* sp. DS11 (Kim et al., 1998) was at 7 to 7.5. *A. niger* NRRL 3135 secreted two different phytases, one with pH optima at 5.5 and 2.5, and the other at 2.0; as such, these enzymes were designated as phyA and phyB, respectively (Howson and Davis, 1983). Phytases of *T. lanuginosus* (Berka et al., 1998) and *A. fumigatus* (Pasamontes et al., 1997b) were optimally active at pH 6.0 to 6.5. The yeast phytases showed optimal activity in the pH range of 4.0 to 5.0 (Nakamura et al., 2000). The

Table 2. The biochemical properties of phytases from various microbes.

Source	MW(kDa)	T_{opt}	pH_{opt}	K_m(mM)	pI	Specificity	Reference
Fungi							
A. fumigatus	75	58	5.0	-	-	-	Mullaney et al. (2000)
A. niger	85	58	2.5 5.0	0.04	4.5	P	Ullah and Gibson (1987)
A. niger SK-57	60	50	5.5, 2.5	0.0187	-	P	Nagashima et al. (1999)
A. niger	-	55	5.5	0.20	4.9	-	Berka et al. (1998)
A. niger	353	55	2.5	0.606	-	P	Vats and Banerjee (2005)
A. oryzae	120–140	50	5.5	0.33	4.15	B	Shimizu (1993)
A. nidulans	77.8	55	5.5	-	-	-	Wyss et al. (1999b)
R. oligosporus	-	55	4.5	0.15	-	-	Sutardi and Buckle (1988)
A. niger ATCC9142	84	65	5.0	0.10	-	B	Casey and Walsh (2003)
R. oligosporus	124	65	5.00	0.010	-	B	Casey and Walsh (2004)
Peniophora lycii	72	50-55	4-4.5	-	3.61	-	Lassen et al. (2001)
Ceriporia sp.	59	55-60	5.5-6.0	-	7.36-8.01	-	Lassen et al. (2001)
Agrocybe pediades	59	50	5.0-6.0	-	4.15-4.86	-	Lassen et al. (2001)
Trametes pubescens	62	50	5.0-5.5	-	3.58	-	Lassen et al. (2001)
Thermomyces lanuginosus	60	65	7.0	0.11	4.7-5.2	B	Berka et al. (1998)
Thermoascus aurantiacus	-	55	-	-	-	-	Nampoothiri et al. (2004)
Rhizomucor pusillus	-	70	5.4	-	-	B	Chadha et al. (2004)
Myceliopthora thermophila	-	37	6.0	-	-	B	Mitchell et al. (1997)
Sporotrichum termophile	456	60	5.5	0.15	4.9	B	Singh and Satyanarayana (2009)
Yeasts							
Saccharomyces cerevisiae	-	45	4.6	-	-	-	Nayini and Markakis (1984)
Schwanomyces castellii	490	77	4.4	0.038	-	B	Segueilha et al. (1992)
Arxula adeninivorans	-	75	4.5	0.25	-	P	Sano et al. (1999)
Candida krusei WZ001[#]	330	40	4.6	-	-	-	Nakamura et al. (2000)
Pichia anomala[#]	64	60	4.0	0.20	-	B	Vohra and Satyanarayana (2002)
P. rhodanensis	-	70-75	4.0-4.5	0.25	-	-	Nakamura et al. (2000)
P. spartinae	-	75-80	4.5-5.0	0.33	-	-	Nakamura et al. (2000)
Bacteria							
Aerobacter aerogens[*]	-	25	4.0-5.0	0.135	-	-	Greaves et al. (1967)
Bacillus sp. DS 11	-	70	7.0	0.55	5.3	P	Kim et al. (1998)
Bacillus subtilis	37	60	7.5	0.04	-	-	Powar and Jagannathan (1982)
B. subtilis (natto)	38	60	6.0–6.5	-	-	-	Shimizu (1992)
B. subtilis	43	55	7.0–7.5	-	6.5	P	Kerovuo et al. (1998)

Table 2. Contd.

B. subtilis	44	55	6.0-7.0	-	5.0	P	Tye et al. (2002)
B. icheniformis	47	65	6.0-7.0	-	5.1	-	Tye et al. (2002)
B. amyloliquefaciens	44	70	7.0–7.5	-	-	-	Kim et al. (1998)
Escherichia coli	42	55	4.5	0.13	6.3-6.5	P	Greiner et al. (1993)
Klebsiella oxytoca	40	55	5.0–6.0	-	-	-	Jareonkitmongkol et al. (1997)
K. aerogenes	700	65	4.5	-	3.7	P	Tambe et al. (1994)
Pseudomonas syringe[*]	47	40	5.5	0.38	-	P	Cho et al. (2003)
Lactobacillus sanfranciscensis[*]	50	45	4.0	-	5.0	B	Angelis et al. (2003)

Phytase location is *intracellular; #Cell bound and in all other cases it is extracellular; B = Broad spectrum, P = Phytate specific.

cell-bound phytase of *Pichia anomala* was maximally activated at pH 4.0 (Vohra and Satyanarayana, 2002), while that for *S. castellii* phytase was at pH 4.4 (Segueilha et al., 1992) and *Arxula adeninivorans* was at pH 4.5 (Sano et al., 1999). Phytases of plant origin have pH optima in the range between 4.0 and 5.6. Recently, alkaline phytase having maximum activity at pH 8.0 was reported from legume seeds (Scott, 1986). Another alkaline phytase was detected in the mature lily pollen that exhibited optimal activity at pH 8.0 (Hara et al., 1985).

Phytases usually show broad substrate spectrum with the highest affinity for phytate. The phytases from *P. anomala*, *Emericella nidulans* and *M. thermophila* phytases exhibited broad substrate specificity, while phytases of *A. niger*, *A. terreus* CBS and *E. coli* were rather specific for phytic acid (Wyss et al., 1999b). Broad substrate specificity was reported for phytases of *S. castellii* (Segueilha et al., 1992) and *S. thermophile* (Singh and Satyanarayana, 2009), while cell-bound phytase from *P. anomala* exhibited broad substrate specificity (Vohra and Satyanarayana, 2002). Only a few phytases have been described as highly specific for phytate such as the alkaline phytases from *B. subtilis* (Powar and Jagannathan,

1982; Shimizu, 1992), *B. amyloliquefaciens* (Kim et al., 1998), lily pollen and cattail pollen (Hara et al., 1985). The acid phytases from *E. coli* (Greiner et al., 1993), *A. niger* and *A. terreus* (Wyss et al., 1999a) had also been reported to be rather specific for phytate.

With the exception of the phytases from *Emericella nidulans* and *Myceliophthora thermophila* (Mitchell et al., 1997), all phytases hitherto studied follow Michaelis-Menten kinetics. In general, phytases from microbial sources exhibit the highest turnover number with phytate, whereas their plant counterparts yield the highest relative rates of hydrolysis with pyrophosphate and ATP (Greiner and Konietzny, 2006). Most of the phytases characterized so far displayed the highest affinity to phytate among all phosphorylated compounds tested. The K_m values of the phytases ranged between 10 and 650 μM (Table 3). Relatively low K_m values have been reported for the phytases from *A. niger* (10 to 40 μM), *A. terreus* (11 to 23 μM), *A. fumigatus* (<10 μM), *Schwanniomyces castellii* (38 μM), *K. aerogenes* (62 μM) and some plant phytases (Greiner and Konietzny, 2006). The K_m and V_{max} values of *S. thermophile* phytase were 0.156 mM and 83.4 U mg^{-1} protein s^{-1} for phytic acid,

respectively (Singh and Satyanarayana, 2009). The catalytic constants for the degradation of phytate by phytases reported so far ranged between <10 (soybean and maize) and 1744 s^{-1} (*E. coli*) [Greiner and Konietzny, 2006]. The kinetic efficiency of an enzyme is validated by means of the k_{cat}/K_m values for a given substrate. The phytase of *E. coli* had a k_{cat}/K_m value of 1.34 x 10^7 M^{-1} s^{-1} (Golovan et al., 2001), which is the highest value reported for any phytase. The turnover number of 6209 s^{-1} and of 4.78 x 10^7 $m^{-1} s^{-1}$ was reported for *E. coli* phytase (Greiner et al., 1993). The k_{cat}/K_m value of the recombinant phytase of *P. anomala* expressed in *Hansenula polymorpha* is 72.5 (μM^{-1} s^{-1}) (Kaur et al., 2010).

Phytases are high molecular weight proteins ranging between 40 and 700 kDa (Table 3). The majority of phytases characterized so far acted like monomeric proteins with molecular masses between 40 and 70 kDa. However, some phytate-degrading enzymes appear to be made up of multiple subunits. Phytase of *S. castellii* has a molecular weight of 490 kDa with a glycosylation of around 31% (Seguilha et al., 1992). The glycosylated protein was tetrameric, with one large subunit (MW 125 kDa) and three identical small subunits (MW 70 kDa). Purified phytase

Table 3. List of commercially available microbial phytases (Modified from Cao et al., 2007).

Company	Phytase source	Production strain	Trademark
AB Enzymes	*Aspergillus awamori*	*Trichoderma reesei*	Finase
Alko Biotechnology	*A. oryzae*	*A. oryzae*	SP, TP and SF
Alltech	*A. niger*	*A. niger*	Allzyme phytase
BASF	*A. niger*	*A. niger*	Natuphos
Biozyme	*A. oryzae*	*A. oryzae*	AMAFERM
DSM	*P. lycii*	*A. oryzae*	Bio-Feed phytase
Fermic	*A. oryzae*	*A. oryzae*	Phyzyme
Finnfeeds International	*A. awamori*	*T. reesei*	Avizyme
Roal	*A. awamori*	*T. reesei*	Finase
Novozyme	*Peniophora lycii*	*A. oryzae*	Ronozyme® Roxazyme®

from *A. fumigatus* revealed a protein with a molecular mass of 60 kDa by SDS-PAGE (Pasamontes et al., 1997a). The molecular masses of the monomeric form of phyA, phyB and acid phosphatase were estimated by SDS-PAGE as 85, 65 and 85 kDa, respectively. An extracellular phytase and an extracellular acid phosphatase were purified from *A. oryzae* K1 and their molecular masses were 60 and 70 kDa, respectively (Shimizu, 1993). The phytase of *A. niger* van Teighem was a 353 kDa homopentameric protein with a monomeric molecular mass of 66 kDa (Vats and Banerjee, 2005), while the phytase of *S. thermophile* is a homopentameric 456 kDa glycosylated protein with a monomeric mass of 90 kDa (Singh and Satyanarayana, 2009), and that of *P. anomala* is a homohexamer with a molecular mass of 390 kDa (Kaur et al., 2010). The rat intestine phytase was reported to be a heterodimer comprising 70- and 90-kDa subunits (Yang et al., 1991). However, the phytases isolated from maize roots (Hubel and Beck, 1996), germinating maize seeds (Laboure et al., 1993), tomato roots (Li et al., 1997), soybean seeds (Hegeman and Grabau, 2001) and *A. oryzae* (Shimizu, 1993) were homodimeric proteins, while a homohexameric structure was proposed for the *A. terreus* enzyme (Yamamoto et al., 1972). Two different forms of phytases have been reported in *K. aerogenes* (Tambe et al., 1994). One, possibly the native enzyme, has an exceptionally large size (700 kDa), and the other, may be a fraction of the native enzyme, exhibits an exceedingly small molecular mass (10 to 13 kDa) with full complement of the activity. Fungal and several plant phytases have been found to be glycosylated with a carbohydrate content of 27.3% (Ullah, 1988). Glycosylation may have an effect on the catalytic properties, the stability or the isoelectric point of an enzyme. The molecular mass and the homogeneity of the purified enzyme from *Bacillus* sp. DS11 were estimated by gel filtration and SDS-PAGE. PAGE under denaturation conditions revealed a single protein band of 44 kDa whose size corresponded well with the molecular mass of 40 kDa obtained by superose-12 column chromatography (Kim et al., 1998). An extracellular phytase of *B. subtilis* (natto) N-77, purified 322-fold by gel filtration and DEAE chromatography had a molecular mass of 36 kDa (Shimizu, 1992), whereas two periplasmic phytases (P_1 and P_2) purified from *E. coli* close to homogeneity, were monomers with a molecular mass of 42 kDa (Greiner et al., 1993).

CRYSTAL STRUCTURE OF PHYTASES

For designing an ideal phytase and its genetic engineering, it is important to have an idea about its structure. Therefore, scientists all over the world are working on this aspect. Recently, the crystal structure of phytase from *Klebsiella* sp. ASR1 has been determined to 1.7 Å resolution using single-wavelength anomalous-diffraction phasing (Bohm et al., 2010). The phytase is different from the *E. coli* phytase in its sequence and phytate degradation pathway, but the overall structure of *Klebsiella* phytase is similar to other histidine-acid phosphatases, such as *E. coli* phytase and human prostatic-acid phosphatase. The stucture of this phytase consisted of two domains (one α and one α/β domain) in which the active site is present in a positively charged cleft between these domains.

The crystal structures of the phytases from *A. niger* (Kostrewa et al., 1997), *E. coli* (Lim et al., 2000) and *B. amyloliquefaciens* (Ha et al., 2000) have been determined. The structures of the *A. niger* and *E. coli* enzyme closely resembled the overall fold of other histidine acid phosphatases. These structures contained a conserved α/β-domain and a variable α-domain and the active site is present at the interface between these domains. This structure also provides the information about substrate binding and the catalytic mechanism. In case of *E. coli* phytase, it was shown that the phosphate is co-ordinated by the two arginine residues of the RHGXRXP-motif, as well as by conserved residues downstream, a further arginine residue and the histidine and aspartate residue of the HD-motif. Furthermore, the histidine residue of the RHGXRXP-motif was shown to be oriented for nucleophilic attack. The phytase from *S. ruminantium* shared no sequence identity with other microbial phytases (Chu et al., 2004). The active site of this phytase is located close to a conserved cysteine-containing (Cys241) P loop. The co-crystallization of *myo*-inositol hexasulfate, with the enzyme revealed that the inhibitor was bound in a pocket slightly away from Cys241 and at the substrate binding site where the phosphate group to be hydrolyzed is held close to the -SH group of Cys241. Crystal structure of *Aspergillus fumigatus* phytase was determined at 1.5 Å resolution to understand the structural basis for its high thermostability (Xiang et al., 2004). However, the overall folding has a resemblance with the structure of other phytases.

Crystal forms I and II were obtained with $CdCl_2$ and $HgCl_2$ and diffracted to 1.5 Å and 2.25 Å resolution, respectively (Lim and Jia, 2002). Hg^{2+} and Cd^{2+} both acted as molecular bridge(s) and played a crucial role in the crystallization of phytase by bridging neighbouring molecules. Despite a lack of sequence similarity, the structure closely resembled the overall folds of other histidine acid phosphatases (Lim et al., 2000). The crystal structure of a thermostable, calcium-dependent and beta propeller type *Bacillus* phytase, complexed with inorganic phosphate, revealed that two phosphates and four calcium ions are tightly bound at the active site (Shin et al., 2001). Mutation of the residues involved in the calcium chelation resulted in severe defects in the enzyme activity. One phosphate ion, chelating all of the four calcium ions, is close to a water molecule bridging two of the bound calcium ions. The enzyme has two phosphate binding sites, the 'cleavage site', which is responsible for the hydrolysis of a substrate, and the 'affinity site', that increases the binding affinity for substrates containing adjacent phosphate groups.

The crystal structure of *A. niger* NRRL3135 phytase determined at 2.5 Å resolution served to specify all active site residues (Tomschy et al., 2000a, b). Using multiple amino acid sequence alignment approach, Gln27 of *A. fumigatus* phytase was identified as likely to be involved in substrate binding and/or release and, possibly, to be responsible for the considerably lower specific activity of *A. fumigatus* phytase as compared to that of *A. terreus* phytase, which has a 'leu' at an equivalent position. Site-directed mutagenesis of Gln27 of *A. fumigatus* phytase to leu, in fact increased the specific activity, and this and other mutations at position 27 yielded an interesting array of pH activity profiles and substrate specificities. A novel bacterial phytase from a *B. amyloliquefaciens* strain was crystallized using the hanging-drop vapour-diffusion

method (Ha et al., 1999). High-quality single crystals of the enzyme in the absence of calcium ions were obtained using a precipitant solution containing 20% 2-methyl-2, 4-pentanediol and 0.1 M MES (pH 6.5). The crystals contain one monomer per asymmetric unit. Phytase has a α/β-domain similar to that of rat acid phosphatase and α-domain with a new fold (Kostrewa et al., 1997).

DIRECTED EVOLUTION AND PROTEIN ENGINEERING OF PHYTASES

The natural enzymes are adapted in a living cell to perform a particular function, but in most cases, they are poorly suited for industrial applications. Protein engineering is a very active area of research for understanding the structure-function relationships of a particular protein (Lehman et al., 2000a, b, c; Tomschy et al., 2000a, b). In recent years, there has been a widespread enthusiasm for 'directed evolution' as a new tool to optimize the properties of an enzyme of interest (Dalboge and Borchert, 2000; Arnold, 2001). Mostly, enzymes are stabilized by the cumulative effects of small improvements at many locations within the protein molecule (Lehman et al., 2000a, b, c; Tomschy et al., 2000a, b; Coco et al., 2001). The engineering of proteins for improved thermostability is an exciting and challenging field because of its applicability for the industrial use of recombinant proteins (Lehman et al., 2000a, b, c; Tomschy et al., 2000a, b).

Rational design principles and directed evolution

The stability of a protein is determined by both local and long-range interactions between the residues (Tomschy et al., 2000a, b). The thermostability of an enzyme can be enhanced by multiple amino acid exchanges, each of which slightly increases the unfolding temperature of the protein. The rational approaches for thermostability engineering involve the comparison of the amino acid sequence of the protein of interest with a more thermostable, homologous counterpart, followed by replacement of selected amino acids (Tomschy et al., 2000a, b). Three-dimensional structure of the protein of interest could be helpful in this regard. The thermostabilization concepts include the introduction of additional disulfide bridges, improvements in the packing of the hydrophobic core, engineering of surface salt bridge networks or α-helix dipole interactions, changes in α-helix propensity and changes in entropy (Haney et al., 1999; Tomschy et al., 2000a, b). All these rational approaches have been used successfully in the engineering of phytases for improved catalytic activity. Site directed mutagenesis of amino acid residue 300 was resulted in a high phytase activity by *A. niger* NRRL 3135 at pH 3.0 to 5.0, while a single mutation (K300E) resulted in an enhanced hydrolysis of phytic acid at pH 4.0 and 5.0. In this study, the basic amino acid residue lysine (K) was replaced by acidic residue. However, this replacement with another basic residue, or an uncharged but polar residue, did not significantly alter the activity at pH 4.0; but a replacement with basic residue arginine (R) lowered the activity over the pH range from 2.0 to 6.0 (Mullaney et al., 2002).

In A. fumigatus, a 3D structure of the native A. niger NRRL 3135 phytase was used to identify non-conserved amino acids that were not associated with increased catalytic activity (Tomschy et al., 2000a). Consequently, they changed the single amino acid residue (Q27), and this displayed a significant effect on specific activity, pH profile and substrate specificity. A. niger NRRL 3135 and A. niger T213 wild phytases displayed a 3-fold difference in specific activity, despite only 12 amino acid residues difference (Tomschy et al., 2000b). Out of these 12 amino acid residues, nine were distantly placed from active site, and therefore, are not responsible for catalytic activity. In the remaining 3 residues, R297Q mutation was found to fully account for this difference in catalytic activity, because out of the 3 single mutants (E89D, H292N and R297Q), 2 double mutants (E89D R297Q and H292N R297Q) and a triple mutant (E89D H292N R297Q) revealed a 3-fold increase in specific activity. This specific activity is close to the wild type. Molecular modeling revealed that R297Q may directly interact with the phosphate group of phytic acid. This presumed ionic interactions caused strong binding of the substrate and product indicating the product release as the rate-limiting step of the reaction, which is responsible for lower specific activity.

When expressed in A. niger, several fungal phytases were susceptible to proteases (Wyss et al., 1999b). N-terminal sequences of the fragments revealed that cleavage invariably occurred at exposed loops on the surfaces of the molecules. Site directed mutagenesis at the protease-sensitive sites of Aspergillus fumigatus (S151N and R151L/ R152N) and Emericella nidulans phytase (K186G and R187R) yielded mutants with reduced susceptibility to proteases, without affecting the specific activity. Based on E. coli phytase crystal structure, substitution of C200N in a mutant seems to eliminate the disulfide bond between the G helix and the GH loop in the α-domain of the protein which might be modulating the domain flexibility, and thereby the catalytic efficiency and thermostability of the enzyme (Rodriguez et al., 2000).

The consensus approach

The consensus approach is based on the hypothesis that at a given position in an amino acid sequence alignment of homologous proteins, the respective consensus amino

acid contributes more than average to the stability of the protein than the non-consensus amino acids (Lehman et al., 2000a, b, c). Consequently, substitution of non-consensus by consensus amino acids may be a possible approach for improving the thermostability of a protein.

Each amino acid of a protein contributes towards its stability. The mutations responsible for thermostability of a protein with a small effect on the protein stability were combined to generate a consensus protein variant that showed enhanced thermostability (Lehman et al., 2000 a, b, c).

Lehman et al. (2000a) used a computer program to calculate an entire consensus sequence from 13 homologous amino acid sequences of wild-type phytases from mesophilic fungi. This phytase showed an identity of 58.3 to 80% with the parent phytases. The recombinant expression of a synthetic gene gave rise to a consensus phytase (consensus phytase-1) that was 15 to 26 °C more thermostable and showing 15 to 22 °C more denaturing temperature than the wild-type. The backbone of this consensus phytase was modified by Lehman et al. (2000b). They modified the catalytic property by replacing a part of the active site with the corresponding residue of A. niger NRRL3135 phytase, which displayed a pronounced difference in specific activity, substrate specificity and pH profile. This exchange of active site resulted in a decrease in denaturing temperature, but the consensus phytase was still more thermostable than its parents. Further addition of wild-type sequences in the alignment resulted in consensus phytase-10, which displayed a further 7.4 °C increase in denaturing temperature. In another approach, the consensus approach was refined by including six more sequences that yielded consensus phytases-10 and -11 with an increase of 7.4 °C in denaturing temperature. Site directed mutagenesis identified some residues showing their effect on protein thermostability. Nonetheless, the combination of these residues resulted in an increase in the denaturing temperature from 88.0 to 90.4 °C.

MULTIFARIOUS APPLICATIONS OF PHYTASES

Amelioration of the nutritional status of foods and feeds

Phytases are useful in food and feed industries, preparation of myo-inositol phosphate intermediates, combating phosphorus pollution and in plant growth promotion (Idriss et al., 2002; Vohra and Satyanarayana, 2003; Vats and Banerjee, 2004; Greiner and Konietzny, 2006; Rao et al., 2009). The major food supplements in animal food are derived from plant sources such as cereals, legumes, soybean, etc. The presence of phytate in plant foodstuffs causes mineral deficiency due to the chelation of metal ions (De Boland et al., 1975). The presence of phytic acid in rapeseed causes Zn, Mg and

Ca deficiency in chickens (Nwokolo and Bragg, 1977).

Canola meal contains 4 to 6% phytic acid, which reduces the nutrition value of the meal. The phytic acid has been shown to bind with multivalent cations, and hence, reduce their bioavailability. The addition of phytase to high phytate containing diets improves the absorption and utilization of phosphorus (Hughes and Soares, 1998). Dietary phytase also improves the nutritive value of canola protein concentrate and decreases phosphorus output in case of rainbow trout (Forster et al., 1999). Similar reports have been documented for different species like rainbow trout (Rodehutscord and Pfeiffer, 1995), channel catfish (Li and Robinson, 1997), African catfish (Van Weerd et al., 1999), common carp (Schafer et al., 1995) and *Pangasius pangasius* (Debnath et al., 2005). Robinson et al. (2002) reported that 250 units of phytase per kilogram of diet could effectively replace dicalcium phosphate supplement in the diet of channel catfish without affecting growth, feed efficiency or bone phosphorus deposition.

Phytic acid is well known to make complexes with various cations as well as with proteins (Wise, 1983). Phytase added to diets improves the bioavailability of copper and zinc in pigs (Adeola et al., 1995) and poultry (Yi et al., 1996). Microbial phytase also improves the apparent absorption of magnesium, zinc, copper and iron in pigs (Selle and Ravindran, 2007). Similar results have also been reported for fishes (Cao et al., 2007). Phytase addition increases the concentration of minerals like magnesium, phosphorus, calcium, manganese and zinc in plasma, bone and the whole body (Vielma et al., 2004). Yan and Reigh (2002) demonstrated that the phytase supplementation improved the retention of calcium, phosphorus and manganese by catfish fed with an all-plant protein diet. The phytase supplementation in the diets significantly improved the digestibility of minerals, total-P, phytate-P and gross energy (Cheng and Hardy, 2002). The experimental studies in animals and humans have shown that phytic acid rich diets can cause zinc deficiency. Phytic acid does not inhibit copper absorption, but has a modest inhibitory effect on manganese absorption (Lonnerdal, 2000).

The treatment of fish feed with phytase was found to improve protein digestibility and retention in fishes (Cheryan, 1980; Storebakken et al., 1998; Papatryphon et al., 1999; Boling et al., 2001; Cheng and Hardy, 2002; Usmani and Jafri, 2002; Vielma et al., 2004; Sajjadi and Carter, 2004; Debnath et al., 2005; Baruah et al., 2005; Ai et al., 2007; Altaff et al., 2008; Hassan et al., 2009). The inclusion of phytase to broilers diets increased the coefficient of phosphorus retention and reduced the presence of this element in poultry birds, thus, indicating a favorable environmental effect (Ahmad et al., 2000; Brenes et al., 2003; Juanpere et al., 2004; Murugesan et al., 2005; Vohra et al., 2006; Ahmadi et al., 2008; Pillai et al., 2009). Microbial phytases positively affected the pigs' performance and their daily gain, and further, the feed

conversion ratios were ameliorated by organic acids (Jongbloed et al., 2000; Walz and Pallauf, 2002; Revy et al., 2005; Kim et al., 2005; Pomar et al., 2008; Akinmusire and Adeola, 2009; Hill et al., 2009).

The role of phytases in dephytinization and bread making

The presence of phytates in plant food stuffs (De Boland et al., 1975) is well known. Moulds commonly used in oriental food fermentation have been examined for their ability to produce phytase. Tempeh is a popular oriental fermented food made from soyabeans inoculated by moulds (*Rhizopus oligosporus*) in the koji process. The digestibility, vitamin contents and flavour of soyabean were improved by the mould fermentation (Fardiaz and Markakis, 1981). Dietary phytase is inactivated during cooking so the phytate digestion is very poor, thereby affecting mineral absorption. The addition of *A. niger* phytase to the flour containing wheat bran increased iron absorption in humans (Sandberg et al., 1996). The use of phytase was suggested for producing low phytin bread. Also, phytic acid has positive effects. It exerts an antineoplastic effect in animal models of both colon and breast carcinomas. The presence of undigested phytate in the colon may protect it against the development of colonic carcinoma (Iqbal et al. 1994).

By adding mould phytases during bread making, dough phytate could be almost completely eliminated. Caransa et al. (1988) reported that phytase supplementation could accelerate the process of steeping required in the wet milling of corn, thereby improving the properties of corn steep liquor. Supplementation of phytase from a thermophilic mould, *S. thermophile*, improved the bread quality with concomitant reduction in phytate (Singh and Satyanarayana, 2008c). Phytase released inorganic phosphate from calcium, magnesium and cobalt phytates (Singh and Satyanarayana, 2010).

The effect of the supplementation of exogenous phytase to four different bread formulations on the bread quality was assessed by Haros et al. (2001a, b). The supplementation of bread with phytase shortened the fermentation period. There was a considerable increase in the specific bread volume, which is an improvement in the crumb texture and the width/height ratio of the bread slice (Knorr et al., 1981). The chapathi dough with reduced phytic acid levels was developed using a mutated strain of the yeast *Candida versatilis* and it resulted in 10 to 45% reduction in phytate levels (Bindu and Varadaraj, 2005). Wheat flour, sesame oil cake and soymilk were efficiently dephytinized by *S. thermophile* phytase with concomitant reduction in phytic acid content and liberating inorganic phosphate (Singh and Satyanarayana, 2006a; 2008a, 2008b). Similarly, the cell-bound phytase of *P. anomala* resulted in dephytinization of soymilk (Kaur and Satyanarayana, 2010).

Semisynthesis of peroxidase

Peroxidases are ubiquitous enzymes that catalyse a wide variety of selective oxidations with hydrogen peroxide as the primary oxidant (van de Velde et al., 2000). The active site of vanadium chloroperoxidase from *Curvularia inaequalis* closely resembled that of the acid phosphatases and the apoenzyme of vanadium chloroperoxidase exhibits phosphatase-like activity (Hemrika et al., 1997). The combination of phytase with vanadate produced an effective semi-synthetic peroxidase. The effect of pH on the vanadate phytase-catalysed oxidation of thioanisole revealed that the pH optimum coincided with that of phytase. Optimisation led to a maximum enatiomeric excess (ee) of 68% obtained in formate buffer at 4.0°C. The vanadium-incorporated phytase was stable for over three days with only a slight decrease in activity.

A cross-linked enzyme aggregate of 3-phytase was transformed into peroxidase by incorporation of vanadate (Correia et al., 2008). The cross-linked aggregate phytase showed similar efficiency and asymmetric induction as the free enzyme. Moreover, the cross-linked aggregate phytase can be reused at least three times without significant loss of activity. Some other acid, phosphatases and hydrolases were tested for peroxidase activity, when incorporated with vanadate ion. Phytases from *Aspergillus ficuum*, *A. fumigatus* and *A. nidulans*; sulfatase from *Helix pomatia*; and phospholipase D from cabbage, catalyzed the enantioselective oxygen transfer reactions when incorporated with vanadium. However, phytase from *A. ficuum* was unique in catalyzing the enantioselective sulfoxidation as compared to others.

Plant growth promotion

Phosphorus deficiency in soil is a major constraint for agricultural production worldwide. Large proportion of soil P exists in the organic form, of which phytic acid is the pre-dominant form. There are a large number of reports explaining the role of phytase in improving the growth of the plants and reducing the phosphorus pollution. A β-propeller phytase from *Bacillus subtilis* was constitutively expressed in tobacco and *Arabidopsis*, and it was shown to be secreted from their roots (Lung et al., 2005). In transgenic tobacco, phytase activities in leaf and root extracts were 7 to 9-fold higher than those in wild-type. A 4 to 6-fold higher extracellular phytase activity had been recorded in transgenic plants. In sterile liquid culture, using 1 mM sodium phytate as the sole P source, the transgenic tobacco lines accumulated 1.7 to 2.2 times more shoot biomass than the wild-type plants after 30 days of growth with concomitant increase (27 to 36%) in shoot P concentration. Similar observations have been recorded in the transgenic *Arabidopsis*, explaining the mobilization of soil phytate into inorganic phosphate for plant uptake (Lung et al., 2005). Yip et al. (2003) showed that the tobacco line transformed with a neutral *Bacillus* phytase exhibited phenotypic changes in flowering, seed development, and response to phosphate deficiency. The transgenic line showed an increase in number of flower and fruit, lesser seed IP6/IP5 ratio, and enhanced growth under phosphate-starvation conditions as compared to the wild type.

The transgenic *Arabidopsis* plants secreted phytase only from roots when grown on a medium under low phosphate conditions (Mudge et al., 2003). The growth rates and shoot P concentrations of plants were similar when grown on the medium containing phytate or phosphate as the P source. Phytase and phosphatases producing fungi were used as seed inoculants, to help attain higher P nutrition of plants in the soils containing high phytate phosphorus (Yadav and Tarafdar, 2003). The efficiency of different organic P compounds' hydrolysis by different fungi indicated that the fungi have enough potential to exploit native organic phosphorus to benefit plant nutrition. Transgenic *Arabidopsis* plant expressing an extracellular phytase from *Medicago truncatula* led to significant improvement in organic phosphorus utilization and plant growth (Xiao et al., 2005). Using phytate as the sole source of phosphorus, dry weight of the transgenic *Arabidopsis* lines were 3.1 to 4.0-fold higher than the control plants and total phosphorus contents were 4.1- to 5.5-fold higher than the control, suggesting the great potential of heterologous expression of phytase gene for improving plant phosphorus acquisition and for phytoremediation. The growth and phosphorus nutrition of *Arabidopsis thaliana* plants supplied with phytate was improved significantly after the introduction of phytase gene from *Aspergillus niger* (Richardson et al., 2001). Li et al. (2007) showed that both wild type *Bacillus mucilaginosus* and transgenic (containing phytase gene) strains promoted the tobacco plant growth under greenhouse study and field experiments.

The plant growth promotory effect of an extracellular phytase of a thermophilic mould, *Sporotrichum thermophile,* has been reported recently (Singh and Satyanarayana, 2010). Both phytase, as well as the mould, promoted the growth of wheat seedlings. The growth and inorganic phosphate content of the plants were better than the control. Sodium phytate (5 mg plant^{-1}) was adequate for liberating enough phosphorus for the growth of the seedlings. The plant growth, root/shoot length and inorganic phosphate content of test plants were better than the control plants. An enzyme dose of 20.0 U plant^{-1} was found to adequately liberate enough amount of inorganic phosphate required for supporting plant growth. The plant growth, root/shoot length and inorganic phosphate content of test plants were higher than the control (Singh and Satyanarayana, 2010). The compost prepared by the combined action of native microflora of wheat straw along with *S. thermophile* promoted the growth of plants. The inorganic phosphate content of the wheat plants was also high as compared to those cultivated on the compost prepared either with only native microflora or *S. thermophile*. These approaches can be applied as a strategy for boosting the productivity in agriculture and horticulture.

Miscelaneous applications

Preparation of myo-inositol phosphates

There is a continuous demand of inositol phosphates and phospholipids, which play an important role in cell signalling pathways (Billington, 1993). Enzymic hydrolysis of phytic acid using *S. cerevisiae* resulted in the production of D-*myo*-inositol 1,2,6-triphosphate, D-*myo*-inositol 1,2,5-triphosphate, L-*myo*-inositol 1,3,4-triphosphate and *myo*-inositol 1,2,3-triphosphate (Siren, 1986). Greiner and Konietzny (1996) prepared inositol 1,2,3,4,5-pentakisphosphate, inositol 2,3,4,5-tetrakisphosphate, inositol 2,4,5-triphosphate and inositol 2,5-biphosphate using immobilized phytase from *E. coli*. Inositol phosphate derivatives can be used as enzyme stabilizers (Siren, 1986), enzyme substrates for metabolic investigation, as enzyme inhibitors and therefore potential drugs, and as chiral building blocks.

Pulp and paper industry

It has been observed that the removal of plant phytic acid could be important in the pulp and paper industry (Liu et al., 1998). A phytase with activity at elevated temperatures could have the potential as a biological agent to hydrolyse phytic acid during pulp and paper processing. This process will not produce any carcinogenic and toxic byproducts. Therefore, the use of phytases in pulp and paper processing could be ecofriendly and would help in the development of cleaner technologies (Liu et al., 1998).

Combating environmental phosphorus pollution

Phosphorus is an essential ingredient in animal and plant production; however, too much or too little P can be a problem both for animal production and the environment. Researchers all over the world are finding ways for poultry to better utilize P, thus increasing productive efficiency and protecting the environment. The ruminants sustain the microflora that enzymatically releases inorganic phosphorus from phytic acid, though, monogastrics such as humans, chickens and pigs produce little or no phytase in the intestine. Hence, the phytic acid phosphorus is unavailable and the phytic acid is excreted in their feaces (Mullaney et al., 2000). Phytic

acid present in the manure of these animals is enzymatically cleaved by soil and water-borne microorganisms. The phosphorus thus released is transported into the water bodies causing eutrophication. This results in oxygen depletion due to excessive algal growth. Pretreatment of animal feed with phytases will increase the availability of inorganic phosphorus, thereby improving the nutritional status of food and also help in combating phosphorus pollution. Phytases are very well known to reduce pollution caused by excess of phosphorus accumulation in soil and water (Nahm, 2002). The excretion of phosphorus can be reduced by 30%, via replacing feed phosphate with phytase and by equally calculated digestible P content. The addition of phytase to the feed of piglets gives positive results in some experiments such as a significant increase in growth rate and feed intake and a significantly better feed conversion ratio in comparison with the conventional feed. The supplementation of phytase in corn and soybean meal diets was additive, significantly improving P digestibility and dramatically decreasing P excretion to reduce the potential impacts of P from pig manure on the environment (Hill et al., 2009).

Microbial phytase supplementation in the diet of fish can overcome this problem. It makes the chelated phosphorus available to fish, and hence, there is less faecal excretion, thereby reducing environmental pollution. The environmental benefits of using this enzyme in fish feed are thus listed:

1. Reduced requirement of the mineral supplements, thereby reducing chances of excess inorganic phosphorus getting into the aquatic system.
2. Reduced organic phosphorus, that is, phytic acid outputs.

Use of phytase in feeds reduces or sometimes eliminates the necessity of mineral supplementation, which also decreases the cost of feeds. Although phytase was first used for environmental reasons, it is now realized that there are a range of other nutritional and health benefits from using these enzymes.

CONCLUSIONS AND FUTURE PERSPECTIVES

Besides effectively tackling phosphorus pollution in the areas of intensive livestock rearing, phytases have considerable potential in commercial applications. The applications of phytases in improving human health and in synthesis of lower inositol phosphates have increasingly attracted attention. A significant progress has been made in phytase research during the last few decades. The phytases, which exhibit desirable activity profile over a broad pH range, excellent thermal stability, and broad substrate specificity, are more promising for commercial exploitation. Modern day technologies (molecular biology and genetics) could be utilized for the development of staple foods with higher and improved bioavailability of the minerals and proteins. Genetic engineering techniques could be employed for the generation of consensus phytases with improved and desirable properties for applications in food and feed industries (Lehman et al., 2000a, b, c). Adding phytase to the animal diets not only improves the bioavailability of proteins and minerals, but also aids in combating environmental phosphorus pollution in the areas of intensive live stock management.

Transgenic plants of corn, rice, barley and soybean with low phytic acid have been generated; and this could be a novel approach for reducing micronutrient malnutrition and animal waste phosphorus. Further research efforts are needed to understand the molecular biology and genetics of phytic acid accumulation during seed development and feasibility and effectiveness of employing this approach at the community level (Mendoza, 2002). The transgenic plants harboring the microbial phytase genes could also be used to improve soil fertilization and nutrient availability to plants. With the collaborative efforts of phytase scientists from different fields, it would be possible to design and develop an ideal phytase for animal nutrition, human health and environmental protection.

REFERENCES

Adeola O, Lawrence BV, Sutton AL, Cline TR (1995). Phytase-induced changes in mineral utilization in zinc-supplemented diets for pigs. J. Anim. Sci., 73: 3384-3391.

Ahmad T, Rasool S, Sarwar M, Haq A, Hasan Z (2000). Effect of microbial phytase produced from a fungus Aspergillus niger on bioavailability of phosphorus and calcium in broiler chickens. Anim. Feed Sci. Technol., 83: 103-114.

Ahmadi A, Tabatabaei MM, Aliarabi H, Saki AA, Siyar SA (2008). Performance and egg quality of laying hens affected by different sources of phytase. Pak. J. Biol. Sci., 11: 2286-2288.

Ai Q, Mai K, Zhang W, Xu W, Tan B, Zhang C, Li H (2007). Effects of exogenous enzymes (phytase, non-starch polysaccharide enzyme) in diets on growth, feed utilization, nitrogen and phosphorus excretion of japanese seabass, Lateolabrax japonicus. Comp. Biochem. Physiol. A Mol. Integr. Physiol., 147: 502-508.

Akinmusire AS, Adeola O (2009). True digestibility of phosphorus in canola and soybean meals for growing pigs: Influence of microbial phytase. J. Anim. Sci., 87: 977-983.

Altaff K, Hassan S, Satyanarayana T (2008). Use of phytase in plant based feed for aquaculture industry: Cost effective and eco-friendly practice: Present scenario and future perspectives. J. Aqua. Biol., 23: 185-193.

Angelis MD, Gallo G, Corbo MR, McSweeney PLH, Faccia M, Giovine M, Gobbetti M (2003). Phytase activity in sourdough lactic acid bacteria: purification and characterization of a phytase from Lactobacillus sanfranciscensis CB1. Int. J. Food Microbiol., 87: 259-270

Arnold FH (2001). Evolutionary protein design. Adv. Protein Chem., 55: 1-438.

Asada K, Tanaka K, Kasai Z (1969). Formation of phytic acid in cereal grains. Ann. N. Y. Acad. Sci., 165: 801-814.

Barrientos L, Scott JJ, Murthy PP (1994). Specificity of hydrolysis of phytic acid by alkaline phytase from lily pollen. Plant Physiol., 106: 1489-1495.

Baruah KP, Sahu AK, Jain NP, Mukherjee KK, Debnath SC (2005). Dietary protein level, microbial phytase, citric acid and their

interactions on bone mineralization of Labeo rohita (Hamilton) juveniles. Aquacult. Res., 36: 803-812.

Berka RM, Rey MW, Brown KM, Byun T, Klotz AV (1998). Molecular characterization and expression of a phytase gene from the thermophilic fungus Thermomyces lanuginosus. Appl. Environ. Microbiol., 64: 4423-4427.

Billington WD (1993). Species diversity in the immunogenetic relationship between mother and fetus: is trophoblast insusceptibility to immunological destruction the only essential common feature for the maintenance of allogeneic pregnancy? Exp. Clin. Immunogenet., 10(2): 73-84.

Bindu S, Varadaraj MC (2005) Process for the preparation of Chapathi dough with reduced phytic acid level. United States Patent Application #20050048165 dated 3/3/05.

Bohm K, Herter T, Müller JJ, Borriss R, Heinemann U (2010). Crystal structure of Klebsiella sp. ASR1 phytase suggests substrate binding to a preformed active site that meets the requirements of a plant rhizosphere enzyme. FEBS J. 277(5): 1284-1296.

Boling SD, Douglas MW, Johnson ML, Wang X, Parsons CM, Koelkebeck KW, Zimmerman RA (2001). The effects of dietry available phosphorus levels and phytase performance of young and older laying hens. Poult. Sci., 79: 224-230.

Brenes A, Viveros A, Arija I, Centeno C, Pizarro M, Bravo C (2003). The effect of citric acid and microbial phytase on mineral utilization in broiler chicks. Anim. Feed Sci. Technol., 110: 201-219.

Cao L, Wang W, Yang C, Yang Y, Diana J, Yakupitiyage A, Luo Z, Li D (2007). Application of microbial phytase in fish feed. Enz. Microb. Technol., 40: 497-507.

Caransa A, Simell M, Lehmussari M, Vaara M, Vaara T (1988). A novel enzyme application in corn wet milling. Starch 40: 409-411.

Casey A and Walsh G (2003). Purification and characterization of extracellular phytase from Aspergillus niger ATCC 9142. Bioresour. Technol., 86(2): 183-188.

Casey A, Walsh G (2004). Identification and characterization of a phytase of potential commercial interest. J. Biotechnol., 110(3): 313-322.

Chadha BS, Gulati H, Minhas M, Saini HS, Singh N (2004). Phytase production by the thermophilic fungus Rhizomucor pusillus. World J. Microbiol. Biotechnol., 20: 105-109.

Cheng ZJ, Hardy RW (2002). Effect of microbial phytase on apparent nutrient digestibility of barley, canola meal, wheat and wheat middlings, measured in vivo using rainbow trout (Oncorhynchus mykiss). Aquacult. Nutr., 8: 271-277.

Cheryan M (1980). Phytic acid interactions in food systems. Crit. Rev. Food Sci. Nutr., 13: 297-335.

Cho JS, Lee CW, Kang SH, Lee JC, Bok JD, Moon YS, Lee HG, Kim SC, Choi YJ (2003). Purification and characterization of a phytase from Pseudomonas syringae MOK1. Curr. Microbiol. 47(4): 290-294.

Chu HM, Guo RT, Lin TW, Chou CC, Shr HL, Lai HL, Tang TY, Cheng KJ, Selinger BL, Wang AH (2004). Structures of Selenomonas ruminantium phytase in complex with persulfated phytate: Dsp phytase fold and mechanism for sequential substrate hydrolysis. Structure (Camb), 12: 2015-2024.

Coco WM, Levinson WE, Crist MJ, Hektor HJ, Darzins A, Pienkos PT, Charles HS, Monticello DJ (2001). DNA shuffling method for generating highly recombined genes and evolved enzymes. Nat. Biotechnol., 19: 354-359.

Correia I, Aksu S, Adao P, Pessoa JC, Sheldon RA, Arends IWCE (2008). Vanadate substituted phytase: Immobilization, structural characterization and performance for sulfoxidations. J. Inorg. Biochem., 102: 318-329

Dalboge H, Borchert TV (2000). Protein engineering of enzymes. Biochim. Biophys. Acta, 1543: 203-455.

De Boland AR, Garner GB, O'Dell BL (1975). Identification and properties of 'phytate' in cereal grains and oilseed products. J. Agri. Food Chem., 23: 1186-1189.

Debnath D, Pal AK, Sahu NP, Jain KK, Yengkokpam S, Mukherjee SC (2005). Effect of dietary microbial phytase supplementation on growth and nutrient digestibility of Pangasius pangasius (Hamilton) fingerlings. Aquacult. Res., 36: 180-187.

Fardiaz D, Markakis P (1981). Degradation of phytic acid in oncom (fermented peanut press cake). J. Food Sci., 46: 523-525.

Forster I, Higgs DA, Dosanjh BS, Rowshandeli M, Parr J (1999). Potential for dietary phytase to improve the nutritive value of canola protein concentrate and decrease phosphorus output in rainbow trout (Oncorhynchus mykiss) held in 11°C fresh water. Aquacult. Nutr. 179: 109-125.

Fu S, Sun J, Qian L, Li Z (2008). Bacillus phytases: present scenario and future perspectives. Appl. Biochem. Biotechnol. 151(1): 1-8.

Golovan SP, Hayes MA, Phillips JP, Forsberg CW (2001). Transgenic mice expressing bacterial phytase as a model for phosphorus pollution control. Nat. Biotechnol. 19: 429-433.

Graf E (1986). Phytic acid-chemistry and application. Minneapolis, The Pillsbury Co Pilatus Press; p. 42-44.

Greaves MP, Anderson G and Webley DM (1967). The hydrolysis of inositol phosphates by Aerobacter aerogenes. Biochim. Biophys. Acta, 132: 412-418.

Greaves MP, Anderson G, Webley DM (1967). The hydrolysis of inositol phosphates by Aerobacter aerogenes. Biochim. Biophys. Acta, 132: 412-418.

Greiner R, Konietzny U (1996). Construction of a bioreactor to produce special breakdown products of phytate. J. Biotechnol., 48: 153-159.

Greiner R, Konietzny U (2006). Phytase for food application. Food Technol. Biotechnol., 44: 125-140.

Greiner R, Konitzny U, Jany KD (1993). Purification and characterization of two phytases from Escherchia coli. Arch. Biochem. Biophys., 303: 107-113.

Ha NC, Kim YO, Oh TK, Oh BH (1999). Preliminary x-ray crystallographic analysis of a novel phytase from a Bacillus amyloliquefaciens strain. Acta Crystallogr. D Biol. Crystallogr., 55: 691-693.

Ha NC, Oh BC, Shin S, Kim HJ, Oh TK, Kim YO, Choi KY, Oh BH (2000). Crystal structures of a novel, thermostable phytase in partially and fully calcium-loaded states. Nat. Struct. Biol., 7: 147-153.

Han YM, Yang F, Zhou AG, Miller ER, Ku PK, Hogberg MG, Lei XG (1997). Supplemental phytases of microbial and cereal sources improve dietary phytate phosphorus utilization by pigs from weaning through finishing. J. Anim. Sci., 75: 1017-1025.

Haney PJ, Badger JH, Buldak GL, Reich CI, Woese CR, Olsen GJ (1999). Thermal adaptation analyzed by comparison of protein sequences from mesophilic and extremely thermophilic Methanococcus species. Proc. Natl. Acad. Sci. USA. 96: 3578-3583.

Hara A, Ebina S, Kondo A, Funagua T (1985). A new type of phytase from Typha latifolia L. Agric. Biol. Chem., 49: 3539-3544.

Harland BF, Morris ER (1995). Phytate: A good or a bad food component. Nutr. Res. 15: 733-754.

Haros M, Rosell CM, Benedito C (2001a). Fungal phytase as a potential breadmaking additive. Europ. Food Res. Technol., 213 (4-5): 317-322.

Haros M, Rosell CM, Benedito C (2001b). Use of fungal phytase to improve breadmaking performance of whole wheat bread. J. Agric. Food Chem., 49(11): 5450-5454.

Hassan S, Altaff K, Satyanarayana T (2009). Use of soybean meal supplemented with cell bound phytase for replacement of fish meal in the diet of juvenile milkfish, Chanos chanos. Pakistan J. Nutr., 8: 341-344.

Hegeman CE, Grabau EA (2001). A novel phytase with sequence similarity to purple acid phosphatases is expressed in cotyledons of germinating soybean seedlings. Plant Physiol., 126: 1598-1608.

Hemrika W, Renirie R, Dekker HL, Barnett P, Wever R (1997). From phosphatases to vanadium peroxidases: A similar architecture of the active site. Proc. Natl. Acad. Sci. USA, 94: 2145-2149.

Hill BE, Sutton AL, Richert BT (2009). Effects of low-phytic acid corn, low-phytic acid soybean meal, and phytase on nutrient digestibility and excretion in growing pigs. J. Anim. Sci., 87: 1518-1527.

Howson SJ, Davis RP (1983). Production of phytate hydrolyzing enzymes by some fungi. Enz. Microb. Technol., 5: 377-382.

Hubel F, Beck E (1996). Maize root phytase (purification, characterization, and localization of enzyme activity and its putative substrate). Plant Physiol., 112: 1429-1436.

Hughes KP, Soares JJH (1998). Efficacy of phytase on phosphate utilization in practical diets fed to striped bass, morone sexatilis. Aquacult. Nutr., 4: 133-140.

Idriss EE, Makarewicz O, Farouk A, Rosner K, Greiner R, Bochow H,

Richter T, Borriss R (2002). Extracellular phytase activity of Bacillus amyloliquefaciens FZB45 contributes to its plant-growth-promoting effect. Microbiology, 148 (Pt 7): 2097-2109.

Iqbal TH, Lewis KO, Cooper BT (1994). Phytase activity in the human and rat small intestine. Gut, 35: 1233-1236.

Irving GCJ, Cosgrove DJ (1971). Inositol phosphate phosphatases of microbiological origin. Some properties of a partially purified bacterial (Pseudomonas sp.) phytase. Aust. J. Biol. Sci., 24: 547-557.

Jareonkitmongkol S, Ohya M, Watanbe R, Takagi H, Nakamori S (1997). Partial purification of phytase from a soil isolate bacterium, Klebsiella oxytoca MO-3. J. Ferm. Bioeng., 83: 393-394.

Jongbloed AW, Mroz Z, Kemme PA (1992). The effect of supplementary Aspergillus niger phytase in diets for pigs on concentration and apparent digestibility of dry matter, total phosphorus, and phytic acid in different sections of the alimentary tract. J. Anim. Sci., 70: 1159-1168.

Jongbloed AW, Mroz Z, van der Weij-Jongbloed R, Kemme PA (2000). The effects of microbial phytase, organic acids and their interaction in diets for growing pigs. Livestock Prod. Sci., 67: 113-122.

Juanpere J, Pérez-Vendrell AM, Brufau J (2004). Effect of microbial phytase on broilers fed barley-based diets in the presence or not of endogenous phytase. J. Animal Feed Sci. Technol., 115: 265-279.

Kaur P, Singh B, Böer E, Straube N, Piontek M, Satyanarayana T, Kunze G (2010). Pphy--a cell-bound phytase from the yeast Pichia anomala: molecular cloning of the gene PPHY and characterization of the recombinant enzyme. J. Biotechnol., 149(1-2): 8-15.

Kaur P, Kunze G, Satyanarayana T (2007). Yeast phytases: Present scenario and future perspectives. Crit. Rev. Biotechnol., 27: 93-109.

Kaur P, Satyanarayana T (2010). Improvement in cell-bound phytase activity of Pichia anomala by permeabilization and applicability of permeabilized cells in soymilk dephytinization. J. Appl. Microbiol. (In press).

Kerovuo J, Lauraeus M, Nurminen P, Kalkkinen N, Apajalahti J (1998). Isolation, characterization, molecular gene cloning and sequencing of a novel phytase from Bacillus subtilis. Appl. Environ. Microbiol., 64: 2079-2085.

Kim JC, Simmins PH, Mullan BP, Pluske JR (2005). The effect of wheat phosphorus content and supplemental enzymes on digestibility and growth performance of weaner pigs. Anim. Feed Sci. Technol., 118: 139-152.

Kim YO, Kim HK, Bae KS, Yu JH, Oh TK (1998). Purification and properties of thermostable phytase from Bacillus sp. DS11. Enz. Microb. Technol., 22: 2-7.

Knorr D, Watkins TR, Carlson BL (1981). Enzymatic reduction of phytate in whole wheat breads. J. Food Sci., 46: 1866-1869.

Kostrewa A, Grueninger-Leitch F, D'Arcy A, Broger C, Mitchell D, vanLoon APGM (1997). Crystal structure of phytase from Aspergillus ficuum at 2.5 A resolution. Nat. Struct. Biol., 4: 185-190.

Laboure AM, Gagnon J, Lescure AM (1993). Purification and characterization of a phytase (myo-inositol-hexakisphosphate phosphohydrolase) accumulated in maize (Zea mays) seedlings during germination. Biochem. J., 295 (Pt 2): 413-419.

Lassen SF, Breinholt J, Ostergaard PR, Brugger R, Bischoff A, Wyss M, Fuglsang CC (2001). Expression, gene cloning and characterization of five novel phytases from four basidomycete fungi: Peniophora lycii, Agrocybe pediades, Ceriporia sp., and Trametes pubescens. Appl. Environ. Microbiol., 67: 4701-4707.

Lehman M, Kostrewa D, Wyss M, Brugger R, D'Arcy A, Pasamontes L, van Loon APGM (2000b). From DNA sequence to improved functionality: Using protein sequence comparisons to rapidly design a thermostable consensus phytase. Protein Eng., 13: 49-57.

Lehman M, Lopez-Ulibarri R, Loch C, Viarouge C, Wyss M, van Loon APGM (2000a). Exchanging the active site between phytases for altering the functional properties of the enzyme. Protein Sci., 9: 1866-1872.

Lehmann M, Pasamontes L, Lassen SF, Wyss M (2000c). The consensus concept for thermostability engineering of proteins. Biochim. Biophys. Acta, 1543: 408-415.

Lei XG, Ku PK, Miller ER, Yokoyama MT (1993a). Supplementing corn-soybean meal diets with microbial phytase linearly improves phytate phosphorus utilization by weanling pigs. J. Anim. Sci., 71: 3359-3367.

Lei XG, Ku PK, Miller ER, Yokoyama MT, Ullrey DE (1993b).

Supplementing corn-soybean meal diets with microbial phytase maximizes phytate phosphorus utilization by weanling pigs. J. Anim. Sci., 71: 3368-3375.

Lei XG, Stahl CH (2001). Biotechnological development of effective phytases for mineral nutrition and environmental protection. Appl. Microbiol. Biotechnol., 57: 474-481.

Li M, Osaki M, Honma M, Tadano T (1997). Purification and characterization of phytase induced in tomato roots under phosphorus-deficient conditions. Soil Sci. Plant Nutr., 43: 179-190.

Li MH, Robinson EH (1997). Microbial phytase can replace inorganic phosphorus supplements in channel catfish lactalurus punctatus. J. World Aquacult. Soc. 28: 402-406.

Li X, Wu Z, Li W, Yan R, Li L, Li J, Li Y, Li M (2007). Growth promoting effect of a transgenic Bacillus mucilaginosus on tobacco planting. Appl. Microbiol. Biotechnol., 74(5): 1120-1125.

Lim D, Golovan S, Forsberg CW, Jia Z (2000). Crystal structures of Escherichia coli phytase and its complex with phytate. Nat. Struct. Biol. 7: 108-113.

Lim D, Jia Z (2002). Heavy metal-mediated crystallization of Escherichia coli phytase and analysis of bridging interactions. Protein Pept. Lett., 9: 359-365.

Liu BL, Jong CH Tzeng YM (1998). Effect of immobilization on pH and thermal stability of Aspergillus ficuum phytase. Enz. Microb. Technol., 25: 517-521.

Lonnerdal B (2000). Dietary factors influencing zinc absorption. J. Nutr., 130(5S Suppl): 1378S-1383S.

Lung SC, Chan WL, Yip W, Wang L, Yeung EC, Lim BL (2005). Secretion of beta-propeller phytase from tobacco and Arabidopsis roots enhances phosphorus utilization. Plant Sci. 169(2): 341-349.

Maga JA (1982). Phytate: Its chemistry, occurrence, food interactions, nutritional significance, and methods of analysis. Crit. Rev. Food Sci. Nutr., 16: 1-48.

Mendoza C (2002). Effect of genetically modified low phytic acid plants on mineral absorption. Int. J. Food Sci. Technol., 37: 759-767.

Mitchell DB, Vogel K, Weimann BJ, Pasamontes L, van Loon APGM (1997). The phytase subfamily of histidine acid phosphatase: Isolation of genes for two novel phytases from fungi Aspergillus terreus and Myceliophthora thermophila. Microbiology, 143: 245-252.

Mudge SR, Frank WS, Richardson AE (2003). Root-specific and phosphate-regulated expression of phytase under the control of a phosphate transporter promoter enables Arabidopsis to grow on phytate as a sole P source. Plant Sci., 165(4): 871-878.

Mullaney EJ, Daly CB, Kim T, Porres JM, Lei XG, Sethumadhavan K, Ullah AHJ (2002). Site-directed mutagenesis of Aspergillus niger NRRL 3135 phytase at residue 300 to enhance catalysis at pH 4.0. Biochem. Biophys. Res. Commun., 297: 1016-1020.

Mullaney EJ, Daly CB, Ullah AH (2000). Advances in phytase research. Adv. Appl. Microbiol., 47: 157-199.

Mullaney EJ, Ullah AH (2003). The term phytase comprises several different classes of enzymes. Biochem. Biophys. Res. Commun., 312: 179-184.

Murugesan GS, Sathishkumar M, Swaminathan K (2005). Supplementation of waste tea fungal biomass as a dietary ingredient for broiler chicks. Bioresour. Technol., 96:1443-1448.

Nahm KH (2002). Efficient feed nutrient utilization to reduce pollutants in poultry and swine manure. Crit. Rev. Env. Sci. Technol., 32: 1-16.

Nakamura Y, Fukuhara H, Sano K (2000). Secreted phytase activities of yeasts. Biosci. Biotechnol. Biochem., 64: 841-844.

Nampoothiri KM, Tomes GJ, Roopesh K, Szakacs G, Nagy V, Soccol CR, Pandey A (2004) Thermostable phytase production by Thermoascus aurantiacus in submerged fermentation. Appl. Biochem. Biotechnol., 118: 205-214.

Nayini NR, Markakis P (1984). The phytase of yeast. Lebensm. Wiss. Technol., 17: 24-26.

NagashimaT, Tange T, Anazawa H (1999). Dephosphorylation of phytate by using Aspergillus niger phytase with a high affinity for phytate. Appl. Environ. Microbiol., 65(10): 4682-4684.

Nwokolo EN, Bragg DB (1977). Influence of phytic acid and crude fiber on the availability of minerals from four protein supplements in growing chicks. Can. J. Anim. Sci., 57: 475-480.

Pandey A, Szakacs G, Soccol CR, Rodriguez-Leon JA, Soccol VT

(2001). Production, purification and properties of microbial phytases. Bioresour. Technol., 77: 203-214.

Papatryphon E, Howell RA, Soares JJH (1999). Growth and mineral absorption by a striped bass morone sexatilis fed a plant feedstuff based diet supplemented with phytase. J. World Aquacult. Soc., 30: 161-173.

Pasamontes L, Haiker M, Henriquez-Huecas M, Mitchell DB,van Loon APGM (1997a). Cloning of the phytases from Emericella nidulans and the thermophilic fungus Talaromyces thermophilus. Biochim. Biophys. Acta, 1353: 217-223.

Pasamontes L, Haiker M, Wyss M, Tessier M, van Loon APGM (1997b). Gene cloning, purification, and characterization of a heat-stable phytase from the fungus Aspergillus fumigatus. Appl. Environ. Microbiol., 63: 1696-1700.

Pillai UP, Manoharan V, Lisle A, Li X, Bryden W (2009). Phytase supplemented poultry diets affect soluble phosphorus and nitrogen in manure and manure-amended soil. J. Environ. Qual., 38: 1700-1708.

Pomar C, Gagne F, Matte JJ, Barnett G, Jondreville C (2008). The effect of microbial phytase on true and apparent ileal amino acid digestibilities in growing-finishing pigs. J. Anim. Sci., 86: 1598-1608.

Powar VK, Jagannathan V (1982). Purification and properties of phytate-specific phosphatase from Bacillus subtilis. J. Bacteriol., 151: 1102-1108.

Quan CS, Difan S, Zhang LH, Wang YJ, Ohta Y (2001). Purification and properties of a phytase from Candida krusei WZ-001. J. Biosci. Bioeng., 94(5): 419-425.

Radcliffe JS, Zhang Z, Kornegay ET (1998). The effects of microbial phytase, citric acid, and their interaction in a corn-soybean meal-based diet for weanling pigs. J. Anim. Sci., 76: 1880-1886.

Ramachandaran S, Krishnan R, Nampoothiri KM, Szackacs G and Pandey A (2005). Mixed substrate fermentation for the production of phytase by Rhizopus spp. using oil cakes as substrates. Process Biochem., 40(5): 1749-1754.

Ramseyer L, Garling D, Hill G, Link J (1999). Effect of dietary zinc supplementation and phytase pre- treatment of soybean meal or corn gluten meal on growth, zinc status and zinc-related metabolism in rainbow trout, Oncorhynchus mykiss. Fish Physiol. Biochem., 20: 251-261.

Rao DE, Rao KV, Reddy TP, Reddy VD (2009). Molecular characterization, physicochemical properties, known and potential applications of phytases: An overview. Crit. Rev. Biotechnol., 29(2): 182-198.

Reddy NR, Sathe SK, Salunkhe DK (1982). Phytases in legumes and cereals. Adv. Food Res., 82: 1-92

Revy PS, Jondreville C, Dourmad JY, Nys Y (2005);. Assessment of dietary zinc requirement of weaned piglets fed diets with or without microbial phytase. J. Anim. Physiol. Anim. Nutr., 90: 50-59.

Richardson AE, Hadobas PA, Hayes JE (2001). Extracellular secretion of Aspergillus phytase from Arabidopsis roots enables plants to obtain phosphorous from phytate. Plant J., 25: 641-649.

Riley WW, Austic RE (1984). Influence of dietary electrolytes on digestive tract pH and acid-base status of chicks. Poult. Sci., 63: 2247-2251.

Robinson EH, Li MH, Manning BB (2002). Influence of dietary calcium, phosphorus, zinc and sodium phytate level on cataract incidence, growth and histopathology in juvenile chinook salmon (Oncorhynchus tshawytscha).Comparison of microbial phytase and dicalcium phosphate for growth and bone mineralization of pond-raised channel catfish, Ictalurus punctatus. J. Appl. Aquacult., 12: 81-88.

Rodehutscord M, Pfeffer E (1995). Effects of supplemental microbial phytase on phosphorus digestibility and utilization in rainbow trout (Oncorhynchus mykiss). Water Sci. Technol., 31: 143-147.

Rodriguez E, Mullaney EJ, Lei XG (2000). Expression of the Aspergillus fumigatus phytase gene in Pichia pastoris and characterization of the recombinant enzyme. Biochem. Biophys. Res. Commun., 268: 373-378.

Roopesh K, Ramachandran S, Nampoothiri KM, Szakacs G, Pandey A (2006). Comparison of phytase production on wheat bran and oilcakes in solid-state fermentation by Mucor racemosus. Bioresour. Technol., 97(3): 506-511.

Sajjadi M, Carter CG (2004). Effect of phytic acid and phytase on feed intake, growth, digestibility and trypsin activity in atlantic salmon

(*Salmo salar*, L.) Aquacult. Nutr., 10: 135-142.

Sandberg AS, Hulthen LR, Turk M (1996). Dietary *Aspergillus niger* phytase increases iron absorption in humans. J. Nutr., 126: 476-480.

Sano L, Fukuhara H, Nakamura Y (1999). Phytase of the yeast *Arxula adeninivorans*. Biotechnol. Lett., 21: 33-38.

Satyanarayana T, Vohra A(2003). A synergistic feed composition to enhance phosphorous availability, assimilation and retention in non-ruminants. Indian Patent No. 197593

Schafer A, Koppe WM, Meyer-Burgdorff KH, Gunther KD (1995). Effects of a microbial phytase on the utilization of native phosphorus by carp in a diet based on soybean meal. Water Sci. Technol., 31: 149-155.

Scott JJ, Loewus FA (1986). A calcium activated phytase from pollen of *Lilium longiflorum*. Plant Physiol., 82: 333-335.

Segueilha L, Lambrechts C, Boze H, Moulin G, Galzy P (1992). Purification and properties of phytase from *Schwanniomyces castellii*. J. Ferment. Bioeng., 74: 7-11.

Selle PH, Ravindran R (2007). Microbial phytase in piggery nutrition. Anim. Feed Sci. Technol., 135: 1-41.

Selle PH, Ravindran R (2008). Microbial phytase in poultry nutrition. Livestock Sci., 113: 99-122.

Sheppy C (2001). The current feed enzyme market and likely trends. In MR Bedford (ed) Enzymes in farm animal nutrition. CABI, Publishing, Oxon, U.K. pp.1-10.

Shieh TR, Ware JH (1968). Survey of microorganism for the production of extracellular phytase. Appl. Microbiol., 16: 1348-1351.

Shimizu M (1992). Purification and characterization of phytase from *Bacillus subtilis* (natto) N-77. Biosci. Biotechnol. Biochem., 56: 1266-1269.

Shimizu M (1993). Purification and characterization of phytase and acid phosphatase produced by *Aspergillus oryzae* K1. Biosci. Biotechnol. Biochem., 57: 1364-1365.

Shin S, Ha NC, Oh BC, Oh TK, Oh BH (2001). Enzyme mechanism and catalytic property of β-propeller phytase. Structure, 9: 851-858.

Singh B, Kaur, P, Satyanarayana, T (2006). Fungal phytases for improving the nutritional status of foods and combating environmental phosphorus pollution. In AK Chauhan, and A Verma (eds.) Microbes: Health and Environment. IK International Publishers, New Delhi, pp. 289-326.

Singh B, Satyanarayana T (2006a). A marked enhancement in phytase production by a thermophilic mould *Sporotrichum thermophile* using statistical designs in a cost-effective cane molasses medium. J. Appl. Microbiol. 101: 344-352.

Singh B, Satyanarayana T (2006b). Phytase production by thermophilic mold *Sporotrichum thermophile* in solid-state fermentation and its application in dephytinization of sesame oil cake. Appl. Biochem. Biotechnol., 133: 239-250.

Singh B, Satyanarayana T (2008a). Improved phytase production by a thermophilic mould *Sporotrichum thermophile* in submerged fermentation due to statistical optimization. Bioresour. Technol., 99: 824-830.

Singh B, Satyanarayana T (2008b). Phytase production by a thermophilic mould *Sporotrichum thermophile* in solid state fermentation and its potential applications. Bioresour. Technol., 99: 2824-2830.

Singh B, Satyanarayana T (2008c). Phytase production by *Sporotrichum thermophile* in a cost-effective cane molasses medium in submerged fermentation and its application in bread. J. Appl. Microbiol., 105: 1858-1865.

Singh B, Satyanarayana T (2009). Characterization of a HAP-phytase from a thermophilic mould *Sporotrichum thermophile*. Bioresour. Technol., 100: 2046-2051.

Singh B, Satyanarayana T (2010). Plant growth promotion by an extracellular HAP-phytase of a thermophilic mold *Sporotrichum thermophile*. Appl. Biochem. Biotechnol., 160(5): 1267-1276.

Siren M (1986). New *myo*-inositol triphosphoric acid isomer. Pat SW 052950.

Sreeramulu G, Srinivasa DS, Nand K, Joseph R (1996). *Lactobacillus amylovorus* as a phytase producer in submerged culture. Lett Appl Microbiol., 23: 385-388.

Storebakken T, Shearer KD, Roem AJ (1998). Availability of protein, phosphorus and other elements in fish meal, soy-protein

concentrate and phytase-treated soy-protein-concentrate-based diets to atlantic salmon, *Salmo salar*. Aquacult. Nutr., 161: 365-379.

Sunita K, Kim YO, Lee JK, Oh TK (2000). Statistical optimization of seed and induction conditions to enhance phytase production by recombinant *Escherichia coli*. Biochem. Eng. J., 5: 51-56.

Sutardi M, Buckle KA (1988). Characterization of extra and intracellular phytase from *Rhizopus oligosporus* used in tempeh production. Int. J. Food Microbiol., 6: 67-69.

Tambe SM, Kakli SG, Kelkar SM, Parekh LJ (1994). Two distinct molecular forms of phytase from *Klebsiella aerogenes*; evidence for unusually small active enzyme peptide. J. Ferment. Bioeng., 77: 23-27.

Tomschy A, Tessier M, Wyss M, Brugger R, Broger C, Schnoebelen L (2000a). Optimization of the catalytic properties of *Aspergillus fumigatus* phytase based on the three-dimensional structure. Protein Sci., 9: 1304-1311.

Tomschy A, Wyss M, Kostrewa D, Vogel K, Tessier M, Hofer S et al. (2000b). Active site residue 297 of *Aspergillus niger* phytase critically affects the catalytic properties. FEBS Lett., 472: 169-172.

Tyagi PK, Tyagi PK, Verma SVS (1998). Phytate phosphorus content of some common poultry feed stuffs. Indian J. Poult. Sci., 33: 86-88.

Tye AJ, Siu FK, Leung TY and Lim BL (2002). Molecular cloning and the biochemical characterization of two novel phytases from *B. subtilis* 168 and *B. licheniformis*. Appl. Microbiol. Biotechnol., 59(2-3): 190-197.

Ullah AHJ (1988). *Aspergillus ficuum* phytase: Partial primary structure, substrate selectivity, and kinetic characterization. Prep. Biochem., 18: 459-471.

Ullah AHJ, Gibson DM (1987). Extracellular phytase (EC 3.1.3.8) from *Aspergillus ficuum* NRRL 3135: purification and characterization. Prep. Biochem., 17: 63-91.

Usmani N, Jafri AK (2002). Influence of dietry phytic acid on growth, conversion efficiency and carcass composition of mrigal *Cirrhinus mrigala* (hamilton) fry. J. World Aquacult. Soc., 33: 199-204.

Van de Velde F, Konemann L, van Rantwijk F, Sheldon RA (2000). The rational design of semisynsheticperoxidascs. Biotechnol. Bioeng., 67: 87-96.

Van Weerd JH, Khalaf KHA, Aartsen FJ, Tijssen PAT (1999). Balance trials with African catfish *Clarias gariepinus* fed phytase-treated soybean meal-based diets. Aquacult. Nutr., 5: 135-142.

Vats P, Banerjee UC (2004). Production studies and catalytic properties of phytases (*myo*-inositolhexakisphosphate phosphohydrolases): An overview. Enz. Microb. Technol., 35: 3-14.

Vats P, Banerjee UC (2005). Biochemical characterisation of extracellular phytase (*myo*-inositol hexakisphosphate phosphohydrolase) from a hyper-producing strain of *Aspergillus niger* van Teighem. J. Ind. Microbiol. Biotechnol., 32: 141-147.

Vielma J, Ruohonen K, Gabaudan J, Vogel K (2004). Top-spraying soybean meal-based diets with phytase improves protein and mineral digestibilities but not lysine utilization in rainbow trout, *Oncorhynchus mykiss* (walbaum). Aquacult. Res., 35: 955-964.

Vohra A, Rastogi SK, Satyanarayana T (2006). Amelioration in growth and phosphate assimilation of poultry birds using cell-bound phytase of *Pichia anomala*. World J. Microbiol. Biotechnol., 22: 553-558.

Vohra A, Satyanarayana T (2002). Statistical optimization of the medium components by response surface methodology to enhance phytase production by *Pichia anomala*. Process Biochem., 37: 999-1004.

Vohra A, Satyanarayana T (2003). Phytases: Microbial sources, production, purification, and potential biotechnological applications. Crit. Rev. Biotechnol., 23: 29-60.

Walz OP, Pallauf J (2002). Microbial phytase combined with amino acid supplementation reduces P and N excretion of growing and finishing pigs without loss of performance. Int. J. Food Sci. Technol., 37: 835-848.

Wise A (1983). Dietry factors determining the biological activities of phytase. Nutr. Abstr. Rev., 53: 791-806.

Wodzinski RJ, Ullah AHJ (1996). Phytase. Adv. Appl. Microbiol., 42: 263-301.

Wyss M, Brugger R, Kronenberger A, Remy R, Fimbel R, Oesterhelt O (1999b). Biochemical characterization of fungal phytases (*myo*-inositolhexakisphosphate-phosphohydrolases): Catalytic properties.

Appl. Environ. Microbiol., 65: 367-373.

Wyss M, Pasamontes L, Friedlein A, Remy R, Tessier M, Kronenberger A (1999a). Biophysical characterization of fungal phytases (*myo-inositolhexakisphosphate-phosphohydrolases*): Molecular size, glycosylation pattern and engineering of proteolytic resistance. Appl. Environ. Microbiol., 65: 359-366.

Xiang T, Liu Q, Deacon AM, Koshy M, Kriksunov IA, Lei XG, Hao Q, Thiel DJ (2004). Crystal structure of a heat-resilient phytase from *Aspergillus fumigatus*, carrying a phosphorylated histidine. J. Mol. Biol., 339: 437-445.

Xiao K, Harrison MJ, Wang ZY (2005). Transgenic expression of a novel *M. truncatula* phytase gene results in improved acquisition of organic phosphorus by *Arabidopsis*. Planta, 222(1): 27-36.

Yadav RS, Tarafdar JC (2003). Phytase and phosphatase producing fungi in arid and semi-arid soils and their efficiency in hydrolyzing different organic P compounds. Soil Biol. Biochem., 35(6): 745-751.

Yamamoto S, Minoda Y, Yamada K (1972). Chemical and physicochemical properties of phytase from *Aspergillus terreus*. Agric. Biol. Chem., 36: 2097-2103.

Yan W, Reigh RC (2002). Effects of fungal phytase on utilization of dietary proteins and minerals and dephosphorylation of phytic acid in the alimentary tract of channel catfish, *Lactalurus punctatus* fed an all plant-protein diet. J. World Aquacult. Soc., 33: 10-22.

Yang WJ, Matsuda Y, Inomata M, Nakagava H (1991). Development and dietary induction of 90 kda subunits of rat intestinal phytase. Biochem. Biophys. Acta, 1075: 83-87.

Yanke LJ, Bae HD, Selinger LB, Cheng KJ (1998). Phytase activity of anaerobic ruminal bacteria. Microbiology, 144 (Pt 6): 1565-1573.

Yi Z, Kornegay ET, Denbow DM (1996). Effect of microbial phytase on nitrogen and amino acid digestibility and nitrogen retention of turkey poults fed corn-soybean meal diets. Poult. Sci., 75: 979-990.

Yip W, Wang L, Cheng C, Wu W, Lung S, Lim BL (2003). The introduction of a phytase gene from *Bacillus subtilis* improved the growth performance of transgenic tobacco. Biochem. Biophys. Res. Commun., 310: 1148-1154.

Yoon SJ, Choi YJ, Min K, Cho KK, Kim JW, Lee SC (1996). Isolation and identification of phytase producing bacterium, *Enterobacter* sp. 4 and enzymatic properties of phytase enzyme. Enz. Microb. Technol., 18: 449-454.

12

DNA microarrays and their applications in medical microbiology

Chijioke A. Nsofor

Department of Biotechnology, Federal University of Technology, Owerri, Imo State, Nigeria.

Rapid diagnosis and treatment of disease is often based on the identification and characterization of causative agents derived from phenotypic characteristics. This can be laborious and time consuming, often requiring many skilled personnel and a large amount of lab space. However, the introduction of nucleic acid amplification techniques into molecular biology has transformed the laboratory detection of pathogens. The progression of the molecular diagnostic revolution currently relies on the ability to efficiently and accurately offer multiplex detection and characterization for a variety of infectious disease pathogens. DNA microarray analysis has the capability to offer robust multiplex detection. Multiple microarray platforms exist, including printed double-stranded DNA and oligonucleotide arrays, in situ-synthesized arrays, high-density bead arrays, electronic microarrays, and suspension bead arrays. The aim of this paper was to review DNA microarray technology, highlighting two major types: the oligonucleotide-based array and the PCR product-based array. Although, the use of microarrays to generate gene expression data has become routine, applications pertinent to microbiology continue to rapidly expand. This review highlights uses of microarray technology that impact diagnostic microbiology, including the detection and identification of pathogens, determination of antimicrobial resistance, epidemiological strain typing, and determination of virulence factors.

Key words: DNA microarray, applications, microbiology.

INTRODUCTION

The large-scale genome sequencing effort and the ability to immobilize thousands of DNA fragments on a surface, such as coated glass slide or membrane, have led to the development of DNA microarray technology (Cassone et al., 2006). An entire microbial genome can be easily represented in a single array, making it feasible to perform genome-wide analysis (Ye et al., 2006; Akondi and Lakshmi, 2013). The two common applications of DNA microarray technology in molecular biology are the exploration of genome-wide transcriptional profiles and the measurement of the similarities or differences in genetic contents among different microbes (Peterson et al., 2010). DNA microarray technology is being used to study many bacterial species ranging from standard laboratory strains and pathogens to environmental isolates

(Murakami et al., 2002).

DNA microarrays are basically a miniaturized form of dot blot, but in a high-throughput format. There are two major types of DNA microarrays; one is the oligonucleotide-based array and the other is the PCR product-based array (Panicker et al., 2004). A DNA microarray experiment consists of array fabrication, probe preparation, hybridization and data analysis (Call et al., 2001). Although the basic array technology is the same, there are fundamental differences in its application to prokaryotes and eukaryotes. For example, total RNA is usually labeled for a bacterial array experiment, while poly(A) RNA is often used for eukaryotic arrays.

Detection of single bacterial genes (for example antibiotic resistance genes or species-specific genes) in diagnostics and in epidemiological studies is typically carried out by PCR, whereas DNA microarrays have been developed to perform a large number of different hybridization experiments simultaneously on a single membrane or glass substrate. They are well-suited to comprehensively investigate and quantitatively compare the expression levels of a large number of genes, but they can also be easily used in qualitative studies to detect selected DNA sequences (Call et al., 2003a; Call et al., 2003b; Perreten et al., 2005).

Simply defined, a microarray is a collection of microscopic features (most commonly DNA) which can be probed with target molecules to produce either quantitative (gene expression) or qualitative (diagnostic) data (Miller and Tang, 2009; Yu et al., 2013). Although other types of microarrays exist, such as protein microarrays (Lopez and Pluskal, 2003; MacBeath, 2002), this review will focus on DNA microarrays.

With the availability of complete genome sequences of many microorganisms, the DNA microarray technology has become a very powerful tool to explore global gene expression profiles and to measure genome-wide differences in genetic contents. Since only the abundance of transcripts or presence and absence of DNA regions are measured with a DNA microarray, interpretation of the array data can be difficult in the absence of other supporting evidence. This is especially true when the physiological events are not well studied. In addition, it is not easy to sort out secondary effects caused by mutations, expression of certain genes, and different growth conditions.

As a result, the greatest impact of this technology will not be realized until it is combined with other high-throughput genomic methods, biochemistry, genetics and physiology. Analysis of a systematically perturbed metabolic network in yeast clearly demon-strates the power of an integrated approach to build, test, and refine a model of a cellular metabolic pathway.

PCR PRODUCT-BASED DNA MICROARRAYS

Primer design

The first step of DNA microarray construction for microbes with known genome sequences is the design of primers to amplify specific regions of interest. In a bacterial genome, there are open reading frames (ORFs) and intergenic regions. ORFs are potential protein coding regions identified by computer analysis or experimentally. The intergenic regions include promoters or regions encoding small RNA molecules which may have regulatory functions (Lease and Belfort, 2000).

An ideal array should contain both ORFs and intergenic regions, although most of the current PCR product-based arrays only contain ORFs. Primer design for the whole genome can be carried out with a computer program

such as PrimeArray (Raddatz et al., 2001). PrimeArray is specifically designed to compute the oligonucleotide primer pairs for genome-scale gene amplification. The simplest way to design primers is to use the beginning and ending regions of a specific ORF. If an ORF is too long (>3.0 kb), primers can be designed to reduce the size of the PCR product. Since there are repeated regions in a genome, amplification of unique sequences is necessary to avoid cross-hybridization. Before proceeding to the making of all the primers for the whole genome, it is important to test a portion of the primer sets to ensure the primer quality and desirable amplification results.

PCR amplification

The goal of the whole genome PCR amplification is to achieve the highest success rate and yield in a high-throughput manner. Conditions for PCR amplification are initially optimized with a few 96-well plates. The optimized conditions vary with genomes since they could have different GC contents and secondary structures. After completion of the first round of PCR for all the plates, further optimization or redesign of primers for failed reactions is necessary.

To reduce chromosomal DNA contamination and increase yield, some arrays are constructed with second round PCR products (Wei et al., 2001). This practice may not be necessary as long as the amount of genomic DNA template remains low. In fact, the second round of amplification could increase the number of reactions that have multiple bands.

PCR product purification

To remove unincorporated nucleotides and primers, it is recommended to purify the PCR product. PCR purification can be done in either 96- or 384-plate format by ethanol precipitation or by commercial purification systems (Millipore, Qiagen, Whatman). The 96-well multiscreen filter plates from Millipore have been found to give excellent DNA product recovery with no significant contamination at a relatively low cost (Hegde et al., 2000).

If the goal is to perform high-throughput analysis for gene discovery, purification of PCR products is not necessary. The binding efficiency of PCR products and the quality of the array may be slightly compromised, but there is a significant saving in time and money

Spotting

The purified PCR products are spotted onto membranes or coated glass slides. DNA microarrays on coated glass slides are prepared by printing DNA products with high-speed robots. The arraying robots can be custom made or purchased from commercial sources. The common problems associated with glass slides are spot morphology, high background, and batch variability. Although there are no perfect slides, aminosilane-coated slides (Corning, Telechem, Amersham Pharmacia Biotech) and poly-L- lysine-coated slides are commonly used. The PCR products are resuspended in an appropriate solution before spotting.

The two common solutions are high-salt buffer (3×SSC) and 50% dimethyl sulfoxide (DMSO). One of the factors in determining which chemistry to use is the type of slides. For CMT-GAPS aminosilane-coated glass microscope slides, DMSO has been found to offer many advantages (Hegde et al., 2000). It is hygroscopic and has a low vapor pressure, which allows DNA to be stored for long periods of time without significant evaporation.

After spotting, it is often necessary to check the quality of the slides and spotting results. This can be routinely performed with SYBR green staining (Battaglia et al., 2000) by hybridizing the array with a Cy-labeled genomic DNA or random 9-mer sequence. The SYBR green dye can be used to measure the amount of DNA in each spot and the spotting integrity, while the genomic DNA hybridization is a better indicator of the hybridization background and the quality of the slides (Ye et al., 2001).

Total RNA labeling

The cDNA probes for array hybridization are synthesized from total RNA by reverse transcriptase. The nucleotides can be labeled with radioisotopes (such as P^{32} orthophosphate) or fluorescent markers. Either random primers or specific primers are used in the reaction. The amount of total bacterial RNA used varies with the organism, stage of growth, type of array and labeling method. Typically, 7 to 15 µg of total RNA in combination with 6 µg of random hexamers generally yields good labeling efficiency and reasonable signal intensity with Cy5 or Cy3 fluor for arrays on glass slides. The incorporation efficiencies of Cy5- and Cy3-labeled nucleotides are not equal.

A two-step labeling procedure using aminoallyl-dUTP is gaining popularity due to the increased labeling efficiency and reduction in dye bias and cost. In this two-step procedure, primary aliphatic amino groups are first incorporated during cDNA synthesis. In the second step, the monofunctional N-hydroxylsuccinimide-activated fluorescent dye (Cy3, Cy5) is coupled to cDNA by chemical reaction with the amino functional groups.

Since the substrate for the reverse transcriptase is identical for all the samples and is less bulky, the two-step method could yield equivalent molarities of labeled probe with higher efficiency than the one-step labeling procedure (Call et al., 2001b). After labeling, it is neces-

sary to remove the unincorporated dyes to reduce background. This can be achieved by conventional DNA purification methods. To insure the probe quality, the labeling efficiency for Cy3 and Cy5 needs to be calculated. The calculation is based on the extinction coefficient with the following formula: (O.D.$_{550}$×dilution factor×total volume)/0.15 for Cy3, or (O.D.$_{650}$×dilution factor×total volume)/0.25 for Cy5. The total amount of incorporated dye obtained is in pmol (Ye et al., 2001).

It is worthwhile to point out that there are other fluorescent markers, for example, Alexa fluor from Molecular Probes (Nsofor et al., 2013) or labeling methods that are available or being developed.

Genomic DNA labeling

Genomic DNA probes can be used for normalization, slide quality control, antibiotics resistance gene profile and comparative genomic studies. Genomic DNA can be labeled by nick translation or by random priming with the Klenow fragment of DNA polymerase. Direct chemical labeling of nucleic acids is also a commonly used method.

For example, the Universal Linkage System (ULS) is a technique for binding any marker group or label to DNA and RNA (Kreatech Diagnostics, Amsterdam, The Netherland). For the random priming method, the genomic DNA samples are first sheared by mechanical means such as nebulization or sonication. DNA fragments within 1-3 kb are collected and labeled by either the one- or two-step labeling procedure. A labeling reaction with 0.5-2 µg of genomic DNA often yields enough probes for a single hybridization experiment.

Hybridization and data acquisition

The amount of probe used for hybridization depends on the array format and labeling method. For arrays on glass slides, a reasonable signal to background ratio can be obtained with probes containing 100-200 pmol of incorporated fluorescent dye. In a typical hybridization reaction, equal amounts of Cy3- and Cy5-labeled probes based on the incorporated dye concentration are combined (Ye et al., 2000). To correct for the difference in labeling efficiency of Cy3 and Cy5, a dye swap procedure is used. In other words, the two samples are labeled with opposite dyes and the resulting probes are hybridized to two different slides.

The overall procedure for a PCR product-based DNA microarray hybridization is basically the same as for a Southern blot except for a few modifications. Before hybridization, most glass slides need to be treated to block or inactivate the non-specific binding sites. The procedure employed depends on the slide type and spotting chemistry. For aminosilane-coated slides, a pre-

prehybridization solution containing 1% BSA, 5×SSC and 0.1% SDS has been found to be effective (Hegde et al., 2000). Usually, the hybridization solution containing the probe is placed onto the array and covered with a cover slip. The glass slide is then placed in a humidified chamber. The temperature of hybridization and washing conditions depend on the GC content of the organism. Bacteria with high GC content require a more stringent washing condition in order to minimize non-specific binding.

After hybridization, the signal intensities of all the spots on a glass slide are captured by scanners (GSI Lumonics, Molecular Dynamics, Genomic Solutions, Axon, and others). For membrane arrays hybridized with P^{32}-labeled probes, a phosphor imaging system (Molecular Dynamics) can be used. Processing of array images involves three steps: spot finding, quantification, and background estimation. These steps are performed with the software provided by the scanner vendors or by other sources. In Nsofor et al., 2013), Applied Precision arrayWoRx scanner for image capturing and Applied Precision SoftWoRx Tracker software for processing and analysis of array images was used.

Data normalization

There are several systematic variables in a DNA microarray experiment that can affect the measurement of mRNA levels making direct comparisons. Sources of the variations include inherent errors from sample handling, slide to slide variation, difference in labeling or hybridization efficiency, and variations during image analysis. These differences are not due to the actual changes in gene expression levels. Normalization is a process of minimizing these variations, establishing a common base for comparison. Normalization can be done within the slide to adjust the dye incorporation efficiency, between the two slides for dye swap experiments and across slides for replicates of the same experiment (Yang et al., 2001). After normalization, the ratio is calculated for each spot on the slide.

OLIGONUCLEOTIDE-BASED DNA MICROARRAYS

Instead of using PCR products, DNA microarrays can be constructed with short oligonucleotides. In the Affymetrix system, Oligonucleotide is synthesized *in situ* on a derivatized glass surface using a combination of photolithography and combinatorial chemistry. The *Escherichia coli* Genome Array system by Affymetrix uses a protocol for the enrichment and labeling of the non-polyadenylated mRNA of prokaryotes. The mRNA is directly labeled so that it represents the natural distribution of RNA species within the sample. No reverse transcription or amplification steps are involved. On the other hand, the enrichment procedure could also poten-

tially alter the mRNA population.

Selinger et al. (2000) have reported the application of a 30-base pair resolution *E. coli* genome array for RNA expression analysis using the Affymetrix system. This array contains on average one 25-mer oligonucleotide probe per 30 base pairs over the entire genome, with one every six bases for the intergenic regions and every 60 bases for 4290 ORFs. Twofold concentration differences can be detected at levels as low as 0.2 mRNA copies per cell. The array also permits the investigation of intergenic regions of the genome.

A system using one optimized 70-mer probe per gene has been developed by Operon Technologies (http://www.operon.com). It is similar to a PCR-based microarray except that no amplification is required. Another type of array has also been fabricated by immobilizing oligonucleotides in a polyacrylamide gel (Proudnikov et al., 1998).

Overall, the oligonucleotide-based DNA microarray has many advantages: (i) no amplification is required, and thus, there are no failed amplifications. It is difficult to obtain a high success rate of amplification for micro-organisms that contain high GC content or complex DNA structure; (ii) there are fewer chances for contamination due to non-specific amplification and mishandling; (iii) There is a reduction in cross-hybridization and an increase in the differentiation of overlapping genes or highly homologous regions. (iv) it is easier to normalize concentrations of oligonucleotides; (v) high-density oligonucleotide arrays enable high coverage of the genome, and thus, allow a precise mapping of the transcriptional regions and identification of alternative promoters. However, the cost of making long oligonucleotides is high. There are a limited number of whole microbial genome arrays that are available in the Affymetrix system (Proudnikov et al., 1998).

DNA microarray database

DNA microarray experiments generate vast amounts of data. The goal of the array database is to allow researchers to retrieve, analyze and visualize the array data. It can also serve as a means to link array data to other information, such as DNA and protein sequences, protein expression profiles and cellular function. In addition, an array database will make it possible to compare gene expression profiles across microbial species.

The EcoReg website (http://www.genomics.lbl.gov/~ecoreg/) was designed to be a repository of primary data (analogous to Genbank) for bacteria transcriptional control processes. It is a bioinformatics database project to facilitate improved understanding and modeling of the transcriptional control of bacteria gene expression.

There are commercial software packages available to facilitate database construction. For example,

GeneSpring Suite™ from Silicon Genetics includes a web database and other tools for data sharing. The SCOUT platform developed by Lion Bioscience is based on a sequence retrieval system (SRS). It integrates the array data analysis package, arraySCOUT, to other analysis tools and databases. They include bioSCOUT for automated gene and genome analysis, pathSCOUT for metabolic pathway analysis, and πSCOUT for analysis of protein-protein interaction. This integration also allows a sharing of data on an enterprise-wide level (Call et al., 2001).

Validation of DNA microarray data

There are three major sources of errors associated with the application of DNA microarrays: initial construction of the print-ready plates, array experiment, and data analysis. During the construction of the print-ready plates, there are handling issues such as plate transfer, which could lead to cross-contamination or other types of mishandling. During PCR amplification, the presence of multiple bands can lead to false results (Call et al., 2003).

Mistakes can also come from the generation of the final gene list, leading to a mismatch between the clone and spot position on the final array. As a result, it is advisable to check the hybridization results against well-characterized genes and internal controls when a new set of spotting plates is made.

The common experimental errors can stem from uneven hybridization, inefficient labeling and problems during RNA preparations. During data analysis, it is difficult to have an accurate ratio calculation when the signal intensity is low. When the fold of induction is low, the result can be misinterpreted. All these issues lead to the need for validation of the array experiment. Supporting evidence can be obtained from enzymatic assays, reporter gene systems, and other direct RNA quantification methods such as quantitative PCR (qPCR), nuclease protection assay, and primer extension. qPCR (Applied   Biosystems, http://www.appliedbiosystems.com/) offers a high-throughput advantage and can be an excellent tool to supplement array analysis. As with any qPCR reaction, proper controls need to be implemented and care must be taken to avoid DNA contamination in RNA samples.

When the results of a DNA microarray experiment were compared with those obtained from a Northern blot, the sensitivity of the DNA microarray was found to be slightly less than that of a Northern blot analysis (Taniguchi et al., 2001). In most genes, the data obtained by the two methods were consistent.

However, in 4 of 46 genes compared, the DNA microarray failed to detect the expression changes that were revealed by the Northern blot. The data demonstrated that DNA microarrays provide quantitative data that are comparable to the Northern blot in general.

APPLICATIONS OF DNA MICROARRAYS IN MEDICAL MICROBIOLOGY

Microbial detection and identification

Perhaps the most promising area in applying DNA microarray technology in medical microbiology is its use for simultaneous assessment of large numbers of microbial genetic targets (Stover et al., 2003; Gentry and Zhou, 2006). Specific microbial gene amplification by either a broad-range or a multiplex PCR prior to micro-array analysis enhances test sensitivity. The amplification of universal microorganism targets by broad-range PCR followed by sequencing analysis has been considered a standard procedure (Tang et al., 1998); however, microarrays have emerged as potential tools for bacterial detection and identification given their high parallelism in screening for the presence of a wide diversity of genes. The most commonly used gene targets have been the 16S bacterial and 28S fungal and intergenic transcribed spacers (ITSs) in rRNA genes, and microarray technology has been incorporated to compensate for the time-consuming sequencing identification procedure (Tang et al., 1998). An oligonucleotide microarray targeting the 16S rRNA gene was developed for the detection of a panel of 40 predominant human intestinal bacterial pathogens in human fecal samples (Wang et al., 2004; Wapner et al., 2012). Assays using broad-range PCR incorporated with microarrays have been shown to allow rapid bacterial detection and identification with positive blood cultures (Anthony et al., 2000; Marlowe et al., 2003). A similar procedure was developed and used for the rapid diagnosis of bloodstream infections caused by common bacterial pathogens in the pediatric and general populations (Shang et al., 2005; Cleven et al., 2006). PCR amplification, in combination with an oligonucleotide microarray, was used to identify *Bacillus anthracis* based on the rRNA ITS region (Nubel et al., 2004; Roh et al., 2012). Several studies reported the use of microarrays to identify pathogenic yeasts and molds by targeting the ITS regions in fungal rRNA genes (Hsiao et al., 2005; Huang et al., 2006). In another study, a DNA microarray was established to detect and identify 14 commonly encountered fungal pathogens in clinical specimens collected from neutropenic patients (Spiess et al., 2007; McLoughlin, 2011).

The key for broad-range PCR amplification followed by microarray identification to work is to target the right gene. It is critical to use a gene "broad" enough so that most related microorganisms can be covered in one amplification reaction. On the other hand, the targeted gene should possess enough polymorphic information to supply sufficient discriminatory power to differentiate and characterize related microorganisms. Degenerate primer sets can be designed to increase the coverage of relatively variable genes. Other universal bacterial genes have been used to detect and identify organisms using microarrays.

For mycobacterial detection and identification, the *gyrB*, *rpoB*, and *katG* genes have been targeted by using microarrays (Fukushima et al., 2003). Microarrays targeting the 23S rRNA and *gyrB* genes for bacterial detection and identification using clinical specimens have been described (Kakinuma et al., 2003; Kostic et al., 2007). In addition to bacterial and myco-bacterial organisms, microarrays following broad-range PCR amplification have been used to detect and identify fungal, parasitic, and viral pathogens (Diaz and Fell, 2005; Korimbocus et al., 2005; Wang, et al., 2005).

Microarrays have also been incorporated with multiplex PCR amplification for the simultaneous detection and identification of a panel of microbial pathogens in a single reaction. Khodakov et al. (2008) described a novel microarray-based approach for the simultaneous identification and quantification of human immunodeficiency virus type 1 (HIV-1) and hepatitis B and C viruses in donor plasma specimens (Khodakov et al., 2008). A microarray technique for the detection and identification of enteropathogenic bacteria at the species and subspecies levels was developed, covering pathogenic *E. coli*, *Vibrio cholerae*, *Vibrio parahaemolyticus*, *Salmonella enterica*, *Campylobacter jejuni*, *Shigella* spp., *Yersinia enterocolitica*, and *Listeria monocytogenes* (You et al., 2008).

A microarray-based multiplexed assay was developed to detect foot-and-mouth disease virus with rule-out assays for two other foreign animal diseases and four domestic animal diseases that cause vesicular or ulcerative lesions that are indistinguishable from those of foot-and-mouth disease virus infection of cattle, sheep, and swine (Lenhoff et al., 2008). Bøving et al. (2009) developed a novel multiplex PCR with product detection by the Luminex suspension array system covering a panel of bacterial and viral pathogens causing meningitis. This system detected and identified nine microorganisms including *Neisseria meningitidis*, *Streptococcus pneumoniae*, *E. coli*, *Staphylococcus aureus*, *L. monocytogenes*, *Streptococcus agalactiae*, herpes simplex virus types 1 and 2, and varicella zoster virus directly from cerebrospinal fluid (Bøving *et al.*, 2009). The ResPlex I system, manufactured by Qiagen (Valencia, CA), was used to detect a panel of bacterial pathogens related to community-acquired pneumonia from tracheal aspirates collected from hospitalized antibiotic-treated children. The data indicated that the ResPlex I system significantly enhanced the pathogen-specific diagnosis of community-acquired pneumonia in children (Deng et al., 2009).

Comparative genomics and microbial typing

Genomic hybridization of a whole genome array detects the presence or absence of similar DNA regions in other microorganisms, allowing genome-wide comparison of their genetic contents. It is an effective way to conduct a comparative genomic study in the absence of complete genome sequences. DNA microarrays have been used to investigate genome differences between *M. tuberculosis*, *M. bovis* and the various Bacille Calmette-Guérin (BCG) daughter strains (Behr et al., 1999) within the species of *H. pylori* and *M. tuberculosis* (Kato-Maeda et al., 2001) among different isolates of *S. pneumoniae* (Hakenbeck et al., 2001). These studies show that DNA microarrays can facilitate a better understanding of the genetic differences between closely related organisms, providing useful information for the identification of virulence factors, exploration of molecular phylogeny, improvement of diagnostics and development of vaccines.

DNA microarray technology is also an excellent way to identify changes in genetic content of the same strain after long-term adaptation or strain optimization. After adaptation for 2000 generations to a stressful high temperature of 41.5°C, *E. coli* was examined on a genome-wide scale for duplication/deletion events by using DNA arrays (Riehle et al., 2001). A total of five duplication and deletion events were detected, providing additional evidence for the idea that gene duplication plays an integral role in adaptation, specifically as a means for gene amplification.

Numerous studies that use DNA microarrays for microorganism typing by taking advantage of its simultaneous detection of a variety of genomes have been reported. The accurate identification and prompt typing of pathogens causing diarrheal diseases are critical for directing clinical intervention, including appropriate antibiotic administration, and facilitating epidemiological investigations. Microarray-based approaches along with other genetic approaches that can be used to support or replace the classical serotyping method for several conventional diarrhea bacterial pathogens have already been offered. The use of microarrays has included *Salmonella*, *Helicobacter*, and *Campyloba cter* species (Fitzgerald et al., 2007; Salama et al., 2000; Volokhov et al., 2003 Willse et al., 2004). PCR followed by a microarray hybridization step has been used for the detection and typing of *E. coli* virulence genes (van Ijperen et al., 2002). A serotype-specific DNA microarray for the identification of clinically encountered *Shigella* and pathogenic *E. coli* strains has being described (Li et al., 2006). Diagnostic microarrays based on the ArrayTube format were devised for virulence determinant detection as well as for protein-based serotyping of *E. coli* (Korczak et al., 2005; Anjum et al., 2007). A novel ArrayTube assay, which incorporates oligonucleotide DNA probes representing 24 of the most epidemiologically relevant O antigens and 47 H antigens, has been described for fast DNA serotyping of *E. coli* (Ballmer et al., 2007). Microarrays have also been used to characterize and type other gastroenteritis-causing viral pathogens including rotavirus, norovirus, and astrovirus (Chizhikov et al., 2002; Honma et al., 2007; Jaaskelainen and Maunula, 2006; Lovmar et al., 2003). Beyond diarrheal illnesses, Pas et al. (2008) reported the comparison of reverse hybridization, microarray, and sequence analysis for hepatitis B virus (HBV) genotyping, suggesting that the InnoLipa HBV genotyping strip assay, a microarray-based system, detected dual infections and was an easy and quick tool for HBV genotyping.

Determination of virulence factors

Many genes associated with virulence are regulated by specific conditions. One way to determine the candidate virulence factors is to investigate the genome-wide gene expression profiles under relevant conditions, such as physiological changes during interaction with the host. A second approach would rely on comparative genomics. In a genome comparison study among H. pylori strains, a class of candidate virulence genes was identified by their coinheritance with a pathogenicity island (Salama et al., 2001). The whole genome microarray of H. pylori was also shown to be an effective method to identify differences in gene content between two H. pylori strains that induce distinct pathological outcomes (Israel et al., 2001). It was demonstrated that the ability of H. pylori to regulate epithelial cell responses related to inflammation depends on the presence of an intact cag pathogenicity island.

Gene expression profiles of drugs, resistance, inhibitors and toxic compounds

Inhibition of a particular cellular process may result in a regulatory feedback mechanism, leading to changes in gene expression patterns. Exploring the gene expression profiles with DNA microarrays may reveal information on the mode of action for drugs, resistance, inhibitors or toxic compounds. DNA microarray hybridization experiments have been conducted in M. tuberculosis to explore the changes in gene expression induced by the antituberculous drug isoniazid (INH) (Wilson et al., 1999). INH selectively interrupts the synthesis of mycolic acids, which are branched β-hydroxy fatty acids. Microarray experiments showed that isoniazid induced several genes that encode proteins that are physiologically relevant to the drug's mode of action, including an operonic cluster of five genes encoding type II fatty acid synthase enzymes and fbpC, which encodes trehalose dimycolyl transferase. Insights gained from this approach may define new drug targets and suggest new methods for identifying compounds that inhibit those targets. In addition to the alternation in gene expression patterns related to the drug's mode of action, drugs can induce changes in genes related to stress responses that are linked to the toxic consequences of the drug. Each type

of compound often generates a signature pattern of gene expression. A database populated with these signature profiles can serve as a guide to elucidate the potential mode of action as well as side effects of uncharacterized compounds.

Another successful application of DNA microarray techniques in medical microbiology is the determination of antimicrobial resistance by simultaneously detecting a panel of drug resistance-related mutations in microbial genomes (Call et al., 2003; Crameri et al., 2007; Hager, 2006; Perreten et al., 2005; Zhu et al., 2007a; Zhu et al., 2007b). The emergence of multidrug-resistant tuberculosis, extensively drug-resistant tuberculosis, and time-consuming phenotypic antimycobacterial susceptibility procedures has stimulated the pursuit of microarray platforms in antituberculosis drug resistance determinations. Oligonucleotide based DNA arrays have been used for parallel species identification and rifampin resistance-related mutations in mycobacteria (Troesch et al., 1999) and, more specifically, for the detection of M. tuberculosis strains that are resistant to rifampin (Yue et al., 2004) or isoniazid, kanamycin, streptomycin, pyrazinamide, and ethambutol (Gegia et al., 2008). Oligonucleotide microarrays were developed to analyze and identify drug-resistant M. tuberculosis strains, and it was found that the results were comparable with those of standard antimicrobial susceptibility testing (Strizhkov et al., 2000; Mikhailovich et al., 2001). A low-cost and -density DNA microarray was designed to detect mutations that confer isoniazid and rifampin resistance in M. tuberculosis isolates. The low-cost and -density array protocol takes 45 min after PCR amplification, with only minimal laboratory equipment required (Aragon et al., 2006). Antonova and colleagues developed a method for the detection and identification of mutations in the M. tuberculosis genome determining resistance to fluoroquinolones by hybridization on biological microchips (Antonova et al., 2008). A recently developed QIAplex system combines a novel multiplex PCR amplification and suspension bead array identification for the simultaneous detection of 24 M. tuberculosis gene mutations responsible for resistance to isoniazid, rifampin, streptomycin, and ethambutol (Gegia et al., 2008).

Microarray-based techniques face several application challenges to determine antimicrobial resistance in the clinical setting. First, genomes of some pathogens continue to mutate under natural and therapeutic selective pressures, which is well demonstrated by HIV-1. An Affymetrix microarray was developed to provide HIV-1 antiretroviral-drug-resistant profiles (Kozal et al., 1996; Vahey et al., 1999; Wilson et al., 2000). The product was discontinued due to rapidly emerging HIV-1 genome mutations. The company now has a comprehensive, high-density microarray available to identify every mutation in resistance-related HIV-1 genomes. Second, molecular mechanisms for many antimicrobial drug resistances remain to be discovered while novel resistance genes and mutations continue to emerge. It takes considerable time and effort to decipher all of the resistance-related mutations and transfer the basic science findings to clinical applications. For *M. tuberculosis*, until such knowledge is available, the currently used phenotypic methods for identifying resistance will continue to play an invaluable role in optimizing the therapy of persons with tuberculosis.

Staphylococcus aureus, including methicillin-resistant *S. aureus* (MRSA), is an important pathogen in hospitals and, increasingly, in communities around the world. Advanced laboratory techniques, including diagnostic microarray analysis, have been sought to rapidly identify staphylococcal isolates and determine antimicrobial susceptibility patterns. DNA microarray analyses of large samples of clinically characterized community-acquired MRSA strains have been reported, which provide broad insights into evolution, pathogenesis, and disease emergence (Koessler et al., 2006; Scherl et al., 2006). DNA microarrays based on the array-tube platform (ClonDiag Chip Technologies, Jena, Germany) have been used for characterizing and genotyping staphylococcal DNA, including their relevant resistance determinants and virulence factors (Monecke et al., 2007; Monecke and Ehricht, 2005; Monecke et al., 2006). Microarrays provide a valuable epidemiological tool for the detailed characterization of MRSA isolates and comparison of strains at a global level (Monecke et al., 2007).

Analyses of microbial evolution and epidemiology

DNA microarrays can be used to explore the variability in genetic content and in gene expression profiles within a natural population of the same or related species and between the ancestor and the descendents. As a result, it provides very rich information on the molecular basis of microbial diversity, evolution and epidemiology. Genomes within the species of *M. tuberculosis* have been compared with a high density oligonucleotide microarray to detect small-scale genomic deletions among 19 clinically and epidemiologically well-characterized isolates (Kato-Maeda et al., 2001).

This study reveals that deletions are likely to contain ancestral genes whose functions are no longer essential for the organism's survival, whereas genes that are never deleted constitute the minimal mycobacterial genome. As the amount of genomic deletion increased, the likelihood that the bacteria will cause pulmonary cavitation decreased, suggesting that the accumulation of mutations tends to diminish their pathogenicity.

CONCLUSION

In conclusion, applications of DNA microarrays for gene expression profiling experiments between two samples appear to be relatively reliable. The array technology,

however, cannot give a reasonable estimation of the actual amount of mRNA. The measurement of relative abundance of particular mRNA species within the same sample needs to be further tested and improved.

Currently, both oligonucleotide and PCR product-based arrays are used for the study of bacterial species. Whether one format will prevail in the future will largely depend on robustness, feasibility (cost and availability of technology) and purpose of the experiments. For example, short oligonucleotide arrays may not be suitable for comparative genomic studies for organisms that are not closely related. Additionally, an individual array could be made of both oligonucleotides and PCR products.

The use of DNA microarrays as a tool for phylogenetic studies and strain identification merits attention. For many organisms, the 16S rRNA approach often fails to truly reflect their genetic potential. This gap can be bridged by comparative genomic methods with whole genome arrays in the absence of genome sequences. It is not difficult to envision the future construction of a DNA array that will contain unique sequences of 16S rRNA, 23S rRNA, and many key functional genes for most of the representative bacterial species. This array could be useful in food, medical, environmental, and agricultural applications.

With the increasing applications of DNA microarrays and generation of enormous quantities of data, the construction of a database and the linking of relevant functional information will be the next important phase of technology development. Centralization of genomic data, including DNA sequences and array results, will be very beneficial to the research community. The construction of EcoReg, EcoCyc and EcoSal web sites for *E. coli* is an excellent starting point and could serve as a model for other prokaryotic microorganisms.

REFERENCES

Akondi KB, Lakshmi VV (2013). Emerging Trends in Genomic Approaches for Microbial Bioprospecting. J. Integr. Biol. 17(2):61-70.

Anjum MF, Mafura M, Slickers P, Ballmer K, Kuhnert P, Woodward MJ, Ehricht R (2007). Pathotyping *Escherichia coli* by using miniaturized DNA microarrays. Appl. Environ. Microbiol. 73:5692-5697.

Anthony RM, TJ Brown, French GL (2000). Rapid diagnosis of bacteremia by universal amplification of 23S ribosomal DNA followed by hybridization to an oligonucleotide array. J. Clin. Microbiol. 38:781-788.

Antonova OV, Gryadunov DA, Lapa SA, Kuz'min AV, Larionova EE, Smirnova TG, Nosova EY, Skotnikova OI, Chernousova LN, Moroz AM, Zasedatelev AS, Mikhailovich VM (2008). Detection of mutations in *Mycobacterium tuberculosis* genome determining resistance to fluoroquinolones by hybridization on biological microchips. Bull. Exp. Biol. Med. 145:108-113

Aragon LM, Navarro F, Heiser V, Garrigo M, Espanol M, Coll P (2006). Rapid detection of specific gene mutations associated with isoniazid or rifampicin resistance in *Mycobacterium tuberculosis* clinical isolates using non-fluorescent low-density DNA microarrays. J. Antimicrob. Chemother. 57:825-831.

Ballmer K, Korczak BM, Kuhnert P, Slickers P, Ehricht R, Hachler H (2007). Fast DNA serotyping of *Escherichia coli* by use of an oligonucleotide microarray. J. Clin. Microbiol. 45:370-379.

Battaglia C, Salani G, Consolandi C, Bernardi LR, De Bellis G (2000). Analysis of DNA microarrays by non-destructive fluorescent staining using SYBR green II. BioTechniques 29:78-81.

Behr MA, Wilson MA, Gill WP, Salamon H, Schoolnik GK, Rane S, Small PM (1999). Small,Comparative genomics of BCG vaccines by whole-genome DNA microarray. Science 284:1520-1523.

Bøving MK, Pedersen LN, Moller JK (2009). Eight-plex PCR and liquid-array detection of bacterial and viral pathogens in cerebrospinal fluid from patients with suspected meningitis. J. Clin. Microbiol. 47:908-913

Call DR, Borucki MK, Loge FJ (2001b). Detection of bacterial pathogens in environmental samples using DNA microarrays. J. Microbiol. Methods. 53: 235-243.

Call DR, Bakko MK, Krug MJ, Roberts MC (2003). Identifying antimicrobial resistance genes with DNA microarrays. Antimicrob. Agents Chemother. 47:3290-3295.

Call DR, Chandler DP, Brockman F (2001a). Fabrication of DNA microarrays using unmodified oligonucleotide probes. BioTechniques 30:368-379.

Call DR, Borucki MK, FJ Loge (2003). Detection of bacterial pathogens in environmental samples using DNA microarrays. J. Microbiol. Methods 53:235-243.

Cassone M, Marco MD, Francesco I, Marco RO, Gian MR and GP (2006). DNA Microarray for Detection of Macrolides Resistance Genes. Antimicrob. Agents Chemother. 50(6):2038-41

Chizhikov V, Wagner M, Ivshina A, Hoshino Y, Kapikian AZ, Chumakov K (2002). Detection and genotyping of human group A rotaviruses by oligonucleotide microarray hybridization. J. Clin. Microbiol. 40:2398-2407

Cleven BE, Palka-Santini M, Gielen J, Meembor S, Kronke M, Krut O (2006). Identification and characterization of bacterial pathogens causing bloodstream infections by DNA microarray. *J. Clin. Microbiol.* 44:2389-2397

Crameri A, Marfurt J, Mugittu K, Maire N, Regos A, Coppee JY, Sismeiro O, Burki R, Huber E, Laubscher D, Puijalon O, Genton B, Felger I, Beck HP (2007). Rapid microarray-based method for monitoring of all currently known single-nucleotide polymorphisms associated with parasite resistance to antimalaria drugs. J. Clin. Microbiol. 45:3685-3691.

Deng J, Zheng Y, Zhao R, Wright PF, Stratton CW, YW Tang. (2009). Culture versus polymerase chain reaction for the etiologic diagnosis of community-acquired pneumonia in antibiotic-pretreated pediatric patients. *Pediatr. Infect. Dis. J.* 28:53-55.

Diaz MR, Fell JW (2005). Use of a suspension array for rapid identification and genotypes of the *Cryptococcus neoformans* species complex. J. Clin. Microbiol. 43:3662-3672

Fitzgerald C, Collins M, van Duyne S, Mikoleit M, Brown T, Fields P (2007). Multiplex,bead-based suspension array for molecular determination of common *Salmonella* serogroups. J. Clin. Microbiol. 45:3323-3334.

Fukushima M, Kakinuma K, Hayashi H, Nagai H, Ito K, Kawaguchi R (2003). Detection and identification of *Mycobacterium* species isolates by DNA microarray. J. Clin. Microbiol. 41:2605-2615

Gegia M, Mdivani N, Mendes RE, Li H, Akhalaia M, Han J, Khechinashvili G, Tang YW (2008). Prevalence of and molecular basis for tuberculosis drug resistance in the Republic of Georgia: validation of a QIAplex system for detection of drug resistance-related mutations. Antimicrob. Agents Chemother. 52:725-729

Gentry TJ, Zhou J (2006). Microarray-based microbial identification and characterization. In: Y. W. Tang and C. W. Stratton (ed.), Advanced techniques in diagnostic microbiology. Springer Science and Business Media, New York, NY. pp. 276-290.

Hager, J. (2006). Making and using spotted DNA microarrays in an academic core laboratory. Methods Enzymol. 410:135-168.

Hakenbeck R, Balmelle N, Weber B, Gardes C, Keck W, de Saizieu A (2001). Mosaic genes and mosaic chromosomes: intra- and interspecies genomic variation of Streptococcus pneumoniae. Infect. Immun. 69:2477-2486.

Hegde P, Qi R, Abernathy K, Gay C, Dharap S, Gaspard R, Hughes JE, Snesrud E, Lee N, Quackenbush J (2000). A concise guide to cDNA microarray analysis. BioTechniques 29:548-562.

Honma S, Chizhikov V, Santos N, Tatsumi M, Timenetsky MD,

Linhares AC, Mascarenhas JD, Ushijima H, Armah GE, Gentsch JR, Hoshino Y (2007). Development and validation of DNA microarray for genotyping group A rotavirus VP4 (P[4], P[6], P[8], P[9], and P[14]) and VP7 (G1 to G6, G8 to G10, and G12) genes. J. Clin. Microbiol. 45:2641-2648

Hsiao CR, Huang L, Bouchara J-P, Barton R, Li HC, Chang TC (2005). Identification of medically important molds by an oligonucleotide array. J. Clin. Microbiol. 43:3760-3768.

Huang A, Li J-W, Shen Z-Q, Wang X-W, M Jin (2006). High-throughput identification of clinical pathogenic fungi by hybridization to an oligonucleotide microarray. J. Clin. Microbiol. 44:3299-3305

Israel DA, Salama N, Arnold CN, Moss SF, Ando T, Wirth HP, Tham KT, Camorlinga M, Blaser MJ, Falkow S, Peek RM Jr. (2001). Helicobacter pylori strain-specific differences in genetic content, identified by microarray, influence host inflammatory responses. J. Clin. Invest. 107: 611-620.

Jaaskelainen AJ, Maunula L (2006). Applicability of microarray technique for the detection of noro- and astroviruses. J. Virol. Methods 136:210-216.

Kakinuma K, Fukushima M, Kawaguchi R (2003). Detection and identification of *Escherichia coli, Shigella,* and *Salmonella* by microarrays using the *gyrB* gene. Biotechnol. Bioeng. 83:721-728

Kato-Maeda M, Rhee JT, Gingeras TR, Salamon H, Drenkow J, Smittipat N, Small PM (2001).Comparing genomes within the species Mycobacterium tuberculosis. Genome Res. 11 547-554.

Khodakov DA, Zakharova NV, Gryadunov DA, Filatov FP, Zasedatelev AS, Mikhailovich VM (2008). An oligonucleotide microarray for multiplex real-time PCR identification of HIV-1, HBV, and HCV. BioTechniques 44:241-246, 248

Koessler T, Francois P, Charbonnier Y, Huyghe A, Bento M, Dharan S, Renzi G, Lew D, Harbarth S, Pittet D, Schrenzel J (2006). Use of oligoarrays for characterization of community-onset methicillin-resistant *Staphylococcus aureus*. J. Clin. Microbiol. 44:1040-1048.

Korczak B, Frey J, Schrenzel J, Pluschke G, Pfister R, Ehricht R, Kuhnert P (2005). Use of diagnostic microarrays for determination of virulence gene patterns of *Escherichia coli* K1, a major cause of neonatal meningitis. J. Clin. Microbiol. 43:1024-1031.

Korimbocus J, Scaramozzino N, Lacroix B, Crance JM, Garin D, Vernet G (2005). DNA probe array for the simultaneous identification of herpesviruses, enteroviruses, and flaviviruses. J. Clin. Microbiol. 43:3779-3787

Kostic T, Weilharter A, Rubino S, Delogu G, Uzzau S, Rudi K, Sessitsch A, Bodrossy L (2007). A microbial diagnostic microarray technique for the sensitive detection and identification of pathogenic bacteria in a background of nonpathogens. Anal. Biochem. 360:244-254

Kozal MJ, Shah N, Shen N, Yang R, Fucini R, Merigan TC, Richman DD, Morris D, Hubbell E, M Chee, Gingeras TR (1996). Extensive polymorphisms observed in HIV-1 clade B protease gene using high-density oligonucleotide arrays. Nat. Med. 2:753-759.

Lease RA, Belfort M (2000). A trans-acting RNA as a control switch in *Escherichia coli*: DsrA modulates function by forming alternative structures. Proc. Natl. Acad. Sci. USA 97:9919-9924.

Lenhoff RJ, Naraghi-Arani P, Thissen JB, J Olivas, Carillo AC, Chinn C, Rasmussen M, Messenger SM, Suer LD, Smith SM, Tammero LF, Vitalis EA, Slezak TR, Hullinger PJ, Hindson BJ, Hietala SK, Crossley BM, McBride MT (2008). Multiplexed molecular assay for rapid exclusion of foot-and-mouth disease. J. Virol. Methods 153:61-69

Li Y, Liu D, Cao B, Han W, Liu Y, Liu F, Guo X, Bastin DA, Feng L, Wang L (2006). Development of a serotype-specific DNA microarray for identification of some *Shigella* and pathogenic *Escherichia coli* strains. J. Clin. Microbiol. 44:4376-4383.

Lopez MF, Pluskal MG (2003). Protein micro- and macroarrays: digitizing the proteome. J. Chromatogr. B Analyt. Technol. Biomed. Life Sci. 787(1):19-27.

Lovmar L, Fock C, Espinoza F, Bucardo F, Syvanen AC, Bondeson K (2003). Microarrays for genotyping human group A rotavirus by multiplex capture and type-specific primer extension. J. Clin. Microbiol. 41:5153-5158.

MacBeath G (2002). Protein microarrays and proteomics. Nat. Genet. 32(Suppl.):526-532.

Marlowe EM, Hogan JJ, Hindler JF, Andruszkiewicz I, Gordon P, Bruckner DA (2003). Application of an rRNA probe matrix for rapid identification of bacteria and fungi from routine blood cultures. J. Clin. Microbiol. 41:5127-5133.

McLoughlin KS, (2011). Microarrays for Pathogen Detection and Analysis. Brief Funct Genomics.10(6):342-353.

Mikhailovich V, Lapa S, Gryadunov D, Sobolev A, Strizhkov B, Chernyh N, Skotnikova O, Irtuganova O, Moroz A, Litvinov V, Vladimirskii M, Perelman M, Chernousova L, Erokhin V, Zasedatelev A, Mirzabekov A (2001). Identification of rifampin-resistant Mycobacterium tuberculosis strains by hybridization, PCR, and ligase detection reaction on oligonucleotide microchips. J. Clin. Microbiol. 39:2531-2540

Miller MB, Tang YW (2009). Basic Concepts of Microarrays and Potential Applications in Clinical Microbiology. Clin. Microbiol. Rev. 22(4):611-633.

Monecke S, Ehricht R (2005). Rapid genotyping of methicillin-resistant Staphylococcus aureus (MRSA) isolates using miniaturised oligonucleotide arrays. Clin. Microbiol. Infect. 11:825-833.

Monecke S, Berger-Bachi B, Coombs G, Holmes A, Kay I, Kearns A, Linde HJ, O'Brien F, Slickers P, Ehricht R (2007). Comparative genomics and DNA array- based genotyping of pandemic Staphylococcus aureus strains encoding Panton-valentine leukocidin. Clin. Microbiol. Infect. 13:236-249.

Monecke S, Slickers P, Hotzel H, Richter-Huhn G, Pohle M, Weber S, Witte W, Ehricht R (2006). Microarray-based characterisation of a Panton-Valentine leukocidin-positive community-acquired strain of methicillin-resistant Staphylococcus aureus. Clin. Microbiol. Infect. 12:718-728.

Murakami S, Nakashima R, Yamashita E, and Yamaguchi A. (2002).Crystal structure of bacterial multidrug efflux transporter AcrB. Nature 419:587-593.

Nsofor CA, Iroegbu CU, Davis MA, Orfe L, Call DR (2013). DNA microarray-based detection of antibiotic resistance genes of human isolates of Escherichia coli in Nigeria. J. Bacteriol. Res. 5(6):68-75

Nubel U, Schmidt PM, Reiss E, Bier F, Beyer W, Naumann D (2004). Oligonucleotide microarray for identification of Bacillus anthracis based on intergenic transcribed spacers in ribosomal DNA. FEMS Microbiol. Lett. 240:215-223.

Panicker G, Call DR, Krug MJ, Bej AK (2004). Detection of Pathogenic Vibrio spp. in shellfish by using multiplex PCR and DNA microarrays. Appl. Environ. Microbiol. 7436-7444

Pas SD, Tran N, de Man RA, Burghoorn-Maas C, Vernet G, Niesters HG (2008). Comparison of reverse hybridization, microarray, and sequence analysis for genotyping hepatitis B virus. J. Clin. Microbiol. 46:1268-1273.

Perreten V, Vorlet-Fawer L, Slickers P, Ehricht R, Kuhnert P, Frey J (2005). Microarray-based detection of 90 antibiotic resistance genes of gram-positive bacteria. J. Clin. Microbiol. 43:2291-2302.

Peterson G, Jianfa B, Nagaraja TG, Sanjeev N (2010). Diagnostic microarray for human and animal bacterial diseases and their virulence and antimicrobial resistance genes. J. Microbiol. Methods 80:223-230

Proudnikov D, Timofeev E, Mirzabekov A (1998). Immobilization of DNA in polyacrylamide gel for the manufacture of DNA and DNA-oligonucleotide microchips. Anal. Biochem. 259:34-41.

Raddatz G, Dehio M, Meyer TF, Dehio C (2001). PrimeArray: genome-scale primer design for DNA microarray construction. Bioinformatics 17:98-99.

Riehle MM, Bennett AF, Long AD (2001).Genetic architecture of thermal adaptation in Escherichia coli. Proc. Natl. Acad. Sci. USA 98:525-530.

Roh SW, Guy CJ, Abell Kyoung-Ho Kim, Young-Do N and Jin-Woo B (2012). Comparing microarrays and nextgeneration sequencing technologies for microbial ecology research. Trends Biotechnol. 28 (6):234-256

Salama N, Guillemin K, McDaniel TK, Sherlock G, Tompkins L, Falkow S (2001). A whole-genome microarray reveals genetic diversity among Helicobacter pylori strains. Proc. Natl. Acad. Sci. USA 97:14668-14673.

Scherl A, Francois P, Charbonnier Y, Deshusses JM, Koessler T, Huyghe A, Bento M, Stahl-Zeng J, Fischer A, Masselot A, Vaezzadeh A, Galle F, Renzoni A, Vaudaux P, Lew D, Zimmermann-Ivol CG, Binz PA, Sanchez JC, Hochstrasser DF, Schrenzel J (2006). Exploring

glycopeptide-resistance in Staphylococcus aureus: a combined proteomics and transcriptomics approach for the identification of resistance-related markers. BMC Genomics 7:296.

Selinger DW, Cheung KJ, Mei R, Johansson EM, Richmond CS, Blattner FR, Lockhart DJ, Church GM (2000). RNA expression analysis using a 30-base pair resolution Escherichia coli genome array. Nat. Biotechnol. 18:1262-1268.

Shang S, Chen G, Wu Y, Du L, Zhao Z (2005). Rapid diagnosis of bacterial sepsis with PCR amplification and microarray hybridization in 16S rRNA gene. Pediatr. Res. 58:143-148

Spiess B, Seifarth W, Hummel M, Frank O, Fabarius A, Zheng C, Morz H, Hehlmann R, Buchheidt D (2007). DNA microarray-based detection and identification of fungal pathogens in clinical samples from neutropenic patients. J. Clin. Microbiol. 45:3743-3753.

Stover AG, Jeffery E, Xu JC, Persing DH (2003). Hybridization array, p. 619-639. In: D. H. Persing, F. C. Tenevor, J. Versalovic, Y. W. Tang, E. R. Unger, D. A. Relman, and T. J. White (ed.), Molecular microbiology: diagnostic principles and practice. ASM Press, Washington, DC.

Strizhkov BN, Drobyshev AL, Mikhailovich VM, Mirzabekov AD (2000). PCR amplification on a microarray of gel-immobilized oligonucleotides: detection of bacterial toxin- and drug-resistant genes and their mutations. BioTechniques 29:844-848, 850-852, 854.

Tang Y-W, Ellis NM, Hopkins MK, Smith DH, Dodge DE, Persing DH. (1998). Comparison of phenotypic and genotypic techniques for identification of unusual aerobic pathogenic Gram-negative bacilli. J. Clin. Microbiol. 36:3674-3679

Taniguchi M, Miura K, Iwao H, Yamanaka S (2001).Quantitative assessment of DNA microarrays—comparison with Northern blot analyses. Genomics 71:34-39.

Troesch A, Nguyen H, Miyada CG, Desvarenne S, Gingeras TR, Kaplan PM, Cros P, Mabilat C (1999). Mycobacterium species identification and rifampin resistance testing with high-density DNA probe arrays. J. Clin. Microbiol. 37:49-55

Vahey M, Nau ME, Barrick S, Cooley JD, Sawyer R, Sleeker AA, Vickerman P, Bloor S, Larder B, Michael NL, Wegner SA (1999). Performance of the Affymetrix GeneChip HIV PRT 440 platform for antiretroviral drug resistance genotyping of human immunodeficiency virus type 1 clades and viral isolates with length polymorphisms. J. Clin. Microbiol. 37:2533-2537.

van Ijperen C, Kuhnert P, Frey J, Clewley JP (2002). Virulence typing of Escherichia coli using microarrays. Mol. Cell. Probes 16:371-378

Volokhov D, Chizhikov V, Chumakov K, Rasooly A (2003). Microarray-based identification of thermophilic Campylobacter jejuni, C. coli, C. lari, and C. upsaliensis. J. Clin. Microbiol 41:4071-4080.

Wang RF, Beggs ML, Erickson BD, Cerniglia CE (2004). DNA microarray analysis of predominant human intestinal bacteria in fecal samples. Mol. Cell. Probes 18:223-234

Wang Z, Orlandi PA, Stenger DA (2005). Simultaneous detection of four human pathogenic microsporidian species from clinical samples by oligonucleotide microarray. J. Clin. Microbiol. 43:4121-4128

Wapner RJ, Christa Lese Martin MD, Brynn Lev, Blake CB, Christine ME (2012). Chromosomal Microarray versus Karyotyping for Prenatal Diagnosis. N. Engl. J. Med. 367:2175-2184.

Wei Y, Lee JM, Richmond C, Blattner FR, Rafalski JA, LaRossa RA (2001) High-density microarray-mediated gene expression profiling of Escherichia coli. J. Bacteriol. 183: 545-556.

Willse A, Straub TM, Wunschel SC, Small JA, Call DR, Daly DS, Chandler DP (2004). Quantitative oligonucleotide microarray fingerprinting of Salmonella enterica isolates. Nucleic Acids Res. 32:1848-1856.

Wilson M, DeRisi J, Kristensen HH, Imboden P, Rane S, Brown PO, Schoolnik GK(1999). Exploring drug-induced alterations in gene expression in Mycobacterium tuberculosis by microarray hybridization. Proc. Natl. Acad. Sci. USA 96: 12833-12838.

Wilson JW, Bean P, Robins T, Graziano F, Persing DH (2000). Comparative evaluation of three human immunodeficiency virus genotyping systems: the HIV-GenotypR method, the HIV PRT GeneChip assay, and the HIV-1 RT line probe assay. J. Clin. Microbiol. 38:3022-3028.

Yang YH, Dudoit S, Luu P, Speed TP (2001). In: Normalization for cDNA Microarray Data. pp. 1-12

Ye RW, Wang T, Bedzyk L, Croker KM (2001). Applications of DNA microarrays in microbial systems. J. Microbiol. Methods 47:257-272.

Ye Y, He X, Szewczyk P, Nguyen T, Chang G (2006). Structure of the multidrug transporter EmrD from *Escherichia coli*. Science 312:741-744.

Ye RW, Tao W, Bedzyk L, Young T, Chen M, Li L (2000). Global gene expression profiles of *Bacillus subtilis* grown under anaerobic conditions. J. Bacteriol. 182: 4458-4465.

You Y, Fu C, Zeng X, Fang D, Yan X, Sun B, Xiao D, Zhang J (2008). A novel DNA microarray for rapid diagnosis of enteropathogenic bacteria in stool specimens of patients with diarrhea. J. Microbiol. Methods 75:566-571

Yu H, David YW, Lee W, Siddaraju MN, Shivaprasad HV, Brian MB, Kamal DM (2013). Microarray Analysis Reveals the Molecular Basis of Antiarthritic Activity of Huo-Luo-Xiao-Ling Dan. Evid. Based Complement. Altern. Med 234-237.

Yue J, Shi W, Xie J, Li Y, Zeng E, Liang L, Wang H (2004). Detection of rifampin-resistant *Mycobacterium tuberculosis* strains by using a specialized oligonucleotide microarray. Diagn. Microbiol. Infect. Dis. 48:47-54

Zhu L-X, Zhang Z-W, Wang C, Yang H-W, Jiang D, Zhang Q, Mitchelson K, Cheng J (2007b). Use of a DNA microarray for simultaneous detection of antibiotic resistance genes among staphylococcal clinical isolates. J. Clin. Microbiol. 45:3514-3521.

Zhu L-X, Zhang Z-W, Liang D, Jiang D, Wang C, Du N, Zhang Q, Mitchelson K, Cheng J (2007a). Multiplex asymmetric PCR-based oligonucleotide microarray for detection of drug resistance genes containing single mutations in *Enterobacteriaceae*. Antimicrob. Agents Chemother. 51:3707-3713.

13

Toward a comprehensive description of microbial processes through mechanistic and intelligent approaches

Pratap R. Patnaik

Institute of Microbial Technology, Sector 39-A, Chandigarh-160036, India.
IMTECH communication No. 048/2007. E-mail: pratap@imtech.res.in.

Microbial processes functioning in bioreactors under realistic conditions are subject to incomplete dispersion and the presence of noise from the environment and within the cells. These factors complicate the development of good quantitative descriptions of microbial reactors. Most analyses have therefore focused on either the intra-cellular or the extra-cellular processes, ignoring or simplifying the other facet. The resulting models are thus useful only for the intended purposes and in limited domains, but they do not include a comprehensive description of all features. These models have employed one or more of three main approaches to develop quantitative descriptions – mechanistic, cellular intelligence (or cybernetic), and artificial intelligence. Models using judicious combinations of two or more methods have wider and more versatile applicability. However, no model has accommodated both intra-cellular and extra-cellular noise in a macroscopic description of a nonideal bioreactor. Based on a review of recent studies, such a conceptual model is presented here. It combines all three approaches in a flexible design.

Key words: Microbial processes, dispersion, noise, modeling approaches, comprehensive description.

INTRODUCTION

Cellular processes are often more complex than chemical processes, even when both generate the same outputs. Many factors contribute to this complexity. At a fundamental level, living cells sustain more complex and intricate networks of reactions than many chemical reactions. Cellular reaction networks are not always completely deciphered, thus underlining their complexity and necessitating simplifying methods to formulate workable kinetic models (Gombert et al., 2000; Varner and Ramkrishna, 1999). Unlike chemical reactions, metabolic reactions are regulated by structurally and functionally complex molecules such as DNA, RNA and enzymes, many of whose concentrations are small but important. The low concentrations also make these molecules sensitive to noise at the genetic level (Kaern et al., 2005), which in turn has an impact on the observed behavior of the process (Haag et al., 2005).

Cellular metabolic processes also respond to environmental changes in ways that are difficult to capture through models constructed on chemical kinetic princeples alone. The lag phase behavior of cultures transferred from one medium to another, the responses to abrupt changes in input streams, and cellular dynamics in the presence of external noise are some examples. Such observations may however be described quantitatively by "intelligent" models. By contrast with the static nature of chemical kinetic models, intelligent models employ either inherent or artificial intelligence. Inherent intelligence is the basis of the so-called cybernetic models (Dhurjati et al., 1985; Patnaik, 2000) which attribute to living cells the ability to retain information, understand it and adjust their responses on the basis of past experience. Artificial intelligence has been invoked largely to model macro-scopic microbial behavior under the influences of noise and spatial variations in a bioreactor. Methods such as artificial neural networks, fuzzy logic and genetic algorithms then provide more faithful representations of varied cellular dynamics than mechanistic models do (Hodgson et al., 2004; Patnaik, 2006a).

While intelligent descriptions of microbial processes in bioreactors have decisive advantages over mechanistic models under realistic conditions, they have limitations too. Cybernetic models tend to be quite complex, often resulting in large sets of differential equations, and sometimes more than one cybernetic goal seems to explain the observed behavior equally well (Patnaik, 2000; Straight and Ramkrishna, 1994). A pivotal feature of cybernetic models is the presence of regulatory key enzymes, but it has not always been possible to establish a correspondence between these and the enzymes actually detected. A common criticism of artificial intelligence methods is that they are too empirical and do not incorporate the physiological features of cellular processes. Consequently, it becomes difficult to provide physical interpretations for the parameters of these models and relate them to kinetic and thermodynamic features (James et al., 2002; Lubbert and Jorgensen, 2001).

Despite their inadequacies, mechanistic models are simple and are derived through biological and physical principles. Therefore they have greater physiological closeness to internal cellular processes; as a result, their parameters can be attributed physical meaning and can be manipulated by introducing mutational, genetic or operational changes (Haag et al., 2005; Hodgson et al., 2004). Given the different strengths of mechanistic models and intelligent models, it might be useful to combine the two approaches to develop composite (or hybrid) models for microbial reactors. These composite models should, in principle, turn out to be adaptive, flexible and self-regulating like intelligent models, but also possess the fundamental biological basis of mechanistic models.

The idea of composite or hybrid models is not new. Many previous studies (Coleman et al., 2003; Galvanauskas et al., 2004; James et al., 2002; Patnaik, 2003a) have combined two or more modeling approaches to optimize and control bioreactors for different microbial cultures. However, no investigation has yet been reported of an approach that includes both intra-cellular and extra-cellular noise as well as cybernetic kinetics and bioreactor nonidealities. Since all these are important features of real microbial processes, this communication provides a perspective of how the development of existing methods of modeling can lead to such a composite descriptive framework.

NONIDEAL FEATURES OF CELLS AND REACTORS

Cellular noise and complexity

Cells synthesize products as a result of the expression of specific genes. Molecules such as DNA, mRNA and proteins are involved in gene expression. These molecules are usually present in concentrations sufficiently low for stochastic effects to become significant (Kaern et al., 2005; Raser and O'Shea, 2005). As a result, the amount of a particular protein that a gene synthesizes fluctuates from cell to cell in a population and with time for each cell. These fluctuations (termed genetic noise) arise from a number of sources but they may be categorized broadly as either (a) intrinsic or (b) extrinsic.

Intrinsic noise refers to fluctuations associated with promoter activation or deactivation and the synthesis and decay of mRNA and proteins. Extrinsic noise pertains to

fluctuations in gene products such as RNA polymerase, ribosomes and certain proteins. Although external to the relevant genes, these fluctuations act on the genes, thus complicating both genetic expression and the identification of the two contributing effects. Nevertheless, Elowitz et al.'s (2002) two-reporter assay provides an elegant method to differentiate and quantify these two types of genetic noise.

While it may be possible to measure intrinsic noise and extrinsic noise and relate them to biochemical parameters (Swain et al., 2002), it still remains difficult to quantify individual repositories of intrinsic noise. Intrinsic noise may be present at any of three levels: (a) individual genes, (b) reaction pathways in a network, and (c) the cells as a whole. Each of these sources affects particular intra-cellular processes but all three interact as shown in Figure 1.

Since cellular processes involve complex networks of reactions regulated at the genetic level, they should be able to withstand stochastic effects so that the cellular machinery can function without being destabilized. In other words, cells should be sufficiently robust to both intrinsic and extrinsic noise. Complexity and robustness are inter-related, and examples abound in biological and ecological systems (Carson et al., 2006). Robustness is the maintenance of specific characteristics of system behavior in the face of perturbations (Carson and Doyle, 2002; Kitano, 2004a; Stelling et al., 2004). Kitano (2004a) has argued that complex evolvable systems are necessarily robust; since evolution is a fundamental trait of living cells, they too are robust.

Many factors contribute to robustness, of which feedback is a prominent example. While positive feedback amplifies fluctuations and negative feedback attenuates them, the former also increases phenotypic diversity in a population of cells (Becsei and Serrano, 2000; Kaern et al., 2005; Rao et al., 2002). In a heterogeneous population, different phenotypes have different survival probabilities under given environmental conditions, and this has important implications for disease control (Balaban et al., 2004). Since both negative and positive feed-back have beneficial as well as detrimental effects, it may be worthwhile to design gene networks that incorporate the helpful features of both; research on HIV (Richman, 2001) and cancer (Kitano, 2004b) indicates that this is possible.

The evolvability of living systems is also a core concept underlying cybernetic models of microbial processes. The cybernetic approach (Dhurjati et al., 1985; Varner and Ramkrishna, 1999) attributes to living cells the ability to learn from their experiences and accordingly respond optimally to changing circumstances. Cybernetic models have not only provided more faithful representations of the dynamic behavior of microbial reactors (Dhurjati et al., 1985; Hodgson et al., 2004; Namjoshi and Ramkrishna, 2001; Patnaik, 2000) but also explained uncommon patterns of behavior that were considered aberrant by mechanistic modeling approaches (Narang et al., 1997;

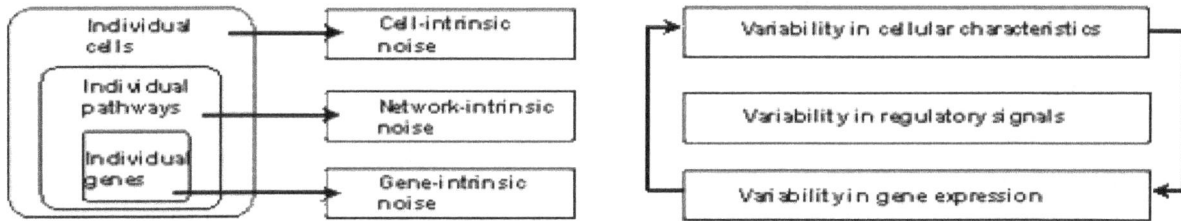

Figure 1. Noise sources inside a cell and their effects on cellular processes. Reproduced from Kaern et al. (2005) with permission from Macmillan Publishers Ltd. © 2005.

Straight and Ramkrishna, 1994). Such models have also helped to increase productivity (Patnaik, 2006a) and to suggest new feeding methods with multiple substrates to promote desired objectives (Patnaik, 2003b).

It is important to realize that the cybernetic concept is consistent with the basic premises of complex evolvable systems (Carson and Doyle, 2002; Kitano, 2004a). Both address the dynamics of complex processes and both are predicated on the Mendelian idea of evolution as a method of survival. Not surprisingly, therefore, the cybernetic framework may be used to suggest genetic modifications that will alter the rates and fluxes of path-ways in a metabolic network so that a desired objective, such as over-expression of an enzyme or suppression of a by-product, is favored (Varner and Ramkrishna, 1998).

Heterogeneity and noise in bioreactors

Populations of cells are usually cultivated in bioreactors under conditions that favor specific objectives. Typical objectives are the growth of the cells themselves, the formation of particular products, and the removal of harmful components from the environment of the cells. To achieve these objectives, it is possible to have elaborate controls of small bioreactors used in the laboratory but practical difficulties and high costs limit both measurements and control in large bioreactors. Therefore, larger reactors are less 'ideal' than small vessels and generate less of the product(s). Two significant nonideal features are: (a) spatial variations within the reactor, as a result of incomplete mixing or dispersion and (b) the influx of noise from the environment. The latter feature is obviously more likely in continuous flow and fed-batch operations; nevertheless, kinetic and thermodynamic considerations often favor the choice of such operating modes (Liden, 2001).

While spatial heterogeneity increases with reactor size, even small bioreactors can have significant gradients on a microscopic scale (Larsson et al. 1996). This makes the optimal positioning of sensors difficult and expensive. Moreover, the presence of spatial variations and the influx of noise from the environment create differences among the cells and in the distribution of fluxes along the pathways of metabolic networks, in the yields of products

and sometimes in the stability of the fermentation. These effects are illustrated by numerous studies of the production of ethanol by *Saccharomyces cerevisiae* in continuous cultures (Garhyan and Elnashaie, 2004; Garhyan et al., 2003; Patnaik, 2005). Recall here that extrinsic noise also creates differences between cells in a population (Kaern et al., 2005; Raser and O'Shea, 2005). Since both extrinsic noise and external (environmental) noise have an impact on the cells, this raises an intriguing question: will the two sources of noise amplify or nullify each other? Depending on the operating conditions, either effect is possible: noise may drive a culture from a monotonic stationary state to an oscillating or a chaotic state and, alternately, proper filtering of noise can restore stable states from chaotic conditions (Garhyan and Elnashaie, 2004; Garhyan et al., 2003; Patnaik, 2006b).

The complexities described above make it difficult to formulate mechanistic models and control policies based on them that are sufficiently accurate, simple, flexible, adaptive, fast and robust. These difficulties have motivated the use of artificial intelligence (AI) methods for online estimations of important variables and optimal control of bioreactors. Artificial neural networks (Gadkar et al., 2005), genetic algorithms (Hodgson et al., 2004) and fuzzy logic (Arnold et al., 2002) have been used in different applications. Although they are superior to classical methods of modeling and control, AI methods are, in effect, 'black box' entities with poor physiological foundations. Not surprisingly, therefore, they are often difficult to train, do not always yield unique models and have limited capability outside their training domains. These weaknesses have generated hybrid models that combine AI or cybernetic models with mechanistic equations. Many applications of hybrid models have been reviewed recently (Galvanauskas et al., 2004). While establishing the usefulness of hybrid approaches, they also reveal areas that are still uncertain and require further inquiry. These are discussed in later sections.

A CONCEPTUAL BASIS FOR MODEL DEVELOPMENT

The formulation of a comprehensive model for a biore-bioreactor begins at the cellular level, integrates this with the extra-cellular environment, establishes a quantitative description of this environment through mass balances, and incorporates non ideal features. The processes inside the cells themselves are sufficiently complex to require a hierarchical portrayal; Figure 2 presents such a portrait. Many studies (Dun and Ellis, 2005; Shimizu, 2002; Wang et al., 2006) have discussed cellular functions from the perspective of Figure 2, so that discussion will not be repeated here. An overriding feature of this conceptual depiction is the emergence of 'omic' structures as building blocks for intra-cellular processes. Genomics describes the primary functions at the level of genes. Ensembles of genes may differ from one genome to another, thus controlling the expression of different proteins through DNA transcription processes (transcript-tomics). The expressed proteins differ in their structures, functions, stability and interactions within individual mole-cules and between molecules; these features are the domain of proteomics.

Macroscopic manifestation of these intra-cellular facets begins at the metabolomic stage, where metabolic regulatory networks and fluxes along pathways constituting these networks are analyzed. The flux distributions are controllable and are critical to the product distributions obtained in microbial cultures. Thus metabolomics provides a vital interface between genome-level processes and those at the bioreactor level. The hierarchical structure of the 'omic' domain continues through the metabolome to the bioreactor, as illustrated in Figure 2 (Ortoleva et al., 2003; Wang et al., 2006). Note that from the microscopic to the macroscopic level the number of variables reduces from up to 10^5 (reflecting the complexity of cellular processes) to less than 10. This reduction is important because practical monitoring and control of a bioreactor can be done efficiently only with a limited number of variables.

Multi-cellular processes that differ so widely in complexity, time scales and the number of variables may understandably be described by more than one approach. Different workers have adopted different approaches and used different assumptions, according to the conditions of the system being studied and the objective. There appear to be five main approaches, with differences and similarities:

i.) Equation-oriented approach.
ii.) Signal-oriented approach.
iii.) Cellular intelligence approach.
iv.) Artificial intelligence approach.
v.) Composite (or hybrid) approach.

These are discussed below, followed by a proposal to combine some of them to formulate a composite comprehendsive model.

EQUATION-ORIENTED MODELING

The underlying premise for equation-oriented models is that biological processes follow the same laws as chemical processes. This implies that, as for chemical reactions, mass balances and kinetic equations can be formulated on the basis of measurements of inputs, outputs and intermediates. We recapitulate here that cellular systems involve reactions inside the cells and transport between (a) the cells and the extra-cellular broth, (b) through the broth itself and (c) sometimes between the broth and the external environment. On this basis two streams of modeling have evolved independently. One relates to the intra-cellular metabolic processes (Gombert and Nielsen, 2000; Varner and Ramkrishna, 1999) and the other pertains to bioprocesses (Bailey, 1998; Lubbert and Jorgensen, 2001). However, a comprehensive description of a microbial system should encompass both streams, and Haag et al. (2005) work illustrates how this may be done.

Consider a set of reactions that follow the stoichiometry

$$\sum_{i=1}^{m} \kappa_{S_i} S_i \rightarrow \sum_{i=1}^{n} \kappa_{P_i} P_i \tag{1}$$

Where S_i is the i-th substrate and P_i the i-th product. In a perfectly mixed continuous flow stirred tank bioreactor, the mass (or molar) balances for each component in the extra-cellular fluid may be written as

$$\frac{d(\bar{c}_{ex} V)}{dt} = \bar{K}_{ex} \bar{r} V + \bar{c}_{in} Q_{in} - \bar{c}_{ex} Q_{out} \tag{2}$$

where $\bar{c}_{ex}$ is the vector of exit concentrations, $\bar{c}_{in}$ the vector of inlet concentrations, Q_{in} the inflow rate, Q_{out} the outflow rate, V the volume of material in the bioreactor and t the elapsed time, $\bar{r}$ contains the rate terms for the concentrations $\bar{c}_{ex}$, and $\bar{K}_{ex}$ is a vector of stoichiometric rate constants.

Normally $Q_{out} = Q_{in}$ and hence V is constant. Then, with the dilution rate defined as $D = Q_{in}/V$, Eqn. (2) becomes

$$\frac{d\bar{c}_{ex}}{dt} = \bar{K}_{ex} \bar{r} + (\bar{c}_{in} - \bar{c}_{ex})/D \tag{3}$$

Haag et al. (2005) also accounted for the common observation that, at any time, some cells are active (or viable) and others are inactive (or dead). Ignoring detailed mechanisms, they considered simply that, overall, dead cells (X_d) arise irreversibly from viable cells (X_v). Then the mass balances for the biomass may be expressed as:

$$\frac{dc_{x_v}}{dt} = \bar{k}_{x_v}^T \bar{r} - Dc_{x_v} = (\mu - k_d)c_{x_v} - Dc_{x_v} \tag{4}$$

$$\frac{dc_{x_d}}{dt} = \overline{k}_{x_d}^T \overline{r} - Dc_{x_d} = k_d c_{x_v} - Dc_{x_d} \qquad (5)$$

Here $\overline{k}_{x_v}$ and $\overline{k}_{x_d}$ contain the respective reaction rate constants, and k_d is the rate constant for the decay of viable cells. Combining Eqns. (4) and (5), the total specific growth rate may be obtained from

$$\frac{dc_{x_{tot}}}{dt} = \frac{dc_{x_v}}{dt} + \frac{dc_{x_d}}{dt} = \mu_{tot} c_{x_v} - Dc_{x_{tot}} = \mu c_{x_v} - Dc_{x_{tot}} \qquad (6)$$

The equality $\mu_{tot} = \mu$ is understandable since only viable cells contribute to the growth. Metabolic reactions inside the cells contribute to changes in the viable cell mass. Therefore, similar to Equation (2), global balances for the intra-cellular metabolites, $\overline{c}_i$, may be written as

$$\frac{d(\overline{c}_i c_{x_v} V)}{dt} = \overline{K}_i \overline{q} c_{x_v} V - \overline{c}_i k_d c_{x_v} V - \overline{c}_i c_{x_v} Q_{out} \qquad (7)$$

The vector $\overline{q}$ contains the metabolic fluxes along the reaction pathways. Equation (7) may be recast in the form of Equation (3) to obtain

$$\frac{d\overline{c}_i}{dt} = \overline{K}_i \overline{q} - \overline{c}_i \mu \qquad (8)$$

Since the control volume of the bioreactor alone is a closed system, the mass flows between the cells and their environment have to be balanced. This leads to

$$\overline{c}_{ex}(t) = \overline{K}_{ex} \overline{r}(\overline{c}_{ex}, \overline{c}_i) + \overline{g}_{ex}(\overline{c}_{ex}, t) \qquad (9)$$

$$\overline{c}_i(t) = \overline{K}_i \overline{q}(\overline{c}_{ex}, \overline{c}_i) - \overline{c}_i \mu(\overline{c}_{ex}, \overline{c}_i) \qquad (10)$$

with $\overline{g}_{ex}$ containing the terms representing exchanges with the environment outside the bioreactor. In the simplest case without gaseous inflow or outflow,

$$\overline{g}_{ex}(\overline{c}_{ex}, t) = [\overline{c}_{in}(t) - \overline{c}_{ex}(t)]D \qquad (11)$$

This model is rigorous but complex. Moreover, it depends on intra-cellular concentrations, which are usually difficult to measure, and it ignores intra-cellular regulatory processes (Dhurjati et al., 1985; Ptnaik, 2000). These limitations have been exposed in studies of hybridoma cultures (Namjoshi et al., 2003; Zupke and Stephanopoulos, 1995).

SIGNAL-ORIENTED APPROACH

As an alternative to the "bag full of enzymes" (Lengeler et al., 1999) approach of equation-oriented mechanism-based modeling, the signal-oriented approach views a cell as a mosaic of function units with signals flowing between them. The content and the nature of flows determine the ultimate functions of a cell. Three biological criteria are used to demarcate these units: (i) the presence of an enzymatic network with a common physiological task, (ii) control of this network at the genetic level by a common regulatory unit, such as an operon or a regulon, usually organized in a hierarchical way, and (iii) the coupling of this regulatory network through a signal transduction system.

Based on these concepts, Kremling and coworkers (Kremling et al., 2000) proposed a signal-oriented description of cellular dynamics. They illustrated their method with *Escherichia coli*. In their application each functional unit is characterized by two "coordinates". The structural coordinate is described by the number and type of inputs and outputs. For example, a functional unit may describe transcription processes connecting a pool of nucleotides with the RNAs. The second coordinate is behavioral and it is expressed by mathematical equations describing the structural object. As might be expected, functional units differ in their complexity and response times.

Kremling et al. (2000) also assigned to each unit an indicator molecule (called an alarmone) whose level of activity controls the activities of superimposed regulatory networks. It is worth remembering this concept in order to draw a parallel with the cybernetic approach to be described later.

Each function may be composed of one or more elemental submodels or model objects. These are basically of three types: (i) substance storages, (ii) substance transformers and (iii) signal transformers. Substance storage devices may either contain genetic information (as in the cases of DNA, RNA and proteins) and may not (e.g. intermediate metabolites). Signal transformers form the central nervous system of the signal-oriented approach. As Kremling et al. (2000) point out; they provide the crucial links between the reception of stimuli, either from inside the cells or from the external environment, and the cellular responses.

Signal transformers also help to differentiate between metabolic flows and signal flows. Metabolic networks comprise metabolic and regulatory subnetworks. The metabolic subnetwork contains the metabolic fluxes whereas the regulatory component describes the signal transduction processes. The signal-oriented approach may be illustrated by the simple example of the synthesis of a protein (Kremling et al., 2000). This requires the processes of transcription, translation and replication, which are modeled as shown in Figure 3. Each process has three components–substance storage, substance transformer and signal transformer. As seen, transcription and translation provide unidirectional signal transfer. The signal transformer of the transcription cascade processes information about DNA sequences, regulatory proteins,

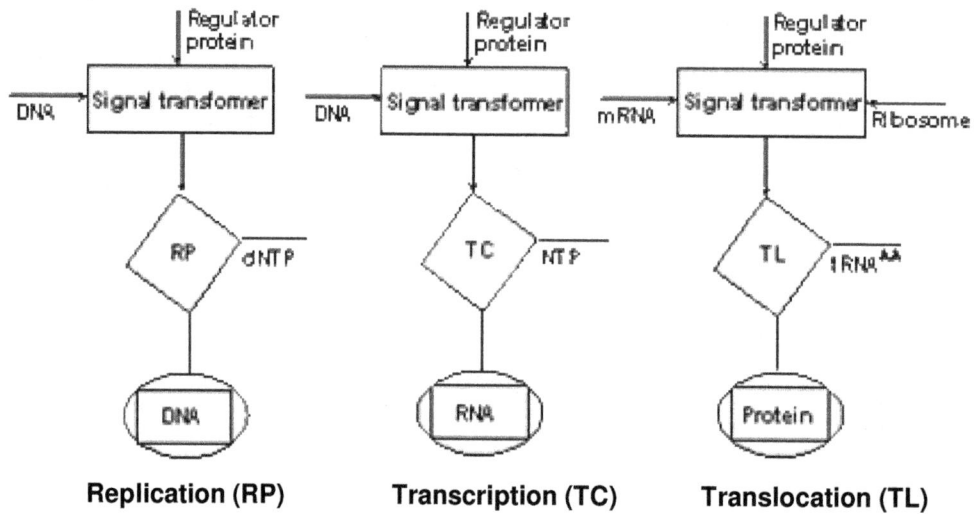

Replication (RP)   **Transcription (TC)**   **Translocation (TL)**

Figure 3. Signal flow depictions of the processes of replication (RP), transcription (TC) and translation (TL) in the synthesis of a protein. Reproduced from Kremling et al. (2000) with permission from Academic Press © 2000.

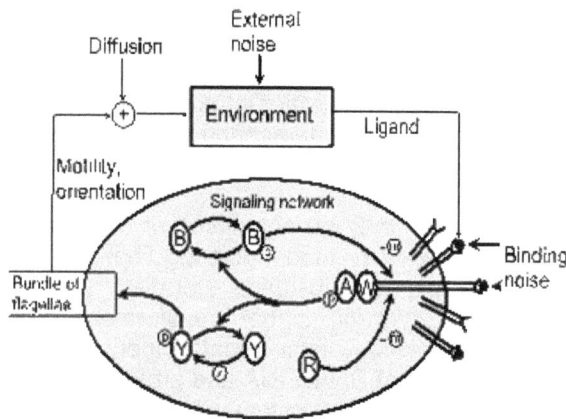

Figure 4. Signal flow diagram of the chemosensory system of *Escherichia coli*. Redrawn from Andrews et al. (2006) © Authors 2006.

etc. and determines the transcription efficiency. The product of transcription - mRNA - is an input to the signal transformer of the translation stage, where the translation efficiency is determined for the synthesis of the final protein.

The chemosensory system of *E. coli* (Figure 4) demonstrates the application of the signal-oriented approach to a more complex problem. Contrary to Fickian diffusion, *E. coli* move up a chemical gradient. A chemotactic path comprises alternate straight-line movements ("runs") and corrective changes of direction ("tumbles"). The movements are effected by long helical flagellae attached to rotary motors embedded in the cell surface. When the motors spin counter-clockwise, they propel the cells along straight paths. Clockwise rotations generate tum-

bles. Chemoreceptors projecting out from the cell surface sense the stimuli and generate a network of signals that eventually control the rotary motors. Details of the chemosensory mechanism are described elsewhere (Andrews et al., 2006; Baker et al., 2006).

Kremling and coworkers preferred a different application that perhaps illustrates better both the strengths and the weaknesses of their methodology. They analyzed the diauxic growth of *E. coli* on a mixture of glucose and lactose (Kremling et al., 2001). While the signal-oriented model could express the observed growth patterns, it also allowed variations in the identification of the functional units and in the signal flow diagrams. In their earlier work (Kremling et al., 2000) they had called this flexibility but the multiplicity of competing designs also make it difficult to reach an optimal representation and a reliable interpretation of the biological processes. Similar to cybernetic models of metabolic processes, to be discussed next, the signal-flow method suffers from a surfeit of parameters, many of which have to be estimated independently, and the non-uniqueness of topologies just mentioned.

CELLULAR INTELLIGENCE APPROACH

Ramkrishna and associates proposed a different perspective of microbial metabolism and growth. Like Kremling and coworkers, they were motivated by the inability of mechanistic models to portray and predict certain features of microbial cultures. For example, mechanistic kinetics accounts for steady state variations with the dilution rate in continuous flow bioreactors by invoking an unproven maintenance term but still fails to handle the transient approach to a steady state. Another case is diauxic growth on mixture of two substrates, where the

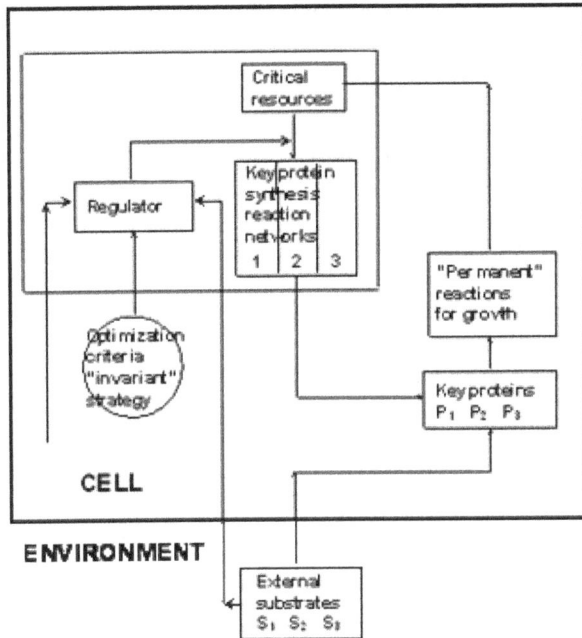

Figure 5. Information flow diagram of a typical cell, showing internal regulatory controls and the effect of the environment. Redrawn from Dhurjati et al. (1985) with permission from John Wiley and Sons Inc., New York ©1985.

mechanistic approach cannot explain why one substrate is ignored until the other is exhausted (Ramkrishna, 2003).

To account for such apparent oddities, Ramkrishna and coworkers (Dhurjati et al., 1985; Straight and Ramkrishna, 1994; Varner and Ramkrishna, 1999) proposed that living cells have internal regulatory controls that enable the cells to exercise judgment in a given set of conditions. An alternate way to describe this is to say that cells possess some rudimentary intelligence that enables them to learn from their experiences and respond to environmental changes in a manner that is most favorable to themselves.

Figure 5 represents a flow sheet of the key stages in a cybernetic modeling framework (Dhurjati et al., 1985). A typical cell contains an "adaptive machinery" that controls metabolic transformations in response to extra-cellular variations. The extra-cellular soup is viewed as a resource pool (of substrates and other nutrients), whose constituents are allocated optimally to different metabolic pathways such that a desired objective (such as cell growth) is maximized. Once the essential proteins are synthesized, a "permanent machinery" carries out the metabolic reactions for replication of cellular material. The third component is a "regulator", and it embodies a crucial feature of cybernetic modeling that distinguishes it from mechanistic modeling. The regulator controls the distribution of resources to achieve the maximization objective referred to above. The cells choose their object-

tives such that their survival is favored at all times. In this sense, the cybernetic approach formalizes the evolutionary approach that Demain (1971) had recognized nearly four decades ago.

Cybernetic modeling begins with the concept of a key enzyme whose synthesis and activity are regulated cybernetically. The utilization of each substrate (in a mixture) is regulated by a corresponding key enzyme. To explain the cybernetic process in simple terms, let n substrates S_1, S_2, ..., S_n contribute to the synthesis of an equal number of proteins P_1, P_2, ..., P_n. Let R_i be the allocation rate of S_i to P_i, and R the total allocation rate. Then $u_i = R_i/R$ is the fractional allocation S_i to P_i. Based on Mandelstam and McQuillen's (1968) work, Dhurjati et al. (1985) considered R to be constant, thereby resulting in the constraint

$$\sum_{i=1}^{n} u_i = \sum_{i=1}^{n} R_i / R = 1 \tag{12}$$

The total rate of production of cell mass, X, is the sum of the concentrations from the individual substrates:

$$\frac{dx}{dt} = -\sum_{i=1}^{n} Y_{si} \frac{ds_i}{dt} \tag{13}$$

where Y_{si} is the yield of cell mass per unit mass of S_i consumed. The rate of consumption of S_i depends, among other factors, on its key enzyme E_i, whose activity, e_i, varies with time. For Monod kinetics,

$$\frac{ds_i}{dt} = -\frac{\mu m_i e_i s_i x}{Y_{si}(K_{si} + s_i)} \tag{14}$$

Here μ_{mi} is the maximum specific growth rate on S_i and K_{si} is a saturation constant. Equation (14) differs from a classic Monod equation by including e_i, whose rate of change is simply the difference between its synthesis and degradation rates:

$$\frac{de_i}{dt} = r_{E_i} - (\beta_i + r_x)e_i \tag{15}$$

The first term represents synthesis and the second is for degradation.

At this point the cybernetic approach invokes its central concept. The activities of the key enzymes are regulated by a set of cybernetic variables λ_i^s such that they are proportional to the returns from the respective enzymes. If r_{ij} is the rate of formation of the i-th product from the j-th resource (or substrate) then:

$$\lambda_{ij}^s = \frac{r_{ij}}{\max_k (r_{ik})} \tag{16}$$

Equation (16) applies to "substitutable" substrates, that is, where any one of a set of substrates may contribute to a synthesis pathway for a product. For "complementary" substrates, where each substrate has a choice of pathways,

$$\lambda_{ij}^c = \frac{r_{ij} / P_{ij}}{\max_k (r_{kj} / P_{kj})} \qquad (17)$$

Expounding on these fundamental ideas, many workers have expanded the cybernetic approach and applied it to different systems. Their studies, like Kremling et al. (2000, 2001) examples of signal-oriented modeling, expose the value as well as some weaknesses of cybernetic descriptions.

By incorporating regulatory controls that enable the cells to utilize information gained from experience and thereby respond intelligently to external conditions, cybernetic modeling is able to overcome the rigidity and the limitations of mechanistic modeling. It has successfully portrayed lag phase behavior, cellular responses to changes in dilution rate, and both diauxic and triauxic growth, features which have been difficult to describe by mechanistic methods (Bapat et al., 2006; Namjoshi and Ramkrishna, 2001; Ramkrishna, 2003). However, as mentioned above, the cybernetic approach too has weaknesses. One weakness is that quite complex models may be required to describe adequately metabolic dynamics of multi-cellular systems, especially under non-ideal conditions. A second difficulty is the inability sometimes to identify a unique cybernetic goal, thus creating uncertainty about the cellular response itself. Recent reviews (Patnaik, 2000, 2001a, 2008) have discussed these aspects in detail and suggested combining cellular intelligence with other methods.

ARTIFICIAL INTELLIGENCE APPROACH

The successes of artificial intelligence (AI) methods in different disciplines and particularly in remote sensing and control, together with the difficulties of obtaining rapid on-line acquisition of intra-cellular data perhaps motiveted the use of AI for microbial processes. Many recent applications provide a testimony to the value of AI in cellular systems.

AI was initially employed in microbial cultures for broadly two purposes (Schugerl, 2001). One is for estimations of time-dependent variables that are difficult or/and expensive to monitor by instrumental methods. The second class of applications was for bioreactor control. The latter use of course required on-line data, provided either by AI methods or by sensory hardware, but it depended also on good models of biological processes. However, it is often difficult to formulate mathematical models that are sufficiently simple, accurate and flexible to be useful under realistic conditions. This difficulty generated a third class of applications of AI, for bioprocess modeling and optimization, and it has also been a driving force for cybernetic models.

The early applications of AI have been reviewed by Patnaik (1998) and by Lubbert and Simutis (1998), whereas more recent work has been discussed by Komives and Parker (2003). These applications have employed different AI methods, notably artificial neural networks, fuzzy logic and genetic algorithms. Often two or more techniques have been used in conjunction, sometimes combined also with classical mathematical models for certain features.

Neural networks seem to be the most favored method to represent microbial processes. They have been used to portray both cell growth and related dynamics (Acuna et al., 1998; Valdez-Castro et al., 2003) of microbial behavior in bioreactors affected by incomplete mixing of the broth and the inflow noise from the environment (Patnaik, 2002), and for early detection of different types of process faults (Vora et al., 1997). However, neural networks, being essentially "black box" devices, suffer from weak physiological links with the biological process, thereby creating difficulties under nonideal conditions (Chen and Rollins, 2000). Other AI methods such as fuzzy logic and genetic algorithms avoid some of these problems but have others of their own.

For instance, the choice of the membership function in fuzzy logic or the fitness function in genetic algorithms is not always known uniquely, and competing candidates may perform equally well within the experimental errors. Nevertheless, both methods have been successful in specific situations. Arnold et al. (2002) study of the adaptation of microorganisms to an inhibiting factor in an industrial brewing process provides a good practical example of the usefulness of fuzzy logic in formalizing the intuitive knowledge of skilled vinegar brewers. Hodgson and coworkers (2005) used genetic programming to reach the interesting conclusion that, for *Saccharopolyspora erythrea* cultures, constrained mathematical forms were superior to flexible unconstrained models, even though no prior knowledge of the fermentation was used.

The strengths and the weaknesses of individual AI methods suggest the possibility of combining them such that the composite model minimizes the overall weakness and/or capitalizes on the strengths. Diverse applications illustrate the validity of this approach for different microbial systems. In one of the early studies, Ye et al. (1994) coupled fuzzy logic with a feed-forward neural network to control β-galactosidase production by recombinant *E. coli*. More complex devices were employed by Coleman et al. (2003) and Fellner et al. (2003). The former maximized the production of green fluorescent protein by *E. coli* through a combination of decision-free analysis, a neural network and a hybrid genetic algorithm. In a novel hybrid network, the latter authors introduced a fuzzy node, a differential equation node and chemometric node into a back propagation network to obtain on-line estimates of diacetyl alcohol in a brewery fermenter.

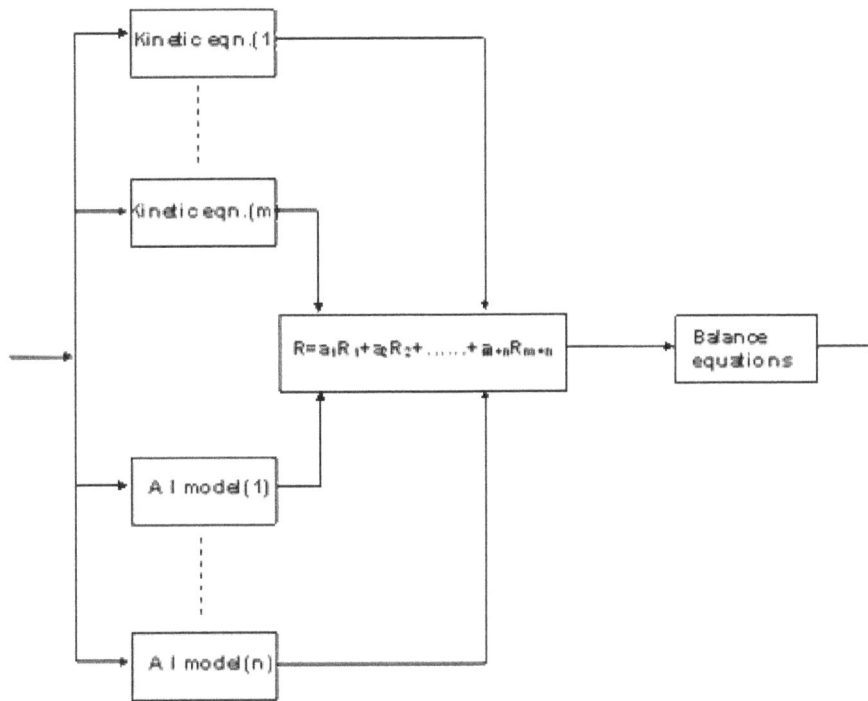

Figure 6. Block diagram of the development of a hybrid model for a microbial reactor. Mechanistic and AI models contribute in a weighted manner to the overall kinetics. The balance equations (for the bioreactor) may be mass balances or AI representations.

All the investigations cited so far have used one or more AI methods but no mechanistic model. While avoiding the weaknesses of mechanistic modeling, they have also lost its physiological relevance. Recent work by Galvanauskas et al. (2004) and by Patnaik (2003a, 2006b) has highlighted the benefits of integrating some mechanistic information with AI methods. Patnaik (2003a, 2006b) has also shown how neural networks may be designed to filter the inflow noise into a bioreactor, an important function for industrial applications.

The basic structure of a hybrid model is portrayed in Figure 6. A set of mechanistic models (e.g. Monod kinetics or substrate inhibition equations) and one or more AI models contribute to the complete kinetic description. Their outputs are fed to a bioreactor model, which may be a set of macroscopic equations or purely AI descriptors or a combination. Figure 6 incorporates Galvanauskas et al. (2004) recommendation to use a weighted combination of the kinetic components but in a more generalized manner. α_j here is the weight assigned to the j-th kinetic model; obviously $\sum_j \alpha_j = 1$ Their values may be determined iteratively.

These composite (or hybrid) neural models still do not account for intra-cellular noise. However, as discussed earlier, noise inside the cells has a vital impact on the metabolic processes (Kaern et al., 2005; Kitano, 2004a; Raser and O'Shea, 2005; Stelling et al., 2004). Moreover,

since substrates supplied from outside participate in metabolic reactions, external noise enters the cells and interacts with internal noise. Although the nature of these interactions are not yet clear, we do have the interesting information that neither source of noise *per se* is entirely detrimental and optimal filtering of either, and preferably both, can indeed enhance cellular functions (Andrews et al., 2006; Patnaik, 2006b; Rao et al., 2002). These observations and the benefits of combining AI methods with equation-oriented models presents the possibility of designing composite architectures that optimally blend cellular intelligence, artificial intelligence and mechanistic models in a framework that links intra-cellular and extra-cellular processes. This concept is discussed below.

CONCEPTUAL DEVELOPMENT OF A COMPREHENSIVE MODEL

Kaern et al. (2005) have provided an elegant exposition of the different sources of noise in a microbial culture and their inter-relationship. Their concept is captured diagrammatically in Figure 1, where the influx of external noise (from the environment) has been added. Environmental noise enters through feed streams, permeates the culture broth and penetrates the cells through the diffusional transfer of substrates. Within the cells, external noise encounters intra-cellular noise. Of the two types of intra-cellular noise, extrinsic noise contributes more substantially to stochasticity in gene expression than intrinsic noise (Elowitz et al., 2002; Kaern et al., 2005; Raser and O'Shea, 2005; Stelling et al., 2004)]. Whereas intrinsic noise causes differences between reporter genes in a single cell, extrinsic noise creates differences between cells. The latter thus has a direct effect on the behavior of a population of cells in a bioreactor and is therefore significant for bioreactor dynamics.

Kaern et al. (2005) and other analysts of cellular noise (Elowitz et al., 2002; Raser and O'Shea, 2005; Rao et al., 2002; Thattai and van Oudenaarden, 2004) concentrated on fluctuations in cellular components and did not consider the impact of environmental noise on these processes. On the other hand, Patnaik (2002, 2003a, 2006b) and others (Coleman et al., 2003; Galvanauskas et al., 2004; Gadkar et al., 2005; James et al., 2002) studying bioreactor problems focused on observable macroscopic variations, either without intra-cellular detail or by lumping molecular-level fluctuations into simple mathematical descriptions. Both approaches have value in their respective spheres but neither is complete, so here we will try to evolve a way to concatenate them to develop a comprehensive description of microbial processes.

The development of even a conceptual strategy needs to recognize the central role of yet another source of fluctuations that interfaces those at the cellular level with macroscopic fluctuations induced by external variations. This is the *binding noise* discussed by Andrews et al. (2006). Although their work pertains to chemotaxis, the idea and their model should be applicable to any cellular system. The mechanism of chemotaxis involves the binding of chemical ligands to receptor clusters protruding from the surfaces of cells. Since small molecules in low concentrations are involved, stochastic fluctuations are associated with this process (Kaern et al., 2005; Raser and O'Shea, 2005). Andrews et al. (2006)] argued that bacterial cells possess an inherent mechanism to filter this noise optimally. To model this mechanism, they relied on Yi et al.'s (2000) observation that the presence of an integral feedback system (Figure 7a) imparts robust adaptation to the cells. Such a mechanism may be modeled by a differentiator in series with a low-pass filter (Figure 7b). In a more general context of any cellular reaction system, the ligand-receptor binding may be replaced by the binding of substrate molecules to active sites on the cell surface, thus retaining the validity of Andrews et al.'s (2006) depiction.

Feedback circuits are the most common device to regulate noise at the genetic level. Negative feedback attenuates fluctuations whereas positive feedback amplifies them. While many examples of feedback regulation in biological systems are known and many of their mechanisms have been identified (Kaern et al., 2005; Raser and O'Shea, 2005; Rao et al., 2002; Thattai and van Oudenaarden, 2004), the construction of gene circuits that that impart the desired feedback features is still at an early stage. However, Becskei et al. (2001) designs of negative

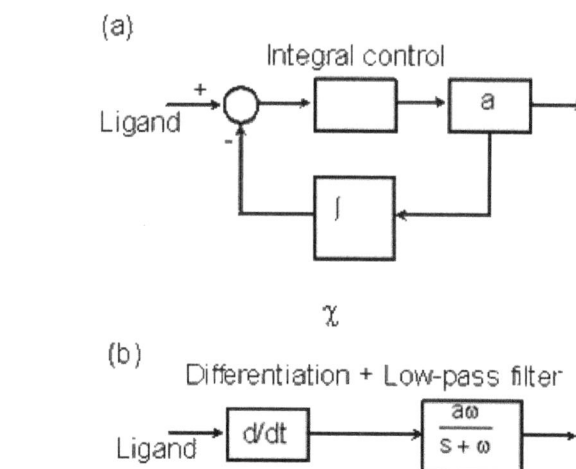

Figure 7. Regulatory system for filtering of binding noise. (a) Schematic representation of the filtering process. (b) Equivalent analog model. Redrawn from Andrews et al. (2006) ©Authors 2006.

negative (Becsei and Serrano, 2000) and positive (Becsei et al., 2001) feed-back modules in *E. coli* to control the fluctuations of a green fluorescence protein used as a marker of gene expression indicate the feasibility of synthetic gene circuits. Interestingly, simple negative feedback corresponds to a low-pass filter, which also forms part of an integrated feedback model (Andrews et al., 2006). This similarity between the two main modes of cellular feedback stabilization makes it convenient to use similar noise filtering methods.

The macroscopic broth in which the cells are immersed is subject to noise from the environment as well as dispersion in the broth itself. Although environmental noise, carried mainly by inlet streams, may be reduced by classical algorithmic filters such as the extended Kalman filter, the low-pass Butterworth filter and the cusum filter, the static nature of these devices restricts their adaptability to varying disturbances, and thus undermines their effectiveness. Neural networks as filtering devices perform much better (Gadkar et al., 2005; Patnaik, 2001b, 2002). However, these networks may be difficult to train, have limited extrapolation capability and have little organic connection with the metabolic processes. These limitations have led to the development of hybrid filters that combine neural networks for some variables with algorithmic filters for others (Galvanauskas et al., 2004; Hodgson et al., 2004; James et al., 2002; Patnaik, 2006b).

The development of hybrid filters seems to have been motivated by the success of similar modules for bioreactors *per se*. While good mixing can be achieved in small laboratory-scale bioreactors, this is difficult in large reactors. The presence of spatial gradients (Larsson et al., 1996; Liden, 2001) and the limitations as well as the

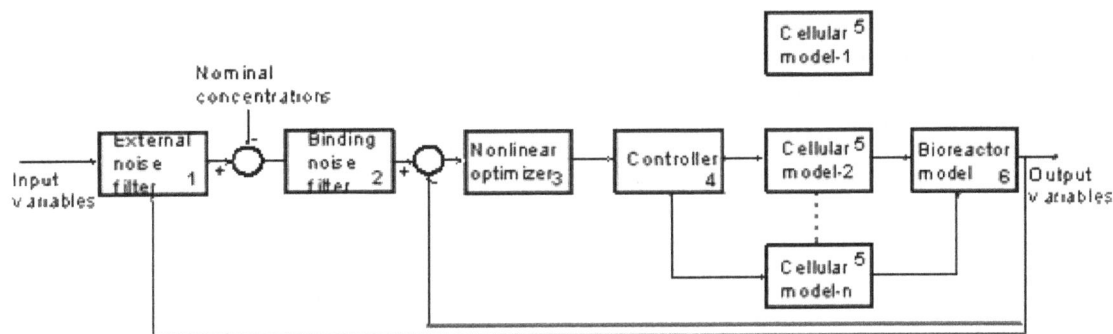

Figure 8. Flow sheet of a conceptual composite model for a microbial system. The model includes both intra-cellular and extra-cellular processes. Previous studies suggest the following representations for the components: Box 1 = auto-associative neural network, Box 2 = Differentiator + Low-pass filter, Box 3 = Optimization algorithm or AI system, Box 4 = Feed-forward neural network, Box 5 = Flexible combination of mechanistic, cybernetic and AI models, and Box 6 = Macroscopic balances + Elman neural network.

costs of instrumental methods make it impractical to have on-line measurement of all important variables (Mandenius, 2004; Sonnleiter, 2000). Off-line measurements can be too slow for rapid data feedback and control, especially in response to disturbances. Moreover, the time-varying nature of external noise and of dispersion in the broth often invalidate conventional kinetic equations developed on the basis of laboratory-scale observations. As a result control strategies that require a good model of a bio-process have limited validity in nonideal situations. Hybrid models overcome these weaknesses by enabling rapid on-line estimations of variables that are difficult to measure and by implementing "intelligent" control policies.

Most hybrid models of microbial processes (Coleman et al., 2003; Galvanauskas et al., 2004; James et al., 2002; Lubbert and Jorgensen, 2001) have used combinations of one or more AI methods with mathematical equations. However, both AI and equation-oriented descriptions ignore internal regulatory controls that cybernetic models recognize. Therefore Patnaik (2006c) recently proposed a conceptual approach that models bioreactors through arrays of neural networks, cybernetic models and AI models. He did not, however, include either binding noise or genetic noise. Since these are not isolated from each other or from extra-cellular noise and dispersion, a more holistic model should embrace all of these. Figure 8 presents a flow sheet for such a model.

Inflow streams are first filtered to prune environmental noise present in them (box 1). This filter generates the same output variables as it receives but with reduced noise. An auto-associative neural network is germane to this requirement and many applications (Patnaik, 2001b, 2003a, 2006b) have shown that it is superior to algorithmic filters and other neural configurations. On entering the cytoplasm, the substrate molecules bind to active sites on the cell surfaces; box 2 in Figure 8 contains the binding noise filter, which may be represented by a differentiator coupled to a low-pass filter (Andrews et al.,

2006; Rao et al., 2002). The cellular reactions, leading to biomass growth and the formation of products, begin after the binding process. This is a critical phase, both biologically and for model building. Kinetic equations derived from observations of ideal laboratory-scale fermentations often cannot be translated directly to large non-ideal bioreactors (Liden, 2001; Lubbert and Jorgensen, 2001). AI and cellular intelligence (cybernetic) models provide more accurate, flexible and faithful descriptions. Since each method has strengths and weaknesses, a judicious combination of more than one approach is often recommended (Arnold et al., 2002; Coleman et al., 2003; Galvanauskas et al., 2004; Hodgson et al., 2004; James et al., 2002; Patnaik, 2003a; Ramkrishna, 2003). The sequence of blocks labeled (5) illustrates this idea. For generality, the microbial process is considered to have an arbitrary number of 'n' variables, each of whose rates of change is described by an appropriate cellular model.

The bioreactor itself may be represented by either macroscopic mass balances or by neural networks or by a combination of the two. If neural networks are used, a recurrent network of the Elman (1990) type turns out to be the best. The reason is that an Elman network has internal feedback loops that mimic the internal recirculation streamlines in a reactor with finite dispersion (Patnaik, 2003a, 2006c). The entire process is under neural control (box 4) because this is more efficient than adaptive PID control (Gadkar et al., 2005; Hisbullah et al., 2002; Patnaik, 2003a). Box (3) is a critical component in that it compares input and output data and thereby adjusts the settings of the controller dynamically. For a PID controller this involves manipulating the gain and the integral and differential time constants, whereas a neural controller requires adjustments of the weights associated with the inter-neuron signal flows. Although an AI routine may be employed for the optimizer, algorithmic multivariable optimization methods may also be acceptable if the variables of interest do not have widely different dynamics.

CONCLUDING REMARKS

Owing to the complexity of the processes involved, quantitative descriptions of cellular systems functioning in realistic situations have focused on either microscopic intracellular phenomena or a macroscopic observable behavior. Either approach generates workable models that are useful for their intended limited purposes but not beyond. Macroscopic models account for dispersion in the culture broth, transport processes and the inflow of noise form the environment, but they tend to ignore or simplify metabolic detail, intra-cellular controls and noise within the cells. Models focused on the cells *per se* describe the latter features in detail while lumping or ignoring their macroscopic manifestations.

Given that cellular processes differ widely, and even a given process behaves differently under different conditions, it is worthwhile to develop comprehensive descriptions that incorporate both intra- and extra-cellular processses under realistic conditions. The resulting model(s) may then be tailored for specific applications. The versatility and usefulness of composite models are suggested by examples of the use of two or more descriptive methodologies to simulate and optimize microbial processes. These approaches are broadly of three types – mechanistic (or equation-oriented), cellular intelligence (or cybernetic) and artificial intelligence.

Based on the topologies of the hybrid models that have been used for microbial reactors, a conceptual framework for a comprehensive composite model is devised. The flow sheet for such a model accommodates (a) fluid dispersion in the broth, (b) environmental noise, (c) genetic noise and (d) ligand or substrate binding noise in a flexible manner. Each phenomenon may described by a suitable modeling approach and, according to each application, one or more of them may be simplified or dispensed with or assigned a suitable weightage. This flexibility also allows the relative contributions of different processes to the overall behavior of a microbial reactor to be adjusttable on-line, a feature that is useful when flux distributions or product patterns or morphology or other relevant characteristics of the cells change as a fermentation progresses.

REFERENCES

Acuna G, Latrille E, Beal C, Corrien G (1998). Static and dynamic neural network models for estimating biomass concentration during thermophilic lactic acid bacteria batch cultures. J. Ferment. Bioeng. 85: 615-622.

Andrews BW, Yi T-M, Iglesias P (2006). Optimal noise filtering in the chemotactic response of Escherichia coli. PloS Comput. Biol. 2: 1407-1418.

Arnold S, Becker T, Delgado A, Emde F, Enenkel A (2002). Optimizing high strength acetic acid bioprocess by cognitive methods in an un- steady state cultivation. J. Biotechnol. 97: 133-145.

Bailey J (1998). Mathematical modeling and analysis is biochemical engineering: past accomplishment and future opportunities. Biotechnol. Prog. 14: 8-20.

Baker MD, Wolanin PM, Stock JB (2006). Systems biology of bacterial chemotaxis. Curr. Opin. Microbiol. 9: 187-192.

Balaban NQ, Merrin J, Chait R, Kowalik L, Leibler S (2004). Bacterial persistence as a phenotypic switch. Science 305: 1622-1625.

Bapat PM, Sohoni SV, Moses TA, Wangikar PP (2006). A cybernetic model to predict the effect of freely available nitrogen substrate on rifamycin B production in complex media. Appl. Microbiol. Biotechnol. 72: 662-670.

Becsei A, Serrano L (2000). Engineering stability in gene networks by autoregulation. Nature 405: 590-593.

Becskei A, Seraphin B, Serrano L (2001). Positive feedback in eukaryotic gene networks: cell differentiation by graded to binary response conversion. EMBO J. 20: 2528-2535.

Carson E, Feng DD, Pons ML, Socini-Sessa R, van Straten G (2006). Dealing with biological and ecological complexity: Challenges and opportunities. Ann. Revs. Control. 30: 91-101.

Carson JM, Doyle J (2002). Complexity and robustness. Proc. Natl. Acad. Sci. USA 99 Suppl. 1: 2538-2545.

Chen VCP, Rollins DK (2000). Issues regarding artificial neural network modeling for reactors and fermenters. Bioprocess Eng. 22: 85-93.

Coleman MC, Buck KK, Block DE (2003). An integrated approach to optimization of Escherichia coli fermentations using historical data. Biotechnol. Bioeng. 84: 274-285.

Demain AL (1971). Overproduction of microbial metabolites and enzymes due to alteration of regulation. Adv. Biochem. Eng./Biotechnol. 1: 113-142.

Dhurjati P, Ramkrishna D, Flickinger C, Tsao GT (1985). A cybernetic view of microbial growth: modeling cells as optimal strategists. Biotechnol. Bioeng. 27: 1-9.

Dun WB, Ellis DI (2005). Metabolomics: current analytic platforms and methodologies. Trends Analyt. Chem. 24: 285-294.

Elman J (1990). Finding structure in time. Cognitive Sci. 14: 1789-1811.

Elowitz M, Levine A, Siggle E, Swain P (2002). Stochastic gene expression in a single cell. Science. 297: 1183-1186.

Fellner M, Delgado A, Becker T (2003). Functional nodes in dynamic neural networks for bioprocess modeling. Bioprocess Biosyst. Eng. 25: 263-270.

Gadkar KG, Mehra S, Gomes J (2005). On-line adaptation of neural networks for bioprocess control. Comput. Chem. Eng. 29: 1047-1057.

Galvanauskas V, Simutis R, Lubbert A (2004). Hybrid process models for process optimization, monitoring and control. Bioprocess Biosyst. Eng. 26: 393-400.

Garhyan P, Elnashaie SSEH (2004). Static/dynamic bifureation and chaotic behavior of an ethanol fermenter. Indl. Eng. Chem. Res. 43: 1260-1273.

Garhyan P, Elnashaie SSEH, Haddad SA, Ibrahim G, Elshishini SS (2003). Exploration and exploitation of bifurcation/chaotic behavior of a continuous fermenter for the production of ethanol. Chem. Eng. Sci. 58: 1479-1496.

Gombert AK, Nielsen J (2000). Mathematical modeling of metabolism. Curr. Opin. Biotechnol. 11: 180-186.

Haag JE, Wouwer AV, Bogaerts P (2005). Dynamic modeling of complex biological systems: a link between metabolic and macroscopic description. Math. Biosci. 193: 25-49.

Hisbullah M, Hussain MA, Ramachandran KB (2002). Comparative evaluation of various control schemes for fed-batch fermentation. Bioprocess Biosyst. Eng. 24: 309-318.

Hodgson BJ, Taylor CN, Ushio M, Leigh JR, Kalganova T, Baganz F (2004). Intelligent modeling of bioprocesses: a comparison of structured and unstructured approaches. Bioprocess Biosyst. Eng. 26: 353-359.

James S, Legge R, Budman H (2002). Comparative study of black-box and hybrid estimation methods in fed-batch fermentation. J. Process Contr. 12: 113-121.

Kaern M, Elston TC, Blake WJ, Collins JJ (2005). Stochasticity in gene expression: from theories to phenotypes. Nat. Revs. Genet. 6: 451-464.

Kitano H (2004a). Biological robustness. Nat. Revs. Genet. 5: 826-837.

Kitano H (2004b). Cancer as a robust system: implications for anticancer therapy. Nat. Revs. Cancer 4: 227-235.

Komives C, Parker RS (2003). Bioreactor state estimation and control. Curr. Opin. Biotechnol. 14: 468-474.

Kremling A, Bettenbrock K, Laube B, Jahreis K, Lengeler JW, Gilles ED

(2001). The organization of metabolic reaction networks. III Application for diauxic growth on glucose and lactose. Metabol. Eng. 3: 362-379.

Kremling A, Jahreis K, Lengeler JW, Gilles ED (2000). The organization of metabolic reaction networks: A signal-oriented approach. Metabol. Eng. 2: 190-200.

Larsson G, Tornquist M, Wernersson ES, Tragardh C, Noorman H, Enfors SO (1996). Substrate gradients in bioreactors: origins and consequences. Bioprocess Eng. 14: 281-289.

Lengeler JW, Drews G, Schlegel HG (1999). Biology of the Prokaryotes. Oxford: Thieme Verlag, Stuttgardt/Blackwell Science.

Liden G (2001). Understanding the bioreactor. Bioprocess Biosyst. Eng. 24: 273-279.

Lubbert A, Jorgensen S (2001). Bioreactor performance: a more scientific approach for practice. J. Biotechnol. 85: 187-212.

Lubbert A, Simutis R (1998). Advances in modeling for bioprocess supervision and control. In: Subramanian G (Ed) Bioseparation and Bioprocessing. Weinheim: Wiley-VCH, Vol. I, Ch.15.

Mandelstam J, McQuillen K (1968). Biochemistry of Bacterial Growth. Oxford: Blackwell.

Mandenius CF (2004). Recent developments in the monitoring, modeling and control of biological production systems. Bioprocess Eng. 26: 347-351.

Namjoshi A, Ramkrishna D (2001). Multiplicity and stability of steady states in continuous bioreactors. Dissection of cybernetic models. Chem. Eng. Sci. 56: 5593-5607.

Namjoshi AA, Hu WS, Ramkrishna D (2003). Unveiling steady state multiplicity in hybridoma cultures. The cybernetic approach. Biotechnol. Bioeng. 81: 80-91.

Narang A, Konopka A, Ramkrishna D (1997). New patterns of mixed substrate utilization during batch growth of Escherichia coli. Biotechnol. Bioeng. 55: 747-757.

Ortoleva P, Berry E, Brun Y, Fan J, Fontus M, Hubbard K, Jaquaman K, Jarymowycz L, Navid A, Sayyed-Ahmad A, Shrief Z, Stanley F, Tuncay K, WeiTzke E, Wu LC (2003). The Karyote physico-chemical genomic, proteomic, metabolic cell modeling system. OMICS J. Integrat. Biol. 7: 269-283.

Patnaik PR (1998). Neural network applications to fermentation processes. In: Subramanian G. (Ed.) Bioseparation and Bioprocessing. Weinheim: Wiley-VCH, I, Ch. 14.

Patnaik PR (2000). Are microbes intelligent beings? An assessment of cybernetic modeling. Biotechnol. Adv. 18: 267-288.

Patnaik PR (2001a). Microbial metabolism as an evolutionary response: The cybernetic approach to modeling. Crit. Revs. Biotechnol. 21: 155-175.

Patnaik PR (2001b). A simulation study of dynamic neural filtering and control of a fed-batch bioreactor under nonideal conditions. Chem. Eng. J. 84: 533-541.

Patnaik PR (2002). Neural optimization of fed-batch streptokinase fermentation in a nonideal bioreactor. Can. J. Chem. Eng. 80: 920-926.

Patnaik PR (2003a). An integrated hybrid neural system for noise filtering, simulation and control of a fed-batch recombinant fermentation. Biochem. Eng. J. 15: 165-175.

Patnaik PR (2003b). Effect of fluid dispersion on cybernetic control of microbial growth on substitutable substrates. Bioprocess Biosyst. Eng. 25: 315-321.

Patnaik PR (2005). Application of the Lyapunov exponent to detect noise-induced chaos in oscillating microbial cultures. Chaos Solitons Fractals 26: 759-765.

Patnaik PR (2006a). Fed-batch optimization of PHB synthesis through mechanistic, cybernetic and neural approaches. Bioautomation 5: 23-38.

Patnaik PR (2006b). Hybrid filtering to rescue stable oscillations from noise-induced chaos in continuous cultures of budding yeast. FEMS Yeast Res. 6: 129-138.

Patnaik PR (2006c). Synthesizing cellular intelligence and artificial intelligence for bioprocesses. Biotechnol. Adv. 24: 129-133.

Patnaik PR (2008). Intelligent models of the quantitative behavior of microbial systems. Food Bioprocess Technol. (in press).

Ramkrishna D (2003). On modeling of bioreactors for control. J. Process. Contr. 13: 581-589.

Rao CV, Wolf DM, Arkin AP (2002). Control, exploitation and tolerance of intracellular noise. Nature 420: 231-237.

Raser JM, O'Shea EK (2005). Noise in gene expression: origins, consequences, and control. Science 309: 2010-2013.

Richman DD (2001). HIV chemotherapy. Nature 410: 995-1001.

Schugerl K (2001). Progress in monitoring, modeling and control of bioprocesses during the last 20 years. J. Biotechnol. 85: 149-173.

Shimizu H (2002). Metabolic engineering – Integrating methodologies of molecular breeding and bioprocess systems engineering. J. Biosci. Bioeng. 94: 563-573.

Sonnleiter B (2000). Instrumentation of biotechnological processes. Advances in Biochemical Engineering/Biotechnology 66:1-64.

Stelling J, Sauer U, Szallasi Z, Doyle FJ III, Doyle J (2004). Robustness of cellular functions. Cell 118: 675-685.

Straight JV, Ramkrishna D (1994). Cybernetic modeling and regulation of metabolic pathways: growth on complementary nutrients. Biotechnol. Prog. 10: 574-587.

Swain PS, Elowitz MB, Siggia ED (2002). Intrinsic and extrinsic contributions to stochasticity in gene expression. Proc. Natl. Acad. Sci. USA 99: 12795-12800.

Thattai M, van Oudenaarden A (2004). Stochastic gene expression in fluctuating environments. Genetics 167: 523-530.

Valdez-Castro L, Baruch I, Barrera-Cortes J (2003). Neural networks applied to the prediction of fed-batch fermentation kinetics of Bacillus thuringiensis. Bioprocess Biosyst. Eng. 25: 229-233.

Varner J, Ramkrishna D (1998). Application of cybernetic models to metabolic engineering: Investigation of storage pathways. Biotechnol. Bioeng. 58: 282-290.

Varner J, Ramkrishna D (1999). Mathematical modeling of metabolic pathways. Curr. Opin. Biotechnol. 10: 146-150.

Vora N, Tambe SV, Kulkarni BD (1997). Counterpropagation neural networks for fault detection and diagnosis. Comput. Chem. Eng. 21: 177-185.

Wang Q, Wu C, Chen T, Chen X, Zhao X (2006). Integrating metabolomics into systems biology to exploit metabolic complexity: strategies and applications in microorganisms. Appl. Microbiol. Biotechnol. 70: 151-161.

Ye K, Jin S, Shimizu K (1994). Fuzzy neural network for the control of high cell density cultivation of recombinant Escherichia coli. J. Ferment. Bioeng. 77: 663-673.

Yi T-M, Huang Y, Simon MI, Doyle J (2000). Robust perfect adaptation in bacterial chemotaxis through integral feed-back control. Proc. Natl. Acad. Sci. USA 97: 4649-4653.

Zupke C, Stephanopoulos G (1995). Intracellular flux analysis in hybridomas using mass balances and in vitro ^{13}C NMR. Biotechnol. Bioeng. 45: 292-297.

14

The role of vaccine derived polioviruses in the global eradication of polio-the Nigeria experience as a case study

Okonko I. O.[1*], Babalola E. T.[1], Adedeji A. O.[1], Onoja B. A.[1], Ogun A. A.[2], Nkang A. O.[1] and Adu F. D.[1]

[1]Department of Virology, Faculty of Basic Medical Sciences, University of Ibadan College of Medicine, University College Hospital (UCH), Ibadan, University of Ibadan, Ibadan, Nigeria. WHO Regional Reference Polio Laboratory, WHO Collaborative Centre for Arbovirus Reference and Research, WHO National Reference Centre for Influenza.
[2]Department of Environmental Health, Faculty of Public Health, University of Ibadan College of Medicine, University College Hospital (UCH) Ibadan, Nigeria

This review reports the role of vaccine derived poliovirus (VDPV) in the global eradication of poliomyelitis. A vaccine derived poliovirus (VDPV) is a rare strain of poliovirus, genetically mutated from the strain contained in OPV. The OPV contains a weakened or attenuated version of poliovirus, activating an immune response in the body. A vaccinated person transmits the weakened virus to others, who also develop antibodies to polio, ultimately stopping transmission of poliovirus in a community. The World Health Assembly in 1988, resolved to eradicate poliomyelitis from the world by the year 2000 and since then, the Global Polio Eradication Initiative (PEI) of the World Health Organization (WHO) has led to a decline in global polio incidence, from an estimated 350,000 cases in 1988 to under 3,500 in the year 2000, with the last remaining global poliovirus reservoirs confined to parts of Southeast Asia and Sub-Saharan Africa. In the African Region (AFRO) of the WHO, eradication strategies were accelerated following supporting resolutions by WHO's Regional Committee for African in 1995 and the organization for African Unity in 1996. Despite the reported success in National Immunization days (NIDs), establishment of acute flaccid paralysis (AFP) surveillance and accelerated efforts to meet the year 2000 targets including "mopping-up" executed in 1999 and subsequent years, Nigeria, the most populous country in Africa, remains one of the major reservoirs for wild poliovirus transmission. Conversely, American region (AMRO) of the WHO was certified as polio-free in 1994 as was the Western Pacific Region (WPRO) in 2000. Recommendations have therefore being presented on ways of evaluating vaccine administration to boost its output in checkmating the increasing waves of paralytic poliomyelitis (including vaccine associated paralytic poliomyelitis-VPP) and prevalence of wild poliovirus in the country. However, there are obstacles to the global eradication which involve among others, vaccine derived polioviruses (VDPVs) in areas with low oral poliovirus vaccine (OPV) coverage. In addition, long term excretion of neurovirulent immunodeficiency-associated vaccine derived polioviruses (iVDPVs) can lead to poliovirus spread to contacts. Overcoming these obstacles is challenging.

Key words: Global eradication, Poliovirus, poliomyelitis, vaccine administration, vaccine derived poliovirus (VDPV).

BACKGROUND OF THE STUDY

Poliomyelitis is an acute infectious systematic viral disease affecting human and is of widely varying severity from a non specific illness to almost irreversible paralysis or death due to polioviruses of serotypes 1, 2 and 3 infections. In rarely severe cases, death is often due to asphyxiationespecially children below the age of five years, it has remained endemic in some parts of the world including Nigeria. Poliomyelitis has been ravaging many developing countries especially those in sub-Sahara Africa. The cumulated number of children with wild poliovirus recorded has been increasing astronomically especially in Africa continent and particularly in Nigeria (White and Fenner, 1994; WHO, 1997, 2000; Abdulraheem and Saka, 2004).

In 1988, the World Health Assembly launched the Global Polio Eradication Initiative, which aimed to use large-scale vaccination with the oral vaccine to eradicate polio worldwide by the year 2000. Although important progress has been made, polio remains endemic in several countries. Also, the current control measures will likely be inadequate to deal with problems that may arise in the post-polio era. A panel convoked by the National Research Council concluded that the use of antiviral drugs may be essential in the polio eradication strategy (WHO, 1997, 2000; Abdulraheem and Saka, 2004; GPE, 2007; Agbeyegbe, 2007).

The World Health Assembly in 1988, resolved to eradicate poliomyelitis from the world by the year 2000 and since then, the Global Polio Eradication Initiative (PEI) of the World Health Organization (WHO) has led to a decline in global polio incidence, from an estimated 350,000 cases in 1988 to under 3,500 in the year 2000, with the last remaining global poliovirus reservoirs confined tom parts of Southeast Asia and Sub-Saharan Africa. The American region (AMRO) of the WHO was certified as polio-free in 1994 as was the Western Pacific Region (WPRO) in 2000. However, there are obstacles to the global eradication which involve among others, vaccine derived polioviruses (VDPVs) in areas with low oral poliovirus vaccine (OPV) coverage. In addition, long term excretion of neurovirulent immunodeficiency-associated vaccine derived polioviruses (iVDPVs) can lead to poliovirus spread by contacts (WHO, 1997, 2000; Abdulraheem and Saka, 2004; Adu et al., 2007; GPEI, 2007; Agbeyegbe, 2007). Overcoming these obstacles is challenging.

In developing countries like Nigeria, poliomyelitis is a public health problem and it was first brought to focus in 1961. In 1984, the annual coverage on immunization in Nigeria was 21% with Nigeria reporting the highest cases of poliomyelitis in Africa region (EPA, 1989). In 1992 alone, about 957 cases of poliomyelitis were reported to the World Health Organization (Adu et al., 1996).

The effort is how to eradicate poliomyelitis through the efficient and purposeful national immunizations programmme. The vaccine, invented by Albert Sabbin, is easier to give, offers much stronger protection and can beneficially "infect" other family members or neighbors, protecting them too. But in rare cases, it can mutate into something resembling wild polio virus, which can paralyze or kill. Ten billion doses of oral vaccine had been given in the last 10 years, so such mutations are presumably extremely unusual (McNeil, 2007). The rational for providing several dose of OPV is to ensure initial seroconversion

against all the types of poliovirus and not to boost waning immunity. Oral polio vaccine is ineffective if given before the age of six month. This is because of the vaccine neutralization by maternal polio antibodies in the baby and also in the breast milk. Nevertheless, several studies show that among breastfed infants, who are fed OPV in the first three days of life, 20 to 40% develop serum antibodies and 30 - 60% excrete vaccine virus (Halsey and Galazka, 1985; Abdulraheem and Saka, 2004).

Polio often circulates undetected; in only one of 200 infections will it cause paralysis, which signals health officials to look for the virus in the area (McNeil, 2007). Outbreaks of vaccine-derived polio are unusual but not unheard of. Individual cases have been known for years. For example, a former lieutenant governor of Virginia was partly paralyzed in 1973, apparently after changing the diapers of his son, who had received oral vaccine (McNeil, 2007). The first spreading outbreak of a vaccine-derived strain, in which 22 children were paralyzed, was detected in 2001 in the Dominican Republic and Haiti (McNeil, 2007). Experts now believe another took place in Egypt in the late 1980s but went unnoticed amid the much larger numbers of wild-type infections. There have been in the Philippines, Madagascar, China and Indonesia (McNeil, 2007). All were eventually eliminated by immunizing more children, and experts argue that the latest outbreaks were able to spread because, until recently, only 30 to 40% of the children in northern Nigeria were vaccinated. About 70% of the children in Nigeria have been vaccinated now (McNeil, 2007). In 2000, the United States switched to inject vaccine made from killed virus, which cannot mutate. But oral drops with the live, weakened version of the virus are still used in most poor countries, including those where the disease has never been eliminated: Nigeria, India, Pakistan and Afghanistan (McNeil, 2007).

Nigeria indeed is fighting an unusual outbreak of polio caused by mutating polio vaccine; the only remedy is to keep vaccinating children there. The ongoing outbreak in 18 northern states of Nigeria's 36 states started in 2006 and was reported in September 2007 issue of Morbidity and Mortality Weekly Review of CDC (Adeija, 2007). This polio outbreak is only appearing in areas where people are refusing to be vaccinated or where there is not enough oral polio vaccine. Heightened immunization campaign for children was a necessity to stop the endemic from spreading (Adeija, 2007). The best way to overcome the outbreak of vaccine-related polio virus is to increase immunization coverage, making sure that all children get the vaccine (Adu, 2007). This review therefore reports the role of vaccine derived poliovirus (VDPV) in the global eradication of poliomyelitis.

Trends in global eradication of polio infection

The goal of global eradication of poliomyelitis by the year 2000 was approved by the World Health Assembly in

*Corresponding author. E-mail: mac2finney@yahoo.com.

1988 and was adopted by the World Health Organization (WHO). The program has two main objectives: no more cases of poliomyelitis caused by wild polioviruses and the absence of wild poliovirus circulation, as evidenced by examining specimens from humans and the environment (Wright et al., 2000). The effort is how to eradicate poliomyelitis through the efficient and purposeful national immunizations programme. The rational for providing several dose of OPV is to ensure initial seroconversion against all the types of poliovirus and not to boost waning immunity. Oral polio vaccine is ineffective if given before the age of Six month. This is because of the vaccine neutralization by maternal polio antibodies in the baby and also in the breast milk. Nevertheless, several studies show that among breastfed infants, who are fed OPV in the first three days of life, 20 to 40% develop serum antibodies and 30 - 60% excrete vaccine virus (Halsey and Galazka, 1985; Abdulraheem and Saka, 2004).

In line with World Health Organization (WHO) global polio eradication initiative (GPEI), Nigeria and other polio endemic countries have designated National Immunization days (NIDs) for mass vaccination campaigns (GPE, 2007; Agbeyegbe, 2007). The deadline of the global polio eradication initiative coordinated by the WHO is approaching rapidly. Eradication may not be accomplished in the year 2000 itself, but it is imminent (Agbeyegbe, 2007).

The success of this eradication programme could not be achieved without the highly effective, live, attenuated oral poliovaccine (OPV), as formulated by Albert Sabin (Sabin and Boulger, 1973). The advantages of this vaccine are numerous, e.g. it confers lifelong humoral and mucosal immunity, it is cheap to produce and easy to administer and it is fairly stable (Melnick, 1996). There are, however, a few disadvantages, the most prominent being the frequent reversion to neurovirulence upon replication in humans (Macadam et al., 1989; Dunn et al., 1990; Georgescu et al., 1994) and the capacity to cause vaccine-associated paralytic poliomyelitis (VAPP), which occurs in approximately one case per 750,000 first doses of OPV (CDC, 1997).

Extensive use of the two available vaccines, the live attenuated oral poliovirus vaccine (OPV), or the Sabin vaccine, and the inactivated poliovirus vaccine, or the Salk vaccine, has dramatically reduced the numbers of poliomyelitis cases caused by wild poliovirus infection (Ward et al., 1993). The used of OPV still remains unresolved and this is perhaps because of the different neutralizing antibody level in different geographical areas. In temperate climates immunity rates to all the three polio serotypes after the three doses is about 95% whereas a developing country like India, has 50 – 60% (John and Jayabal, 1972). Five doses of OPV are required to get to 90 – 95% seroconversion rates (John, 1967). The prevention of further cases of poliomyelitis is essential if this disease is to be eradicated. This is best achieved by an oral polio vaccine (Sabin) manufactured with all three types of attenuated live virus. At least two; and preferably

three (and sometimes more) doses should be given to all children and to all babies from birth onwards.

The prevention of further cases of poliomyelitis is essential if this disease is to be eradicated. This is best achieved by an oral polio vaccine (Sabin) manufactured with all three types of attenuated live virus. At least two; and preferably three (and sometimes more) doses should be given to all children and to all babies from the age of three months onwards. Intensive immunization campaigns are necessary in the developing countries of the world. Epidemics of paralytic poliomyelitis in the developing countries of the tropics and subtropics have, in fact, shown a threefold increase in the past 10 years and are continuing to increase. Nationwide immunization campaigns are therefore an urgent necessity for all developing countries and, once started, must continue if future epidemics are to be prevented.

Current trends in global polio eradication

In 1981 a WHO-initiated collaborative study on various markers for the intratypic differentiation of polioviruses was conducted (WHO, 1981). The serum neutralization test with cross-absorbed antisera was shown to be superior to all other tests in the study. Since then, new developments in microbiological diagnostics and molecular virology have added new possibilities for the rapid and reliable intratypic differentiation of poliovirus isolates. Cross-absorbed intratype-specific polyclonal rabbit antisera (PAbs) are currently used in an enzyme-linked immunosorbent assay (ELISA) format (Osterhaus et al., 1983; Glikmann et al., 1984). Differences in the antigenic structure between vaccine-related and wild-type poliovirus strains reflect differences in the viral RNA.

It is evident that towards the end of the eradication campaign, the role of the virus diagnostic laboratories is increasing, as all poliovirus isolations from paralytic cases require virological follow-up to elucidate the source of the agent (WHO, 2004b). That is, whether the paralysis is due to a wild-type infection or to VAPP, or to infection with a non-polio enterovirus causing polio-like symptoms (Melnick, 1984; Hayward et al., 1989). Since beginning of 2001, all polio isolates from WHO laboratory network are tested by two Intratypic differentiation (ITD) methods; one antigenic and one genomic. All isolates that are Sabin-like in one test but not Sabin-like in the other are sequenced in VP1.

Recently, fears arose about the successful eradication of polio despite a reduction in polio endemic countries from 125 in 1988 to 7 in 2002, and a decline in reported cases of 1919 annually from about 350,000 (WHO, 2004f) globally during the same period. In 2003, polio cases in the world were n the increase and Nigeria went from being one of the seven countries with endemic polio to reporting the highest number of polio cases in the world (WHO, 2003). New cases of polio in nine hitherto polio-free countries resulted from wild poliovirus genetically linked to the poliovirus endemic in Nigeria (WHO, 2004d), while 7 of the countries-Benin, Burkina Faso, Cameroon, Chad, Cote d'Ivoire, Ghana and Togo were in West African, the other 2 were the Central African Republic and Botswana in the Southern region of Africa (Agbeyegbe, 2007).

The outbreak in Botswana reinforced fears of a global outbreak of polio. The present state of global travel presents a favorable environmental condition for the spread of infectious disease (CDC, 2000c). A remaining challenge to World Health Organization (WHO) Global Polio Eradication Initiative is the continued circulation of wild poliovirus in Nigeria (CDC, 2005b, 2006a, b, 2007). Wild poliovirus type 1 has spread from reservoirs in northern Nigeria (and southern Niger) to 18 previously polio-free countries in 2002 - 2005, from Guinea in the west to Indonesia at the southeastern rim of Asia, re-sulting in over 1200 cases associated with imported virus since 2002 (CDC, 2006b; Adu et al., 2007; Agbeyegbe, 2007).

Moreover, northern Nigeria remains by far the world's largest reservoir for wild poliovirus type 3, with 244 polio wild poliovirus type 3 cases reported in 2005 and 278 wild type 3 cases reported in 2006 (CDC, 2007). Coverage by supplementary immunization with NIDs has also been low and insufficient to block widespread wild poliovirus transmission in northern Nigeria (CDC, 2005b, 2007).

In 2003, some states in Nigeria opted out of the program, disrupting the campaign. Thereafter, Nigeria surpassed India reporting the highest polio cases globally (WHO, 2003; Agbeyegbe, 2007). The 2003 - 2004 polio vaccination boycotts in Nigeria threatened the global polio eradication initiative of having a polio-free world by 2005. Encouraged by the global eradication of small pox from the world in 1980, WHO launched the global polio eradication initiative in 1988 seeking to eradicate polio from the world by the year 2000 (Agbeyegbe, 2007).

By the end of 2003, Nigeria accounted for 45% of all global cases of polio and 70% of all cases in 2004. Within ten months, twelve polio free countries confirmed polio cases resulting from a poliovirus genetically linked to that endemic in northern Nigeria (WHO, 2004). The situation appeared to have been fully resolved when Kano the last state holding out resumed polio vaccination in July 2004 (Agbeyegbe, 2007). Despite the severe challenges, a major milepost to poliovirus eradication has been the disappearance of indigenous wild poliovirus type 2 which was last found in West Africa in 1997, and last found anywhere in the world (in Uttar Pradesh, India) in 1999 (WHO, 2001). Thus, in Nigeria, as in other parts of the world, the only current exposure to poliovirus type 2 is from use of the live, attenuated oral poliovaccine (OPV).

However, in recent years a new dimension of risk was identified with the discovery of highly divergent vaccine-derived polioviruses (VDPVs) (Adu et al., 2007). In September 2002, a type 2 VDPV was isolated from an incompletely immunized 21-month-old boy from Plateau state, in north central, Nigeria. The VDPV isolate differed from the Sabin type 2 (Sabin 2) OPV strain at 2.5% of VP1 nucleotides, and had $3D^{pol}$ sequences derived from a source distinct from any of the Sabin strains (Adu et al., 2007).

Over the past 4 years, there have been localized outbreaks of cVDPV in Hispaniola, Philippines and Madagascar. Retrospective studies have confirmed occurrence of cVDPV in Belarus, Pland, Egypt etc. All cVDPV occurred in an environment of poor OPV coverage. Currently, the most urgent priority is to stop wild poliovirus circulation in Nigeria and other countries in Africa, and strenuous efforts are being made to close the large gaps in immunity to poliovirus in the continent (CDC, 2006b, 2007). Type 2 polioviruses are most readily controlled because of the high immunogenicity of the Sabin 2 component of OPV and the marked tendency of Sabin 2-related viruses to spread to and minimize contacts (Sutter et al., 2004). OPV has been the vaccine of choice for the more than 190 countries which have eliminated polio. OPV remains the only vaccine used by the Global Polio Eradication Initiative to interrupt all wild poliovirus transmission, globally (GPEI, 2008).

Vaccine Derived Polioviruses (VDPVs)

VDPVs are defined as poliovirus isolates having >1% nucleotide sequence divergence in the ~900 nucleotide (nt) region encoding the major capsid protein, VP1 (CDC, 2005a; Kew et al., 2005). This definition follows from the rapid rate of nucleotide sequence evolution in poliovirus (~1% per year) and the normal period of poliovirus excretion of less than 3 months (Alexander et al., 1997; Kew et al., 2005). A vaccine derived polioviruses (VDPVs) is a rare strain of poliovirus, genetically mutated from the strain contained in OPV. The OPV contains a weakened or attenuated version of poliovirus, activating an immune response in the body. A vaccinated person transmits the weakened virus to others, who also develop antibodies to polio, ultimately stopping transmission of poliovirus in a community. Between October 2001 and April 2002, five cases of acute flaccid paralysis associated with vaccine-derived poliovirus (VDPV) type 2 isolates were reported in the southern province of the Republic of Madagascar (Rousset et al., 2003). All cVDPVs occur in an environment of poor or low routine/mass immunization OPV coverage, poor sanitation, tropical condition and crowding (WHO, 1997, 2000; Abdulraheem and Saka, 2004; Adu et al., 2007; GPE, 2007; Agbeyegbe, 2007).

Similarities of VDPV to wild polioviruses

i.) Capacity for sustained person-to-person transmission.
ii.) Significant paralytic attack rate.
iii.) Highly neurovirulent in transgenic mouse model.
iv.) Replicates at 39.5°C.
v.) Undergoes recombination with non-polio entero-

viruses (NPEVs) during circulation

Categories of VDPVs

Two well defined categories of VDPVs have been recognized: (1) immunodeficiency-associated VDPVs (iVDPVs) associated with chronic poliovirus infections (Bellmunt et al., 1999; Halsey et al., 2004; Kew et al., 1998; Khetsuriani et al., 2003; MacLennan et al., 2004; Minor, 2001; WHO, 2004a), and (2) circulating VDPVs (cVPDVs) associated with polio outbreaks in areas with low rates of OPV coverage (Kew et al., 2002, 2005; Liang et al., 2006; Rousset et al., 2003; Shimizu et al., 2004; Yang et al., 2003). A third category, ambiguous VDPVs (aVDPVs) include clinical isolates from patients with on recognized immunodeficiency and not associated with an outbreak (Cherkasova et al., 2002; Georgescu et al., 1997; Korotkova et al., 2003), and environmental isolates whose ultimate sources have not been identified (Blomqvist et al., 2004; CDC, 2005a; Shulman et al., 2000).

Immunodeficiency-associated Vaccine derived polioviruses (iVDPVs)

iVPDVs refers to vaccine derived polioviruses which is associated with chronic poliovirus infections (Adu et al., 2007). This can be excreted over prolonged periods (>6 months) by a small proportion of immunodeficient persons exposed to OPV. World Health Organization iVDPV registry identified only 30 persons excreting iVDPV since the introduction of OPV in 1961 - 1962. Persons with B-cell immunodeficiencies are mostly at risk for iVDPV infections. The first reports of iVDPV came from high-income developed countries (e.g. the USA, Western Europe and Japan) with high OPV coverage (Bellmunt et al., 1999; Halsey et al., 2004; Kew et al., 1998; Khetsuriani et al., 2003; MacLennan et al., 2004; Minor, 2001; WHO, 2004a).

Recent reports include middle-income countries while no iVDPVs have been reported from low-income develop countries where survival rates for persons with B-cell deficiencies are low. Although OPV is not recommended for immunodeficiency patients, that is often inadvertently administered because certain primary immunodeficiencies [e.g. common variable immunodeficiency (CVID) develop later in life]. Exposure is usually from receipts of OPV, though 3 of the known iVDPV infections occurred in immunized persons (Bellmunt et al., 1999; Halsey et al., 2004; Kew et al., 1998; Khetsuriani et al., 2003; MacLennan et al., 2004; Minor, 2001; WHO, 2004a).

Circulating Vaccine-derived Polioviruses (cVDPVs)

This also refers to vaccine-derived polioviruses that are associated with polio outbreaks in areas with low rates of OPV coverage (Adu et al., 2007). Circulating cVDPV cause polio outbreaks during extensive circulation in populations with poor vaccine coverage and hygiene. Low vaccination coverage increases the proportion of non-immune persons in a population; this increases the potential for VDPVs to circulate. cVDPVs have produced several localized polio outbreaks, episodes in different countries such as Belarus (type 2, 1965 - 66), Poland (type 3, 1968), Egypt (30 cases of type 2, 1988 - 1993), Hispaniola (25 cases of type 1, 2000 - 2001), Philippines (3 cases of type 1, 2001), Madagascar (5 cases of type 2, 2002) and 8 independent outbreaks in 8 countries; Egypt, Haiti, Dominican Republic of Congo, Philippines, Madagascar, China, Indonesia and Cambodia have been associated with cVDPVs. A single isolates of vaccine/nonvaccine recombinant type 2 VDPVs were obtained under similar epidemiologic conditions in Peru in 1983 and Parkistan in 2000 (Kew et al., 2005). Prospective and retrospective studies of > 3,600 Sabin isolates have detected 1 additional drifted virus in immunodeficient patient in Argentina (EPI/SEARO/WHO, 2007). The largest documented outbreak (46 polio cases) occurred in the Indonesian Island of Madura (Bellmunt et al., 1999; Halsey et al., 2004; Kew et al., 1998; Khetsuriani et al., 2003; MacLennan et al., 2004; Minor, 2001; WHO, 2004a).

Genetic studies stored isolates suggest that a type 2 cVDPV circulated endemically in Egypt for 10 years (approximately from 1983 to 1993). Outbreaks of cVDPV have been associated with all 3 poliovirus serotypes. 2 independent type 2 cVDPV outbreaks occurred in Madagascar in 2002 and 2005 possibly signaling a higher potential for the emergence of type 2 cVDPVs (Adu et al., 2007).

In very rare instances, the virus in the vaccine can mutate into a form that can paralyze. When the virus regains the ability to circulate, it is called a circulating vaccine-derived poliovirus cVDPV). As with naturally occurring polioviruses, the only protection against cVDPV is full vaccination. The spread of a cVDPV shows that too many children remain under-immunized (GPEI, 2008). A fully immunized population will be protected from all strains of poliovirus, whether wild or vaccine derived (Bellmunt et al., 1999; Halsey et al., 2004; Kew et al., 1998; Khetsuriani et al., 2003; MacLennan et al., 2004; Minor, 2001; WHO, 2004a).

Ambiguous Vaccine-derived Polioviruses (aVDPVs)

This also refers to vaccine derived polioviruses that are associated with clinical isolates from patients with on recognized immunodeficiency and environmental isolates whose ultimate sources have not been identified, and not associated with an outbreak (Blomqvist et al., 2004; CDC, 2005a; Cherkasova et al., 2002; Georgescu et al., 1997; Korotkova et al., 2003; Shulman et al., 2000). There are reports of highly evolved VDPV that do not easily fit the above classifications. For instance, such strains were isolated from immunocompetent patients with vaccine-associated paralytic poliomyelitis (VAPP), a healthy contact of the VAPP case, and from sewage without any known source of excretion, they are called aVDPV. The sewage isolates have similar genetic and antigenic properties as iVDPVs, but measures to identify the infected persons have been unsuccessful (Bellmunt et al., 1999; Halsey et al., 2004; Kew et al., 1998; Khetsuriani et al., 2003; MacLennan et al., 2004; Minor, 2001; WHO, 2004a).

In addition, the finding that the Plateau/02 VDPV isolates had non-capsid sequences derived from nonvaccine enterovirus sources by Adu et al. (2007), suggests that it had been in circulation, as recombination with species C enteroviruses frequently but not invariably (Liang et al., 2006) occurs with circulating wild polio-viruses (Liu et al., 2003) and cVPDVs (Arita et al., 2005; CDC, 2005a; Kew et al., 2002; Rousset et al., 2003; Shimizu et al., 2004; Yang et al., 2003), but has not been observed with iVPDVs (Bellmunt et al., 1999; Kew et al., 1998, 2005; Martin et al., 2000; Yang et al., 2005).

Vaccine/nonvaccine recombinants are occasional found, but are rare among isolates with limited (<1%) capsid sequence divergence from the parental Sabin strain (Arita et al., 2005; Kilpatrick et al., 2004). Seque-nces of iVPDV isolates, by contrast, frequently have many mixed-base positions, indicative of multiple iVDPV lineages infecting the immunodeficient patients (Kew et al., 1998; Martin et al., 2000; Yang et al., 2005). As till date, Plateau/02 was not confirmed as cVDPV because no related VDPV isolates have been found in Nigeria or elsewhere, the VDPV was found in an area where conditions favour VDPV emergence and spread (Adu et al., 2007).

BIOLOGICAL PROPERTIES OF VDPVs

i.) Have capacity to cause paralytic polio in humans.
ii.) Potential or demonstrated capacity for sustained circulation.
iii.) VDPVs have lost key attenuation mutations and resemble WPVs biologically.
iv.) All known cVDPVs but no iVDPVs are recombinants with no structural protein sequences derived from species C enteroviruses, a property associated with poliovirus circulation.
v.) Most VDPVs are antigenic variants of Sabin strains, but antigenic evolution to be faster in iVDPVs than in cVDPVs.

The frequency of VDPVS occurrence is represented in Table 1.

Implications of circulating vaccine-derived poliovirus

A circulating vaccine-derived poliovirus is a rare strain of

poliovirus, genetically changed from its original strain contained in Oral Polio Vaccine (OPV). The emergence of a vaccine-derived poliovirus that can circulate in the population shows that too many children remain under-immunized. A fully-immunized population with OPV will be protected from all strains of poliovirus, whether wild or VDPVs. Although quite rare, cVDPVs are not a new phenomenon and have occurred in various parts of the world. In the past 10 years worldwide: over 10 billion doses of OPV have been administered to more than 2 billion children; 9 cVDPV outbreaks have occurred in 9 countries, in communities with low OPV coverage, resulting in under 200 polio cases; during that period, more than 33,000 children were paralyzed by wild poliovirus while over 3.5 million polio cases were prevented by OPV. cVDPVs in the past have been rapidly stopped with 2 - 3 rounds of high-quality immunization campaigns with OPV. The solution is the same for all polio outbreaks: immunize every child several times with OPV to stop polio transmission, regardless as to the origin (GPEI, 2008).

The Nigerian VDPVS experience: when and how it started

Nigeria has the highest number of polio cases, accounting for 61% of global polio cases and 95% of cases in Africa according to the disease surveillance unit of the World Health Organization (WHO) (Adeija, 2007). Sixty-nine children in Nigeria have been partially paralyzed after weakened viruses from polio vaccines were inadvertently transmitted to people in unvaccinated regions in the north of the country (Adeija, 2007). According to McNeil (2007), officials of the World Health Organization (WHO) fear that news of the outbreak will be a new setback for eradication efforts in northern Nigeria, where vaccinations were halted in 2003 for nearly a year because of rumors that the vaccine sterilized Muslim girls or contained the AIDS virus. During that lull, polio spread to many new countries, although most have snuffed out the small outbreaks that resulted. Seventy (70) of Nigeria's last 1,300 polio cases stemmed from a mutant vaccine virus rather than "wild type" virus, which causes most polio (McNeil, 2007).

The emergence of a circulating vaccine-derived poliovirus in Nigeria reaffirms that not enough children are protected from poliovirus (wild or vaccine-derived) and that much more must be done to reach all children with vaccine. Of the 69 children with cVDPV in Nigeria, 40% had never been vaccinated; 87% were under-vaccinated (three or fewer doses). Consistent with global recommendations, three rounds of trivalent OPV (the recommended vaccine for the type of cVDPV in Nigeria) were conducted in northern Nigeria after the first case was confirmed in 2006. The first round was conducted in November 2006 another in January 2007 and a further

Table 1. Frequency of VDPVS occurrence.

Country Episodes of cVDPVs in:	Type	Year	No. cases
Belarus	Type 2	1965-66	
Poland	Type 3	1968	
Egypt	Type 2	1988-1993	30
Hispaniola	Type 1	2000-2001	25
Philippines	Type 1	2001	3
Madagascar	Type 2	2002	5
Nigeria	Type 2	2006-2007	69
Nigeria: Case study			
States	**Type**	**Year 2006**	**Year 2007**
High risk states in Nigeria	cVDPV-2	**No. cases**	**No. cases**
Bauchi	cVDPV-2	1	2
Borno	cVDPV-2	4	2
Kebbi	cVDPV-2	1	1
Sokoto	cVDPV-2	0	1
Zamfara	cVDPV-2	1	1
Very high risk states in Nigeria			
Kaduna	cVDPV-2	2	5
Kano	cVDPV-2	23	13
Katsina	cVDPV-2	1	4
Jigawa	cVDPV-2	6	1

Source: EPI/SEARO/WHO (2007) and Adu (2008)

round in March 2007. These three rounds of immune-zation have reduced by more than half the number of cVDPV transmission strains and the geograp-hical extent of the virus. In September 2007, an additional dose of trivalent vaccine was administered to children in the 13 high risk northern states, including those where the cVDPV continued to circulate.

In 2000 Bauchi/01 originally identified as Sabin 2 was isolated from Bauchi state, detected by REC primers (Kilpatrick et al., 1996, 1998) to be abnormal Sabin. Sequence studies showed <1% nucleotide change in VP1 but a vaccine/nonvaccine recombination in the capsid 3D region (Adu et al., 2003). In 2002 Plateau/02 VDPV was isolated from incompletely immunized 21 months old child. It was initially identified as Sabin by ITD. Sequences studies showed 22/903 (2.5%) nucleo-tide change in the VP1 region. Sequences upstream 620 of the 5'-UTR and downstream 5840 3C region derived from species C enterovirus unrelated to OPV. Ironically the VDPV did not circulate (Adu et al., 2007). The Plateau/02 VDPV isolated in Kanke LGA of Plateau state, North Central Nigeria. Plateau OPV coverage is 36% while Kanke LGA OPV coverage is 22%. It was quite possible that this VDPV was circulating but was not detected by surveillance (Recombination with type C enterovirus).

Type 2 cVDPV was observed in Nigeria 2005 - 2007, cVDPVs with 5 – 9 nt. changes in 2005 in Kaduna, Lagos, Sokoto and Zamfara States. A retrospective

observation was made on cVDPVs in 2006 and 2007. Noticed by clustering of Sabin 2 cases, not flagged by ELISA screen and 5 - 21 nt. differences from Sabin 2; the most divergent VDPV has 2.3% VP1 change (21/903 nt). There were circulation of many independent lineages of type 2 VDPV in Nigeria in 2005 - 2007; this was detected in 9 states. NIE type 2 cVDPV is characterized by multiple lineages, 9 states in Northern Nigeria and one isolate from Lagos, Southern Nigeria; 70 VP1 sequences (Adu, 2008). Determinants of neurovirulence and markers of attenuation have reverted or were changed or removed by recombination, 5 - 21 nt. differences from Sabin 2 in VP1. Most cVDPV are recombinants with human entero-virus C 3D coding region. The outbreak is significant, especially considering the low attack rate of type 2 polio compared to type 1. A new VDPV screening method is essential for type 2 and type 3 VDPVs, all cVDPVs were NSL in VP1 region using real-time PCR VDPV assay. Most preVDPV and VDPVs were NSL in 3D region using the real-time PCR VDPV assay. Type 2 VDPV has been detected in combination with wild polio in a few cases, and there are orphan cVDPVs. Most have recombined with wild polio/species C enteroviruses in the 3D region, typical of cVDPV. In 2006 and 2007, all the VDPVs except one found high risk (Bauchi, Borno, Kebbi, Sokoto, and Zamfara) and very high risk (Kaduna, Kastina, Jigawa and Kano) northern states where immun-ization coverage has been low. Aftermath of the OPV controversy: hidden rejection/Non-compliance (Adu, 2008).

Information on all cVDPVs in 2006 - 2007, including the cases in Nigeria have been available publicly since April 2007, and have been included in presentations at various polio eradication and global laboratory network meetings (GPEI, 2008). Reports on both the work of the global lab network and on VDPVs in general have been issued as standard every year. Since introduction of monovalent oral polio vaccine against type 1 (mOPV1) in Nigeria, wild poliovirus type 1 has declined: 58 cases have been reported this year as compared to 846 last year. Type 1 polio, which has caused international outbreaks, has a higher paralytic attack rate than the two other types and is the eradication effort's primary target (GPEI, 2008).

Wild poliovirus remains a greater threat to children in Nigeria than vaccine-derived virus. Since 2005, Nigeria has reported over 2000 polio cases due to wild poliovirus. In that same period, there have been 69 cases due to circulating vaccine-derived poliovirus (GPEI, 2008). Summarily, specific areas and children are still being missed by IPDs (particularly in hard to reach and non-compliance states); states in the polio high risk states are not making enough progress in RI outside the IPDs. More effort is needed to sustain the growth in RI services country-wide. Special attention to enhance the population immunity in the polio high risk states to check the outbreak of cVPDV is urgently required.

Lesson learned from Nigeria and Hispaniola Episodes

Nigeria continues to improve its polio immunization activities, both supplementary and routine to stop all polio transmission, including the cVDPV. The critical issue is to achieve high coverage during these activities by reaching all children. The cVDPVs in Nigeria are due to type 2 poliovirus, which was eliminated in the wild in 1999. It is the most responsive of the 3 types of poliovirus to OPV. Previous type 2 cVDPVs were detected in Madagascar in 2002 and 2005 and in Egypt in the 1980 - 90s. Enhancing routine immunization with trivalent OPV (targeting all 3 types of polio) in the northern states is the key to maintaining immunity against type 2 polio, as monovalent OPVs are increasingly used to eradicate type 1 and type 3 wild polioviruses.

i.) Keep in mind that vaccine-derived polioviruses do exist.
ii.) Maintain AFP indicators at pre-eradication levels.
iii.) Maintain high coverage in every district and WHO laboratory networks.
iv.) Maintain NIDs until adequate coverages are reached everywhere.

Factors enhancing VDPVs emergence

High use of OPV during mass campaigns

The use of 2 highly efficient vaccines, the Sabin live oral polio vaccine (OPV) and the Salk inactivated polio vaccine (IPV) resulted in a dramatic decrease in poliovirus morbidity and led to the virtual disappearance of wild polioviruses from most of the world.

IPV: This consists of formalin inactivated wild-type polioviruses. It does not induce adequate immunity in the gastrointestinal tract and does not prevent cryptic virus circulation in communities. Life long immunity to poliomyelitis can be induced with a single dose of inactivated polio vaccine (IPV) administered at 5 or 7 months of age (Salk, 1984).

OPV: This consists of live attenuated poliovirus strains of 3 serotypes (I, II and III). It induces adequate immunity in the gastrointestinal tract and in population with high vaccination coverage, prevents virus circulation. The major shortcoming of OPV is its ability to cause rare cases of vaccine-associated paralytic poliomyelitis in vaccine recipients and unimmunized or non-adequately immunized contact persons.

Prolonged circulation of vaccine-derived poliovirus increases the likelihood of its reversion to a neurovirulent strain that could eventually assume the transmission characteristics of wild polioviruses. Most developed, polio-free countries in recent years have switched to the exclusive of IPV, but the great majority of children throughout the rest of the world are still being vaccinated with OPV. In countries that have already achieved eradication of poliomyelitis but are still using OPV, the only source of rare cases of paralytic poliomyelitis is the vaccine itself (GPEI, 2007, 2008). The risk of VDPV is higher in countries that have already achieved eradication compared to countries having cases of wild poliovirus.

Low vaccination coverage (by increasing non-immune population), including factors contributing to low coverage

Also low OPV coverage resulting into low levels of population immunity favored the selection and transmission of vaccine derived variants with biological properties indistinguishable from those of wild polioviruses in countries such as Nigeria, Afganistan, Somalia and Parkistan. Although quite rare, cVDPVs are not a new phenomenon and have occurred in various part of the world. Nonetheless, the benefits of oral polio vaccine far outweigh the risk of a cVDPVs (Kew et al., 2005).

Human altitudes and error

Human altitudes toward vaccination for example, false rumors in 2003, that the polio vaccine was unsafe leading to the shut down vaccination campaigns in northern Nigeria, which likely contributed to the outbreak of cVDPV, the largest outbreak to date of poliomyelitis caused by VDPV. Ninety nine cases were reported from October 2005 to August 2007, with two additional related cases in neighbouring Niger. There were allegations that the polio vaccine can spread HIV (Kapp, 2004), which further strengthened the boycott and weakened the public's confidence in the vaccine (Agbeyegbe, 2007).

Also, it is a well-known and established fact that no vaccine is entirely safe, and even then, there exists the possibility of human error during administration of vaccine (Clements et al., 1999). However, despite questions of safety, vaccines have continually contributed to disease prevention and control (Plotkin and Plotkin, 1999). Strained relations between the West and Muslims in Northern Nigeria, some of whom identified positively with acts of terrorism (Nigeria World, 2001) provided an enabling environment for advocating boycott of polio vaccination (Agbeyegbe, 2007). Furthermore, deaths from a Pfizer drug trial had created lingering uncertainty in northern Nigeria about the safety of Western health initiative (BBC, 2004).

Political instability

In several countries particularly Afganistan and Somalia, part of Parkistan and Ethiopia political instability or armed conflicts make vaccination logistically difficult and unpredictable. In addition to this internal politics in the 2003 immunization boycott, were ramifications from the international political arena. Anti-western sentiments have increased among Northern Muslims fundamentalist following the September 11, 2001 attacks and America's war on terrorism (Freedom house, 2004; Agbeyegbe, 2007). Given the distrust and growing antagonism towards America, the involvement of the West in a program that benefits Muslims was viewed with suspicion (Dhimmi Watch, 2004).

Cultural and religious objections

Many Nigerians are blaming the outbreak on vaccination efforts; an attitude experts fear may ruin previous gains in eradicating polio in the country. Most of the anti-immunization campaigns in Nigeria have been predominantly Muslim north of Nigeria, and a number of Muslim clerics have been quoted in the Nigerian media as claiming that vaccines are dangerous and cause sterility or illness (Adeija, 2007).

Cultural and religious objections under vaccinations efforts, resulting in persistently low immunity in the population and consequently, a high incidence of vaccine-derived poliomyelitis. The 2003 boycott of national polio vaccination campaigns by some northern Nigerian states threatening the impending success of the global initiative (Agbeyegbe, 2007). Agbeyegbe (2007) examined the role of religion in the boycott, as reflected in the questioning of

the authenticity and the safety of polio vaccine by the Supreme Council of Sharia in Nigeria (SCSN). The influential Islamic preachers by raising questions on immunization and the safety of the OPV laid foundation for the boycott of the northern states of Nigeria from the national immunization programme (Gamii, 2004; Agbeyegbe, 2007) while some predominantly Muslim states in northern, Nigeria implemented the boycott (The Guardian, 2004). Due to the intensity of the religious divide, Muslims dismissed earlier efforts by the Federal Government to assure northern Muslims of the safety of the vaccines as not being credible since Muslims leaders had called vaccination unsafe (Csmonitor, 2004). Religion was however, a proximal factor in the 2003 boycott since principal advocates, SCSN, Jama'atu Nasril Islam (JNI) are Islamic organizations, and their primary concern was for Muslims in Northern Nigeria to boycott the immunization exercise (Ebonugwo and Ndiribe, 2004; Africa Action, 2004; Ishr, 2004; Agbeyegbe, 2007).

Increasingly, some Muslims in the north are resisting compliance with some UNICEF and UN conventions being questioned for their incompatibility with the application of Sharia (Science in Africa, 2004). Whereas the undercurrents between Muslims and Christians in Nigeria as well as Western donors may have been sufficient to begin the controversy on the polio vaccine, other factors helped to sustain it. In October 2003, the Organization of Islamic Countries (OIC) issued a statement calling Muslim nations to speed up polio elimination in their countries. This position by the OIC counters claims that immunizations are an anti-Islamic practice (Marshall, 2003; Agbeyegbe, 2007).

Government negligence

The inability of the Nigerian government to acknowledge the risk involved in vaccination however negligible raises doubt about the sincerity of the government, and positions the boycotts of polio vaccination proponents as a more reliable source of information. The government does not appear to have positioned itself as a credible authority to implement immunization programs. No vaccine is fully safe, a perfectly potent and without risk of administering error (Clements et al., 1999). In 1991, WHO recommended post-immunization surveillance for any nations that implement national immunization programs (WHO, 1991). Beyond monitoring for adverse effects, some countries have established compensation schemes for injuries that result from vaccination (BBC, 2004). Such efforts by government do not necessarily raise suspicion about immunization programs but may rather raise the credibility of government with the public.

Lack of risk communication

Lack of risk communication is one of the factors affecting

polio eradication especially in Nigeria. Although, the boycott of immunization is no longer in effect, low participation during vaccination may persist reflecting a failure to implement risk communication (Agbeyegbe, 2007). The silence of the government over the alleged report by JNI and widely in the media that the government acknowledged the use of contaminated vaccines but claimed that the contaminated batch had been completely used, could be interpreted as indicative of the accurateness of the report. To address such situations, risk communication is increasingly becoming important in public health (Rudd et al., 2003; Rothman and Kiviiniemi, 1999).

Risk communication offers a two-way communication process that presents the expert opinions based on scientific facts to the public, and acknowledges the fears and concerns of the public, seeking to rectify knowledge gaps that foster misrepresentation of risk (Leiss, 2004; Aakko, 2004; Frewer, 2004). The information delivered to the public over the immunization boycott period was capable of affecting their risk perception. Public perception of risk shaped by variables such as bias, values, beliefs and experience may disregard facts and rational reasoning, and be entirely subjective. Individuals perceive risks which they are familiar with (polio cases) and have control over (refuse to submit to vaccination) as more acceptable than unfamiliar risk situations (infertility) over which they have no control over which the individual can exercise no control (compulsory vaccination) (Renn, 2004; Aakko, 2004; Frewer, 2004; Ropeik, 2004).

Prolonged circulation of vaccine virus (among existing gaps of non-immune population), including factors contributing to it

Tropical climate

Results of some studies indicate that circulation of vaccine-derived polioviruses in temperate countries may be limited in time and it is behaved that tropical climates are more conducive than ones to prolonged vaccine virus circulation reason (Abdulraheem and Saka, 2004; Adu et al., 2007; GPEI, 2007; Agbeyegbe, 2007). The World Health Organization has identified as the highest priority research evaluation of persistent vaccine-derived poliovirus circulation in tropical countries

Levels of sanitation

Low rates of OPV coverage coupled with poor sanitation, tropical conditions, and the prior eradication of the corresponding serotype of wild poliovirus are said to be the main risks factors for cVDPV emergence (Kew et al., 2005).

Population densities

Also, low rates of OPV coverage coupled with over-crowding is said to be one of the main risks factors for cVDPV emergence (Kew et al., 2005).

Conclusion and Recommendations

As with the other epidemics in Egypt and Hispaniola, VDPV circulated in a province of Madagascar with low OPV coverage (CDC, 2001; Kew et al., 2002). Because a high OPV coverage rate helps prevent the circulation of both VDPVs and wild PVs, obtaining and maintaining high rates of immunization coverage are essential (Wood et al., 2000). Moreover, two recombinant VDPV lineages in Madagascar indicate that recombination is frequent between OPV and cluster C enteroviruses. Similar recombinant VDPVs have been implicated in the epidemics in Hispaniola and in the Philippines (CDC, 2001a, b). Determining whether the neurovirulence and transmissibility of these VDPVs could be the result of the recombination with non-polio enteroviruses is important. These VDPVs have major implications for the cessation of immunization with OPV after certification that wild PV has been eradicated.

Most polio-free countries no longer immunize against polio. Thus, the global population was vulnerable to polio spreading from Nigeria. The resurgence of polio at a global level would not only mean the $3 billion investment in eradication efforts will be over with, it will also leave the world at the beginning of another eradication effort requiring more financial commitment (WHO, 2004f). The most efficient means to close the immunity gaps in Nigeria and other countries in Africa is by implementation of high quality NIDs, coupled with strengthened routine immunization and sensitive AFP and poliovirus surveillance (Adu et al., 2007). Negotiation for days of tranquility (the suspension of hostilities in order to allow for vaccination) will have to remain an integral part of polio eradication activities for the foreseeable future.

The above mentioned factors have caused great setbacks in the eradication of polioviruses-both the wild types and the vaccine-derived. Nevertheless, only 4 countries where the virus remains endemic-Nigeria, India, Parkistan, and Afghanistan-account for 93% of the world's cases of poliomyelitis, unlike all other countries, they have never succeeded in interrupting the transmission of wild poliovirus. Furthermore, these countries are also faced with the problem of vaccine-derived poliovirus. Wild poliovirus remains a greater threat to children in Nigeria than vaccine-derived virus (GEPI, 2008).

Researchers were preparing vaccine alternatives to combat polio outbreak. Monovalent vaccines (which protect against only one of the three strains) should be used in the areas where the virus is still circulating. But nationwide, the polyvalent vaccine (which protects against all three strains of poliovirus) should be aggressively used and coverage increased (Adu, 2007).

Immunization remains a widely used public health intervention and one of the most cost effective prevention and control measure of poliomyelitis (WHO, 2004e). The smallpox vaccine made possible the global eradication of smallpox (Dittman, 2001). Rates of OPV coverage sufficient to block wild poliovirus circulation, will also be sufficient to prevent cVDPV emergence and spread (Adu et al., 2007). However, once wild poliovirus circulation has stopped, high rates of poliovaccine coverage must be maintained until global cessation of OPV use (WHO, 2004c).

The findings by many authors, that vaccine-derived polioviruses may circulate under suitable conditions prevents an additional challenge to efforts to eradicate polio worldwide. The threshold rates of vaccine coverage needed to suppress circulation of VDPVs are unknown but probably vary by poliovirus serotype and environmental factors (e.g. population density, levels of sanitations and climate) (Kew et al., 2005). However, when OPV coverage rates are sufficient to prevent circulation of wild polioviruses, they probably are sufficient to prevent circulation of VDPVs.

The ability to rapidly distinguish cVPDVs from iVDPVs will assume greater importance in the post-OPV era (WHO, 2004c), when any poliovirus detected 6 months after cessation of OPV use might signal continued poliovirus transmission. During this critical phase, virologic findings, particularly nueclotide sequence information, will be needed to test for links between any sporadic poliovirus infections that may be detected (Adu et al., 2007). Sustainable establishment of a global integrated virologic surveillance network and enhanced molecular characterization surveillance will provide greater insights into its circulation than would be possible with strictly epidemiologic tools. Tracking individual virus lineages will help in knowing the patterns of transmission and for cultural monitoring the progress of eradication effort. Studies on will continue, as risks of cVDPV may change over time. Therefore, intensive virologic surveillance will be especially important in populous high-risk areas, such as Nigeria, which have been major reservoirs for wild poliovirus circulation (Adu et al., 2007). Intensifying polio eradication strategies, all children 0 - 15 years must be sufficiently immunized for high population immunity as well as bringing up a scientifically sound strategy to discontinue global eradication.

Although the polio immunization boycott in the northern states of Nigeria has now been fully suspended in Nigeria, and all 36 states of the nation are implementing the polio vaccination program, the risk of a global outbreak originating from Nigeria may remain. Sharing appropriate information can correct the public's misconceptions on OPV and increase its responsiveness to vaccination programs. In order to achieve this, the authorities should use risk communication strategies (Agbeyegbe, 2007). The successful use of polio vaccine (OPV) has helped to bridge the gap and this has contributed immensely to the limit of spread of the disease. In the study, it was found that most of the mother and guardian did not know how many OPV

(booster doses) their children have received. This indirectly showed the efficient coverage of national immunization programme. In view of these, concerted effort should be made by the government and WHO towards complete eradication of poliomyelitis. Parents and guardians should be educated on the need to allow their children to receive the normal schedule of OPV and booster doses given during national immunization programmes.

REFERENCES

Aakko E (2004). Risk communication, risk perception, and public health. WMJ 103: 25-27.

Abdulraheem IS, Saka MJ (2004). The immune response to polio virus after natural infection and immunization with oral polio vaccine: panoramic view of the issues and problems. Nig. Med. Pract. 46 (3): 50-55.

Adeija A (2007). Vaccine-derived polio spreads in Nigeria. Science and DevelopmentNetwork.SciDev.Net.http://www.scidev.net/News/News/index.cfm?fuseaction-readNews&itemid=3958&language=1.

Adu FD. (2007). In: Vaccine-derived polio spreads in Nigeria, Adeija A. (ed.). Science and Development Network. SciDev.Net. http://www.scidev.net/News/News/index.cfm?fuseaction-readNews&itemid=3958&language=1

Adu FD (2008). VDPVs: Nigeria Experience. Power Point Presentation delivered at EPI/SEARO/WHO. p. 1-6.

Adu FD, Iber J, Bukbuk D, Cumede N, Yang S, Jorba J, Sule WF, Yang C, Burris C, Pallansch M, Harry T, Kew O (2007). Isolation of recombinant type 2 vaccine-derived poliovirus (VDPV) from a Nigerian child. Virus Res. 127:17-25

Adu FD, Iber J, Harry T, Burns C, Oyedele O, Adeniji JA, Osei-Kwasi M, Kilpatrick D, Tomori O, Kew O (2003). Some genetic characteristics of Sabin-like poliovirus isolated from acute flaccid paralysis in Nigeria. Afr. J. Biotechnol. 2: 460-464.

Adu FD, Odemuyiwa SO, Tomori O (1996). Circulation of poliovirus among risk groups in Ibadan, Nigeria. Transactions Royal Society of Tropical Medicine & Hygiene 90:126–127.

Africa Action (2004). Nigeria: Obasanjo on Sharia crisis. http://www.africaaction.org/docs00/nig003. htm. retrieved.

Agbeyegbe L (2007). Risk communication: The over-looked factor in the Nigeria polio immunization boycott crisis. Nig. Med. Pract. 51(3): 40-44.

Arita M, Zhu SL, Yoshida H, Yoneyama T, Miyamura T, Shimizu H (2005). A Sabin 3-derived poliovirus recombinant contained a sequence homologous with indigenous human enterovirus species C in the viral polymerase coding region. J. Virol. 79: 12650-12657.

BBC News (2004). Africa/Nigeria Muslims oppose polio vaccination. http://news.bbc.co.uk/2/hi/africa/ 2070634.stm.

Bellmunt A, May G, Zell R, Pring-Akerblom P, Verhagen W, Heim A (1999). Evolution of poliovirus type 1 during 5.5 years of prolonged enteral replication in an immunodeficient patient. Virol. 265: 178-184.

Blomqvist S, Savolainen C, Laine P, Hirttio P, Lamminsalo E, Penttila E, Joks S, Roviainen M, Hovi T (2004). Characterization of highly evolved vaccine-derived poliovirus type 3 isolated from sewage in Estonia. J. Virol. 78: 4876-4883.

Centers for Disease Control and Prevention (2001a). Circulation of a type 2 vaccine-derived poliovirus—Egypt, 1982–1993. MMWR Morb. Mortal. Wkly. Rep. 50:41–42.

Centers for Disease Control and Prevention (2001b). Acute flaccid paralysis associated with circulating vaccine-derived poliovirus—Philippines, 2001. Morb. Mortal. Wkly. Rep(MMWR). 50:874–5.

Centers for Disease Control and Prevention (CDC) (1997). Paralytic poliomyelitis-United States, 1980-1994. Morbidity and Mortality Weekly Report 46: 79-83.

Centers for Disease Control and Prevention (CDC) (2006). Enterovirus surveillance--United States, 2002-2004. Morbidity and Mortality Weekly Report (MMWR) 17: 55(6): 153-156.

Centers for Disease Control and Prevention. (CDC) (2000b).

Enterovirus surveillance United States, 1997–1999. Morbidity and Mortality Weekly Report 49: 913–916.

Centre for Disease Control and Prevention (CDC) (2000a). Progress towards poliomyelitis eradication: African Region. J. Am. Med. Assoc. (JAMA) 14:1781-1782.

Centre for Disease Control and Prevention (CDC) (2000c). Preventing emerging infectious diseases-A strategy for the 21st century. CDC, Atlanta. pp. 3.

Centre for Disease Control and Prevention (CDC) (2005a). Laboratory surveillance for wild and vaccine-derived polioviruses, January 2004-June 2005. Morbidity and Mortality Weekly Report (MMWR) 54:958-961.

Centre for Disease Control and Prevention (CDC) (2005b). Progress towards poliomyelitis eradication-Nigeria, January 2004-July 2005. Morb. Mort. Wkly. Rep. (MMWR). 54:958-961

Centre for Disease Control and Prevention (CDC) (2006a). Progress towards interruption of wild poliovirus transmission-worldwide, January 2005-March 2006. Morbidity and Mortality Weekly Report (MMWR) 55:458-462.

Centre for Disease Control and Prevention (CDC) (2006b). Resurgence of wild poliovirus type 1 transmission and consequences of importation-21 countries, 2002-2005. Morbidity and Mortality Weekly Report (MMWR) 55:145-150.

Centre for Disease Control and Prevention (CDC) (2007). Progress towards poliomyelitis eradication-Nigeria, 2005-2006. Morbidity & Mortality Weekly Report (MMWR) 56:278-281.

Cherkasova EA, Korotkova EA, Yakovenko ML, Ivanova OE, Eremeeva TP, Chumakov KM, Agol VI (2002). Long-term circulation of vaccine-derived poliovirus that causes paralytic disease. J. Virol. 76: 6791-6799.

Clements CJ, Evans G, Dittman S, Reeler AV (1999). Vaccine safety concerns everyone. Vaccine S90-94.

CS Monitor. 2004. Nigerian Islamist veto vaccines. http://csmonitor.com/2004/0330/p06s01woaf.htm.

Dhimmi W (2004). Sharia vs. polio in Nigeria. http://www.jihadwatch.org/dhimmiwatch/archives/ 000957.php.

Dittman S. 2001. Vaccine safety: risk communication-a global perspective. Vaccine 19: 2446-2456.

Dunn G, Begg NT, Cammack N, Minor PD (1990). Virus excretion and mutation by infants following primary vaccination with live oral poliovaccine from two sources. J. Med. Virol. 32: 92-95.

Ebonugwo M, Ndiribe O (2004). polio vaccine controversy: between politics and science. Vanguard, Lagos, Nigeria.

EPI/SEARO/WHO (2007). Circulating Vaccine Derived Polioviruses (cVDPV). Power Point Presentation delivered at EPI/SEARO/WHO. pp. 1-2.

Frewer L (2004). The public and effective risk communication. Toxicol. Lett. 149: 391-397.

Gamii News (2004). The Islamic perspective of immunization. http://www.gamii.com/NEWS1602.htm.

Georgescu MM, Balanant J, Macadam A, Otelea D, Combiescu M, Combiescu AA, Crainic R, Delpeyroux F (1997). Evolution of the Sabin type 1 poliovirus in humans: characterization of strains isolated from patients with vaccine-associated paralytic poliomyelitis. J. Virol. 71: 7758-7768.

Georgescu MM, Delpeyroux F, Tardy-Panit M, Balanant J, Combiescu, M, Combiescu, AA, Guillot S, Crainic R (1994). High diversity of poliovirus strains isolated from the central nervous system from patients with vaccine-associated paralytic poliomyelitis. J. Virol. 68: 8089-8101.

Glikmann G, Moynihan M, Petersen I, Vestergaard BF (1984). Intratypic differentiation of poliovirus strains by enzyme-linked immunosorbent assay (ELISA): poliovirus type 1. Dev. Biol. Stand.55: 199–208.

Global Polio Eradication (GPE) (2007). Circulating Vaccine-derived polio virus. Page available at URL-http://www.polioeradication.org.

Global Polio Eradication Initiative (GPEI) (2008). Circulating Vaccine-derived polio virus: Implications of circulating vaccine-derived poliovirus. Page available at URL-http://www.polioeradication.org.

Halsey N, Galazka A (1985). The efficacy of DPT and oral poliomyelitis immunization schedules initiated from birth to 12 weeks of age. Bulletin of World Health Organization 63: 1151-1169.

Halsey NA, Pinto J, Espinosa-Rosales F, Faure-Fontenia MA, da Silva

E., Khan AJ, Webster ADB, Minor PD, Dunn G, Asturias E, Hussain H, Pallansch MA, Kew OM, Winkelstein J, Sutter R, Polio Project Team (2004). Search for poliovirus carriers in persons with primary immune deficiency diseases in the United States, Mexico, Brazil, and the United Kingdom. Bulletin of WHO 82: 3-8.

Hayward JC, Gillespie SM, Kaplan KM, Packer R, Pallansch M, Plotkin S, Schonberger LB (1989). Outbreak of poliomyelitis-like paralysis associated with enterovirus 71. Pediatr. Infect. Dis. J. 8: 611-616.

Ishr Organization (2004). Sharia in Nigeria. http://www.ishr.org/activities/campaigns/stoning/ sharianigeria.html.

Kapp C (2004). Nigerian states again boycott polio-vaccination drive. The Lancet p. 709.

Kew OM, Morris-Glasgow V, Landaverde M, Burns C, Shaw J, Garib Z, Andre J, Blackman E, Freeman CJ, Jorba J, Sutter RW, Tambini G, Venczel L, Pedreira C, Laender F, Schimizu H, Yoneyama T, Miyamura M, de Quadros C (2002). Outbreak of poliomyelitis in Hispaniola associated with circulating type 1 vaccine derived poliovirus. Sci. 296:356-359.

Kew OM, Mulders MN, Lipskaya GY, da Silva EE, Pallansch MA (1995). Molecular epidemiology of polioviruses. Seminars in Virol. 6: 401-414.

Kew OM, Sutter RW, de Gourville EM, Dowdle WR, Pallansch MA (2005). Vaccine-derived polioviruses and the endgame strategy for global polio eradication. Annu. Rev. Microbiol. 59: 587-635.

Kew OM, Sutter RW, Nottay B, McDonough M, Prevots DR, Quick L, Pallansch M (1998). Prolonged replication of a type 1 vaccine-derived poliovirus in an immunodeficient patient. J. Clin. Microbiol. 36: 2893-2899.

Khetsuriani N, Prevots DR, Quick L, Elder ME, Pallansch M, Kew O, Sutter RW (2003). Persistence of vaccine-derived polioviruses among immunodeficient persons with vaccine-associated paralytic poliomyelitis. J. Infect. Dis. 188:1845-1852

Kilpatrick DR, Ching K, Iber J, Campagnoli R, Freeman CJ, Mishrik N, Pallansch MA, Kew OM (2004). Multiplex PCR method for identification of recombinant vaccine-related polioviruses. J. Clin. Microbiol. 42: 4313-4315.

Kilpatrick DR, Nottay B, Yang SJ, da Silva E, Penaranda S, Pallansch MA, Kew OM (1998). Serotype-specific identification of polioviruses by PCR using primers containing mixed-base or deoxyinosine residues at positions of codon degeneracy. J. Clin. Microbiol. 36: 352-357.

Kilpatrick DR, Nottay B, Yang SJ, Mulders MN, Holloway BP, Pallansch MA, Kew OM (1996). Group-specific identification of polioviruses by PCR using primers containing mixed-base or deoxyinosine residues at positions of codon degeneracy. J. Clin. Microbiol. 34: 2990-2996.

Korotkova EA, Park R, Cherkasova EA, Lipskaya GY, Chumakov KM, Feldman E, Kew OM, Agol VI (2003). Retrospective analysis of a local cessation of vaccination against poliomyelitis: a possible scenario for the future. J. Virol. 77: 12460-12465.

Leiss W (2004). Effective risk communication practice. Toxicol. Lett. 149:399-404.

Liang X, Zhang Y, Xu W, Wen N, Zou S, Lee LA, Yu J (2006). An outbreak of poliomyelitis caused by type 1 vaccine-derived poliovirus in China. J. Infect. Dis. 194: 545-551.

Liu H-M, Zheng D-P, Zhang L-B, Oberste MS, Kew OM , Pallanch MA (2003). Serial recombination during circulation of type 1 wild-vaccine recombinant polioviruses in China. J. Virol. 77:10994-11005.

Macadam A, Arnold C, Howlett J, John A, Marsden S, Taffs F, Reeve P, Hamada N, Wareham K, Almond J, Cammack N, Minor PD (1989). Reversion of the attenuated and temperature sensitive phenotypes of the Sabin type 3 strain of poliovirus in vaccinees. Virol. 172: 408-414.

MacLennan C, Dunn G, Huissoon AP, Kumararatne DS, Martin J, O'Leary P, Thompson RA, Osman H, Wood P, Minor P, Wood DJ, Pillay D (2004). Failure to clear persistent vaccine-derived neurovirulent poliovirus infection in an immunodeficient man. Lancet 363: 1509-1513.

Marshal SJ (2003). Islamic states renew commitment to eradicate polio. Bulletin of WHO 18: 918.

Martin J, Dunn G, Hull R, Patel V, Minor PD (2000). Evolution of the Sabin strain type 3 poliovirus in an immunodeficient patient during the entire 637-day period of virus excretion. J. Virol. 74: 3001-3010.

McNeil Jr. DG (2007). Polio in Nigeria Traced to Mutating Vaccine. An

American Empirical Pictures by Wes Anderson.

Melnick JL (1984). Enterovirus type 71 infections : a varied clinical pattern sometimes mimicking paralytic poliomyelitis. Rev. Infect. Dis. 6: S387-S390.

Melnick JL (1996). Enteroviruses : polioviruses, coxsackieviruses, echoviruses, and newer enteroviruses. In: Fields Virology, 3rd edn, pp. 655-712. Edited by B. N. Fields, D. M. Knipe & P. M. Howley. Philadelphia : Lippincott±Raven.

Minor P (2001). Characteristics of poliovirus strains from long-term excretors with primary immunodeficiencies. Dev. Biol. 105: 75-80

Nigerian World Letters and Viewpoints (2001). The 'Talibans' of Northern Nigeria. http://nigeriaworld.com/letters/2001/oct/241.html.

Osterhaus ADME, van Wezel A, Hazendonk T, Uytdehaag F, van Asten J, van Steenis B (1983). Monoclonal antibodies to polioviruses.Comparison of intratypic strain differentiation of poliovirus type 1 using monoclonal antibodies versus cross absorbed antisera. Intervirol. 20: 129–136.

Osterhaus ADME, van Wezel AL, Hazendonk AG, UytdeHaag FCGM, van Asten JAAM, van Steenis G (1983). Monoclonal antibodies to polioviruses. Comparison of intratypic strain differentiation of poliovirus type 1 using monoclonal antibodies versus cross-absorbed antisera. Intervirol. 20: 129-136.

Plotkin SL, Plotkin SA (1999). A short history of vaccination. In: Plotkin SA, Orenstein WA eds. Vaccines 3rd edition, Philadelphia, PA: Saunders, pp. 1-27.

Renn O (2004). Perception of risks. Toxicol. Lett. 149: 405-413.

Ropeik D (2004). The consequences of fear. EMBO reports 5: S56-S60.

Rothman AJ, Kiviniemi MT (1999). Treating people with information: an analysis and review of approaches to communicating health risk information. J. National Cancer Inst. Monograph 25: 44-51.

Rousset D, Rakoto-Andrianarivelo M, Razafindratsimandresy R, Randriamanalina B, Guillot S, Balanant J, Mauclere P, Delpeyroux F (2003). Recombinant vaccine-derived poliovirus in Madagascar. Emerg. Infect. Dis. 9: 885-887.

Rudd RE, Comings JP, Hyde JN (2003). Leave no one behind: improving health and risk communication through attention to literacy. J. Health Commun. 8:104-115.

Sabin AB, Boulger LR (1973). History of Sabin attenuated poliovirus oral live vaccine strains. J. Biol. Stand. 1: 115-118.

Salk J (1984) One dose immunizations against paralytic poliomyelitis using a non non infectious vaccine. Rev. Infect. Dis. 6: S444-S450.

Science Africa (2004). Nigeria Muslim suspicion of polio vaccine lingers on. http://www.scienceafrica.co .za/2004/march/polio.htm.

Shimizu H, Thorley B, Paladin FJ, Brussen KA, Stambos V, Yuen L, Utama A, Tano Y, Arita M, Yoshida H, Yoneyama T, Benegas A, Roesel S, Pallansch M, Kew O, Miyamura T (2004). Circulation of type 1 vaccine-derived poliovirus in the Philippines in 2001. J. Virol. 78: 13512-13521.

Shulman L, Manor J, Handsher R, Delpeyroux F, McDonough M, Halmut T, Silberstein I, Alfandari J, Quay J, Fisher T, Robinov J, Kew O, Crainic R, Mendelson E (2000). Molecular and antigenic characterization of a highly evolved derivative of the type 2 oral poliovaccine strain isolated from sewage in Israel. J. Clin. Microbiol. 38: 3729-3734.

Sutter RW, Kew OM, Cochi SL (2004). Poliovirus vaccine-live. In: Plotkin SA, Orenstein WA. Eds., Vaccines, 4th ed. WB Saunders Company, Philadelphia, pp. 651-705.

The Guardian (2004). UNICEF deplores rejection of polio vaccine, Zamfara, others adamant. The Guardian, Lagos, Nigeria.

Ward NA, Milstien JB, Hull HF, Hull BP, Kim-Farley RJ (1993). The WHO-EPI initiative for the global eradication of poliomyelitis. Biol. 21:327–333.

White DO, Fenner FJ (1994). Family Picornaviridae. In: Medical Virology 4th Edition. Academic press New York. pp. 381-488.

Wood DJ, Sutter RW, Dowdle WR (2000). Stopping poliovirus vaccination after eradication: issues and challenges. Bull. World Health Organization 78: 347–57.

World Health Organization (WHO). (1981). Markers of poliovirus strains isolated from cases temporally associated with the use of live poliovirus vaccine.

World Health Organization (WHO). (2004b). Polio Laboratory Manual, 4th edition. World Health Organization, Geneva.

World Health Organization (WHO). (1997). Field Guide for supplementary activities aimed at achieving polio eradication, 1996 revision, WHO/EPI/GEN/95.01 Rev.1, World Health Organization, Geneva, Switzerland p. 153.

World Health Organization (WHO). (1997). Polio: the beginning of the end Geneva: World Health Organization, Geneva, Switzerland.

World Health Organization (WHO). (2001). Transmission of wild poliovirus type 2-apparent global interruption. Wkly Epidemiol. Rec. 76: 95-97.

World Health Organization (WHO). (2003). WHO says polio spreading from Nigeria page at URL-http://www.washingtonpost.com/wp-dyn/articles/A63438-2003Oct22. html.

World Health Organization (WHO). (2004a). Laboratory surveillance for wild and vaccine-derived polioviruses, January 2003-June 2004. Wkly. Epidemiol. Rec. 79: 393-398.

World Health Organization (WHO). (2004c). Progress towards global poliomyelitis eradication: preparation for the oral polioviruses vaccine cessation era. Wkly. Epidemiol. Rec. 79: 349-356.

World Health Organization (WHO). (2004d). Polio reported in Botswana page at URL-http://www.who.int./mediacentre/notes/2004/np11/en/print.html.

World Health Organization (WHO). (2004e). The history of vaccination-diseases and vaccines. http://www.childrensvaccine.com/files/WHO-vaccine-history.pdf.

World Health Organization (WHO). (2004f). Poliomyelitis. http://www.who.int./mediacentre/factsheets/

Wright PF, Kim Farley RJ, de Quadros CA, Robertson SE, Scott RM. Ward NA, Henderson RH (1991). Strategies for the global eradication of poliomyelitis by the year 2000. New Engl. J. Med. 325:1774–1779.

Wyatt HV (1973). Poliomyelitis in hypogammaglobulinemias. J. Infect. Dis. 128: 802-806.

Yang CF, Chen HY, Jorba J, Sun HC, Yang SJ, Lee HC, Haung YC, Lin TY, Chen PJ, Shimizu H, Nishimura Y, Utama A, Pallansch MA, Miyamura T, Kew OM, Yang JY (2005). Intratypic recombination among lineages of type 1 vaccine-derived poliovirus emerging during chronic infection of an immunodeficient patient. J. Virol. 79: 12623-12634.

Yang CF, Naguib T, Yang SJ, Nasr E, Jorba J, Ahmed N, Campagnoli R, van der Avoort H, Shimizu H, Yoneyama T, Miyamura T, Pallansch MA, Kew OM (2003). Circulation of endemic type 2 vaccine-derived poliovirus in Egypt, 1983 to 1993. J. Virol. 77: 8366-8377.

15

The HIV-based host derived R7V epitope; functionality of antibodies directed at it and the predicted implications for prognosis, therapy or vaccine development

Bremnæs Christiane and Meyer Debra

Department of Biochemistry, Faculty of Natural and Agricultural Sciences, School of Biological Sciences, University of Pretoria, Pretoria 0002.

Host protein beta-2 microglobulin (β2m) is incorporated into the HIV-1 coat during budding. Individuals not progressing to AIDS produce antibodies directed to an epitope contained in β2m which is designated R7V. These antibodies increased with duration of HIV-infection in non-progressor patients and protected against HIV replication. Purified R7V antibodies neutralized different HIV-1 isolates and did not bind to human cells. In individuals progressing to AIDS or using antiretroviral treatment, a lower prevalence of R7V antibodies was observed. This review summarizes findings on the R7V epitope and antibodies directed at it. Suggestions are also made as to necessary research on R7V which may clarify the importance of this epitope in HIV therapy, prognosis or vaccine development.

Key words: R7V, epitope, antibodies, β2m, HIV, ELISA.

Table of content

INTRODUCTION

HIV, the retrovirus known for disabling CD4 expressing T cells and causing immune system dysfunction, incorporates host cell proteins into its envelope during budding. These virus incorporated host proteins can be bystanders, assist in the life cycle and viral ability to avoid immune system detection, while some of these proteins retain their functional ability or engage in responses that are detrimental to the pathogen. The host immune response to these cellular proteins when they are presented as virus-associated particles allows room for investigating them as prognostic markers, therapeutic tools or potential vaccine candidates. Existing AIDS prognostic markers like CD4 cell count and viral load are not infallible and better therapies against HIV is still under development because of shortcomings in existing treatment regimes. In addition, the uneven record of success of HIV vaccine strategies leaves room for considering host proteins incorporated by the virus as a radically different means of generating protective immunity.

Beta-2 microglobulin (β2m) is one of the host proteins incorporated into the envelope of HIV-1 and under discussion in this review are the possible uses of antibodies to R7V, an epitope within this protein. R7V is incorporated into the envelope of HIV-1 and located on the exterior surface (Le Contel et al., 1996). This epitope has been suggested as possible vaccine target (Le Contel et al., 1996; Galéa et al., 1999 a and b; Chermann, 2001; Haslin and Chermann, 2007b) and antibodies induced by it as prognostic (Galéa et al., 1996; Chermann, 2001; Ravanini et al., 2007; Kouassi et al., 2007; Sanchez et al., 2008) or therapeutic markers (Haslin and Chermann, 2002, 2004 and 2007b; Haslin et al., 2007a). Data that lead to these suggestions are reviewed here. Because reports on R7V are limited (only 11 articles in peer reviewed journals and a number of patents), data contained in posters (4 posters reviewed by a scientific committee and presented at international conferences) as well as personal communications with researchers (Webber, 2009) are also referenced to provide a complete picture of work done on this epitope thus far.

Cellular proteins in HIV

Cellular or host proteins can be incorporated by HIV-1

*Corresponding author. E-mail: debra.meyer@up.ac.za. Tel: +27 12-420-2300. Fax: +27 86-638-1904.

Abbreviations: AIDS, acquired immunodeficiency syndrome; **ARV,** antiretroviral; **β2m,** beta-2 microglobulin; **CMV,** cytomegalovirus; **CTL,** cytotoxic T-lymphocyte; **ELISA,** enzyme-linked immunosorbent assay; **HAART,** highly active anti-retroviral therapy; **HIV,** human immunodeficiency virus; **HLA,** human leukocyte antigen; **LTNP,** long term non-progressor; **MAbs,** monoclonal antibodies; **MHC,** major histocompatibility complex; **OD,** optical density.

either on its surface or inside the viral lipid envelope (Ott, 2008; Figure in Galéa et al., 1999b). Some host proteins present in the virus retain their functional ability and can affect infectivity, tropism and pathogenesis. According to Ott (2008) there are three possible ways in which HIV-1 incorporates cellular proteins inside or on its surface. Most of the cellular proteins incorporated by the virus are taken up as simple bystanders because of their close proximity during the budding process. The presence of these non-specifically incorporated proteins provides important information about the local environment where HIV-1 assembles and the site of budding. The second possible mechanism of incorporation involves host proteins that act as partners in the assembly and budding process and these are incorporated by interacting with one of the viral proteins. Lastly, HIV-1 can hijack cellular proteins by incorporating them for a post-assembly step where they act as captives helping the virus replicate or evade the immune system (Ott, 1997 and 2008). Selected cellular proteins detected in HIV-1 as well as their role in favour of the virus are detailed by Ott, 1997 and 2008. Since β2m is part of the major histocompatibility complex (MHC) class I, the assumption is that this protein is incorporated into HIV's envelope in the same manner as MHC class I. From information provided in the review by Ott, 2008 it seems that HIV incorporates this protein as a bystander as there is no evidence that a specific HIV protein acts as a binding partner to bring the protein complex into the viral particle.

Beta-2 microglobulin

β2m is a 12 kDa, 99 amino acid protein which is non-covalently bound to the 45 kDa heavy chain of the MHC class I molecule (Roitt et al., 2001) where it is essential for expression of this molecule. MHC class I plays a central role in the immune system, is omnipresently expressed and binds peptide antigens for presentation to the CD8[+] T-cells (Rosano et al., 2005) during the initiation of a cellular immune response. β2m is also associated with CD 1 (a protein related to MHC class I) for presentation of lipids, glycolipids and lipid antigens to T-cells (Roitt et al., 2001). Because β2m is expressed in almost all nucleated cells (Arthur et al., 1992), it can be found in all potential virus target cells. This protein along with neopterin and other serum and cellular markers that correlate with clinical progression of HIV disease (Fahey et al., 1990; Hofmann et al., 1990; Melmed et al., 1989) when found free in a variety of physiological fluids (Rosano et al., 2005), also serve as indicators of the degree of immune activation.

Beta-2 microglobulin as vaccine target

Arthur et al. (1992); Ott (1997) and Haslin et al. (2002) confirmed β2m as one of the cellular proteins incorporated into the surface of HIV-1. Monoclonal antibodies (MAbs) directed to β2m were able to immunoprecipitate intact viral particles and inhibit the life-cycle of HIV (Arthur et al., 1992). These data suggests β2m to be an integral and functioning part of the HIV-1 surface, involved in the process of HIV-infection and pathogenesis (Hoxie et al., 1987; Devaux et al., 1990; Corbeau et al., 1990; Arthur et al., 1992; Le Contel et al., 1996). β2m is immunogenic only when exposed on the viral surface and not when it is part of the human leukocyte antigen (HLA)/MHC class I at the host cell surface or when it is part of circulating β2m. MAbs against β2m reacts with free urinary β2m (Liabeuf et al., 1981) but not with the protein when it is associated with cell surface proteins.

To prevent binding between HIV-1 and a host cell, it may be worth considering targeting β2m in vaccine design and possibly avoiding the problems associated with the variability of viral proteins (Galéa et al., 1999b; Haslin et al., 2002). However it is only possible to consider such a vaccine approach if epitopes in the cellular determinant used will only be exposed when it is carried away by the extracellular infectious agent, or the epitopes are nonimmunogenic in their natural presenta-tion by the cell and is modified when it is presented at the surface of the virion (Chermann et al., 2000). This appears to be the case with epitopes in β2m. Studies demonstrating an inability of the potential vaccine antigen to induce autoimmune responses need to be designed. β2m based vaccine formulas appear to be under investigation as is evident in the references used in the patent by Chermann et al. (2000).

Most HIV protein-vaccine strategies to date focused on the variable viral envelope proteins, especially the V3 loop of gp120. Hewer and Meyer (2004 and 2005) hypothesized that exploiting the advantageous properties of V3 loop peptides and at the same time acounting for the variability of this region would aid in developing an effective vaccine component for inducing broadly cross reactive neutralizing antibodies. Addressing variability by novel synthesis induced viral strain specific immune responses with some amount of cross-reactivity which underscores the need for circumventing viral variability by other means.

The R7V epitope of beta-2 microglobulin

A seven amino acid epitope in β2m was shown to be present at the surface of divergent HIV isolates by Le Contel et al. in 1996. These authors studied several short overlapping peptides derived from β2m for their ability to reverse the neutralizing action of MAbs directed to the protein. Among the tested peptides, the heptamers R7V (Arg-Thr-Pro-Lys-Ile-Gln-Val), S7K (Ser-Gln-Pro-Lys-Ile-Val-Lys) and F7E (Phe-His-Pro-Ser-Asp-Ile-Glu) were efficient in reversing the action of these MAbs, with R7V being the most efficient. The R7V peptide consists of

fewer hydrophobic (PIV) than hydrophilic (RTKQ) amino acids. The tertiary structure of isolated human β2m is described as an antiparallel β-barrel fold (Rosano et al., 2005). Multiple Protein Sequence Analysis programs, including DeepView/Swiss-Pdb Viewer 3.7 (SP5), describe percentages of the 7 amino acids in the R7V peptide as either a random coil or an extended strand which are regions that could form part of a β-barrel (PDB entry 1LDS; Trinh et al., 2002).

Polyclonal antibodies detected using an R7V peptide as antigen

Antibodies directed to R7V were detected in HIV-infected individuals, primarily asymptomatic and long term infected patients naïve of treatment. In studies done by Galéa et al. (1996) and Sanchez et al. (2008), R7V antibodies were present in the majority of asymptomatic patients (HIV seropositive for 5 - 12 years, stage A1-A2 according to the 1993 CDC classification, Galéa et al., 1996; class A according to the 1993 CDC classification and naïve of highly active antiretroviral therapy, HAART, Sanchez et al., 2008) and long-term infected patients (infected for more than 10 years without HAART, Sanchez et al., 2008). The highest prevalence of R7V antibodies were found in non-progressor patients compared to progressors who go on to develop AIDS with associated CD4 cell decreases and viral load increases (Galéa et al., 1996; Sanchez et al., 2008, Table 1). R7V antibodies were also found in individuals who were not infected with HIV but at lower levels (Galéa et al., 1996; Sanchez et al., 2008, Table 1). This is interesting because R7V is believed to only be immunogenic when incorporated into the HIV envelope and the virus was absent in these patients. These antibodies could be cross-reactive or perhaps the HIV negative individuals were infected by other enveloped viruses containing host derived R7V-like epitopes (more detail in paragraphs to follow).

The study done by Galéa and colleagues (1996) used an "in-house" enzyme-linked immunosorbent assay (ELISA) and the authors calculated concentrations of R7V antibodies from a standard curve while Sanchez et al. (2008) used the anti-R7V ELISA from Ivagen (Bernis, France) and determined the antibody ratio by normalizing the optical density (OD) value for the sample with the OD value for the internal calibrator. The Ivagen ELISA developed for detection of R7V in human serum and/or plasma of individuals confirmed as being seropositive for HIV-infection is not commercially available.

The Sanchez et al. (2008) study contained several groups; A (201 HIV negative, 160 HIV positive on treatment and 88 HIV positive untreated patients from USA), B (177 asymptomatic and 131 symptomatic HIV positive patients from USA) and C (45 untreated Italian non-progressor patients infected with HIV-1). From the

Table 1. Summary of studies reporting on the presence of R7V antibodies in HIV-infected and uninfected individuals. HIV-1 subtypes listed were inferred from locations where samples were collected since this information was not always stated.

Percent individuals	R7V	antibody	producing	Total	HIV-1 subtype/country	References
HIV Positive vs. HIV negative						
%	n	%	n			
53.7	95	32.0	69	164	-	Galéa et al., 1996[3,4]
53.2	248	5.5	201	449	(B) USA	Sanchez et al., 2008[4,7]
33.5	507	3.0 [2]	201	708	(B) USA	Haslin and Chermann, 2007b[1,3,4]
9.1[7]	33	0.0	10	43	(B) Turkey	Ergünay et al., 2008[4,7]
Asymptomatic vs. Symptomatic						
%	n	%	n			
59.0	63	38.0	8	71	(B) Italy	Ravanini et al., 2007[1,4]
59.1	22	13.1	61	83	(A) Cameroon	Tagny et al., 2007[1,4]
42.0	36	9.0	53	89	(A) Ivory Coast	Kouassi et al., 2007[1,4]
64.4[5]	177[5]	35.1	131	308	(B) USA	Sanchez et al., 2008[4]
35.7[6]	29[6]	31.8[6]	22[6]	51	-	Galéa et al., 1996[2,3,4]
Treatment naïve vs. Treatment						
%	n	%	n			
67.0	45	35.0	17	62	(B) Italy	Ravanini et al., 2007[1,4]
56.8	88	21.3	159	247	(B) USA	Haslin and Chermann, 2007b[1,3,4]
38.0	50	2.0	50	100	(A) Ivory Coast	Kouassi et al., 2007[1,4]
71.6	88	43.1	160	248	(B) USA	Sanchez et al., 2008[4]
CD4 cell count:						
> 200 cells/µl vs. 0-200 cells/µl						
%	n	%	n			
59.1	22	13.1	61	83	(A) Cameroon	Tagny et al., 2007[1,4]
Years of HIV-infection (treatment naïve):						
		%	n			
< 5 years		14.3	-	-	(B) USA	Haslin and Chermann, 2007b[1,3,4]
5 - 10 years		50.0	-	-		
> 10 years		68.3	-	-		
0 - 5 years		~ 40.0	-	-	(B) USA	Sanchez et al., 2008[4]
5 - 10 years		~ 69.0	-	-		
10 - 20 years		~ 80.0	-	-		
< 5 years		68.0	41	62[8]	(B) Italy	Ravanini et al., 2007[1,4]
5 - 10 years		31.0	13			
≥ 10 years		63.0	8			

third study group Sanchez et al. (2008) noticed a direct correlation between R7V antibody ratio and viral load. On the other hand, no correlation between the R7V antibody ratios and the CD4 T-cell count was detected. Sanchez et al. (2008) also observed a higher prevalence of R7V anti-bodies in untreated patients (71, 6% n = 88) compared to patients on HAART (43.1% n = 160), Table 1. When treatment was considered successful, these authors re-

Table 1. Continues.

Ethnic groups:	%	n			
African Americans	49.5	106	501[2]	(B) USA	Haslin and Chermann, 2007b[1,3,4]
Haitians	38.0	100			
Asians	31.6	95			
Caucasians	26.2	84			
Indians	25.4	67			
Hispanics	22.4	49			
Sex:	%	n			
Males	55.0	61	106	(B) USA, African American	Haslin and Chermann, 2007b[1,3,4]
Females	44.4	45			

[1]Conference proceedings. 3rd South African AIDS Congress, Durban, June 2007.
[2]Calculation errors: The authors stated that 10 of 201 subjects equal 3%. However, 10 of 201 subjects equal 5%. Adding all the ethnic groups equals 501, 507 is stated as the total number by the authors.
[3]Galéa et al. (1996) and Haslin and Chermann (2007b) performed neutralizing assays with the R7V antibodies.
[4]The majority of studies were performed using the anti-R7V ELISA from Ivagen (Bernis, France) with the exception of Galéa et al. (1996) who used an "in-house" ELISA.
[5]Additional results in Sanchez et al. (2008): Long-term infected (more than 10 years) and naïve of treatment. Presence of R7V antibodies: ~ 80%, n not mentioned.
[6] Error: n = 29 stated in the text and n = 28 stated in the summary. Additional results in Galéa et al. (1996): presence of R7V antibodies in progressors: ~ 9% (n = 44).
[7]In Ergünay et al. (2008) all positive patients were on treatment. In Sanchez et al. (2008) 160 patients were on treatment and 88 patients were naïve of treatment.
[8]A sample number of 63 are stated in the poster. 18 of 63 patients were not on treatment.

ported a total disappearance in R7V antibodies for 77% (n = 21) of the patients. During antiviral treatment there is a decrease in newly formed virus particles budding from the host cell, it is therefore possible that fewer viruses containing the R7V epitope are produced in treated individuals. Sanchez et al. (2008) hypothesized that the R7V epitope may no longer be exposed on the virus and thus not visible to the host immune system after successful treatment and therefore suggested that the anti-R7V ELISA from Ivagen was better adapted to the detection of R7V antibodies in asymptomatic patients, still naïve of treatment.

Ergünay and colleagues (2008) following a smaller study (33 HIV positive and 10 HIV negative compared to the 160 patients on HAART in the Sanchez et al., 2008 study) also reported the presence of R7V antibodies in HIV-infected individuals on HAART (Table 1). Only 9.1% of 33 HIV positive individuals on treatment exhibited R7V antibodies and there was no correlation between the presence of the antibodies and disease progression. The Ergünay et al. (2008) report is in Turkish and the English abstract did not provide a description of the patients' disease status or duration of infection. It is difficult to directly compare the study done by Ergünay et al. (2008) and Sanchez et al. (2008) because the latter study used an HIV positive test group including patients both on HAART and not on treatment whereas the HIV positive test group in Ergünay`s study (2008) were all on HAART. The overall agreement between the works done by these

groups is a decrease in prevalence of R7V antibodies in the presence of treatment.

A study reporting on the prevalence of the R7V antibodies in HIV-1 infected individuals on HAART was done by Professor Lynne Webber from the Department of Medical Virology, University of Pretoria, South Africa and presented at the HIV and AIDS Research symposium at the University of Pretoria in February 2009 (Webber, 2009). The prognostic applicability of the anti-R7V ELISA from Ivagen (Bernis, France) was examined using a cohort of 25 HIV-infected patients on HAART. Nine participants were classified as long term non-progressors (LTNPs), defined as patients free of HIV-1 related disease and displaying stable CD4+ T-lymphocyte counts (> 200 cells/μl for more than 10 years). The results indicated that 40% of all the HIV-infected patients and 56% of the LTNPs tested positive for R7V antibodies. Twenty eight percent of all the HIV-infected patients and 44% of the LTNPs were considered doubtful (data collected could not be classified as either positive or negative for R7V antibodies).

R7V Antibody and cross-reactivity

Data collected by Sanchez et al. (2008) demonstrated cross-reactivity of antibodies from individuals infected with other enveloped viruses when using the anti-R7V percent positive results in this study were limited. It is possible that individuals not infected with HIV but infected with other enveloped viruses which may also incorporate β2m in their membrane and also expose the R7V epitope, could produce antibodies to this epitope. Sanchez et al. (2008) observed that a few individuals uninfected with HIV or infected with the enveloped viruses causing mono-nucleosis or rubella gave positive R7V antibody results using the Ivagen ELISA. Three of thirteen individuals (23.0%) infected with mononucleosis and six of eleven (54.5%) individuals infected with rubella responded positive for R7V antibodies (antibodies in the sera were able to bind the R7V antigen in an ELISA). Higher sample numbers are obviously needed to validate these data. Also these antibodies were not tested for an ability to neutralize or precipitate HIV-1 (which are properties of actual R7V antibodies) so it is possible that this response was due to some interference or cross-reactive antibo-dies. There is evidence that cytomegalovirus (CMV, McKeating et al., 1987) and HTLV (Hoxie et al., 1987) incorporate β2m. Since R7V is part of β2m this could mean that individuals infected with viruses other than HIV could test positive for R7V antibodies. However the acquisition of β2m by HIV and CMV differs. CMV acquires β2m after budding from the cell (in a non-HLA-like manner) but still binds MAbs to β2m (Grundy et al., 1987; Tysoe-Calnon et al., 1991; Le Contel et al., 1996). There is not enough evidence to form an opinion on whether CMV (containing β2m) could not induce an R7V like antibody response during infection.

Data collected using assays other than ELISA

Most studies referred to here present conclusions based on ELISA data only (where an R7V peptide was used as antigen). Whether the polyclonal antibodies believed to be R7V antibodies could neutralize or precipitate HIV-1 is not always indicated. Neutralization or precipitation of HIV-1 serves as a means of verifying an actual R7V antibody response. Better validation would be to purify the antibodies before use in either assay. Neither Galéa et al. (1996), Sanchez et al. (2008) nor Ergünay et al. (2008) performed precipitation assays. Of these three studies only Galéa et al. (1996) demonstrated that the presence of R7V antibodies correlated with neutralization of various divergent HIV strains. A study done by Xu et al. (2002) used both ELISA and precipitation to investigate the prevalence of R7V antibodies in HIV-infected patients. However, that data is not referred to in this review because the article is in Chinese and the English abstract does not give information about sample numbers or percentages of R7V antibodies. Since R7V antibodies neutralized divergent HIV-1 strains and this neutralization was reversed by addition of R7V peptide (Le Contel et al., 1996). Galéa et al. (1996) hypothesized that R7V antibodies found in patients could have equivalent neu-

tralizing activity as that observed *in vitro*. Extensive experiments with the two main target cells of HIV-1 infection, peripheral blood lymphocytes and blood derived macrophages, have shown that different T-lymphotropic as well as macrophage-tropic HIV-1 strains was neutralized by β2m MAbs (Le Contel et al., 1996).

Evidence for R7V as potential vaccine target or therapeutic tool

Purified R7V antibodies from sera of rabbits injected with the peptide as well as in sera of non-progressor patients (HIV-1 subtype of infection virus not mentioned) precipitated and neutralized HIV-1 subtypes A, B, C, D and F (Galéa et al., 1999a). The results obtained by Galéa et al. (1999a) is supported by the studies done by Xu et al. (2002) who stated that R7V antibodies were found to inhibit the replication of HIV. Xu et al. (2002) also suggested that R7V antibodies prevent the virus from entering target cells by interfering with the binding of HIV to the co-receptors (CCR5 or CXCR4) of the target cell. In addition, human R7V IgG neutralized virus strains resistant to antiviral drugs and inhibited infection of cells by laboratory as well as primary viral isolates (Galéa et al., 1999b). Furthermore, studies done by Haslin and Chermann (2007b, poster presentation) showed R7V antibodies from HIV-1 subtype B infected individuals to be able to neutralize HIV-1 subtype D. Collectively the presence of R7V antibodies in serum from HIV-infected individuals from different geographic areas suggests that R7V is naturally immunogenic and escapes variability and flexibility observed with the viral proteins in the HIV envelope (Galéa et al., 1999a; Chermann, 2001). These studies suggest R7V-like epitopes to have a potential role in an HIV/AIDS vaccine and the R7V antibodies in the treatment of patients in failure of HAART.

Production of R7V antibodies with antiviral properties has been demonstrated by Haslin and Chermann (2004) and Haslin et al. (2007a). The antibodies were produced by infection of insect cells with a recombinant baculovirus in which the gene corresponding to the R7V antibody was introduced after isolation from EBV-immortalized B lymphocytes of non-progressor patients. These antibodies have been shown to be able to neutralize various clades of HIV-1 including drug-resistant viruses and should therefore be taken into consideration as therapeutic tools (Haslin and Chermann, 2004; Haslin et al., 2007a). There are several studies reporting on the prevalence of the R7V antibodies in HIV-1 infected individuals which were presented at the 3rd South African AIDS Congress in Durban in June 2007 (Ravanini et al., 2007; Haslin and Chermann, 2007b; Tagny et al., 2007; Kouassi et al., 2007). Ravanini et al. (2007) studied a group of HIV- infected patients living in Italy including 63 symptomatic (A class CDC, eighteen patients on antiretroviral, ARV, therapy) and 8 symptomatic patients (B or

C class CDC, 5 patients on ARV therapy). Haslin and Chermann (2007b) studied 507 HIV-infected (on HAART or not and 201 uninfected individuals from USA. The study done by Tagny et al. (2007) was conducted with a group of HIV-infected individuals naïve of ARV therapy and living in Cameroon including 22 asymptomatic (A1 and A2 class CDC 1993) and 61 symptomatic (A3, B or C class CDC 1993) patients. Lastly, 100 HIV-infected individuals (50 naïve of HAART and 50 on HAART) living in the Ivory Coast were used in the study done by Kouassi et al. (2007). Thirty six patients were at the clinical stage A according to CDC and classified asymptomatic and 53 patients classified symptomatic (CDC stage not stated). None of these four studies report on precipitation assays or any other validation study to confirm that the antibodies detected were R7V antibodies. These four studies are discussed in Table 1 with some additional observations described below. Haslin and Chermann (2007b) observed that R7V antibodies from HIV-1 subtype B infected patients were able to neutralize HIV-1 subtype D and that higher titres of R7V antibody plasma were more effective at neutral-lizing HIV-1, suggesting that the neutralizing potential most likely was due to R7V antibodies. A positive correlation between R7V antibody ratio and viral load for a group of asymptomatic patients naïve of ARV therapy (n = 45) was observed by Ravanini et al. (2007) and supported by similar observations by Sanchez et al. (2008).

A vaccine component should induce neutralizing anti-bodies and/or a cellular immune response depending on how it is presented to the immune system. Evidence summarized in this review demonstrates that R7V as antigen in animal studies induced neutralizing antibodies in rabbits (Galéa et al., 1999a). Recombinant MAbs to R7V neutralized viral isolates (Haslin and Chermann, 2004; Haslin et al., 2007a) and naturally produced R7V antibodies isolated from humans did the same (Galéa et al., 1999a and 1999b; Haslin and Chermann, 2007a). In addition, MAbs to β2m, the parent protein of R7V, also neutralized multiple viral strains (Arthur et al., 1992). This evidence supports large scale studies on either protein or peptide for induction of protective humoral immune res-ponses. These studies should perhaps be preceded by studies demonstrating the extent to which β2m or R7V induce autoimmune responses.

R7V antibodies and autoimmunity

A lot still remains to be done with regards to the peptide R7V and what it could mean for prognosis of disease, therapeutics or HIV vaccine development. According to the patent of Chermann et al. (2006) vaccine research has been done with formulations of R7V conjugated to carrier proteins such as KLH and BSA. Because the epitope is host derived one may assume a concern of the possibility that R7V antibodies may initiate an autoimmune respon-

se. However, work by Galéa et al. (1996 and 1999a and b), Haslin and Chermann (2002) and Haslin et al. (2007a) suggest this not to be the case. Data are not shown in these papers but according to the discussion there appears to be no self recognition by R7V anti-bodies; no binding to the surface of human cells by purified R7V antibodies from either patients or immunized rabbits (Galéa et al., 1999b). Nor did the recombinant monoclonal R7V antibody made by Haslin et al. (2007a) bind to human cells. In addition, individuals with naturally high levels of R7V antibodies did not exhibit any auto-immune diseases (Galéa et al., 1996 and 1999b). This suggests that R7V antibodies produced inside an individual were virus specific and therefore not a problem to the host. If there are no autoimmune antibodies it could be due to the fact that the R7V peptide is presented to MHC class II (for antibody production) in the context of viral molecules. R7V anti-bodies would therefore not recognize a corresponding epitope on the surface of the host cells. Also, as men-tioned before, the R7V epitope is only visible to the immune system when this epitope is contained in HIV and not in the natural protein.

R7V and cellular immunity

People exposed to HIV mount an immune response which in some individuals slows down disease progre-ssion. Strong neutralizing antibodies (Watkins et al., 1996; Pilgrim et al., 1997; Richman et al., 2003; Wei et al., 2003) and strong cytotoxic T-lymphocyte (CTL) res-ponses (Wagner et al., 1999; Cao et al., 1995; Klein et al., 1995) have been detected in LTNPs. This means that immune responses slow down disease progression and is certainly of value for prognosis of disease and vaccine development. An HIV vaccine should presumably induce humoral and cellular responses. Neutralizing antibodies as part of a humoral response should eliminate cell-free virus. Cellular immune responses (e.g. CD8+ CTLs) on the other hand should remove already infected cells that escape antibody-mediated neutralization (Lemckert et al., 2004). If R7V is to be considered in vaccine research, knowledge of possible cellular immune response epitopes within it is of importance. CTL responses are mostly related to viral core proteins (Buseyne et al., 1993; Cao et al., 1997; Nakamura et al., 1997; McAdam et al., 1998), but it has been observed with some viral envelope proteins as well (Pinto et al., 1995). R7V (even though it is derived from a protein associated with and important for MHC class I presentation) is recognized as a foreign antigen by the host and induces a humoral immune response as is evident in all the reports on R7V antibodies. It would be interesting to know whether virus associated R7V-like epitopes are capable of inducing a cellular immune response especially given its β2m origins and the role of this protein in MHC class I presentation of antigen. MHC class I molecules are known to bind epitopes ranging from eight to eleven amino acids (Karim and Karim, 2005) and therefore might bind the seven amino acid R7V peptide for presentation during natural infection. No HIV incorporated host proteins have yet been implicated in CTL responses nor has this been exhaustively investi-gated.

CONCLUSION

R7V antibodies have been suggested as a potential therapeutic tool (Haslin and Chermann, 2002, 2004 and 2007b; Haslin et al., 2007a) since 2002. To date no studies report on passive infusion of animal models (or human volunteers) with these R7V antibodies. Nor have recombinant R7V antibodies (Haslin and Chermann, 2004; Haslin et al., 2007a) been used in in vivo therapeutic studies. Passive infusion of animals (Binley et al., 2000; Mascola et al., 2000 and 2003; Mascola, 2002) or humans (Armbruster et al., 2004) with HIV antibodies is not uncommon. Suggestions of a synthetic R7V peptide being considered as a vaccine has been made since 1996 (Le Contel et al., 1996; Galéa et al., 1999a and b; Chermann, 2001; Haslin and Chermann, 2007b). No studies evaluating the in vivo value of R7V antibodies have yet been reported. By this we mean eliciting in vivo neutralizing antibodies in animal models (using synthetic R7V as antigen) and challenging this response with live virus. Subsequently the literature also does not report on any R7V based (phase I/II) vaccine trials. It appears that vaccine development studies have been done and some of it has been published as patents (Chermann et al., 2000 and 2006). Because the prevalence of the R7V antibodies was shown to correlate with non-progression to AIDS it has been postulated that patients who have elevated levels of R7V antibodies have a lower likelihood of progression to AIDS. Further studies are needed to clarify the use of R7V antibodies as possible prognostic markers. The presence of these antibodies in uninfected individuals needs to be clarified as well since it has implications for the former statement. Finally, whether the R7V epitope or the entire β2m protein is incorporated by other or all enveloped viruses has implications for the use of the epitope or antibodies to it in prognosis or therapy. Limited sample number studies commenting on possible cross-reactivity between R7V antigen and antibodies from other enveloped viruses exist but needs to be expanded. There are similarities between HIV and HTLV and there are reports of non-random incorporation of host proteins by the latter which further supports clarifying the extent to which β2m or R7V is incorporated by other viruses.

R7V and antibodies to the epitope holds promise but more extensive and clarifying autoimmunity and cross-reactivity studies are needed to maintain optimism on its use. That a host epitope when incorporated in a viral envelope is immunogenic has been sufficiently demonst-rated. What remains to be rigorously shown is the usefulness of the immunogenicity (challenging the neu-tralisation response with live virus) after autoimmunity

due to these host antigens have been shown to not be a concern.

ACKNOWLEDGEMENTS

This work was supported by the Faculty of Natural and Agricultural Sciences of the University of Pretoria.

REFERENCES

Armbruster C, Stiegler GM, Vcelar BA, Jäger W, Köller U, Jilch R, Ammann CG, Pruenster M, Stoiber H, Katinger HWD (2004). Passive immunization with the anti-HIV-1 human monoclonal antibody (hMAb) 4E10 and the hMAb combination 4E10/2F5/2G12. J. Antimicrob. Chemother. 54: 915-920.

Arthur LO, Bess JW, Sowder II RC, Benveniste RE, Mann DL, Chermann JC, Henderson LE (1992). Cellular proteins bound to immunodeficiency viruses: implications for pathogenesis and vaccines. Sci. 258: 1935-1938.

Binley JM, Clas B, Gettie A, Vesanen M, Montefiori DC, Sawyer L, Booth J, Lewis M, Marx PA, Bonhoeffer S, Moore JP (2000). Passive infusion of immune serum into simian immunodeficiency virus-infected rhesus macaques undergoing a rapid disease course has minimal effect on plasma viremia. Virol. 270: 237-249.

Buseyne F, McChesney M, Porrot F, Kovarik S, Guy B, Riviere Y (1993). Gag-specific cytotoxic T lymphocytes from human immunodeficiency virus type 1-infected individuals: gag epitopes are clustered in three regions of the p24gag protein. J. Virol. 67: 694-702.

Cao Y, Qin L, Zhang L, Safrit J, Ho DD (1995). Virologic and immunologic characterization of long-term survivors of human immunodeficiency virus type 1 infection. N. Eng. J. Med. 332(4), 201-228.

Cao H, Kanki P, Sankale L, Dieng-Sarr G, Mazzara P, Kalams A, Korber B, Mboup S, Walker BD (1997). Cytotoxic T-lymphocyte cross-reactivity among different human immunodeficiency virus type 1 clades; implications for vaccine development. J. Virol. 71: 8615-8623.

Chermann JC, Le Contel C, Galéa P (2000). Immunogenic compositions comprising peptides from β-2-microglobulin. United States Patent 6113902.

Chermann JC (2001). A brief reflection on the development of human retrovirology: the past, the present and the future. J. Hum. Virol. 4: 289-295.

Chermann JC, Le Contel C, Galéa P (2006). Vaccine against infectious agents having an intracellular phase, composition for the treatment and prevention of HIV infections, antibodies and method of diagnosis. United States Patent Application Publication US 2006/0073165 A1.

Corbeau P, Devaux CA, Kourilsky F, Chermann JC (1990). An early postinfection signal mediated by monoclonal anti-β2 microglobulin antibody is responsible for delayed production of human immunodeficiency virus type 1 in peripheral blood mononuclear cells. J. Virol. 64: 1459-1464.

Devaux C, Boucraut J, Poirier G, Corbeau P, Rey F, Benkirane M, Perarnau B, Kourilsky F, Chermann JC (1990). Anti- β2-microglobulin monoclonal antibodies mediate a delay in HIV1 cytopathic effect on MT4 cells. Res. Immunol. 141: 357-372.

Ergünay K, Altinbaş A, Calic Başaran N, Unal S, Us D, Karabulut E, Ustaçelebi S (2008). Investigation of anti-R7V antibodies in HIV-infected patients under highly active antiretroviral therapy. Mikrobiyol. Bul. 42(3):413-419.

Fahey JL, Taylor JMG, Detels R, et al. (1990). The prognostic value of cellular and serological markers in infection with human immunodeficiency virus type 1. N. Eng. J. Med. 322: 166-72.

Galéa P, Le Contel C, Chermann JC (1996). Identification of a biological marker of resistance to AIDS progression. Cell. Pharmacol. AIDS Sci. 3: 311.

Galéa P, Le Contel C, Chermann JC (1999a). A novel epitope R7V common to all HIV-1 isolates is recognized by neutralizing IgG found in HIV-infected patients and immunized rabbits. Vaccine 17: 1454.

Galéa P, Le Contel C, Coutton C, Chermann JC (1999b). Rationale for a vaccine using cellular-derived epitope presented by HIV isolates Vaccine 17: 1700.

Grundy JE, McKeating JA, Griffiths PD (1987). Cytomegalovirus strain AD169 binds β_2 microglobulin in vitro after release from cells. J. Gen. Virol. 68: 777-784.

Haslin C, Chermann JC (2002). Anti-R7V antibodies as therapeutics for HIV-infected patients in failure of HAART. Curr. Opin. Biotechnol. 13: 621-624.

Haslin C, Chermann JC (2004). Therapeutic antibodies a new weapon to fight the AIDS virus. Spectra. Biol. 141: 51-53. Article in French.

Haslin C, Lévêque M., Ozil A., Cérutti P, Chardès T, Chermann JC, Duonor-Cérutti M (2007a). A recombinant human monoclonal anti-R7V antibody as a potential therapy for HIV infected patients in failure of HAART. Hum. Antibodies 16(3-4): 73-85.

Haslin C, Chermann, JC (2007b). Neutralizing anti-R7V Antibodies in United States Human Immunodeficiency Virus type 1-infected patients: their role in disease non-progression. Poster presented at the 3rd SAAIDS Congress, Durban 5-8 June 2007. Online: http://www.ivagen.com/Doc/Poster%20&%20Abstract%20-%20Durban%202007%20-%20Urrma.pdf. URL active: April 24, 2008, 11:38 PM.

Hewer R, Meyer D (2004). Peptide immunogens designed to enhance immune responses against human immunodeficiency virus (HIV) mutant strains; a plausible means of preventing viral persistence. J. Theor. Biol. 233: 85-90.

Hewer R, Meyer D (2005). Evaluation of a synthetic vaccine construct as antigen for the detection of HIV-induced humoral responses. Vaccine 23: 2164-2167.

Hofmann B, Wang Y, Cumberland WG, Detels R, Bozorgmehri M, Fahey JL (1990). Serum $\beta2$ microglobulin level increases in HIV infection: relation to seroconversion, CD4 T cell fall and prognosis. AIDS 4: 207-214.

Hoxie JA, Fitzharris TP, Youngbar PR, Matthews DM, Rackowski JL, Radka SF (1987). Nonrandom association of cellular antigens with HTLV-III virions. J. Hum. Imunol. 18: 39-52.

Karim S, Karim Q (2005). HIV/AIDS in South Africa. Cambridge University press.

Klein MR, Van Baalen CA, Holwerda AM, Kerkhof Garde SR., Bende R., Keet IP, Eeftinck-Schattenkerk JK, Osterhaus AD, Schuitemaker H, Miedema F (1995). Kinetics of Gag-specific cytotoxic T lymphocyte responses during the clinical course of HIV-1 infection: a longitudinal analysis of rapid progressors and long term asymptomatics. J. Exp. Med. 181: 1365-1372.

Kouassi M'Bengue A, Kolou MR, Kouassi B, Crezoit Yapo A, Ekaza E, Prince DM, Kouadio K, Dosso M (2007). Detection of R7V antibodies in HIV patients living in sub-Saharan countries: Case of Abidjan in Ivory Coast in 2006. Poster presented at the 3rd SAAIDS Congress, Durban 5-8 June 2007.

Le Contel C, Galéa P, Silvy F, Hirsh I, Chermann JC (1996). Identification of the $\beta2m$ derived epitope responsible for neutralization of HIV isolates. Cell. Pharmacol. 3:68-73.

Lemckert AAC, Goudsmit J, Barouch DH (2004). Challenges in the search for an HIV vaccine. Eur. J. Epidemiol. 19: 513-516.

Liabeuf A, Le Borgne de Kaouel C, Kourilsky FM, Malissen B, Manuel Y, Sanderson AR (1981). An antigenic determinant of human $\beta2$-microglobulin masked by the association with HLA heavy chains at the cell surface: Analysis using monoclonal antibodies. J. Immunol. 127: 1542-1548.

Mascola JR, Stiegler G, VanCott TC, Katinger H, Carpenter CB, Hanson CE, Beary H, Hayes D, Frankel S, Birx DL, Lewis MG (2000). Protection of macaques against vaginal transmission of a pathogenic HIV-1/SIV chimeric virus by passive infusion of neutralizing antibodies. Nature Med. 6: 207-210.

Mascola JR (2002). Passive transfer studies to elucidate the role of antibody-mediated protection against HIV-1. Vaccine 20: 1922-1925.

Mascola JR, Lewis MG, VanCott TC, Stiegler G, Katinger H, Seaman M, Beaudry K, Barouch DH, Korioth-Schmitz B, Krivulka G, Sambor A, Welcher B, Douek DC, Montefiori DC, Shiver JW, Poignard P, Burton DR, Letvin NL (2003). Cellular immunity elicited by human immunodeficiency virus type 1/simian immunodeficiency virus DNA vaccination does not augment the sterile protection afforded by passive infusion of neutralizing antibodies. J. Virol. 77: 10348-10356

McAdam S, Kaleebu P, Krausa P, Goulder N, French B, Collin T, Blanchard J, Whitworth J, McMichael A, Gotch F (1998). Cross-clade recognition of p55 by cytotoxic T lymphocytes in HIV-1 infection. AIDS, 12: 571-579.

McKeating JA, Griffiths PD, Grundy JE (1987). Cytomegalovirus in urine specimens has host β_2 microglobulin bound to the viral envelope: a mechanism of evading the host immune response? J. Gen. Virol. 68: 785-792.

Melmed RN, Taylor JMG, Detels R, Bozorgmehri M, Fahey JL (1989). Serum neopterin changes in HIV-infected subjects: indicator of significant pathology, CD4 T cell changes, and the development of AIDS. J. Acquired Immune Defic. Syndr. 2: 70-76.

Nakamura Y, Kameoka M, Tobiume M, Kaya M, Ohki K, Yamada T, Ikuta K (1997). A chain section containing epitopes for cytotoxic T, B and helper T cells within a highly conserved region found in the human immunodeficiency virus type 1 Gag protein. Vaccine 15: 489-496.

Ott DE (1997). Cellular proteins in HIV virions. Rev. Med. Virol. 7: 167-180.

Ott DE (2008). Cellular proteins detected in HIV-1. Rev. Med. Virol. 18: 159-175.

Pilgrim AK, Pantaleo G., Cohen OJ, Fink LM, Zhou JY, Zhou JT, Bolognesi DP, Fauci AS, Montefiori DC (1997). Neutralizing antibody responses to human immunodeficiency virus type 1 in primary infection and long-termnonprogressive infection. J. Infect. Dis. 176: 924-932.

Pinto LA, Sullivan J, Berzofsky JA, Clerici M, Kessler HA, Landay AL, Shearer GM (1995). ENV-specific Cytotoxic T Lymphocyte Responses in HIV Seronegative Health Care Workers Occupationally Exposed to HIV-contaminated Body Fluids. J. Clin. Invest. Inc. 96: 867-876.

Ravanini P, Quaglia V, Crobu MG, Nicosia AM, Fila F (2007). Use of anti-R7V antibodies testing as a possible prognostic marker of slow progression in HIV infected patients naïve of treatment. Poster presented at the 3rd SAAIDS Congress, Durban 5-8 June 2007. Ref. n. 682.

Richman DD, Wrin T, Little SJ, Petropoulos C.J (2003). Rapid evolution of the neutralizing antibody response to HIV type 1 infection. Proc. Natl. Acad. Sci. USA 100:4144-4149 Epub 2003 Mar 18.

Roitt I, Brostoff J, Male D (2001). Immunology, Sixth edition. Mosby.

Rosano C, Zuccotti S, Bolognesi M (2005). The three-dimensional structure of $\beta2$ microglobulin: Results from X-ray crystallography. Biochim. Biophys. Acta. 1753: 85-91.

Sanchez A, Gemrot F, Da Costa Castro JM (2008). Development and studies of the anti-R7V neutralizing antibody ELISA test: A new serological test for HIV seropositive patients. J. Immunol. Method. 53-60.

Tagny Tayou C, Ndembi N, Moudourou S, Mbanya D (2007). The anti-R7V antibody and its association to clinico-biological status of HIV-1 positive individuals in Yaoundé, Cameroon. Poster presented at the 3rd SAAIDS Congress, Durban 5-8 June 2007.

Trinh CH, Smith DP, Kalverda AP, Phillips SE, Radford SE (2002). Crystal structure of monomeric human $\beta2$-microglobulin reveals clues to its amyloidogenic properties. Proc. Natl. Acad. Sci. U.S.A. 99: 9771-9776.

Tysoe-Calnon VA, Grundy JE, Perkins SP (1991). Molecular comparison of the beta 2-microglobulin binding site in class I major-histocompatibility-complex alpha-chains and proteins of related sequences. Biochem. J. 277: 359-369.

Wagner R, Leschonsky B, Harrer E, Paulus C, Weber C, Walker BD, Buchbinder S, Wolf H, Kalden JR, Harrer T (1999). Molecular and functional analysis of a conserved CTL epitope in HIV-1 p24 recognized from a long-term nonprogressor: Constraints on immune escape associated with targeting a sequence essential for viral replication. J. Immunol. 162: 3727-3734.

Watkins BA, Buge S, Aldrich K, Davis AE, Robinson J, Reitz MS, Robert-Guroff M (1996). Resistance of human immunodeficiency virus type 1 to neutralization by natural antisera occurs through single amino acid substitutions that cause changes in antibody binding at multiple sites. J. Virol. 70: 8431-8437.

Webber L (2009). Prevalence of anti-R7V antibodies in a cohort of HIV-infected South African patients on HAART. Presentation at the HIV and AIDS Research Symposium at the University of Pretoria, 26-27 February 2009 and personal communications.

Wei X, Decker JM, Wang S, Hui H, Kappes JC, Wu X, Salazar-Gonzales JF, Salazar MG, Kilby JM, Saag MS, Komarova NL, Nowak MA, Hahn BH, Kwong PD, Shaw GM (2003). Antibody neutralization and escape by HIV-1. Nature 422: 307-312.

Xu X, Xing H, Gong W, Chen H, Si C, Wang Y, Chermann JC (2002). Preliminary investigation on the relation between clinical progress and anti-small monomolecular peptides antibody in individual infected with HIV. Zhonghua Shi Tan He Lin Chuang Bing Du Xue Za Zhi. 16(3): 86-287. Article in Chinese.

16

Standard Review Cold-active microbial Lipases: a versatile tool for industrial applications

Babu Joseph[1]*, Pramod W. Ramteke[1], George Thomas[2], and Nitisha Shrivastava[1]

[1]Department of Microbiology and Microbial Technology, College of Biotechnology and Allied Sciences, Allahabad Agricultural Institute- Deemed University, Allahabad 211 007, Uttar Pradesh, India.
[2]Department of Molecular Biology and Genetic Engineering, College of Biotechnology and Allied Sciences, Allahabad Agricultural Institute- Deemed University, Allahabad 211 007, Uttar Pradesh, India.

Lipases are a class of enzymes which catalyze the hydrolysis of long chain triglycerides and constitute the most important group of biocatalysts for biotechnological applications. Cold active lipases have lately attracted attention as a result of their increasing use in the organic synthesis of chiral intermediates. Due to their low optimum temperature and high activity at very low temperatures, which are favorable properties for the production of relatively frail compounds. Cold active lipases are today the enzymes of choice for organic chemists, pharmacists, biophysicists, biochemical and process engineers, biotechnologists, microbiologists and biochemists. The present review describes various industrial applications of cold active microbial lipases in the medical and pharmaceuticals, fine chemical synthesis, food Industry, domestic and environmental applications.

Key words: Biocatalysts, cold active lipase, enzymes, industrial application, lipolytic, Psychrophiles.

Table of Contents

INTRODUCTION

Lipolytic enzymes are currently attracting an enormous attention because of their biotechnological potential (Benjamin and Pandey, 1998). They constitute the most important group of biocatalysts for biotechnological applications. Furthermore, novel biotechnological applications have been successfully established using lipases for the synthesis of biopolymers and biodiesel, the production of enantiopure pharmaceuticals, agro-chemicals, and flavour compounds (Jaeger and Eggert, 2002). The chemo-, regio- and enantio-specific behavior of these enzymes caused tremendous interest among scientists and Industrialists (Saxena et al., 2003). The knowledge of cold adapted lipolytic enzymes in industrial applications is increasing at a rapid and exciting rate. Unfortunately, the studies on cold adapted lipases are incomplete and scattered. Till date no attempt has been undertaken to organize this information. Hence, an over- of this biotech-

*Corresponding author. E-mail: babuaaidu@yahoo.co.in.

Table 1. Bacteria producing cold adapted lipases.

Microorganism	Sources	Reference
Acinetobacter sp. strain No. 6	Siberian tundra soil	Suzuki et al., 2001
Acinetobacter sp. strain No. O_{16}	Ns	Breuil and Kushner, 1975
Achromobacter lipolyticum	Ns	Khan et al., 1967
Aeromonas sp. strain No. LPB 4	Sea sediments	Lee et al., 2003
Aeromonas hydrophila	Marine habitat	Pemberton et al., 1997
Bacillus sphaericus MTCC 7526	Gangothri Glacier (Western Himalayas)	Joseph, 2006
Microbacterium phyllosphaerae MTCC 7530		
Moraxella sp.	Antarctic habitat	Feller et al., 1990
Morexella sp TA144	Antarctic habitat	Feller et al., 1991
Photobacterium lipolyticum M37	Marine habitat	Ryu et al., 2006
Pseudoalteromonas sp. wp27	Deep sea sediments	Zeng et al., 2004
Pseudoalteromonas sp.	Antarctic marine	Giudice et al., 2006
Psychrobacter sp.		
Vibrio sp.		
Pseudomonas sp. strain KB700A	Subterranean environment	Rashid et al., 2001
Pseudomonas sp. B11-1:	Alaskan soil	Choo et al., 1998
Pseudomonas P38	NS	Tan et al., 1996
Pseudomonas fluorescens	Refrigerated milk samples	Dieckelmann et al., 1998
Pseudomonas fluorescens	Refrigerated food	Andersson 1980
Pseudomonas fluorescens	Refrigerated human placental extracts	Preuss et al., 2001
Pseudomonas fragi strain no. IFO3458	BCCM™/LMG2191^T Bacteria collection, Universiteit Gent, Belgium	Alquati et al., 2002
Pseudomonas fragi Strain no. IFO 12049	Ns	Aoyama et al., 1988
Psychrobacter sp. wp37	Deep sea sediments	Zeng et al., 2004
Psychrobacter okhotskensis sp.	Sea coast	Yumoto et al., 2003
Psychrobacter sp. Ant300	Antarctic habitat	Kulakovaa et al., 2004
Psychrobacter immobilis strain B 10	Antarctic habitat	Arpigny et al., 1997
Serratia marcescens	Raw milk	Abdou, 2003
Staphylococcus aureus	Ns	Alford and Pierce, 1961
Staphylococcus epidermidis	Frozen fish samples	Joseph et al., 2006

Ns Not specified.

nologically and industrially important enzyme and its applications has been collected and compiled from the information available in the literature. From the limited number of available reports on cold active lipases, it is clear that most of the studies were focused on isolation, purification and characterization for industrial applications of these enzymes followed by gene cloning, expression and sequencing. A worldwide initiative has been taken up for exploring cold active lipase producing microorganisms and their industrial applications.

Sources of cold active lipases

Cold adapted lipases are largely distributed in microorganisms existing at low temperatures nearly 5°C. Although a number of lipase producing sources are available, only a few bacteria and yeast were exploited for the production of cold adapted lipases (Joseph, 2006). Attempts have been made from time to time to isolate cold adapted lipases from these microorganisms having high activity at low temperatures. A list of various cold adapted lipase

producing psychrophillic and psychro-trophic bacteria are presented in Table 1.

Microbial enzymes are often more useful than enzymes derived from plants or animals because of the great variety of catalytic activities available, the high yields possible, ease of genetic manipulation, regular supply due to the absence of seasonal fluctuations and rapid growth of microorganisms on inexpensive media. Microbial enzymes are also more stable than their corresponding plant and animal enzymes and their production is more convenient and safer (Wiseman, 1995). Various studies showed that a high bacterial count has been recorded as high as 10^5 and 10^6 ml^{-1} in water column and in the sea ice respectively (Sulivan and Palmisano, 1984; Delille, 1993). Cold adapted bacterial strains were isolated mostly from Antarctic and Polar regions, which represents a permanently cold (0 ± 2°C) and constant temperature habitat. Another potential source of cold active lipases is deep sea bacteria. A marine bacterium *Aeromonas hydrophila* growing at a temperature range between 4 and 37°C was found to produce lipolytic enzyme (Pemberton et al., 1997).

Table 2. Fungi producing cold adapted lipases.

Microorganism	Sources	Reference
Aspergillus nidulans	Ns	Mayordomo et al., 2000
Candida antarctica	Antarctic habitat	Patkar et al., 1993;
		Uppenberg et al., 1994a;
		Uppenberg et al., 1994b
		Patkar et al., 1997
		Koops et al.,1999;
		Zhang et al., 2003;
		Siddiqui and Cavicchioli, 2005
C. lipolytica	Frozen food	Alford and Pierce, 1961
Geotrichum candidum	Frozen food	Alford and Pierce, 1961
Pencillium roqueforti	Frozen food	Alford and Pierce, 1961
Rhizopus sp.	Frozen food	Coenen et al., 1997
Mucor sp.	Frozen food	Coenen et al., 1997

Ns - Not specified

However, 16S rDNA sequence of isolated cold adapted *A. hydrophila* exhibited the highest similarity to that of a marine bacterium *A. hydrophila* (95% homology) and showed the same characteristics. This isolate could grow at temperature 4, 10, 20, and 30 °C but not at temperature above 30 °C (Lee et al., 2003). Few bacterial genera have been isolated and characterized from deep-sea sediments where temperature is below 3 °C. They include *Aeromonas* sp. (Lee et al., 2003); *Pseudoalteromonas* sp. and *Psychrobacter* sp. (Zeng et al., 2004); *Photobacterium lipolyticum* (Ryu et al., 2006). Permanently cold regions such as glaciers and mountain regions are another habitat for psychroplillic lipase producing microorganisms (Joseph, 2006). The soil and ice in Alpine region also harbor psychrophillic microorganisms, which produces cold active lipases. In addition to all these permanently cold regions, there are many other accessible and visible soil and water, which become cold both diurnally and seasonally from which cold active lipase producing microbes can be isolated using appropriate low temperature techniques.

The wide spread use of refrigeration to store fresh and preserved foodstuffs provides a great diversity of nutrient rich habitat for some well known psychrotolerant food spoilage microorganisms. Bacterial genera including *Pseudomonas fragi* (Aoyama et al., 1988; Alquati et al., 2002), *Pseudomonas fluorescens* (Dieckelmann et al., 1998) and *Serratia marcences* (Abdou, 2003), which produce cold active lipases, were isolated from refrigerated milk and food samples. Further, the genes encoding for cold active lipases were isolated and cloned into mesophilic bacteria (*Escherichia coli*) as host organism and used for their expression. However, the review of Gerday et al. (1997) revealed an extremely unstable condition for the expression of cold adapted lipases within their host (Feller et al., 1990; Feller et al., 1991). In the other

reports related to expression studies, the stability of gene encoding lipase production in the host is not clear.

Even though many psychrophilic and psychrotrophic bacteria produce lipases, it is clear that only a few lipolytic fungus was reported to produce cold active lipases (Table 2). An extensive research has been carried out in the cold active lipase of *Candida. antarctica* compared to the other psychrophillic fungi. *Candida lipolytica, Geotrichum candidum* and *Pencillium roqueforti* have also been isolated from frozen food samples and reported to produce cold active lipases (Alford and Pierce, 1961). Psychrotropic lipolytic moulds viz., *Rhizopus* sp. and *Mucor* sp. cause havoc with milk and dairy products and soft fruits (Coenen et al., 1997).

Structural modifications for cold activity

Cold adapted lipases probably are structurally modified by an increasing flexibility of the polypeptide chain enabling an easier accommodation of substrates at low temperature. The fundamental issues concerning molecular basis of cold activity and the interplay between flexibility and catalytic efficiency are of important in the study of structure-function relationships in enzymes.

Such issues are often approached through comparison with the mesophilic or thermophilic counterparts, by site directed mutagenesis and 3D crystal structures (Narinx et al., 1997; Wintrode et al., 2000). The molecular modeling of *Pseudomonas immobilis* lipase revealed several features of cold-adapted lipases (Arpigny et al., 1997). A very low proportion of arginine residues as compared to lysine, a low content in proline residues, a small hydrophobic core, a very small number of salt bridges and of aromatic-aromatic interactions are the possible features of lipase for cold adaptation. Similarly the weakening of hydrophobic clusters, the dramatic decrease (40%) of the

Proline content and of the ratio Arg/Arg+Lys makes lipases active at low temperature (Gerday et al., 1997). Moreover when compared to the dehalogenase from *Xanthobacter autotrophicus*, the cold lipase displays a very small number of aromatic – aromatic interactions and of salt bridges. The location of some salt bridges which are absent in the cold lipase seems to be crucial for the adaptation to cold. A large amount of charged residues exposed at the protein surface, have been detected in the cold active lipase from *Pseudomonas fragi* (Alquati et al., 2002). They also observed a reduced number of disulphide bridges and of prolines in loop structures. Arginine residues were distributed differently than in mesophilic enzymes, with only a few residues involved in stabilizing intramolecular salt bridges and a large proportion of them exposed at the protein surface that may contribute to increased conformational flexibility of the cold-active lipase. In addition to this, the structural factors possibly involved in cold adaption are increased number and clustering of glycine residues (providing local mobility), lower number of ion pairs and weakening of charge-dipole interactions in α helices (Georlette et al., 2004; Gomes and Steiner, 2004). The substitution of Glycine with proline by mutation caused a shift of the acyl chain length specificity of the enzyme towards short chain fatty acid esters and enhanced themostability of the enzyme (Kulakovaa et al., 2004). A mutation in the lid region of catalytic triad of cold active lipases from *P. fragi* improved substrate selectivity and thermostability (Santarossa et al., 2005). Introduction of polar residues in the surface exposed lid might be involved in improved substrate specificity and protein flexibility. The sequence alignment study of cold active lipase from *P. lipolyticum* showed three amino acid residues (Ser174, Asp236 and His312) constitute the active site and RG residues (Arg236 and Gly91) making an oxyanion sequence (Ryu et al., 2006). It is understood that the catalytic cavity of the psychrophillic lipase is characterized by high plasticity. These structural adaptations may confer on the enzyme a more flexible structure, in accord with its low activation energy and its low thermal stability. The above discussions may help to obtain information for insights into the molecular mechanisms of cold adaptation and thermolability of cold active lipases and low thermostability and unusual properties like chemo-, regio- and enantiospecificities.

Applications of cold active lipases

The ability of psychrophilic enzymes to catalyze reactions at low or moderate temperature offers a great industrial and biotechnological potential (Gomes and Steiner, 2004). The 'cold activity' (that is, high catalytic activity at low temperatures) and thermolability of lipases might be the key to success in some of their applications. These applications include their use as catalyst for organic synthesis of unstable compounds at low temperature.

Efficient binding of substrate by the enzyme is mediated by the nature and strength of weak interactions which are of two types. Interactions formed with a negative modification of enthalpy and hence are exothermic (Van der Waals interactions, Hydrogen bonds, electrostatic inter-actions) and interactions formed (at least within the low and moderate temperature ranges) with the positive, modification of enthalpy, and thus are endothermic (hydrophobic interactions). The former destabilized by an increase of temperature, whereas the later will tend to be stabilized by moderately high temperatures (Georlette et al., 2004). The cold active lipases from cold adapted microorganisms and their potential applications have been examined (Choo et al., 1998). With the increasing interest in psychrophiles (microorganisms growing at 0°C or lower sometimes restricted to organisms that cannot grow above 20°C) and their applications, cold active lipases will represent a larger share of industrial enzyme market in the coming years. The cold active lipases offer novel opportunities for biotechnological exploitation based on their high catalytic activity at low temperature.

The cold enzymes along with the producing microorganisms cover a broad spectrum of biotechnological applications (Table 3). They include additives in deter-gents (cold washing), additives in food industries (fermentation, cheese manufacture, bakery, meat tenderizing), environmental bioremediations (digesters, composting, oil degradation or xenobiotic biology applications and molecular biology applications), bio-transformation and heterologous gene expression in psychrophilic hosts to prevent formation of inclusion bodies (Feller et al., 1996). A number of relatively straightforward reasons for applications of cold active enzymes in biotechnology have been mentioned by various authors (Rusell, 1998; Margesin and Schinner, 1999; Ohigiya et al., 1999; Gerday et al., 2000; Cavicchioli et al., 2002). Cold active lipases have low thermal stability favorable for some purposes eg: heat liabile lipase can be inactivated by treatment for short periods of time at relatively low temperatures after being used for the processing of food and other materials (Margesin et al., 2002). Thus materials can be prevented from damage during heat inactivation. However, the number of present uses is limited and likely to reflect the state of the field, which has not yet developed as rapidly when compared to the field of thermophilic enzymes. Nevertheless, despite the difficulties with predication, some important advances have been made (Cui et al., 1999). Cold active lipases could be a good alternative to mesophilic enzymes in brewing industry and wine Industries, cheese manufacturing, animal feed supplements, and so on (Collins et al., 2002). The biotechnological potential of cold adapted lipases is in protein polymerization and gelling in fish flesh, improvement in food texture, perfumery and optically active ester synthesis (Cavicchioli and Siddiqui, 2004).

Table 3. Biotechnological applications of cold adapted lipases

Field of Application	Purpose	Reference
Medical and Pharmaceutical application	Synthesis of Aryl aliphatic glycolipids	Otto et al., 2000
	Ethyl esterification of docosahexaenoic acid to Ethyl docosahexaenoate (EtDHA)	Shimada et al., 2001
	Synthesis of citronellol laurate from citronellol and lauric acid	Ganapati and Piyush, 2005
Fine chemical synthesis	Optically active ester synthesis	Anderson et al., 1998
	Ester synthesis, desymmetrization and production of peracids	Zhang et al., 2003
	Organic synthesis of chiral intermediates	Gerday et al., 2000
	Synthesis of butyl caprylate in n-heptane	Tan et al. 1996
	Synthesis of butyl lactate by transesterification	Pirozzi and Greco, 2004
	synthesis of amides	Slotema et al., 2003
Food Industry	Protein polymerization and gelling in fish flesh, improvement in food texture, flavor modification	Cavicchioli and Siddiqui, 2004
	Production of fatty acids and interesterification of fats	Jaeger and Eggert, 2002
Domestic application	Detergents and cold water washing	Gerday et al., 2000 Joseph, 2006
	Production of α-butylglucoside lactate by transesterification for cosmetics	Bousquet et al., 1999
	Conversion of degummed soybean oil to biodiesel fuel	Watanabe et al., 2002
	Synthesis of lipase-catalyzed biodiesel	Chang et al., 2004
Environmental application	Degradation of lipid wastes	Ramteke et al., 2005
	Bioremediation and bioaugumentation	Gerday et al., 2000; Suzuki et al., 2001; Lee et al., 2003
	Removal of solid and water pollution by hydrocarbons, oils and lipids	Margesin et al., 2002

Medical and pharmaceutical application

Enantioselective interesterification and transesterification have great significance in pharmaceutical for selective acylation and deacylation (Stinson, 1995). Lipases are important in application in pharmaceuticals in transesterificatrion and hydrolysis reaction. They play a prime role in production of specialty lipids and digestive aids (Vulfson, 1994). The alteration of temperature during the esterification reaction drastically changes the enantiomeric values and also the stereopreference (Yasufuku et al., 1996). Lipases play an important role in modification of monoglycerides for use as emulsifiers in pharmaceutical applications (Sharma et al., 2001).

Psychrophiles producing cold active lipases may be a good source for polyunsaturated fatty acids for the pharmaceutical industry. It is because of their excellent capability for specific regioselective reactions in a variety of organic solvents with broad substrate recognition makes lipases as an important biocatalyst in biomedical applications (Margesin et al., 2002). A preparation of optically active amines that was intermediate in the preparation of pharmaceuticals and pesticides, which involved in reac-

ting stereospecific N- acylamines with lipases, preferably from Candida antarctica or Pseudomonas sp. (Smidt et al., 1996). In an attempt to determine the substrate specificity for lipases, alkyl esters of 2-arypropionic acids, a class of non-steroidal anti-inflammatory drugs, were hydrolyzed with Caenorhabditis rugosa lipase in which all transformations were highly enantioselective (Botta et al., 1997).

Synthesis of fine chemicals

Some of the industrially important chemicals manufactured from fats and oils by chemical processes could be produced by lipases with greater rapidity and better specificity under mild conditions (Sih and Wu, 1989; Vulfson, 1994). The use of industrial enzymes allows the technologists to develop processes that more closely approach the gentle, efficient process in nature. Some of these processes using cold active lipase from C. antarctica have been patented by pharmaceutical, chemical and food industries. C. antarctica, one of 154 species of the genus Candida belongs to the phylum Ascomycota. It is alkali tolerant yeast found in the sediment of Lake Vanda, Antarctica (Joseph, 2006). The two lipase variants from this organism viz., Lipase A and Lipase B have proven of particular interest to the researchers. Martinelle and Hult (1995) conducted the comparative studies on the interfacial activation of these lipases A and B with *Humicola lanuginose* lipase. The characterization of lipase A for substrate specificity and its utility as biocatalyst was reported (Kirk and Christensen, 2002). Further, the kinetics of acyl transfer reactions in organic media catalyzed by lipase B were studied (Martinelle and Hult, 1995). Anderson et al. (1998) determined the applications of lipase B in organic synthesis and the enantioselectivity of lipase for some synthetic substrates. Rotticci et al. (1998) proposed the molecular recognition of alcohol enantiomers by lipase B. The use of lipase B for the preparation of optically active alcohols was also determined (Rotticci et al., 2001). Studies revealed that size as a parameter for solvent effects on lipase B enantioselectivity. The evaluation of lipase as catalyst in different reaction media for hydrolysis of tributyrin as reaction model has been reported (Salis et al., 2003). The amidase activity of lipase B and structural feature of the substrates were reported (Torres et al., 2006). The performance of lipase B in the enantioselectivity esterifiction of ketoprofen (Ong et al., 2006) and the improvement of enantioselectivity of lipase (fraction B) via adsorption on polyethylenimine-agarose (Torres et al., 2006) were studied recently. The structure and activity of lipase B of *C. antarctica* in ionic liquids (van Rantwijk et al., 2006) and the applications of lipase B in organic synthesis has been reported (Anderson et al., 1998). Shimada et al. (2001) attempted the ethyl esterification of docosahexaenoic acid (DHA) in an organic solvent-free system using *C. antarctica* lipase, which acts strongly on DHA and ethanol. About 88% esterification was attained by shaking the mixture of DHA / ethanol (1:1, mol/mol) and 2 %wt immobilized *C. antarctica* lipase B at 30°C for 24 h. Use of lipase B from *C. antarctica* for the preparation of optically active alcohols has been reported (Rotticci et al., 2001).

Lipase produced by a psychrotroph, *P. fluorescens* P38, was found to catalyze the synthesis of butyl caprylate in n-heptane at low temperatures. The optimum yield of ester synthesis was 75% at 20°C with an organic phase water concentration of 0.25% (v/v). The results are discussed in terms of the structural flexibility of psychrotroph-derived lipase and the activity of this enzyme within a nearly anhydrous organic solvent phase (Tan et al., 1996).

Watanabe et al. (2002) found that the crude soybean oil did not undergo methanolysis with immobilized *C. antarctica* lipase but degummed oil did. The substance that was removed in the degumming step was estimated to inhibit the methanolysis of soybean triacylglycerols (TAGs). Methanolysis is the displacement of alcohol from an ester by methanol in a process similar to hydrolysis, except that methanol is employed instead of water. Met-

hanol is widely used because of its low cost and its physical and chemical advantages. The main components of soybean gum are phospholipids (PLs), and soybean PLs actually inhibited the methanolysis reaction. Indeed, three-step methanolysis successfully converted 93.8% degummed soybean oil to its corresponding methyl esters, and could be reused for 25 cycles without any loss of the activity. This process widely reduced viscosity of triglycerides, thereby enhancing the physical properties the lipase of renewable fuels to improve engine performance. Lipase from *C. antarctica* has been evaluated as catalyst in different reaction media for hydrolysis of tributyrin as reaction model (Salis et al., 2003).

Applications in food Industry

In the food industry, reaction needs to be carried out at low temperature in order to avoid changes in food ingredients caused by undesirable side-reaction that would otherwise occur at higher temperatures. Lipases have become an integral part of the modern food industry. The use of enzymes to improve the traditional chemical processes of food manufacture has been developed in the past few years. Stead (1986) and Coenen et al. (1997) stated that, though microbial lipases are best utilized for food processing, a few, especially psychrotrophic bacteria of *Pseudomonas sp.* and a few moulds of *Rhizopus* sp. and *Mucor* sp. caused havoc with milk and dairy products and soft fruits. Cold active lipase from *Pseudomonas* strain P38 is widely used in non-aqueous biotransformation for the synthesis of n-heptane of the flavoring compound butyl caprylate (Tan et al., 1996). Immobilized lipases from *C. antarctica* (CAL-B), *C. cylindracea* AY30, *H. lanuginosa*, *Pseudomonas* sp. and *Geotrichum candidum* were used for the esterification of functionalized phenols for synthesis of lipophilic antioxidants in sunflower oil Buisman et al. (1998).

Domestic applications

The most commercially important field of application for hydrolytic lipases is their addition to detergents, which are used mainly in household and industrial laundry and in household dishwashers. Godfrey and West (1996) reported that about 1000 t of lipases are sold every year in the area of detergents. The use of cold active lipase in the formulation of detergents would be of great advantage for cold washing that would reduce the energy consumption and wear and tear of textile fibers (Feller and Gerday, 2003). The industrial dehairing of hides and skin at low temperature using psychrophilic protease or keratinase would not only save energy but also reduce the impacts of toxic chemicals used in dehairing.

Enzymes can reduce the environmental load of detergent products since they save energy by enabling a lower wash temperature to be used; allow the content of other often less desirable chemicals in detergents. Addition of cold active lipsase in detergent become biodegradable, leaving no harmful residues, have no negative impact on sewage treatment processes and do not present a risk to aquatic life. Commercial preparations used for the desizing of denim and other cotton fabrics contain both alpha amylase and lipase enzymes. Lipases are stable in detergents containing protease and activated bleach systems. Lipase is an enzyme, which decomposes fatty stains into more hydrophilic substances that are easier to remove than similar non-hydrolyzed stains (Fuji et al., 1986). The low temperature active lipase can be added to detergents to hydrolyze oily stains at the temperature of tap water to reduce energy consumption and protect the color of fabrics (Feller and Gerday, 2003).

The other common commercial applications for detergents is in dish washing, clearing of drains clogged by lipids in food processing or domestic/industrial effluent treatment plants (Bailey and Ollis, 1986). The use of cold active lipase as a liquid leather cleaner (Kobayashi, 1989) and as an ingredient in bleaching composition (Nakamura and Nasu, 1990) has been reported. Similarly its use in decomposition of lipid contaminants in dry-cleaning solvents (Abo, 1990), contact lens cleaning (Bhatia, 1990), degradation of organic wastes on the surface of exhaust pipes, toilet bowls, etc. (Moriguchi et al., 1990) have been reported. The removal of dirt/cattle manure from domestic animals by lipases and cellulases (Abo, 1990), washing, degreasing and water reconditioning by using lipases along with oxidoreductases, which allows for smaller amounts of surfactants and operation at low temperatures (Novak et al., 1990) have been proposed. The lipase component causes an increase in detergency and prevents scaling. The cleaning power of detergents seems to have peaked; all detergents contain similar ingredients and are based on similar detergency mechanisms. To improve detergency, modern types of heavy duty powder detergents and automatic dishwasher detergents usually contain one or more enzymes, such as protease, amylase, cellulase and lipase (Ito et al., 1998).

Environmental applications

There are number of uses of the cold active enzymes, presently it is conceivable that they could be used for environmental bioremediation e.g., as a biodegradable means of treating an oil spill such as that which occurred by the Exon Valdese in Arctic water. Bioremediation for waste disposal is a new avenue in lipase biotechnology. Cheng et al. (1997) characterized cold-adapted organophosphorus acid anhydrolases for application in the efficient detoxification of pesticide and nerve agents. According to Buchon et al. (2000), cold adapted lipases have great potential in the field of wastewater treatment, bioremediation in fat contaminated cold environment, active compounds synthesis in cold condition. Further, more efforts are needed in identifying and cloning of nov-

el lipase genes for environmental applications. Suzuki et al. (2001) identified a psychrotrophic strain of the genus *Acinetobacter* strain No. 6, producing an extracellular lipolytic enzyme that efficiently hydrolyzed triglycerides, such as soybean oil during bacterial growth even at 4 °C. The strain degraded 60% of added soybean oil (initial concentration, 1% w/v) after cultivation in LB medium at 4 °C for 7 days. The psychrophilic microorganisms as well as their enzymes have been proposed as alternative to physicochemical methods for bioremediation of solids and waste waters polluted by hydrocarbons, oils and lipids (Margesin et al., 2002). Belousova and Shkidchenko (2004) isolated 30 strains including *Pseudomonas* sp. and *Rhodococcus* sp. capable of oil degradation at 4-6 °C and maximum degradation of masut and ethanol benzene resins were observed in *Pseudomonas* sp. and maximum degradation of petroleum oils and benzene resins were observed in *Rhodococcus* sp. Further, they stated that the introduction of psychrotrophic microbial degraders of oil products into the environment is most important in the contest of environmental problems in temperate regions. Ramteke et al. (2005) stated that in temperate regions, large seasonal variations in temperature reduce the efficiency of microorganisms in degrading pollutants such as oil and lipids. The lipase active at low and moderate temperature may also be ideal for bioremediation process.

Future outlook

In spite of the importance of cold active lipases, studies on the mechanisms of production of microbial lipases and the role of lipidic substances used as inducers in lipase production are scanty. Cold active lipases represent an extremely versatile group of bacterial extracellular enzymes that are capable of performing a variety of important reactions, thereby presenting a fascinating field for future research. The understanding of structure-function relationships will enable researchers to tailor new lipases active at low temperatures for biotechnological applications. Developments in research are expected from interchange of experiences between biochemists, geneticists and biochemical engineers. Wide and constant screening of new microorganisms for their lipolytic enzymes at low temperature will open novel and simpler routes for the synthetic processes. Consequently, this may pave new ways to solve biotechnological and environmental problems.

REFERENCES

Abdou AM (2003). Purification and Partial Characterization of Psychrotrophic *Serratia marcescens* Lipase. J. Dairy Sci. 86:127–132.

Abo M (1990). Method of purifying dry-cleaning solvent by decomposing liquid contaminants with a lipase, World Organization Patent 9,007,606.

Alford JA, Pierce DA (1961). Lipolytic activity of microorganisms at low and intermediate temperatures. Activity of microbial lipases at temperatures below 0 °C. J. Food Sci. 26:518-524.

Alquati C, De Gioia L, Santarossa G, Alberghina L, Fantucci P, Lotti M (2002). The cold-active lipase of *Pseudomonas fragi*: Heterologous expression, biochemical characterization and molecular modeling. Eur. J. Biochem. 269: 3321-3328.

Anderson EM, Larsson KM, Kirk O (1998). One biocatalyst – many applications: the use of *Candida antarctica* B-lipase in organic synthesis. Biocatalysis Biotransform. 16:181-204.

Andersson RE (1980). Microbial lipolysis at low temperatures. Appl. Environ. Microbiol. 39: 36-40.

Aoyama S, Yoshida N, Inouye S (1988). Cloning, sequencing and expression of the lipase gene from *Pseudomonas fragi* IFO-12049 in *E. coli*. FEBS Lett. 242: 36–40.

Arpigny JL, Lamotte J, Gerday C (1997). Molecular adaptation to cold of an Antarctic bacterial lipase. J. Mol. Catal. B Enzy. 3: 29-35.

Bailey JE, Ollis DF (1986). Applied enzyme catalysis. In: Biochemical Engineering fundamentals. 2nd edn. Mc Graw-Hill, New York, pp. 157–227.

NI, Shkidchenko AN (2004). Low temperature degradation of oil products differing in the extent of condensation. Appl. Biochem. Microbiol. 40: 262-265.

Benjamin S, Pandey A (1998). *Candida rugosa* lipases: Molecular biology and Versatility in Biotechnology. Yeast. 14: 1069-1087.

Bhatia RP (1990). Contact lens cleaning composition containing an enzyme and a carboxylvinyl polymer. United States Patent 4,921,630,

Botta M, Cernia E, Corelli F, Manetti F, Soro S (1997). Probing the substrate specificity for lipases II. Kinetic and modeling studies on the molecular recognition of 2-arylpropionic esters by *Candida rugosa* and *Rhizomucor miehei* lipases. Biochem. Biophys. Acta. 1337: 302–310.

Bousquet MP, Willemot RM, Monsan P, Boures E (1999). Lipase catalyzed α-butylglucoside lactate synthesis in organic solvent for dermo-cosmetic application. J. Biotechnol. 68: 61.

Breuil C, Kushner DJ (1975). Lipase and esterase formation by psychrophilic and mesophilic *Acinetobacter* species. Can. J. Microbiol. 21: 423–433

Buchon L, Laurent P, Gounot AM, Guespin MJF (2000). Temperature dependence of extrcellular enzyme production by Psychotrophic and Psychrophilic bacteria. Biotechnol. Lett. 22: 1577-1581.

Buisman GJH, Helteren CTW, Kramer GFH, Veldsink JW, Derksen JTP, Cuperus FP (1998). Enzymatic esterifications of functionalized phenols for the synthesis of lipophilic antioxidants. Biotechnol. Lett. 20: 131-136.

Cavicchioli R, Saunders KS, Andrews D, Sowers KR (2002). Low-temperature extremophiles and their applications. Curr. Opin. Biotechnol. 13: 253-261.

Cavicchioli R, Siddiqui KS (2004). Cold adapted enzymes. In: A Pandey, C Webb, CR Soccol, C Larroche (eds) Enzyme Technology, Asiatech Publishers, India, pp. 615-638.

Chang H, Liao H, Lee C, Shieh C (2004). Optimized synthesis of lipase-catalyzed biodiesel by Novozym 435. J. Chem. Tech. Biotech. 80: 307 – 312.

Cheng TC, Liu L, Wang B, Wu J, De Frank JJ, Anderson DM, Rastogi VK, Hamilton AB (1997). Nucleotide sequence of a gene encoding an organophosphorous nerve agent degrading enzyme from *Alteramonas haloplanktis*. J. Ind. Microbiol. Biotechnol. 18: 49-55

Choo DW, Kurihara T, Suzuki T, Soda K, Esaki N (1998). A cold-adapted lipase on an Alaskan psychrotroph, *Pseudomonas* sp. strain B11-1: Gene Cloning and Enzyme Purification and Characterization. Appl. Environ. Microbiol. 64: 486-491.

Coenen TMM, Aughton P, Verhagan H (1997). Safety evaluation of lipase derived from *Rhizopus oryzae*: Summary of toxicological data. Food Chem. Toxicol. 35: 315–22.

Collins T, Meuwis MA, Stals I, Claeyssens M, Feller G, Gerday C (2002).. A novel Family 8 Xylanase. Functional and Physicochemical Characterization. J. Biol. Chem. 277: 35133-35139.

Cui Y, Wei D, Yu J (1999). Lipase-catalyzed esterification in organic solvent to resolve racemic naproxen. Biotechnol. Lett. 19: 865-868.

Delille D (1993). Seasonal changes in the abundance and composition of marine heterotrophic bacterial communities in an Antarctic coastal area. Polar Biol. 12: 205-210.

Dieckelmann M, Johnson LA, Beacham IR (1998). The diversity of lipases from psychrotrophic strains of *Pseudomonas*: A Novel Lipase from a highly lipolytic strain of *Pseudomonas fluorescens*. J. Appl. Microbiol. 85: 527–536.

Feller G, Gerday C (2003). Psychrophilic enzymes: Hot Topics in Cold Adaption, Nat. Rev. Microbiol. 1:200-208.

Feller G, Narinx E, Arpigny JL, Aittaleb M, Baise E, Geniot S, Gerday C (1996). Enzymes from Psychrophilic Organisms, FEMS Microbiol. Rev. 18: 189 -202.

Feller G, Thiry M, Arpigny JL, Gerday C (1991). Cloning and Expression in *Escherichia coli* of Three Lipase encoding genes from the Psychrotrophic Antarctic strain *Moraxella* TA144. Gene. 102:111–115.

Feller G, Thiry M, Arpigny JL, Mergeay M, Gerday C (1990). Lipases from Psychrotrophic Antarctic bacteria. FEMS Microbiol. Lett. 66: 239–244.

Fuji T, Tatara T, Minagwa M (1986). Studies on Application of lipolytic enzyme in Detergency I. J. Am. Oil Chem. Soc. 63: 796-799.

Ganapati DY, Piyush SL (2005). Lipase catalyzed transesterification of methyl acetoacetate with n-butanol J. Mol. Cat. B Enzy. 32: 107-113.

Georlette D, Blaise V, Collins T, Amico SD, Gratia E, Hoyoux A, Marx JC, Sonan G, Feller G, Gerday C (2004). Some like it cold: biocatalysis at low temperatures, FEMS Microbiol. Rev. 28: 25-42.

Gerday C, Aittaleb M, Arpigny JL, Baise E, Chessa JP, Garssoux G, Petrescu I, Feller G (1997). Psychrophilic Enzymes: a Thermodynamic Challenge. Biochem. Biophys. Acta. 1342: 119-131.

Gerday C, Aittaleb M, Bentahir M, Chessa JP, Claverie P, Collins T, D'Amico S, Dumont J, Garsoux G, Georlette D, Hoyoux A, Lonhienne T, Meuwis M, Feller G (2000). Cold adapted enzymes: from fundamentals to biotechnology. Trend. Biotechnol. 18: 103-107.

Giudice AL, Michaud L, de Pascale D, Domenico MD, di Prisco G, Fani R, Bruni V (2006). Lipolytic activity of Antarctic Cold adapted marine bacteria. J. Appl. Microbiol. 101: 1039-1048.

Godfrey T, West S (1996). The application of enzymes in industry. In: T Godfrey, J Reichelt (eds) Industrial enzymology, 2nd (edn). The Nature Press, New York, p. 512.

Gomes J, Steiner W (2004). The Biocatalytic Potential of Extremophiles and Extremozymes. Food Technol. Biotechnol. 42: 223-235.

Ito S, Kobayashi T, Ara K, Ozaki K, Kawai S, Hatada Y (1998). Alkaline detergent enzymes from Alkalophiles: Enzymatic Properties, Genetics and Structures. Extremophiles. 2: 185 -190.

Jaeger KE, Eggert T (2002). Lipases for biotechnology. Curr. Opin. Biotechnol. 13(4): 390–397.

Joseph B (2006) Isolation, purification and characterization of cold adapted extracellular lipases from psychrotrophic bacteria: Feasibility as laundry detergent additive. Ph.D Thesis. Allahabad Agricultural Institute – Deemed University, Allahabad, India.

Joseph B, Ramteke PW, Kumar PA (2006). Studies on the enhanced production of extracellular lipase by *Staphylococcus epidermidis*. J. Gen. Appl. Microbiol. 52: 315-320.

Khan IM, Dill CW, Chandan RC, Shahani KM (1967). Production and properties of the extracellular lipase of *Achromobacter lipolyticum*. Biochem.Biophys. Acta. 132: 68-77.

Kirk O, Christensen MW (2002). Lipases from *Candida antarctica*: Unique biocatalysts from a unique origin. Organic Process Research and Development. 6: 446-451.

Kobayashi H (1989). Liquid leather cleaners, Japanese Patent 1: 225-700.

Koops BC, Papadimou E, Verheij HM, Slotboom AJ, Egmond MR (1999). Activity and stability of chemically modified *Candida antarctica* lipase B absorbed on solid supports. Appl. Microbiol. Biotechnol. 52:791-796

Kulakovaa L, Galkina A, Nakayamab T, Nishinob T, Esakia N (2004). Cold-active esterase from *Psychrobacter* sp. Ant300: Gene cloning, Characterization, and the Effects of Gly Pro substitution near the active site on its catalytic activity and stability. Biochemica. Biophysica. Acta. 1696: 59– 65

Lee HK, Min JA, Sung HK, Won HS, Byeong CJ (2003). Purification and Characterization of Cold Active Lipase from Psychrotrophic *Aeromonas* sp. LPB 4. J. Microbiol. 41: 22-27.

Margesin R, Feller G, Gerday C, Rusell N (2002). Cold adapted Microorganisms: Adaptation strategies and Biotechnological Potential. In: Bitton (ed). The Encyclopedia of Environmental Microbiology, John Wiley & Sons, New York, pp 871-885.

Margesin R, Schinner F (1999). Cold-Adapted Organisms—Ecology,

Physiology, Enzymology and Molecular Biology. Springer-Verlag, Berlin, p. 416.

Martinelle M, Hult K (1995). Kinetics of acyl transfer reactions in organic media catalyzed by *Candida antarctica* lipase B. Biochemical Biophysica. Acta. 1251: 191-197.

Mayordomo I, Randez Gil F, Prieto JA (2000). Isolation, Purification and Characterization of a cold-active lipase from *Aspergillus nidulans*. J. Agric. Food Chem. 48: 105-109.

Moriguchi H, Hirata J, Watanabe T (1990). Microorganism based agent for treatment of organic wastes, Japanese Patent 2: 105,899.

Nakamura K, Nasu T (1990). Enzyme containing bleaching composition, Japanese Patent 2: 208,400.

Narinx E, Baise E, Gerday C (1997). Subtilisin from Psychrophilic Antarctic bacteria: characterization and site directed mutagenesis of residues possible involved in the adaptation to cold. Protein Eng. 10: 1271-1279.

Novak J, Kralova B, Demnerova K, Prochazka K, Vodrazka Z, Tolman J, Rysova D, Smidrkal J, Lopata V (1990). Enzyme agent based on lipases and oxidoreductases for washing, degreasing and water reconditioning, European Patent 355,228.

Ohgiya S, Hoshino T, Okuyama H, Tanaka S, Ishizaki K (1999). Biotechnology of enzymes from cold adapted microorganisms. In: R Margesin, F Schinner (eds) Biotechnological applications of cold adapted organisms, Springer- Verlag, Berlin Heidelberg, pp 17- 34.

Ong AL, Kamaruddin AH, Bhatia SW, Long WS, Lim ST, Kumari R (2006). Performance of free *Candida antarctica* lipase B in the enantioselective esterification of (*R*)-ketoprofen. Enzym. Microb. Technol. 39: 924-929.

Otto Y, Sawamoto T, Hasuo M (2000). Tributyrin specifically induces a lipase with a preference for the *sn-2* position of triglyceride in *Geotrichum* sp. F0401B, Biosci. Biotechnol. Biochem. 64: 2497-2499.

Patkar SA, Bjorking F, Zundel M, Schulein M, Svendsen A, Heldt Hansen HP, Gonnsen E (1993). Purification of two lipases from *Candida antarctica* and their inhibition by various inhibitors. Ind. J. Chem. 32: 76-80.

Patkar SA, Svendsen A, Kirk O, Groth IG, Borch K (1997). Effect of mutation in non-consensus Thr-X-Ser-X-Gly of *Candida antarctica* lipase B on lipase specificity, specific activity and thermostabilty. J. Mol. Catal. B Enzym. 3: 51–54.

Pemberton JM, Stephen PK, Radomir S (1997). Secreted enzymes of *Aeromonas*. FEMS Microbiol. Lett. 152:1-10.

Pirozzi D, Greco GJ (2004). Activity and stability of lipases in the synthesis of butyl lactate. Enzym. Microb. Technol. 34: 94-100.

Preuss J, Kaiser I, Gehring U (2001). Molecular characterization of a phosphatidylcholine - hydrolyzing phospholipase C. Eur. J. Biochem. 268: 5081-5091.

Ramteke PW, Joseph B, Kuddus M (2005). Extracellular lipases from anaerobic microorganisms of Antarctic. Ind. J. Biotech. 4:293-294.

Rashid N, Yuji S, Satoshi E, Haruyuki A, Tadayuki I (2001). Low-temperature lipase from psychrotrophic *Pseudomonas* sp. Strain KB700A. Appl. Environ. Microbiol. 67: 4064 - 4069.

Rotticci D, Haffner F, Orrenius C, Norin T, Hult K (1998). Molecular recognition of sec-alcohol enantiomers by *Candida antarctica* lipase B. J. Mol. Catal. B Enzy. 5: 267-272.

Rotticci D, Ottosson J, Norin T, Hult K (2001). *Candida antarctica* lipase B: A tool for the preparation of optically active alcohols. Methods in Biotechnol. 15: 261-276.

Russell RJM, Gerike U, Danson MJ, Hough DW, Tayalor GL (1998). Structural Adaptations of the cold active citrate synthase from an Antarctic Bacterium, Struture. 6: 351–361.

Ryu HS, Kim HK, Choi WC, Kim MH, Park SY, Han NS, Oh TK, Lee JK (2006). New cold-adapted lipase from *Photobacterium lipolyticum* sp. nov. that is closely related to filamentous fungal lipases. Appl. Microbiol. Biotechnol. 70: 321–326.

Salis A, Svensson I, Monduzzi M, Solinas V, Adlercreutz P (2003). The atypical lipase B from *Candida antarctica* is better adapted for organic media than the typical lipase from *Thermomyces lanuginosa*. Biochem. Biophys. Acta. 1646: 145-151.

Santarossa G, Lafranconi PG, Claudia A, Luca DG, Lilia A, Piercarlo F, Lotti M (2005). Mutations in the "lid" region affect chain length specificity and thermostability of a *Pseudomonas fragi* lipase. FEBS Lett. 579: 2383-2386.

Saxena RK, Sheoran A, Giri B, Davidson WS (2003). Purification strategies for microbial lipases. J. Microbiol. Meth. 52: 1–18.

Sharma R, Chisti Y, Banerjee UC (2001). Production, purification, characterization and applications of lipases. Biotechnol. Adv. 19: 627-662.

Shimada Y, Watanabe Y, Sugihara A, Baba T, Ooguri T, Moriyama S, Terai T, Tominaga Y (2001). Ethyl esterification of docosahexaenoic acid in an organic solvent-free system with immobilized *Candida antarctica* lipase. J. Biosci. Bioeng. 92: 19-23.

Siddiqui KS, Cavicchioli R (2005). Improved thermal stability and activity in the cold-adapted lipase B from *Candida antarctica* following chemical modification with oxidized polysaccharides. Extremophiles. 9: 471–476.

Sih CJ, Wu SH (1989). Resolution of enantiomers via biocatalysts. Topics Stereochem. 19: 63–125.

Slotema WF, Sandoval G, Guieysse D, Straathof AJJ, Marty A (2003). Economically pertinent continuous amide formation by direct lipase-catalyzed amidation with ammonia. Biotechnol. Bioeng. 82: 664 – 669.

Smidt H, Fischer A, Fischer P, Schmid RD (1996). Preparation of optically pure chiral amines by lipase catalyzed enantioselective hydrolysis of *N*-acyl-amines. Biotechnol. Tech. 10: 335-338.

Stead R (1986). Microbial Lipases their characteristics, role in food spoilage & industrial uses. J. Dairy Res. 53: 481-505.

Stinson SC (1995). Fine and intermediate chemical markers emphasize new products and process. Chem. Eng. News. 73: 10-26

Sullivan CW, Palmisano AC (1984). Sea ice microbial communities: Distribution, abundance and diversity of ice bacteria in Mc Murdo Soil, Antarctica, in 1980. Appl. Environ. Microbiol. 47:788-795.

Suzuki T, Nakayama T, Kurihara T, Nishino T, Esaki N (2001). Cold-active lipolytic activity of psychrotrophic *Acinetobacter* sp. strain no. 6. J. Biosci. Bioeng. 92: 144–148.

Tan S, Owusu ARK, Knapp J (1996). Low temperature organic phase biocatalysis using cold-adapted lipase from psychrotrophic *Pseudomonas* P38. Food Chem. 57: 415-418.

Torres R, Ortiz C, Pessela BCC, Palomo JM, Mateo C, Guisan JM, Fernandez LR (2006). Improvement of the enantioselectivity of lipase (fraction B) from *Candida antarctica* via adsorpiton on polyethylenimine-agarose under different experimental conditions. Enzym. Microb. Technol. 39: 167-171.

Uppenberg J, Hansen MT, Patkar S, Jones TA (1994b). The sequence, crystal structure determination and refinement of two crystal forms of lipase B from *Candida Antarctica*. Struct. 2 293-308.

Uppenberg J, Patkar S, Bergfors T, Jones TA (1994a). Crystallization and preliminary X-ray studies of lipase B from *Candida antarctica*. J. Mol. Biol. 235: 790-792.

van Rantwijk F, Secundo F, Sheldon RA (2006). Structure and activity of *Candida antarctica* lipase B in ionic liquids. Green Chem. 8: 282–286.

Vulfson EN (1994). Industrial applications of lipases. In: P Wooley, SB Petersen (eds) Lipases, Cambridge, Great Britain: Cambridge University Press. p. 271.

Watanabe Y, Shimada Y, Sugihara A, Tominaga Y (2002). Conversion of degummed soybean oil to biodiesel fuel with immobilized *Candida antarctica* lipase. J. Mol. Catal. B Enzy. 17: 151-155.

Wintrode PL Miyazaki K, Arnold FH (2000). Cold adaptation of a mesophilic subtilisin-like protease by laboratory evolution. J. Biol. Chem. 275: 31635–31640.

Wiseman A (1995). Introduction to Principles. In: A Wiseman (ed) Handbook of enzyme biotechnology. 3rd ed. Padstow, Cornwall, UK: Ellis Horwood Ltd. T.J. Press Ltd., pp 3–8.

Yasufuku Y Ueji S (1996). Improvement five fold of enantioselectivtiy for lipase catalyzed esterification of a bulky substrate at 57 degree in organic solvent. Biotechnol. Tech. 10: 625-628

Yumoto I, Hirota K, Sogabe Y, Nodasaka Y, Yokota Y, Hoshino T (2003). *Psychrobacter okhotskensis* sp. nov., a lipase-producing facultative psychrophile isolated from the coast of the Okhotsk Sea, Int. J. Syst. Evol. Microbiol. 53: 1985–1989.

Zeng X, Xiao X, Wang P, Wang F (2004). Screening and Characterization of psychrotrophic lipolytic bacteria from deep sea sediments. J. Microbiol. Biotechnol. 14: 952-958.

Zhang N, Suen WC, Windsor W, Xiao L, Madison V, Zaks A (2003). Im-

Improving tolerance of *Candida antarctica* lipase B towards irreversible thermal inactivation through directed evolution.   Prot. Eng. 16: 599–605.

17

Engineered pathogenesis related and antimicrobial proteins weaponry against *Phytopthora infestans* in potato plant: A review

Bengyella Louis[1]* and Pranab Roy[2]

[1]Department of Biotechnology, University of Burdwan, Golapbag More-713104, Burdwan, West Bengal, India.
[2]Department of Biochemistry, University of Yaoundé I, Cameroon, BP-812, Cameroon.

Phytopthora infestans (Mont.) de Bary, causal organism of late blight disease is referred to as the most destructive specific pathogen of potato (*Solanum tuberosum* L.). Casualties usually go beyond mere plant destruction, due to its flaring ability to also demolish scientific concerted efforts in establishing novel combat techniques. With high capacity to overcome control measures, it stands at par, and can simply be referred to as a potato-scientist tormentor. This retrospective work examines the role of pathogenesis related proteins and antimicrobial proteins in transgenic potato vis-à-vis *P. infestans* and prospect for associative introduction and overexpression of synergistically resistance conferring genes as a critical step for developing viable transgenic potatoes. The exploitation of enhanced pathogen-inducible promotor overexpressing key pathogenesis related proteins (PR) exemplified by PR-5 (especially osmotin and thaumatin-like (TL) proteins), PR-12 (defensins such as alfAFP (alfalfa antifungal peptide), *Nicotiania megalosiphon* defensins (NmDef02)), PR-13 (the thionins); and antimicrobial encoders such as StEREBP1 (*Solanum tuberosum* ethylene responsive element binding proteins), HEWL (hen egg white lysozyme), CAP (cationic antimicrobial peptide), Barnase cytotoxic protein and oxidative burst through glucose oxidase (GO) have all been incorporated in transgenic potato with variable successes. Apparently, engineering potato with an array of 'transgenes construct' of selected pathogenesis related proteins and antimicrobial proteins may provide a chance to terminate both the dissemination of *P. infestans* and consequent emergence of new strains, terminating the concept of a new agrochemical for *P. infestans* referred to as 'oomicides'.

Key words: Pathogenesis related proteins, osmotin, thaumatin-like (TL) protein, alfalfa antifungal peptide (alfAFP), *Nicotiania megalosiphon* defensins (NmDef02), cationic antimicrobial peptide (CAP), hen egg white lysozyme (HEWL), glucose oxidase (GO).

INTRODUCTION

Natural selection ensured the survival and multiplication of the most viable species. Modernization in genetic engineering followed by classical and modern biotechnology paved the way for selectable artificial evolution, with direct benefits. Today, plants with genomes acquiring single or multiple incorporated genes from a different species are referred to as 'transgenic'; if introgression is through *in vitro* recombinant DNA technology (RDT). Precisely, genetically modified organism (GMO) comprises those created by natural recombination processes or natural selection; whereas transgenic organisms emerge solely from *in vitro* RDT.

The main aims of creating transgenic plants were varied and diversified. The first was obtaining a net gain in productivity, indispensable to sustain 11 billion people as projected as the world population (Moneret, 1998) in 2050. Achievability was through the creation of plant species resistant to biotic and abiotic challenging factors. The second main aim was based on improving the

*Corresponding author. E-mail: bengyellalouis@yahoo.fr.

nutritional content, especially in protein, high carotene (vitamin A), high tocopherol (vitamin E) and other micro and macro nutrients for most food amongst which potatoes, rice, wheat and tomato topped the list (Prabhu, 2010).

Late blight disease is the most important devastator of potatoes in the world (Edwards, 1956; Song et al., 2003). The gravity of its destruction is traced back from the Great Irish potato famine in the 19th century resulting in over a million Irish starved to death and many more emigrating (Edwards, 1956; Kamoun, 2001). Costs of losses and crop protection against the late blight are estimated at US$3.25 billion per annum worldwide (Fry and Goodwin, 1997b). Furthermore, Germany during the Second World War went on famine since all copper supplies were commandeered for war efforts; hence copper fungicides for potato production was not available, *Phytopthora infestans* largely damaged the crops (Carefoot and Sprout, 1967). Into the 21st century, no concrete solution to either curb the spread or eradicate the pathogen has been found despite continued efforts of scientist. Worst still, resistance to chemical fungicide further complicate the issue, giving rise to the idea of an eventual 'oomicides' creation (Gover, 2001); which is very illusive. The emergence of mating types (A1 and A2) of *P. infestans*, undergoing sexual reproduction provides an insight into the pathogenicity of the organism to colonize new terrain (Drenth et al., 1994); this implies production of recombinant strains. The pathogen population in Europe is now highly diverse and there is more evidence of sexual reproduction in several European countries (Drenth et al., 1994; Flier et al., 2003). The incidence of this pathogen has been controlled notably by agronomic practices that include crop rotation, agrochemicals (Fry and Goodwin, 1997b; Shattock, 2002) and by breeding wild species that contain resistance conferring genes (Lara et al. 2006). The use of agrochemicals poses many dangers that include harmful effects on the ecosystem, increased farmer production cost; on the other hand, breeding programs are necessarily time consuming. Moreover, re-emergence of new strains, slow time consuming breeding method and the ploidy nature of potato, led to the introduction of RDT for improvement against biotic and abiotic stress factors. This paper aims to expose the most successful advancement of genetic engineering, in solving the invincibility of *P. infestans* through transgenic potatoes. Bringing out the implications of pathogenesis related proteins and some antimicrobial proteins to the spotlight, usually involved in local resistance and systemic acquired resistance (SAR) responses in conventional plants.

Pathogenesis related proteins (PR) as weaponry for transgenic potatoes against *P. infestans*

The operational definition of PRs is that of polypeptides with relatively low molecular weights (of 10 to 40 Kilo Dalton (KDa)) that accumulate extracellular infected plant tissue, exhibit high resistance to proteolytic degradation, and often, but not always, possess extreme isoelectric points (Van Loon, 1985). Since their discovery in 1970, 17 families has been identified and reclassified based on amino acids sequences, serological relationship and/or enzymatic or biological activity (Van Loon et al., 2006).

One of the most exploitable pathogenesis related proteins used in transgenic potatoes is osmotin, a 24 KDa protein. This PR-5 family member had been a hotspot for biotechnologist, since it was suspected to play an important role in enhancing the level of resistance to secondary challenges by pathogens, a phenomenon referred to as SAR. The first evidence that osmotin and other PR-5 family had antifungal properties came from the works of Vigers et al. (1991); indicating the N-terminal sequence of zeamatin, an antifungal protein from corn, was very similar to osmotin. *In vitro* assays, demonstrated that osmotin had antifungal activity against a variety of fungi, including *P. infestans, Candida albicans, Neurospora crassa, and Trichoderma reesei* (Woloshuk et al., 1991; Vigers et al., 1992). Stress con-ditions such as NaCl, desiccation, ethylene, wounding, abscisic acid, tobacco mosaic virus, fungi, and UV light were tipped to be inducers of this protein (LaRosa et al., 1992). The large spectrum of cues, both abiotic and biotic gave, good indications that; osmotin gene could always be activated under field conditions. The constitutive expression of osmotin led to enhancement of potato resistance to late blight (Liu et al., 1993); its role in resistance consist of causing sporangia lysis of *P. infestans* (Woloshuk et al., 1991), plant protection against osmotic stress (Kononowicz et al., 1992) and freezing tolerance (Hon et al., 1995). It was previously unveiled that, at high concentration, it causes the lysis of hyphae tips of fungi (Vigers et al., 1991). A more conclusive antifungal potential of osmotin was shown by Abada et al. (1996).

They concluded that osmotin induces spore lysis, inhibit spore germination or reduce its viability in seven fungal species that exhibited some degree of sensitivity in hyphal growth inhibition tests. These broad spectral mechanisms of action validated osmotin gene, as potential sentinel candidate against *P. infestans* in transgenic potatoes. Overexpression of PR-5 (or thaumatin-like (TL) proteins) in potato delayed development of disease symptoms of *P. infestans* (Liu et al., 1994*) in vitro*, whereas trials in transgenic potato plants overexpressing antisense PR-5 did not exhibit any higher susceptibility (Zhu et al., 1996). Curiously, a basic PR-5 has been identified on the cell wall of *P. infestans* (Jeun and Buchenauer, 2001), implying this pathogen is equally armed and definitely ruling out the use of PR-5 thaumatin-like proteins and osmotin 'single-transgene-construct' (STC) in genetically engineered potatoes since both belongs to the PR-5 family.

Defensins (PR-12 family), like thionins (PR-13 family) are cysteine rich-PR proteins (of about 15 kDa) which distinctively confer resistance with a unique mechanism of action; consisting of pores formation within cell membrane resulting in membrane disruption and consequently leading to cell death. Many defensins exhibit antimicrobial activity (Thevissen et al., 2007), while some are apparently void of antifungal and antibacterial potentials (Liu et al., 2006).

A well studied defensins is Rs-AFP2 (*Raphanus sativus* antifungal protein-2), exploited in tomatoes. As a consequence of its activities, Gao et al. (2000) using a gene for cysteine-rich defensins from alfalfa seeds alfAFP (alfalfa antifungal peptide)-gene regulated by a ^{35}S promoter in transgenic potato, revealed an imposing resistance against *Verticillium dahliae*, *Alternaria solani* and *Fusarium culmorum* but not to *P. infestans*. This lack of activity suggested that, defense gene expressions are pathogen specific.

This further strengthens the need for broad synergistic introgression of transgenes, to constitutively express defense proteins in a given transgenic plant if broad viable resistance is the main aim of RDT. Previously, *Nicotiania megalosiphon* was shown to be highly resistant to *Peronospora hyoscymi* f.sp *tabacina* and oomycetes which causes foliar diseases in tobacco (Borrás-Hidalgo et al., 2010). In similar operational approach of RDT, Portieles et al. (2010) reported that engineered transgenic potato and tobacco over-expressing *N. megalosiphon* defensins (NmDef02) gene successfully conferred high-level resistance both under greenhouse and field conditions against *P. infestans* and *P. hyoscymi* f.sp *tabacina*.

An exceptional breakthrough in improving transgenic potato broad-spectrum resistance to *P. infestans* was initiated in 2007. Lee et al. (2007) identified *Solanum tuberosum* L. ethylene responsive element binding proteins (StEREBP1) to be cold-inducible, playing important regulatory functions in plant development, as well as environmental stress and defense responses (Song et al., 2003).

With the continuous effort to improve potato resistance to phytopathogens, especially oomycetes, transgenic potato lines overexpressing StEREBP1 gene from potato (*S. tuberosum* L.) generated by *Agrobacterium tumefaciens*-mediated transformation was shown to exhibit intense resistance to *P. infestans* (Seok-Jun et al., 2009).

Reported absence of observed symptoms on leaves of this transgenic potato prior to infection with *P. infestans* was glaring indications of resistance conferred by introgressed gene. This case suggested that, enhance resistance to the phytopathogens could be due to up-regulated expression of stress-response genes encoding pathogenesis related proteins; that may have been induced by overexpression of StERESP1 (Seok-Jun et al., 2009).

Some engineered antimicrobial proteins (AP) against *Phytopthora infestans* in transgenic potatoes

The quest to eradicate *P. infestans* had prompted the introgression of animal genes in potatoes. Such intro-gressed animal genes and genes products are known to interact intracellularly, but their activities in an alien cellular environment of transgenic potato can not be predicted precisely in terms of allergenicity and toxicity (Sharma, 2010), raising concerns of consumer safety. The use of hen egg white lysozyme (HEWL) gene, tested both in transgenic potatoes and tobacco, elaborating a certain degree of resistance to several bacteria and chitin fungi such as *Botrytis cinerea*, *Verticillium alboatrium* and *R. solani* (Trudel et al., 1995; reviewed in Grover (2003)) was an obvious demonstration on the enormous power of rDNDt in resolving the late blight disease palaver.

Optimism at the horizon for eradicating late blight disease came from four synthetic cationic peptides pep6, pep7, pep11 and pep20 having *in vitro* inhibitory activities against *P. infestans* and *A. solani* (Ali and Reddy, 2000); however, lack of *in vivo* bioassay, leaves doubts on their antifungal potential against late blight causal. A chimaeric gene is a gene constructed by combining coding sequence from one gene with the regulatory sequence of another gene. This ability to create synthetic recombinant and combinatorial variant proteins encoders offers an opportunity to rDNAt to engineer resistance to a range of phytopathogens simultaneously (Dhekney et al., 2007). This concept was exploited in transgenic potatoes. Engineered chimaera cationic antimicrobial peptide (CAP), composed of cecropin-A at the N-terminus and modified melittin-C terminus; constitutively expressed in transgenic potatoes and has been a powerful source of resistance to *Phytophthora cactorum*, *Erwinia carotovora* and *Fusarium solani*, but not *P. infestans* [reviewed in Grover (2003) and Osusky et al. (2004)]. Chimaeric genes (or synthetic genes) were also introduced into two potato (*S. tuberosum* L.) cultivars, 'Desiree' and 'Russet Burbank' characterized by stable introgression and expression as confirmed by reverse transcription (RT)-PCR and with the recovery of biologically active peptide. Yield and morphology of transgenic 'Desiree plants and tubers was unaffected (Haggag, 2008). Barnase (a cytotoxic protein having RNase activity) is naturally present in *Bacillus amyloliguefaciens*. It's introgression in potatoes led to severe inhibition of *P. infestans* spores (Strittmatter et al., 1995; reviewed in Grover (2003)). Once again, the absence of field studies created a big vacuum on the applicability due to the harsh legislation regulating transgenic field trials.

Expression of gene product which functions by releasing elicitors that regulate plant defenses are highly envisaged in transgenic potatoes. This elicitor includes hydrogen peroxides (H_2O_2), salicyclic acids (SA) and ethylene (C_2H_4). Oxidative burst giving rise to hydrogen peroxide (H_2O_2) generally induced PR, phytoalexins, other relative SAR agents and hypersensitive reactions, conferring broader microbial inhibition. rDNAt exploits the ability of glucose oxidase (GO) hydrogen peroxide and gluconic acid producing potentials. In this sphere, *Aspergillus niger* GO gene incorporation into potato proved to reduce considerably the vulnerability of the plant to *P. infestans, E. carotovora* subspecies *carotovora* and *Verticillim dahlia* (Wu et al., 1995). In this trend, the expression of tobacco catalase, an enzyme with SA-binding activity, in transgenic potato enhanced tolerance to *P. infestans* (Yu et al., 1999).

A resistance gene (R-genes) product acts by mediating rapid localized cell death or hypersensitive responses (Rowland et al., 2005). RB-gene is a race independent resistant gene to late blight (Naess, 2000), discovered in wild species of potato called *Solanum bulbocastanum,* mapped to chromosome 8 (Jones and Simko, 2005). The resistant RB-gene was subsequently cloned and found to belong to the largest class of *R*-genes, encoding proteins with a nucleotide binding site and leucine-rich repeats (Song et al., 2003; van der Vossen et al., 2003). Revelation of RB-gene from the wild type *S. bulbocastanum Dunal* subsp. *bulbocastanum* through traditional breeding instilled hope within breeders. This new optimism was short-lived by gene-drag in this ploidy plant. However, Jones and Simko (2005) and Lara et al. (2006) revealed that, transgenic potato lines containing the RB-gene exhibited strong late blight resistance, compared to backcrossed progenies derived from breeding programs; or those derived from somatic hybrids between *S. tuberosum* and *S. bulbocastanum.*

The performance of transgenic potato was further evaluated by Halterman et al. (2008) using late blight resistance gene RB. This RB-transgenic potato exhibited a strong foliar resistance under greenhouse conditions to *P. infestans,* but with no noticeable tuber resistance to the phytopathogen. Moreover, Halterman et al. (2008) noted that all RB transgenic lines exhibited an increase in resistance to *P. infestans* compared with their non-transgenic counterparts. Hence, RDT approach seems logical in this heterozygous and vegetatively propagated crop (Park et al. 2009).

DISCUSSION

Under-, normal- and over-expressions are the three expressive modes of 'transgenes-construct' so far exploited by biotechnologist in transgenic plants depending on the cis-regulatory elements used for controlling the transgenes. Most success stories in engineered genes are stringently through transgenes-construct over-expression. For most if not all, biotechnologists are switching by choice towards overexpressing 'transgenes-construct' to confer broad-spectrum plant defense, mediated by pathogenesis related proteins and other antimicrobial proteins. However, outcome from successful

defense enhancement due to introgression greatly varies, depending mostly on the targeted phytopathogenic agent susceptibility and the mode of expression of 'transgenes-construct'. Generally, introgression of traits from wild relatives of potato has been difficult, and had mostly been restricted to disease resistance traits (Hermsen, 1994). In an attempt to explain some of the failures seen so far in transgenic potatoes, the most noticeable one is the orientational use of some key amino acids. Transgenes (of PR or AP) in potato genome expression products requires tryptophan, which is also a building block for plant cell wall lignifications; the first line of plant defense. Nicholson and Hammerschidt (1992) showed a decrease in polyphenolic compounds, such as lignin, in transgenic potato tubers due to redirection of tryptophan in transgenic plants through expression of tryptophan decarboxylase; which in turns renders tissues more susceptible to *P. infestans.*

The quality and tenor of rDNAt and the resistance it confers to transgenic potato is so far from satisfaction. In most experiment, STC has been used with shortcomings. From our investigation the following reasons can be suggested to explain the invincibility of *P. infestans* regarding the already created transgenic potato:

1) Antimicrobial proteins- and PR-phytopathogens interactions are specific - that is, gene-for-gene model response. Hence, STC narrowly confer resistance to a particular biotic factor, not a broad-spectrum resistance.
2) *P. infestans* encodes some of the introgressed STC proteins, as in the case of PR-5 TL proteins (Jeun and Buchenauer, 2001). Meaning the pathogen is immune to some of the STC proteins.
3) Misclassification of *P. infestans* as 'fungus' actually placed vital research on wrong track (Govers F (2001). This revelation compromised most previous work geared at eradicating late blight disease.
4) rDNAt method of gene insertion in any recipient DNA is rather a random process which might lead to: a) silencing of the inserted STC itself or surrounding genes at the site of insertion (Azahaguvel et al. 2006); b) genetic instabilities in other genes of interest and c) finally activating some evolutionary silenced genes (Sharma, 2010).
5) Finally, lack of field trials and *in vivo* assays probably due to restrictions pose a challenge to the deployment of field trials, and *in vivo* assays in some countries generally slows down the usefulness of rDNAt in the field of transgenic potatoes.

Of late, a disturbing sequel tormenting first evidence that *P. infestans* can defeat an R-protein through inhibition of recognition of the corresponding effectors protein by Halterman and Chen (2010) is a landmark message to all stakeholders, advocating the use and deployment of the new options transgenic potatoes offers. Halterman and Chen (2010) reports deeply compromise future breeding of RB varieties for the search of broad-spectrum resistance, whereas RDT report from findings of Portieles et al. (2010) gives hope in conferring broad-spectrum resistance against *P. infestans*. Moreover, the scarcity of resistance traits in potatoes or their exhaustive usage in broad-spectrum resistance breeding is an indication; mobilization of genes from other sources through recombinant DNA technology is indispensible. A new approach of using an array of PR and AP encoding genes in 'chimeric-transgenes-construct' is the way forward in conferring a sustainable and durable resistance to potato plants. This biotechnological concept not only improves yield to feed more people, it reduce inputs in terms of chemical fungicide application on the part of the farmers; but also, consumers benefit from nutritionally enriched plants free from chemical residues.

ACKNOWLEDGEMENTS

We thank TWAS (Third World Academic of Sciences) from developing countries, DBT (Department of Biotechnology) and the Government of India for their fellowship support. Also, University of Yaoundé I, Cameroon and The Roger Peller Foundation are acknowledged for their documentation support.

REFERENCES

Abada LR, D'Urzo MP, Liua D, Narasimhan ML, Reuveni M, Zhua JK (1996). Antifungal activity of tobacco Osmotin has specificity and involves plasma membrane permeabilization. Plant Sci., 118: 11-23.

Ali GS, Reddy AS (2000). Inhibition of fungal and bacterial plant pathogens by synthetic peptides: in vitro growth inhibition. Interaction between peptides and inhibition of disease progression. Mol. Plant–Microbe Interact., 13: 847-859.

Azahaguvel P, Vidya D, Sharm A, Varshney RK (2006). Methodologgical advancement in molecular markers to delimit the gene(s) from crop improvement. In floriculture, ornamental and plant biotechnology. Advances and tropical issues. Teixera de silva. J. (ed) Global Science Book, London, UK, pp. 460-469.

Borrás-Hidalgo O, Thomma B, Silva Y, Chacon O, Pujol M (2010). Tobacco blue mould disease caused by *Peronospora hyoscyami* f.sp *tabacina*. Mol. Plant Pathol., 11: 13-18.

Carefoot GL, Sprout ER (1967). Famine in the Wind. Angus and Robertson Ltd Academic Publishers, London, UK.

Dhekney SA, Li ZT, Aman VM, Dutt M, Gray DJ (2007). Genetic transformation of embryonic cultures and recovery of transgenic plants in *Vitis vinifera*, *Vitis rotandifolia* and *Vitis* hybrids. ACTA Hortic., 738: 743-748.

Drenth A, Tas ICQ, Govers SB (1994). DNA fingerprinting uncovers a new sexually reproducing population of *Phytophthora infestans* in the Netherlands. F. Eur. J. Plant Pathol., 100: 97-107.

Edwards RD, Williams TD (1956). "The Great Famine: Studies in Irish History 1845-1852", Lilliput Press, London, UK.

Flier WG, Grünwald NJ, Kroon LPNM, Sturbaum AK, Garay-Serrano E, Lozoya-Saldana H, Fry WE, Turkensteen LJ (2003). The population structure of *Phytophthora infestans* from the Toluca Valley of Central Mexico suggests genetic differentiation between populations from cultivated potato and wild *Solanum* spp. Phytopathology, 93: 382-390.

Fry WE, Goodwin SB (1997b). Resurgence of the Irish potato famine fungus. Bioscience, 47: 363-370.

Gao AG, Hakimi SM, Mittanck CA, Wu Y, Woerner BM, Stark DM, Shah

DM, Liang J, Rommens CMT (2000). Fungal pathogen protection in potato by expression of a plant defensins peptide. Nature Biotech., 18: 1307-1310

Govers F (2001). Misclassification of pest as 'fungus' puts vital research on wrong track. Nature. 411(7): 633

Grover A, Gowthaman R (2003). Strategies for development of fungus resistant transgenic plants. Current science, 84(3): 330-340.

Haggag WM (2008). Biotechnological aspects of plant resistant for fungal disease management. American-Eurasian J. Sustain. Agric., 2(1): 1-18.

Halterman DA, Chen Y, Jiraphan S, Berduo-Sandoval J, Amilcar SR (2010). Competition between *Phytophthora infestans* Effectors Leads to Increased Aggressiveness on Plants Containing Broad-Spectrum Late Blight Resistance. PLoS ONE. 5: 5, e10536.

Halterman DA, Kramer LC, Wielgus S, Jiang J (2008). Performance of transgenic potato containing the late blight resistance gene *RB*. Plant Dis., 92: 39-343.

Hermsen JGT (1994). Introgression of genes from wild species: In Potato genetics. Bradshaw and Mackay (eds). CABI, Wallingford Academic Publishers, London, UK. pp. 515–538.

Hon WC, Griffith M, Mlynarz A, Kwok YC, Yang DSC (1995). Antifreeze proteins in winter rye are similar to pathogenesis-related proteins. Plant Physiol., 109: 879-889.

Jeun YC, Buchenauer H (2001). Infection structures and localization of pathogenesis related protein AP24 in leaves of tomato plants exhibiting systemic acquired resistance against *Phytophthora infestans* after pre-treatment with 3-aminobutyric acid or tobacco necrosis virus. J. Phytopathol., 149: 141-153.

Jones R, Simko IE (2005). Resistance to late blight and other fungi: In genetic improvement of *Solanaceous* crops. Razdan M, and Mattoo A (eds) The Potato Science Publishers Inc. USA., 1: 397-412.

Kamoun S (2001). Nonhost resistance to Phytophthora: novel prospects for a classical problem. Curr. Opinion Plant Biol., 4: 295-300.

Kononowicz AK, Nelson DE, Singh NK, Hasegawa PM, Bressan RA (1992). Regulation of the osmotin gene promoter. Plant Cell, 4: 513-524.

Lara MC, Groza HI, Wielgus SM, Jiang J (2006). Marker-Assisted Selection for the Broad-Spectrum Potato Late Blight Resistance Conferred by Gene RB Derived from a Wild Potato Species. Crop Sci., 46: 589-594.

LaRosa PC, Chen Z, Nelson DE, Singh NK, Hasegawa PM, Bressan RA (1992). Osmotin gene expression is posttranscriptionally regulated. Plant. Physiol., 100: 409-415.

Lee U, Ripflorido I, Hong S, Lurkindale JE, Waters E, Vierling (2007). "The Arabidopsis ClpB/Hsp100 family of proteins: Chaperones for stress and chloroplast development". Plant J., 49: 115-127.

Liu D, Kashchandra GR, Hasegawa PM and Bressan RA (1993). Osmotin overexpression in potato delays development of disease symptoms. Proc. Nati. Acad. Sci. USA., 91: 1888-1892.

Liu YJ, Cheng CS, Lai SM, Hsu MP, Chen CS, Lyu PC (2006). Solution structure of the plant defensins VrD1 from mung beans and it possible role in insecticidal activity against bruchids. Proteins, 63: 777-786.

Liu D, KG Raghothama, PM Hasegawa and RA Bressan, 1994. Osmotin over expression in potato delays development of disease symptoms. Proceedings National Acad. Sci. USA. 91: 1888-1892.

Moneret VDA (1998). Le risque allergique des aliments transgeniques: vrai ou faux probleme? Revue Francaise d'Allergologie et d'Immunologie Clinique, 38(8): 693-699.

Naess SK (2000). Resistance to late blight in *Solanum bulbocastanum* is mapped to chromosome 8. Theoretical appl. Gen., 101: 697-704.

Nicholson RL, Hammer-schmit R (1992). Phenolic compounds and their role in disease resistance. Ann. Rev.Phytopathol., 30: 369-389.

Osusky M, Zhou G, Osuska L, Hancock RE, Kay WW and Mishra S (2004). Transgenic potatoes expressing a novel cationic peptide are resistant to late blight and pink rot. Nature Biotechnol., 18: 1162-1166.

Park TH, Vleeshouwers VG, Jacobsen E, Van Der Vossen (2009). Molecular breeding for resistance to *Phytophthora infestans* (Mont.) de Bary in potato (*Solanum tuberosum* L.): a perspective of cisgenesis resistance mediated by the broad-spectrum resistance gene RB. Phytopathology. Plant Breeding, 128: 109-117.

Portieles R, Ayra C, Gonzales E, Gallo A, Rodriguez R, Chacón O, López Y, Rodriguez M, Castillo J, Pujol M, Enriquez G, Borroto C, Trujillo L, Thomma BPHJ, Borras-Hidalgo O (2010). NmDef02, novel antimicrobial gene isolated from *Nicotiania megalosiphon* confers high-level pathogen resistance under greenhouse and field conditions. Plant Biotechnol. J., 8: 678-690.

Prabhu KV (2010). Genetic Modify Technology. Genetically modified versus traditional crops; How do they differ? Biotechnews. 5(2): 57-60.

Rowland O, Ludwig A, Merrick CJ, Baillieul F, Tracy FE, Durant WE, Yoshioka HY (2005). Functional analysis of Avr9/CF-9 rapidly elicited genes identifies a protein kinase, ACIk1, that is essential for full CF-9-dependent disease resistance in tomato. Plant Cell, 17: 295-310.

Seok-Jun M, Ardales EY, Shin D, Han S, Lee HE, Park SR (2009). The EREBP-gene from *Solanum tuberosum* confers resistance against Oomycetes and a bacterial pathogen in transgenic potato and tobacco plants. Fruits, Vegetable and Cereal Science and Biotechnology 3. Global Sci. Books, 1: 72-79.

Sharma RP (2010). Genetic Modify Technology. Responsible and purposeful use. Biotechnews. 5(2): 68-71.

Shattock RC (2002). *Phytophthora infestans*: Populations, pathogenicity and phenylamides. Pest Manage. Sci., 58: 944-950.

Song J, Bradeen JM, Naess SK, Raaaseh JA, Wielgus SM, Haberlach GT, Liu J, Kuang H (2003). Gene AB cloned from *Solanum tuberosum L* confers broad spectrum resistance to potato Late blight. Proceedings National Acad. Sci. USA., 100: 9128-9133.

Strittmatter G, Janssens J, Opsomer C, Botterman J (1995). Inhibition of fungal disease development in plants by engineering controller cell death. Biotechnology, 13: 1085-1089.

Thevissen K, Kristensen HH, Thomma BPHJ, Cammue BPA, Francois IEJA (2007). Therapeutic potential of antifungal plant and insects defensins, Drug Discov. Today. 12: 966-971.

Trudel J, Potvin C, Asselin A (1995). Secreted hen lysozyme in transgenic tobacco: Recovery of bound enzyme and in vitro growth inhibition of plant pathogens. Plant Sci., 106: 55-62.

van der Vossen, Sikkema E, Hekkert A, Gros J, Stevens P, Muskens M, Wouters D, PereiraStiekema A, Allefs S (2003). An ancient *R* gene from the wild potato species *Solanum bulbocastanum* confers broad-spectrum resistance to *Phytophthora infestans* in cultivated potato and tomato. Plant J. 36: 867-882.

van Loon LC, Rep M, Pieterse CMJ (2006). Significance of inducible defense related proteins in infected plants. Annual Review of Phytopathology, 44: 1-28.

van Loon LC (1985). Pathogenesis-related proteins. Plant Mol. Biol., 4: 111-116.

Vigers AJ, Roberts WK, Selitrenikoff CP (1991). A new family of antifungal proteins. Mol. Plant. Microb. Interact., 4: 315-323.

Vigers AJ, Wiedemann S, Roberts WK, Legrand M, Selitrennikoff C, Fritig B (1992). Thaumatin-like pathogenesis related proteins are antifungal. Plant. Sci., 83: 155-161.

Woloshuk CP, Muelenhoff JS, Sela-Buurlage M, Elzen PJM, Cornelissen BJC (1991). Pathogen-induced proteins with inhibitory activity towards *Phytopthora infestans*. Plant Cell. 3: 619-628.

Wu G, Shortt BJ, Lawrence EB, Levine EB, Fitzsimmons KC, Shah DM (1995). Disease resistance conferred by expression of a gene encoding H_2O_2-generating glucose oxidase in transgenic potato plants. Plant Cell, 7: 1357-1368.

Yu D, Xie Z, Chen C, Fan S, Chen Z (1999). Expression of tobacco class II catalase genes activates the endogenous homologous gene and is associated with disease resistance in transgenic potato plants. Plant Mol. Biol., 39: 477-488.

Zhu B, Chen THH, Li PH (1996). Analysis of late-blight disease resistance and freezing tolerance in transgenic potato plants expressing sense and antisense genes for an osmotin-like protein. Planta, 198: 70-77.

Permissions

All chapters in this book were first published in BMBR, by Academic Journals; hereby published with permission under the Creative Commons Attribution License or equivalent. Every chapter published in this book has been scrutinized by our experts. Their significance has been extensively debated. The topics covered herein carry significant findings which will fuel the growth of the discipline. They may even be implemented as practical applications or may be referred to as a beginning point for another development.

The contributors of this book come from diverse backgrounds, making this book a truly international effort. This book will bring forth new frontiers with its revolutionizing research information and detailed analysis of the nascent developments around the world.

We would like to thank all the contributing authors for lending their expertise to make the book truly unique. They have played a crucial role in the development of this book. Without their invaluable contributions this book wouldn't have been possible. They have made vital efforts to compile up to date information on the varied aspects of this subject to make this book a valuable addition to the collection of many professionals and students.

This book was conceptualized with the vision of imparting up-to-date information and advanced data in this field. To ensure the same, a matchless editorial board was set up. Every individual on the board went through rigorous rounds of assessment to prove their worth. After which they invested a large part of their time researching and compiling the most relevant data for our readers.

The editorial board has been involved in producing this book since its inception. They have spent rigorous hours researching and exploring the diverse topics which have resulted in the successful publishing of this book. They have passed on their knowledge of decades through this book. To expedite this challenging task, the publisher supported the team at every step. A small team of assistant editors was also appointed to further simplify the editing procedure and attain best results for the readers.

Apart from the editorial board, the designing team has also invested a significant amount of their time in understanding the subject and creating the most relevant covers. They scrutinized every image to scout for the most suitable representation of the subject and create an appropriate cover for the book.

The publishing team has been an ardent support to the editorial, designing and production team. Their endless efforts to recruit the best for this project, has resulted in the accomplishment of this book. They are a veteran in the field of academics and their pool of knowledge is as vast as their experience in printing. Their expertise and guidance has proved useful at every step. Their uncompromising quality standards have made this book an exceptional effort. Their encouragement from time to time has been an inspiration for everyone.

The publisher and the editorial board hope that this book will prove to be a valuable piece of knowledge for researchers, students, practitioners and scholars across the globe.

List of Contributors

Olusola Abayomi Ojo
Department of Microbiology, Lagos State University, Badagry Expressway, P.O. BOX 12142 Ikeja, Lagos, Nigeria

Rosfarizan Mohamad
Department of Bioprocess Technology, Faculty of Biotechnology and Biomolecular Sciences, Universiti Putra Malaysia, 43400 UPM Serdang, Selangor, Malaysia

Mohd Shamzi Mohamed
Department of Bioprocess Technology, Faculty of Biotechnology and Biomolecular Sciences, Universiti Putra Malaysia, 43400 UPM Serdang, Selangor, Malaysia

Nurashikin Suhaili
Department of Bioprocess Technology, Faculty of Biotechnology and Biomolecular Sciences, Universiti Putra Malaysia, 43400 UPM Serdang, Selangor, Malaysia

Madihah Mohd Salleh
Faculty of Bioscience and Bioengineering, Universiti Teknologi Malaysia, Skudai, 81310 Johor, Malaysia

Arbakariya B. Ariff
Department of Bioprocess Technology, Faculty of Biotechnology and Biomolecular Sciences, Universiti Putra Malaysia, 43400 UPM Serdang, Selangor, Malaysia

M. N. N. N. Shikongo-Nambabi
Department of Food Science and Technology, Neudamm Campus, University of Namibia, Windhoek, Namibia

N. P Petrus
Department of Animal Science, Neudamm Campus, University of Namibia, Windhoek, Namibia

M. B. Schneider
Faculty of Agriculture and Natural Resources, University of Namibia, Namibia

Fatma Abdelaziz Amer
Microbiology and Immunology Department, Faculty of Medicine, Zagazig University, Zagazig, Egypt

Eman Mohamed El-Behedy
Microbiology and Immunology Department, Faculty of Medicine, Zagazig University, Zagazig, Egypt

Heba Ali Mohtady
Microbiology and Immunology Department, Faculty of Medicine, Zagazig University, Zagazig, Egypt

Heping Zhou
Department of Biological Sciences, Seton Hall University, 400 South Orange Avenue, South Orange, New Jersey

Kalyan K. Mondal
Division of Plant Pathology, Indian Agricultural Research Institute, New Delhi 110 012, India

V. Shanmugam
Institute of Himalayan Bioresource Technology, Council of Scientific and Industrial Research, Palampur 176 062, Himachal Pradesh, India

C. O. Adenipekun
Department of Botany, University of Ibadan, Ibadan, Nigeria

R. Lawal
Department of Botany, University of Ibadan, Ibadan, Nigeria

Rodney Hull
School of Molecular and Cell Biology, University of the Witwatersrand, Wits, 2050, South Africa

Rodrick Katete
School of Molecular and Cell Biology, University of the Witwatersrand, Wits, 2050, South Africa

Monde Ntwasa
School of Molecular and Cell Biology, University of the Witwatersrand, Wits, 2050, South Africa

Priti Maheshwari
Faculty of Arts and Science, Department of Biological Sciences, 4401, University Drive, University of Lethbridge, Lethbridge, Alberta, T1K 3M4, Canada

Sarika Garg
Max Planck Unit for Structural Molecular Biology, c/o DESY, Gebaüde 25b, Notkestrasse 85, D- 22607 Hamburg, Germany

Anil Kumar
School of Biotechnology, Devi Ahilya University, Khandwa Road Campus, Indore – 452001, India

Mohammad Asgharzadeh
Biotechnology Research Center, Tabriz University of Medical Sciences, Tabriz, Iran

Hossein Samadi Kafil
Department of Bacteriology, School of Medicine, Tarbiat Modares University, P.O. Box: 14115-33, Tehran, Iran

Bijender Singh
Department of Microbiology, Maharshi Dayanand University, Rohtak- 124 001, India

Gotthard Kunze
Leibniz-Institut für Pflanzengenetik und Kulturpflanzenforschung (IPK), Corrensstr. 3, D-06466 Gatersleben, Germany

T. Satyanarayana
Department of Microbiology, University of Delhi South Campus, Benito Juarez Road, New Delhi-110 021, India

Chijioke A. Nsofor
Department of Biotechnology, Federal University of Technology, Owerri, Imo State, Nigeria

Pratap R. Patnaik
Institute of Microbial Technology, Sector 39-A, Chandigarh-160036, India

I. O. Okonko
Department of Virology, Faculty of Basic Medical Sciences, University of Ibadan College of Medicine, University College Hospital (UCH), Ibadan, University of Ibadan, Ibadan, Nigeria. WHO Regional Reference Polio Laboratory, WHO Collaborative Centre for Arbovirus Reference and Research, WHO National Reference Centre for Influenza

E. T. Babalola
Department of Virology, Faculty of Basic Medical Sciences, University of Ibadan College of Medicine, University College Hospital (UCH), Ibadan, University of Ibadan, Ibadan, Nigeria. WHO Regional Reference Polio Laboratory, WHO Collaborative Centre for Arbovirus Reference and Research, WHO National Reference Centre for Influenza

A. O. Adedeji
Department of Virology, Faculty of Basic Medical Sciences, University of Ibadan College of Medicine, University College Hospital (UCH), Ibadan, University of Ibadan, Ibadan, Nigeria. WHO Regional Reference Polio Laboratory, WHO Collaborative Centre for Arbovirus Reference and Research, WHO National Reference Centre for Influenza

B. A. Onoja
Department of Virology, Faculty of Basic Medical Sciences, University of Ibadan College of Medicine, University College Hospital (UCH), Ibadan, University of Ibadan, Ibadan, Nigeria. WHO Regional Reference Polio Laboratory, WHO Collaborative Centre for Arbovirus Reference and Research, WHO National Reference Centre for Influenza

A. A. Ogun
Department of Environmental Health, Faculty of Public Health, University of Ibadan College of Medicine, University College Hospital (UCH) Ibadan, Nigeria

A. O. Nkang
Department of Virology, Faculty of Basic Medical Sciences, University of Ibadan College of Medicine, University College Hospital (UCH), Ibadan, University of Ibadan, Ibadan, Nigeria. WHO Regional Reference Polio Laboratory, WHO Collaborative Centre for Arbovirus Reference and Research, WHO National Reference Centre for Influenza

F. D. Adu
Department of Virology, Faculty of Basic Medical Sciences, University of Ibadan College of Medicine, University College Hospital (UCH), Ibadan, University of Ibadan, Ibadan, Nigeria. WHO Regional Reference Polio Laboratory, WHO Collaborative Centre for Arbovirus Reference and Research, WHO National Reference Centre for Influenza

Bremnæs Christiane
Department of Biochemistry, Faculty of Natural and Agricultural Sciences, School of Biological Sciences, University of Pretoria, Pretoria 0002

Meyer Debra
Department of Biochemistry, Faculty of Natural and Agricultural Sciences, School of Biological Sciences, University of Pretoria, Pretoria 0002

Babu Joseph
Department of Microbiology and Microbial Technology, College of Biotechnology and Allied Sciences, Allahabad Agricultural Institute- Deemed University, Allahabad 211 007, Uttar Pradesh, India

Pramod W. Ramteke
Department of Microbiology and Microbial Technology, College of Biotechnology and Allied Sciences, Allahabad Agricultural Institute- Deemed University, Allahabad 211 007, Uttar Pradesh, India

George Thomas
Department of Molecular Biology and Genetic Engineering, College of Biotechnology and Allied Sciences, Allahabad Agricultural Institute- Deemed University, Allahabad 211 007, Uttar Pradesh, India

Nitisha Shrivastava
Department of Microbiology and Microbial Technology, College of Biotechnology and Allied Sciences, Allahabad Agricultural Institute- Deemed University, Allahabad 211 007, Uttar Pradesh, India

Bengyella Louis
Department of Biotechnology, University of Burdwan, Golapbag More-713104, Burdwan, West Bengal, India

Pranab Roy
Department of Biochemistry, University of Yaoundé I, Cameroon, BP-812, Cameroon